第2版

动物解剖涂色书

SAUNDERS
VETERINARY ANATOMY
COLORING BOOK
2/E

编著 〔加〕巴尔吉特・辛格（Baljit Singh）

主译 曹 静 陈耀星

译者（按姓氏笔画排序）

王 璐 王子旭 白 雪 任文姬 佀蓉蓉

陈耀星 徐嘉璐 曹 静 曹鹤馨

ELSEVIER

北京科学技术出版社

Elsevier (Singapore) Pte Ltd.
3 Killiney Road,
#08-01 Winsland House I,
Singapore 239519
Tel: (65) 6349-0200; Fax: (65) 6733-1817

ISBN-13: 978-1-4557-7684-9

This translation of Saunders Veterinary Anatomy Coloring Book, Second Edition by Baljit Singh, was undertaken by Beijing Science and Technology Publishing Co.,Ltd of China and is published by arrangement with Elsevier (Singapore) Pte Ltd.
Saunders Veterinary Anatomy Coloring Book, Second Edition by Baljit Singh由 北京科学技术出版社有限公司进行翻译，并根据北京科学技术出版社有限公司与爱思唯尔（新加坡）私人有限公司的协议约定出版。

《动物解剖涂色书》（第2版）（曹静，陈耀星主译）
ISBN: 978-7-5714-1121-3

著作权合同登记号：图字01—2018—7105号

图书在版编目（CIP）数据

动物解剖涂色书：第2版 / (加) 巴尔吉特·辛格(Baljit Singh) 编著；曹静, 陈耀星主译. — 北京 :北京科学技术出版社, 2020.11
书名原文: Saunders Veterinary Anatomy Coloring Book, 2/E
ISBN 978-7-5714-1121-3

Ⅰ.①动… Ⅱ.①巴… ②曹… ③陈… Ⅲ.①动物解剖学-图谱 Ⅳ.①Q954.5-64

中国版本图书馆CIP数据核字（2020）第164776号

责任编辑：张真真
责任校对：贾 荣
图文制作：华 艺
责任印制：吕 越
出 版 人：曾庆宇
出版发行：北京科学技术出版社
社　　址：北京西直门南大街16号
邮政编码：100035
网　　址：www.bkydw.cn
ISBN 978-7-5714-1121-3

电　　话：0086-10-66135495（总编室）
0086-10-66113227（发行部）
印　　刷：河北鑫兆源印刷有限公司
开　　本：889mm × 1194mm 1/16
字　　数：220千字
印　　张：18.5
版　　次：2020年11月第1版
印　　次：2020年11月第1次印刷

定　　价：**98.00元**

译者前言

本书是由加拿大卡尔加里大学兽医学院院长巴尔吉特·辛格教授编写的*Saunders Veterinary Anatomy Coloring Book*第2版的中文译本。全书含400余张插图，按部位、分层次详细介绍了犬、猫、马、猪、反刍动物、鸟类、小型哺乳动物和爬行动物的局部解剖学结构，内容涉及头部、颈腹部、颈部、背部、脊柱、胸部、腹部、骨盆、生殖器官、前肢以及后肢。

本书通过对解剖结构填涂颜色的形式，使学生或相关专业从业者加深了对动物解剖结构的辨识，以及对不同动物解剖结构的比较。设计思路新颖，使原本枯燥的解剖学知识变得生动有趣。

参加本书翻译工作的有中国农业大学动物医学院曹静副教授、陈耀星教授和王子旭高级实验师，中国农业大学动物医学院研究生王璐、曹鹤馨、徐嘉璐、任文姬、白雪和佀蓉蓉。最后由曹静副教授、陈耀星教授和王子旭高级实验师校对统稿。在付梓之际，感谢为本书翻译和审校工作努力付出的所有人。感谢北京科学技术出版社张真真编辑及其同事对本书的支持和精心编校，使书稿符合出版要求。感谢他们与我们之间的默契合作！

翻译工作是一项艰苦的工程。尽管我们在中文译本中努力真实地反映原著内容，并纠正了原版书的一些错误，但鉴于译校者水平有限，书中如有误译之处，敬请读者批评指正。

全体译者

2020年9月于北京

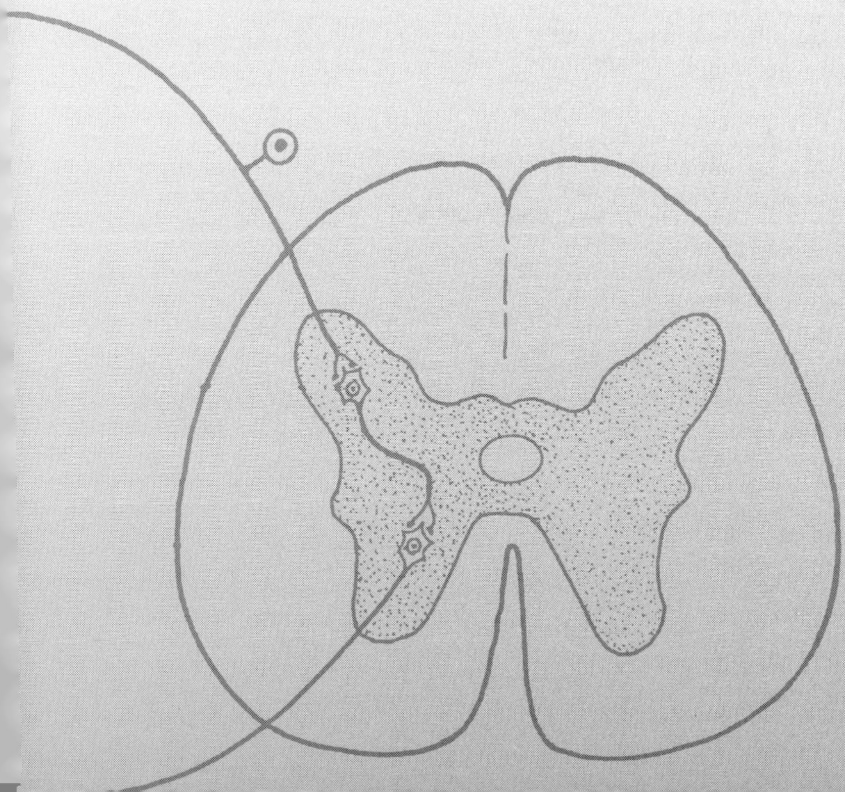

原著前言

欢迎来到动物涂色的世界！这本书共有400余张插图供你学习和涂色，帮助你熟练掌握动物解剖学知识。无论你是动物医学、兽医学专业的学生，还是动物医学从业者，抑或是动物爱好者，这本工具书将通过有趣的涂色方法帮助你学习、复习动物解剖学的内容。

《动物解剖涂色书》（第2版）涵盖犬、猫、马、猪、反刍动物、鸟类及异宠等小型哺乳动物和爬行动物的解剖结构。前7章按机体的7个部位划分，即头部和颈腹部，颈部、背部和脊柱，胸部，腹部，骨盆和生殖器官，前肢，以及后肢。第8章异宠是这一版最新补充的一章，涵盖了你在学习和工作中、日常生活中可能会遇到的各种小型哺乳动物和爬行动物。

《动物解剖涂色书》（第2版）是学习动物解剖学的理想伴侣，互动式练习让你的学习效率更高。本书在考前、课前，甚至在给下一个病例看病之前，都可以使用。

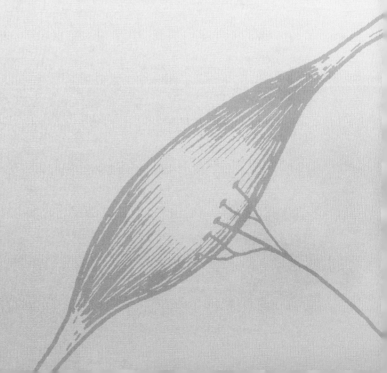

如何使用本书

本书每一页都有一个动物图标，便于你明确本页中的动物。每页页面包含一段简短的文字用于描述动物机体的结构特征和视图方向，以及清晰、易于上色的插图。插图中带编号的引线标识了要填涂的结构，并有编号列表相对应。你可以创造出你自己的“颜色编码”，用相同的颜色填涂插图的方框编码、解剖结构和插图下方或旁边列表中相应的编号框。此外，部分页面设置了一些问题，帮助你温故而知新。

图片来源

Ackerman N, Aspinall V: *The Complete Textbook of Veterinary Nursing, ed 2,* Oxford, 2011, Butterworth-Heinemann.

Colville TP, Bassert JM: *Clinical Anatomy and Physiology for Veterinary Technicians,* ed 3, St Louis, 2015, Mosby.

Dyce KM, Sack WO, Wensing CJG: *Textbook of Veterinary Anatomy,* ed 4, St Louis, 2010, Saunders.

Evans HE, de Lahunta A: *Guide to the Dissection of the Dog,* ed 7, St Louis, 2010, Saunders.

Jepson L: *Exotic Animal Medicine: A Quick Reference Guide,* Oxford, 2009, Saunders.

Longley L: *Saunders Solutions in Veterinary Practice: Small Animal Exotic Pet Medicine,* Philadelphia, 2011, Saunders.

Mader DR: *Reptile Medicine and Surgery,* ed 2, St Louis, 2006, Saunders.

Mitchell M, Tully TN: *Manual of Exotic Pet Practice,* St. Louis, 2008, Saunders.

Quesenberry K, Carpenter JW: *Ferrets, Rabbits, and Rodents,* ed 3, St Louis, 2012, Saunders.

Silverman S, Tell L: *Radiology of Rodents, Rabbits and Ferrets,* St Louis, 2005, Saunders.

目录

1 头部和颈腹部 …………………………… 1

犬

图1-1 犬颅骨的范围 ………………………… 2
图1-2 犬颅骨，背侧观 ……………………… 3
图1-3 犬颅骨，腹侧观 ……………………… 4
图1-4 舌骨器和悬吊于犬颅骨颞部的喉 … 5
图1-5 犬口腔和舌 …………………………… 6
图1-6 犬舌和咽部肌肉 ……………………… 7
图1-7 犬唾液腺 ……………………………… 8
图1-8 犬单齿（A）、外侧（B）和内侧（C）咀嚼肌 …………………… 9
图1-9 犬齿板及成釉器的发育 ………… 10
图1-10 犬喉的横断面 …………………… 11
图1-11 犬脑-垂体-外周器官轴 ……… 12
图1-12 犬头部动脉 ……………………… 13
图1-13 犬中间神经元反射弧 ………… 14
图1-14 犬脊髓内纤维传导路径 ……… 15
图1-15 犬脑，背侧观 ………………… 16
图1-16 犬脑，腹侧观 ………………… 17
图1-17 犬脑，正中矢状面 …………… 18
图1-18 犬脑干，图示成年哺乳动物神经核团 ………………… 19
图1-19 犬视觉和瞳孔反射通路模式图… 20
图1-20 椎体系（实线）和椎体外系（虚线）连接图 ……………… 21
图1-21 犬脑腹侧面动脉 ……………… 22
图1-22 犬面神经分布图 ……………… 23
图1-23 剖开犬眼球，显示眼球壁三层结构 ……………………… 24
图1-24 犬眼前部断面 ………………… 25
图1-25 犬左眼后方的肌肉 …………… 26
图1-26 犬右耳，后面观模式图 ……… 27
图1-27 犬皮肤感觉神经末梢 ………… 28
图1-28 犬毛囊附属结构 ……………… 29
图1-29 犬头部浅层解剖图 …………… 30
图1-30 犬眼眶和翼腭窝的解剖图，外侧观 ……………………… 31
图1-31 犬头部的主要动脉（灰色）和静脉（黑色） ………………… 32
图1-32 犬颈部的第5颈椎横断面 ……… 33
图1-33 犬头颈部主要淋巴结的引流范围 ……………………… 34
图1-34 犬颈部浅层神经，外侧观 …… 35
图1-35 犬颈部静脉，腹侧观 ………… 36
图1-36 分离的幼犬颅骨，腹侧观 …… 37
图1-37 犬咽部肌和舌肌，左外侧观 … 38
图1-38 犬左眼眼外肌及其作用 ……… 39
图1-39 犬面神经和三叉神经的浅表分支 ……………………… 40
图1-40 犬颈总动脉分支，外侧观 …… 41
图1-41 从中央到右侧下颌骨的犬头部肌肉、神经和唾液腺 ……… 42
图1-42 犬正中矢状面的脑膜和脑室 ……………………………… 43
图1-43 犬脑、颅神经和脑干，腹侧观 ……………………………… 44

猫

图1-44 猫头部浅层解剖图 …………… 45
图1-45 猫头部深层解剖图 …………… 46

马

图1-46 马喉软骨外侧观（A）和马喉内部肌肉（B） ………………… 47
图1-47 马左眼前部，从后向前观 …… 48
图1-48 马颅骨，外侧观 ……………… 49

图1-49　马颅骨背侧的脑、额窦、上颌窦的投影 …………………… 50
图1-50　马下切齿的结构 ……………… 51
图1-51　马深层咀嚼肌和右侧二腹肌 … 52
图1-52　马的咽、软腭和舌骨的肌肉 … 53
图1-53　马眼眶的解剖图 ……………… 54
图1-54　马左侧听小骨，前内侧观 …… 55
图1-55　马颈部第4颈椎横断面 ………… 56
图1-56　马头部主要动脉 ……………… 57

牛

图1-57　牛颅骨和下颌骨 ……………… 58
图1-58　牛颈部横断面 ………………… 59
图1-59　牛右眼 ………………………… 60
图1-60　牛颅骨基部和舌与咽、喉的连接… 61
图1-61　牛左颈总动脉分支 …………… 62

猪

图1-62　猪头部的浅层解剖图 ………… 63
图1-63　猪颅骨旁正中面 ……………… 64
图1-64　4周龄猪头部正中面 ………… 65
图1-65　猪腭及舌的发育 ……………… 66
图1-66　猪颈腹侧横断面 ……………… 67
图1-67　猪头部和颈部的淋巴中心 …… 68

鸟类

图1-68　鸟类的头骨 …………………… 69
图1-69　家禽颈部解剖图，腹侧观 …… 70

不同动物的解剖结构比较

图1-70　不同动物的舌和会厌，背侧观… 71
图1-71　犬、猪、牛和马的大唾液腺 … 72
图1-72　不同动物的垂体的正中切面，垂体的前末端位于左侧 ……… 73
图1-73　犬和猪的第3眼睑 …………… 74

2　颈部、背部和脊柱 ……………………75

犬

图2-1　犬脊柱的横断面 ……………… 76
图2-2　犬的颈椎和胸椎 ……………… 77
图2-3　犬脊神经背根和脊髓节段 ……… 78
图2-4　硬膜掀开的犬脊髓末端示意图 … 79
图2-5　犬颈部和胸部的腹侧肌 ………… 80
图2-6　犬背部在第1腰椎的横断面 …… 81
图2-7　犬寰枕关节和寰枢关节相关的肌肉，外侧观 ……………………… 82
图2-8　犬交感神经和副交感神经的外周分布 …………………………… 83
图2-9　犬的轴上肌 …………………… 84

马

图2-10　马胸骨和肋软骨，外侧观 …… 85
图2-11　马头部浅层解剖图 …………… 86

牛

图2-12　牛近轴侧中胚层的分割 ……… 87
图2-13　牛尾部椎管及其内容物 ……… 88
图2-14　牛椎丛-奇静脉系统与主要静脉的连接 ……………………… 89

鸟类

图2-15　鸟类尾脂腺，背侧观 ………… 90

不同动物的解剖结构比较

图2-16　犬（A）、马（B）、牛（C）、猪（D）的甲状腺 ………………… 91

3　胸部 ……………………………………93

犬

图3-1　心房和心室分区示意图 ……… 94
图3-2　犬的心血管循环模式图 ……… 95
图3-3　犬躯干深层肌 ………………… 96
图3-4　犬躯干最深层肌 ……………… 97
图3-5　犬胸部和前肢间的肌肉悬吊 …… 98
图3-6　犬胸膜和心包膜的分布 ……… 99
图3-7　犬心腔纵切面 ………………… 100
图3-8　犬心包膜 ……………………… 101
图3-9　犬心脏神经和相关神经节，左外侧观 ………………………… 102
图3-10　犬动脉壁的组成 …………… 103
图3-11　犬主动脉弓的分支 ………… 104
图3-12　犬胎儿（A）及胎儿出生后（B）

的循环系统 …… 105
图3-13 犬腰荐部淋巴引流，腹侧观 … 106
图3-14 犬胸腔底壁血管 …… 107
图3-15 犬颈部和胸部肌肉，外侧观 … 108
猫
图3-16 猫心脏和肺的左、右侧体表投影 …… 109
马
图3-17 马心脏和肺在左、右胸壁的投影 …… 110
牛
图3-18 牛胸腔，左外侧观 …… 111
图3-19 牛心脏的基部，背侧观 …… 112
图3-20 牛肺叶及支气管树模式图，背侧观 …… 113
猪
图3-21 心内膜管融合后的猪胚胎 …… 114
图3-22 猪心脏原位图 …… 115
图3-23 猪胸腔淋巴中心，左外侧观 … 116
鸟类
图3-24 鸟类飞行肌肉的解剖图，腹侧观 …… 117
图3-25 鸟类右肺（腹中线观）及相关气囊 …… 118
图3-26 鸟类肾脏及其邻近的血管和神经，腹侧观 …… 119

4 腹部 …… 121
犬
图4-1 犬消化器官示意图 …… 122
图4-2 犬腹直肌鞘横断面 …… 123
图4-3 犬腹股沟管及盆膈，左外侧观…… 124
图4-4 犬腹部横断面示意图 …… 125
图4-5 犬腹腔靠近中央的切面，显示腹膜的分布 …… 126
图4-6 犬腹腔动脉分布，腹侧观 …… 127
图4-7 犬切除大网膜后的腹部器官，腹侧观 …… 128
图4-8 犬会阴部的肌肉 …… 129
图4-9 犬肠系膜前动脉和肠系膜后动脉的分布，背侧观 …… 130
图4-10 犬门静脉构造的部分示意图，背侧观 …… 131
图4-11 犬肝脏的脏面 …… 132
图4-12 犬肝脏的发育 …… 133
图4-13 犬肠管在形成肠袢时的发育 … 134
图4-14 犬腹腔和盆腔器官的淋巴引流 …… 135
图4-15 犬腹腔神经节和神经丛，腹侧观 …… 136
图4-16 犬躯干的第11胸椎水平横断面 …… 137
图4-17 犬躯干的第12胸椎水平横断面 …… 138
图4-18 犬第1腰椎（A）和第4或第5腰椎（B）水平的腹部横断面 … 139
图4-19 犬肠道的血液供应，腹侧观…… 140
图4-20 犬（雄性）腹部肌肉和腹股沟深层解剖图，左侧观 …… 141
图4-21 犬腹膜的折转，矢状面 …… 142
图4-22 犬肝脏，脏面…… 143
图4-23 伴有主要吻合支的犬腹腔动脉和肠系膜前动脉分支 …… 144
马
图4-24 附着于骨盆和耻前腱的马腹部肌肉 …… 145
图4-25 马腹部肌肉及其骨骼附着点 … 146
图4-26 马脾，脏面 …… 147
图4-27 马胃和十二指肠的内部结构 … 148
图4-28 从右侧观察的马肠管 …… 149
图4-29 马大肠和肾脏的位置，背侧观 …… 150
图4-30 马脾、胃、胰腺和肝脏的局部解剖图，腹后侧观 …… 151

牛

图4-31　牛腰椎神经与腰椎横突的关系 ………………………… 152

图4-32　牛腹腔脏器的局部解剖图 …… 153

图4-33　牛网膜囊，后面观 …………… 154

图4-34　牛肠道，右侧观 …………… 155

图4-35　牛的恒齿，上颌（A）和下颌（B）………………………… 156

猪

图4-36　公猪腹股沟管，前面观 ……… 157

图4-37　猪升结肠的发育，左外侧观 ………………………… 158

图4-38　猪主要的腹腔动脉和淋巴结 ………………………… 159

图4-39　猪的恒齿，上颌（A）和下颌（B）………………………… 160

鸟类

图4-40　肝脏、胃和小肠向前掀开可见禽类的胃肠道，腹侧观 … 161

不同动物的解剖结构比较

图4-41　置于同一平面的犬（A）、马（B）和牛（C）的胃肠胃 … 162

图4-42　犬和猫（A）、猪（B）、牛（C）、马（D）的大肠 …… 163

图4-43　肠的横断面（一般解剖） …… 164

5　骨盆和生殖器官 ………………… 165

犬

图5-1　犬的荐骨，前面观 …………… 166

图5-2　犬胚胎早期的矢状面 ………… 167

图5-3　犬泌尿系统及生殖器官 ……… 168

图5-4　犬睾丸发育的三个阶段 ……… 169

图5-5　犬肾叶 ………………………… 170

图5-6　犬睾丸和附睾纵切面 ………… 171

图5-7　犬卵巢活动的不同功能阶段 …… 172

图5-8　雌性犬生殖道的血液供应 …… 173

图5-9　犬胚胎外膜的形成 …………… 174

图5-10　髋关节水平的犬骨盆横断面 … 175

图5-11　雄性犬外生殖器官的深层解剖图 ………………………… 176

图5-12　犬右侧腰荐神经和左侧动脉，腹侧观 ………………………… 177

图5-13　犬右侧臀部神经、动脉和肌肉，外侧观 ………………… 178

猫

图5-14　雄性猫原位生殖器官，左外侧观 ……………………… 179

马

图5-15　母马后腹部和盆腔器官原位图，正中面观 ……………………… 180

图5-16　由马荐椎和腰椎后部发出的腹侧支形成腰间神经丛，腹侧观… 181

图5-17　马的右侧卵巢、输卵管和子宫角，外侧观 ……………………… 182

牛

图5-18　牛腹主动脉后部分支和局部淋巴结 ……………………… 183

图5-19　妊娠母牛后腹部和骨盆的旁正中面 ……………………… 184

图5-20　牛盆腔血管和神经的内侧面 ……………………… 185

图5-21　母牛生殖器，背侧观 ……… 186

图5-22　牛生殖道血供示意图，腹侧观… 187

图5-23　牛阴茎及其肌肉，后外侧观 … 188

图5-24　从胎儿乳头顶端开始的发育中的导管系统 ……………………… 189

图5-25　牛的静脉和乳房淋巴引流 …… 190

图5-26　剖开的公牛阴囊，前面观 …… 191

猪

图5-27　2.5厘米的猪胚胎腹顶部，腹侧观和外侧观 ……………… 192

图5-28　来自两个猪原基（后肾索和输尿管芽）的后肾发育 ……… 193

图5-29　猪阴茎游离端（横断面） …… 194

6 前肢 …………………………………… 195

犬

图6-1 犬肱骨，前面观 ………………… 196
图6-2 犬左侧尺骨和左侧桡骨 ………… 197
图6-3 犬肩部与臂前浅层肌肉 ………… 198
图6-4 犬左前肢肩臂肌肉，外侧观和内侧观 …………………………… 199
图6-5 犬爪的轴向切面 ………………… 200
图6-6 犬左前肢前臂部肌肉，外侧观和内侧观 ………………………… 201
图6-7 位于肩关节远端的犬左前肢横断面 ………………………………… 202
图6-8 犬左前臂浅层静脉 ……………… 203
图6-9 犬左前肢近腕部横断面 ………… 204
图6-10 犬右前肢大动脉局部解剖图，内侧观 ………………………… 205
图6-11 犬长骨血液供应 ……………… 206
图6-12 犬左侧膝关节，前面观 ……… 207
图6-13 犬黏液囊（A）和腱鞘（B）的横断面 ………………………… 208
图6-14 犬左前肢骨骼，肌肉附着，外侧观 ………………………… 209
图6-15 犬前肢主要伸肌和屈肌 ……… 210
图6-16 犬右前肢神经分布 …………… 211
图6-17 犬右前肢尺神经分布，内侧观 ………………………… 212
图6-18 犬右前肢静脉示意图，内侧观 ………………………… 213

马

图6-19 马浅层肌肉和静脉 …………… 214
图6-20 马胸部腹侧浅层肌肉 ………… 215
图6-21 马躯干与前肢相连的深层肌肉 ………………………… 216
图6-22 马肩部和肘关节相关肌肉，外侧观 ………………………… 217
图6-23 马右肩和右臂内侧面的神经、血管和肌肉 …………………… 218
图6-24 马左前肢深层肌肉，内侧观 … 219
图6-25 马左前肢远端骨骼，背侧观 … 220
图6-26 马蹄部轴面 …………………… 221
图6-27 马右前肢主要动脉，掌侧观…… 222
图6-28 马左肘横断面 ………………… 223

牛

图6-29 牛左前臂正中横断面 ………… 224
图6-30 经外侧趾劈开的牛蹄矢状面 … 225
图6-31 牛前蹄内侧趾矢状面 ………… 226
图6-32 产生角质的牛蹄真皮，远轴面及底面观 ……………………… 227
图6-33 牛前肢的主要静脉 …………… 228
图6-34 牛右前蹄主要神经，外侧观和背侧观 ……………………… 229

猪

图6-35 猪左前脚，后内侧观 ………… 230

鸟类

图6-36 家禽左翼部骨骼，部分向外侧伸展 ………………………… 231
图6-37 侧向伸展的家禽左翼部腹侧浅层结构 ……………………… 232

不同动物的解剖结构比较

图6-38 腕骨 ………………………… 233
图6-39 甲、爪和蹄示意图 ………… 234

7 后肢 …………………………………… 235

犬

图7-1 犬右脚骨骼，背侧观 ………… 236
图7-2 犬左膝关节，前面观（A~C）…… 237
图7-3 犬后肢和大腿肌肉，外侧观（A）和内侧观（B）………………… 238
图7-4 犬左小腿肌肉，外侧观和内侧观 ………………………………… 239
图7-5 犬左大腿横断面 ……………… 240
图7-6 犬左后肢 ……………………… 241
图7-7 犬左后肢浅层肌肉，内侧观 …… 242
图7-8 犬骨盆和左后肢的肌肉附着，

内侧观 …… 243

图7-9 犬右后肢动脉模式图，内侧观 …… 244

图7-10 犬右侧股部与小腿动脉和神经，外侧观 …… 245

马

图7-11 马左侧股骨，前面观和外侧观 …… 246

图7-12 马左侧胫骨和腓骨，前面观和外侧观 …… 247

图7-13 马左后肢骨，外侧观 …… 248

图7-14 马股，内侧观 …… 249

图7-15 马左膝关节，前面观 …… 250

图7-16 马的跗关节黏液囊、腱鞘和关节囊 …… 251

图7-17 马右后肢主要动脉，跖侧观 …… 252

图7-18 马右后蹄部神经 …… 253

牛

图7-19 牛荐坐韧带，左外侧观 …… 254

图7-20 牛左后肢肌肉，外侧观 …… 255

图7-21 牛左腿的横断面 …… 256

图7-22 牛右后肢大静脉 …… 257

图7-23 牛右后肢神经 …… 258

图7-24 牛左跖骨横断面 …… 259

猪

图7-25 猪后肢淋巴结，外侧观 …… 260

不同动物的解剖结构比较

图7-26 跗骨示意图 …… 261

8 异宠 …… 263

图8-1 鼬科动物白鼬的骨骼 …… 264

图8-2 鼬科动物白鼬的内部解剖图，腹侧观 …… 265

图8-3 兔的骨骼 …… 266

图8-4 雌性野兔内脏解剖图，腹侧观 …… 267

图8-5 野兔头骨横断面显示的下颌、上颌及单个牙齿 …… 268

图8-6 野兔的阴道 …… 269

图8-7 雌性豚鼠内脏解剖图，腹侧观 …… 270

图8-8 仓鼠科动物仓鼠（雌性）内脏解剖图，腹侧观 …… 271

图8-9 仓鼠科动物蒙古沙鼠（雄性）内脏解剖图，腹侧观 …… 272

图8-10 猬科动物非洲刺猬（雌性）内脏解剖图（腹侧观），以及用于卷曲身体和控制脊柱的肌肉 …… 273

图8-11 鼠科动物大鼠（雌性）内脏解剖图，腹侧观 …… 274

图8-12 蛇科动物蛇内脏解剖图，腹侧观 …… 275

图8-13 蛇科动物蛇心脏，背侧观 …… 276

图8-14 蜥蜴科动物草原巨蜥内脏解剖图，腹侧观 …… 277

图8-15 蜥蜴科动物蜥蜴（雌性）的泄殖腔解剖图 …… 278

图8-16 海龟科动物乌龟内脏解剖图，腹侧观，以及胸甲和甲壳的命名 …… 279

图8-17 青蛙的内脏解剖图，腹侧观 …… 280

图8-18 非鳄目爬行动物的一般循环 …… 281

1

头部和颈腹部

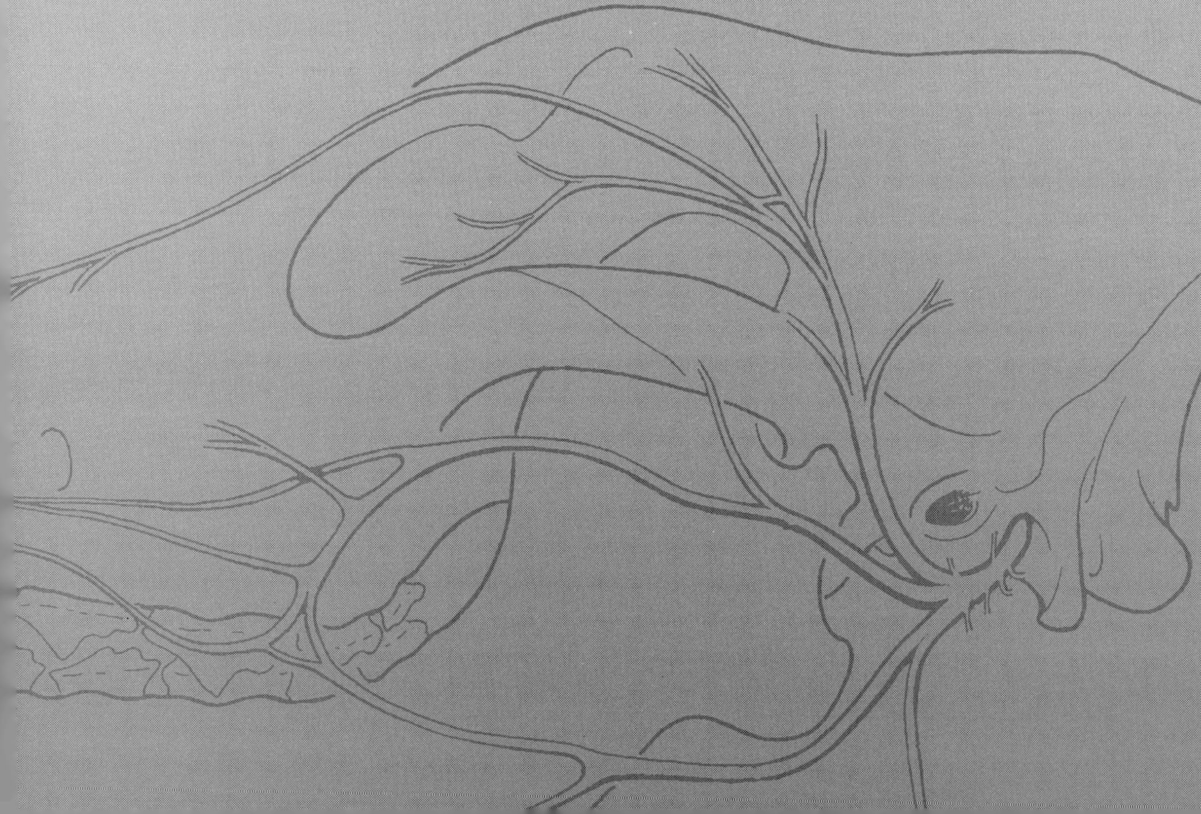

犬

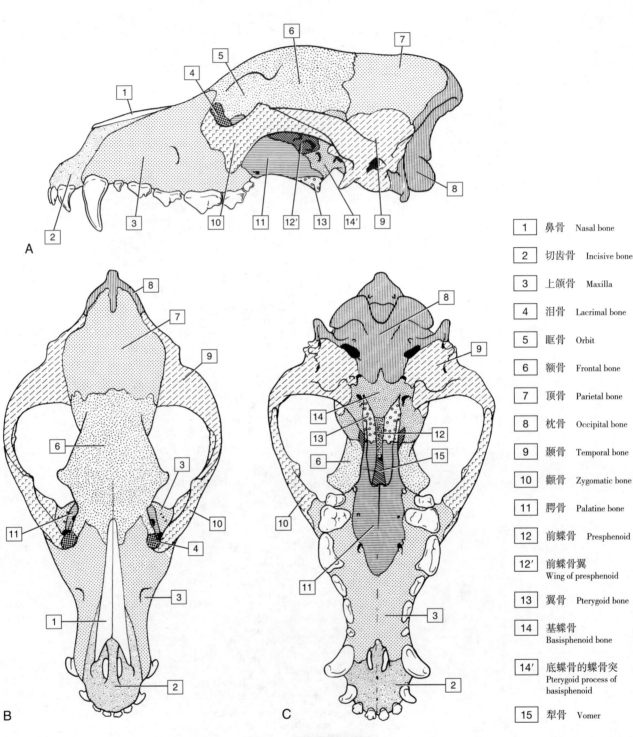

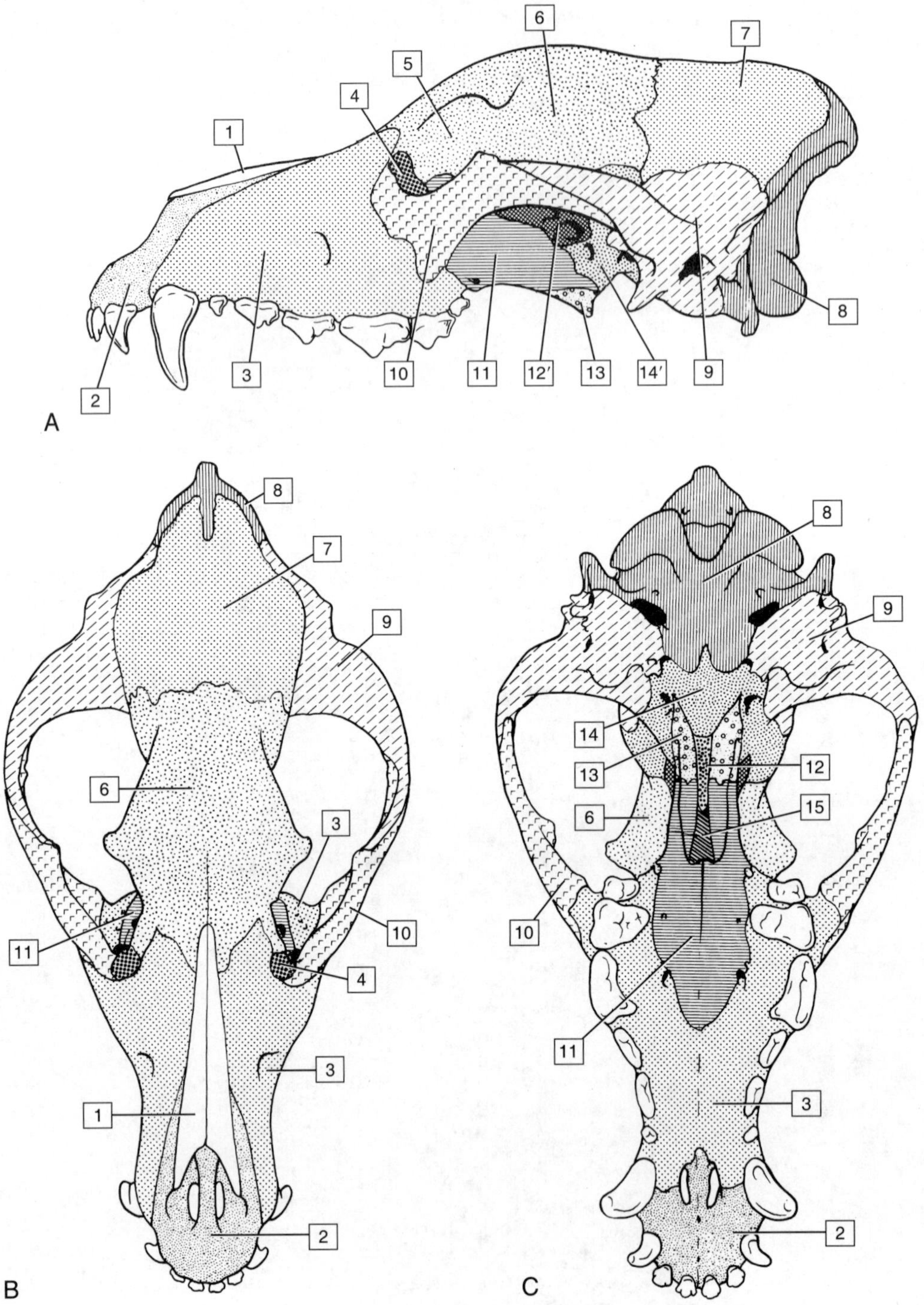

1	鼻骨	Nasal bone
2	切齿骨	Incisive bone
3	上颌骨	Maxilla
4	泪骨	Lacrimal bone
5	眶骨	Orbit
6	额骨	Frontal bone
7	顶骨	Parietal bone
8	枕骨	Occipital bone
9	颞骨	Temporal bone
10	颧骨	Zygomatic bone
11	腭骨	Palatine bone
12	前蝶骨	Presphenoid
12′	前蝶骨翼	Wing of presphenoid
13	翼骨	Pterygoid bone
14	基蝶骨	Basisphenoid bone
14′	底蝶骨的蝶骨突	Pterygoid process of basisphenoid
15	犁骨	Vomer

图1-1　犬颅骨的范围

A. 犬颅骨侧面观；B. 犬颅骨背侧观；C. 犬颅骨腹侧观

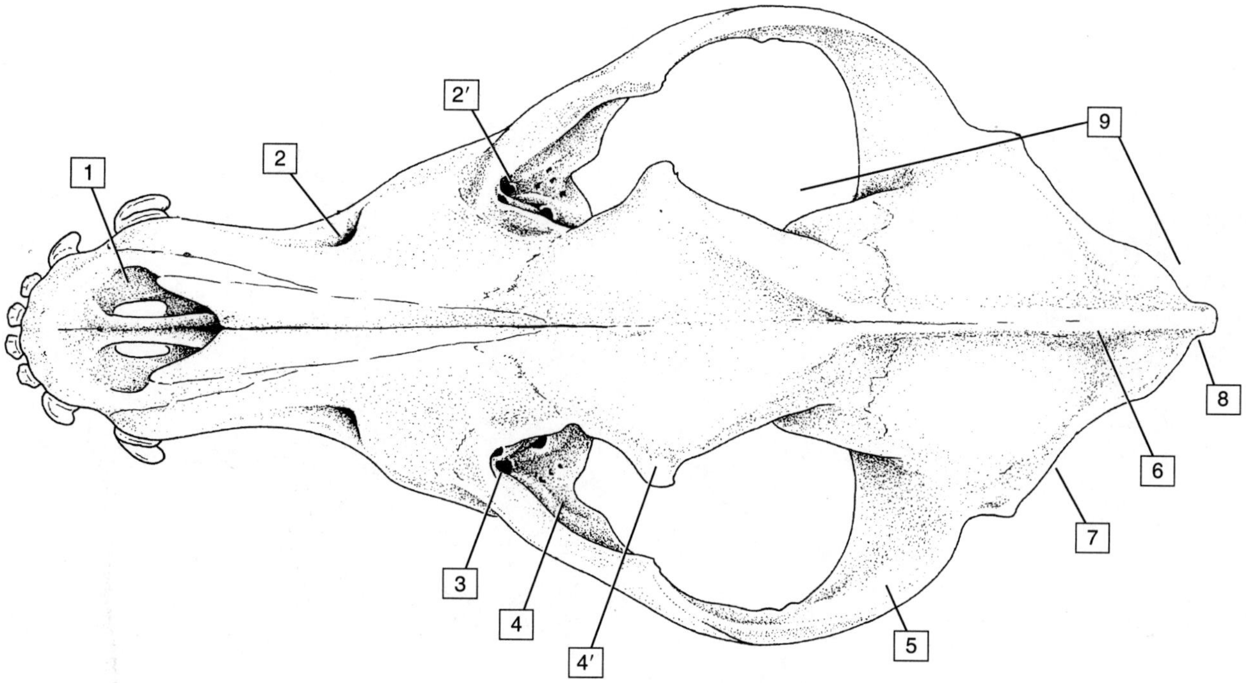

1 鼻孔 Nasal aperture	5 颧弓 Zygomatic arch
2 上颌孔 Maxillary foramen	6 外矢状嵴 External sagittal crest
2′ 眶下孔 Infraorbital foramen	7 颈嵴 Nuchal crest
3 泪囊窝 Fossa for lacrimal sac	8 枕外隆凸 External occipital protuberance
4 眶骨 Orbit	9 颅骨 Cranium
4′ 额骨颧突 Zygomatic process of frontal bone	

图1-2 犬颅骨，背侧观

请画出枕骨、顶骨、腭骨和上颌骨的轮廓。

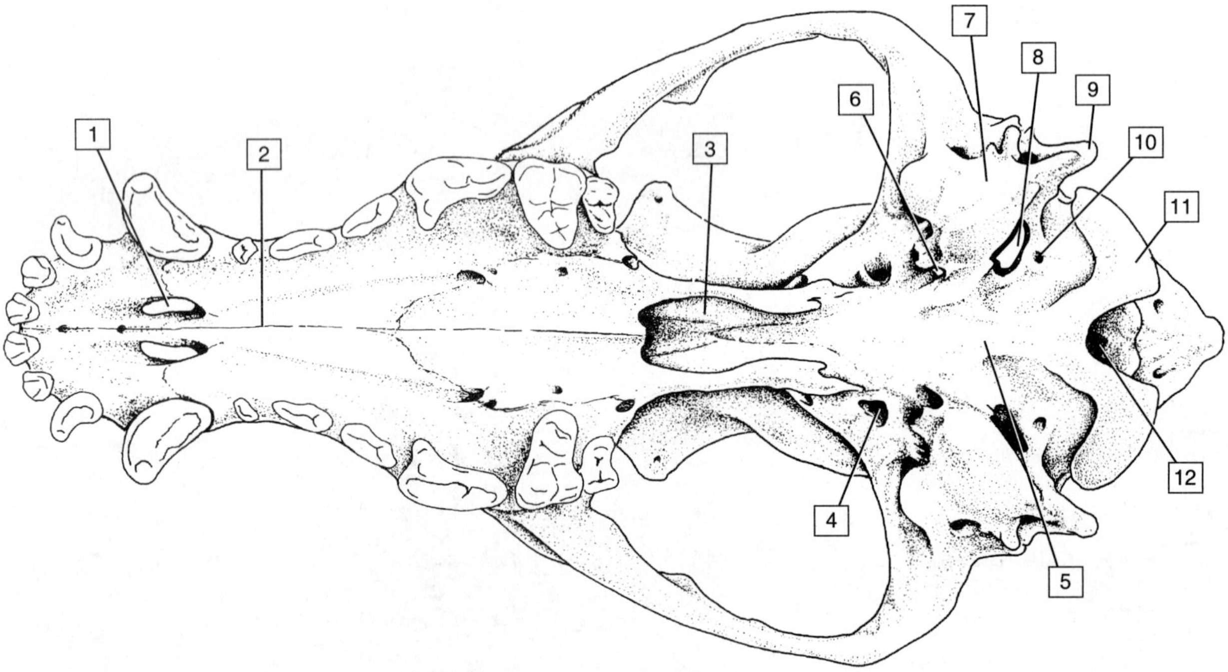

1	腭裂 Palatine fissure	7	鼓泡 Tympanic bulla
2	硬腭 Hard palate	8	颈静脉孔 Jugular foramen
3	鼻后孔 Choanal region	9	髁旁突 Paracondylar process
4	卵圆孔 Oval foramen	10	舌下神经管 Hypoglossal canal
5	颅底 Base of cranium	11	枕髁 Occipital condyle
6	破裂孔 Foramen lacerum	12	枕骨大孔 Foramen magnum

图1-3 犬颅骨，腹侧观

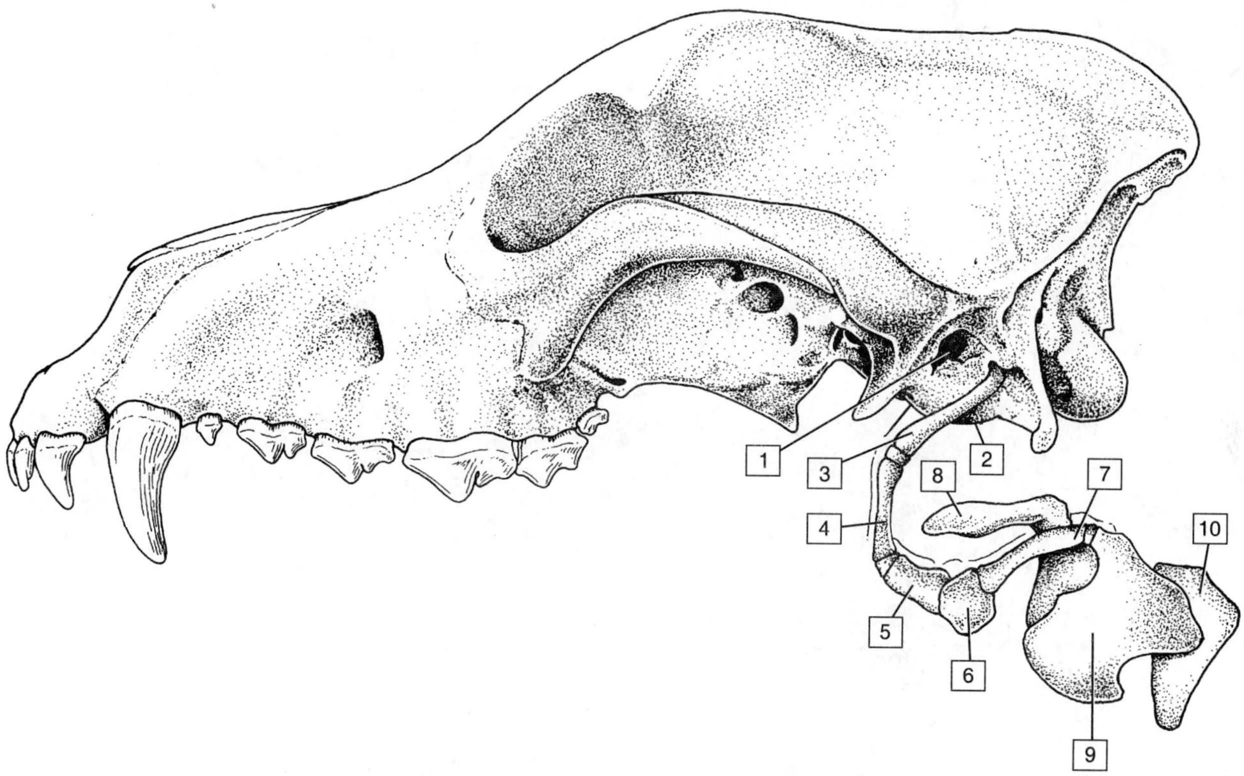

1	外耳道 External acoustic meatus	6	基舌骨 Basihyoid
2	鼓泡 Tympanic bulla	7	甲状舌骨肌 Thyrohyoid
3	茎突舌骨 Stylohyoid	8	会厌软骨 Epiglottic cartilage
4	上舌骨 Epihyoid	9	甲状软骨 Thyroid cartilage
5	角舌骨 Ceratohyoid	10	环状软骨 Cricoid cartilage

图1-4 舌骨器和悬吊于犬颅骨颞部的喉

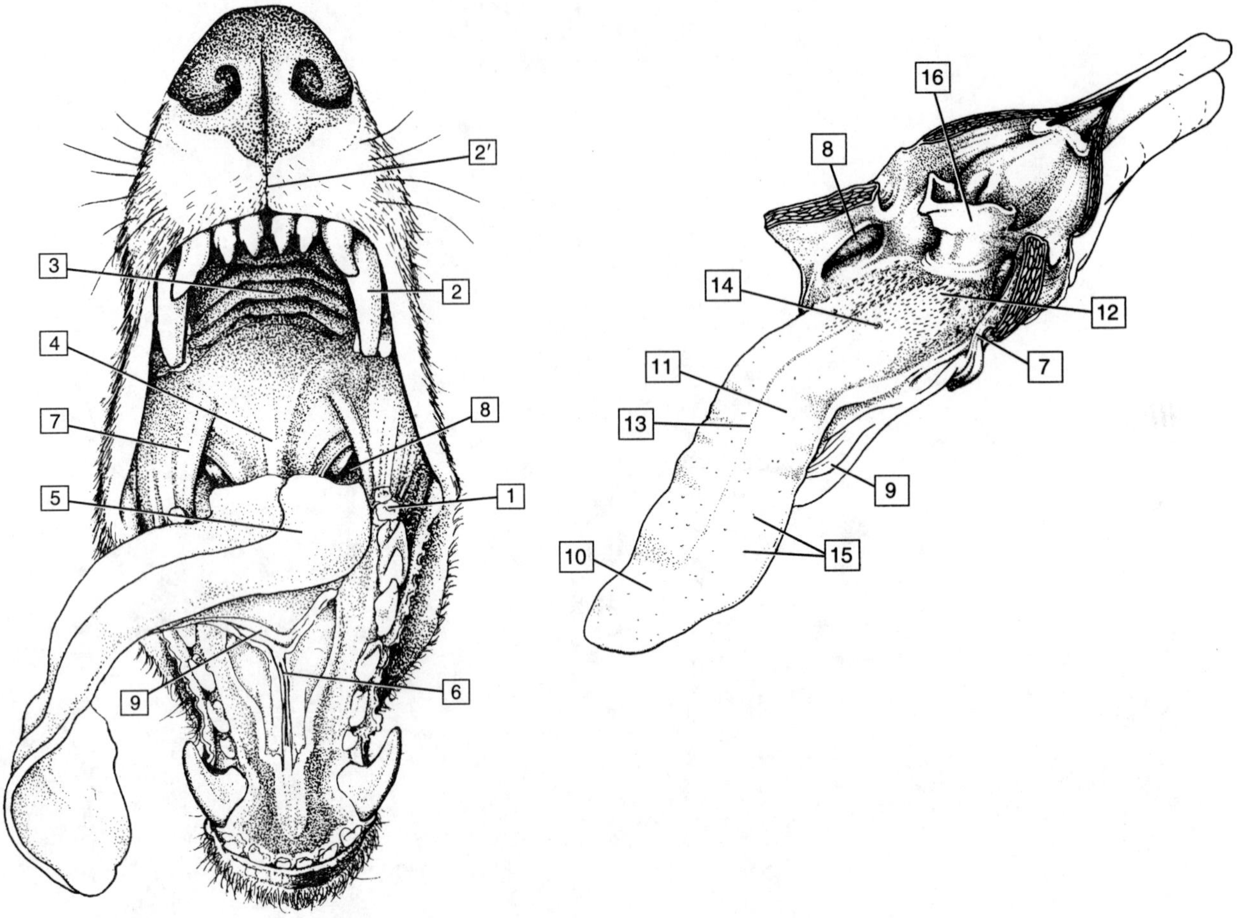

1	前庭	Vestibule
2	犬齿	Canine tooth
2′	鼻镜	Philtrum
3	硬腭	Hard palate
4	软腭	Soft palate
5	舌	Tongue
6	舌下肉阜	Sublingual caruncle
7	腭舌弓	Palatoglossal arch
8	腭扁桃体	Palatine tonsil
9	舌系带	Frenulum
10	舌尖	Apex of tongue
11	舌体	Body of tongue
12	舌根	Root of tongue
13	舌正中沟	Median groove
14	轮廓乳头	Vallate papilla
15	菌状乳头	Fungiform papillae
16	会厌	Epiglottis

图1-5　犬口腔和舌

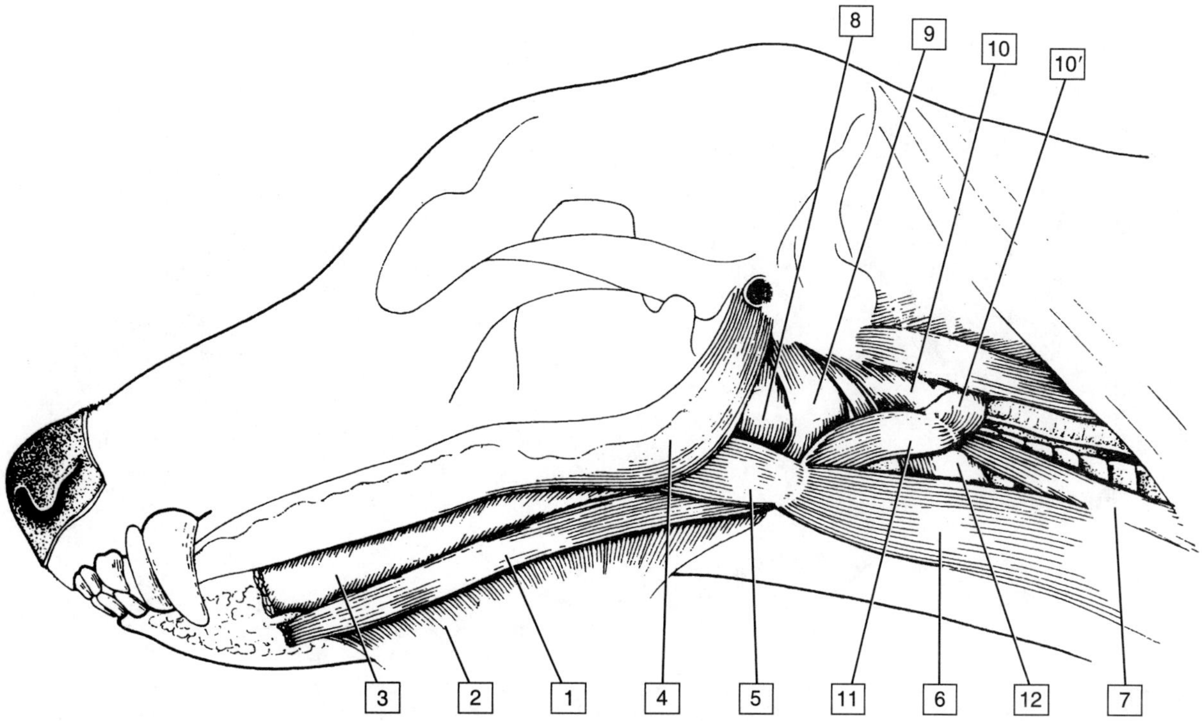

1 颏舌骨肌 Geniohyoideus

2 下颌舌骨肌 Mylohyoideus

3 颏舌肌 Genioglossus

4 茎突舌肌 Styloglossus

5 舌骨舌肌 Hyoglossus

6 胸骨舌骨肌 Sternohyoideus

7 胸骨甲状肌 Sternothyroideus

8 舌骨咽肌（两部分） Hyopharyngeus (two parts)

9 舌骨咽肌（两部分） Hyopharyngeus (two parts)

10 甲咽肌 Thyropharyngeus

10′ 环咽肌 Cricopharyngeus

11 甲状舌骨肌 Thyrohyoid

12 环甲肌 Cricothyroideus

图1-6 犬舌和咽部肌肉

请写出支配 3 ~ 5 肌肉的神经。

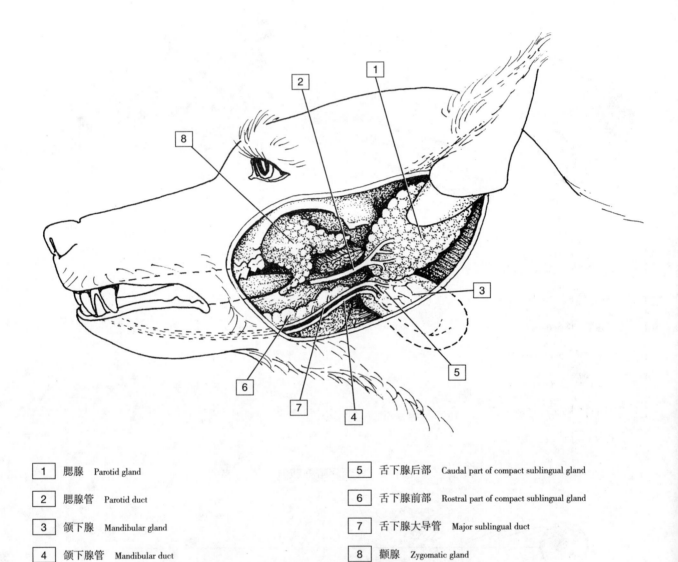

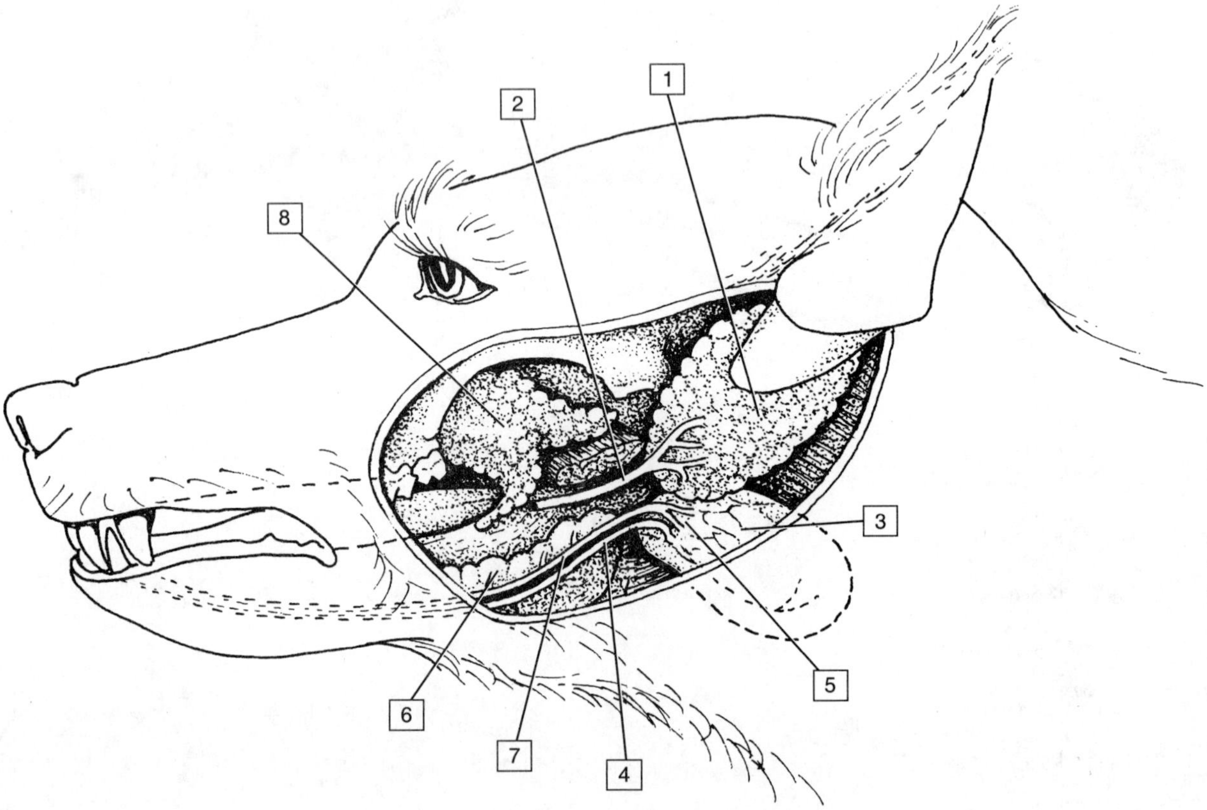

1	腮腺 Parotid gland	5	舌下腺后部 Caudal part of compact sublingual gland
2	腮腺管 Parotid duct	6	舌下腺前部 Rostral part of compact sublingual gland
3	颌下腺 Mandibular gland	7	舌下腺大导管 Major sublingual duct
4	颌下腺管 Mandibular duct	8	颧腺 Zygomatic gland

图1-7 犬唾液腺

1 牙釉质 Enamel

2 牙质 Dentine

3 牙骨质 Cement

4 牙髓 Pulp

5 牙根尖孔 Apical foramen

6 牙周韧带 Periodontal ligament

7 牙槽 Socket (alveolus)

8 牙龈 Gum

9 颞肌 Temporalis

10 咬肌 Masseter

11 二腹肌的前部和腹部 Rostral and belly of digastricus

12 下颌舌骨肌 Mylohyoideus

13 翼内肌 Medial pterygoid

14 翼外肌起点 Origin of lateral pterygoid

15 舌 Tongue

16 下颌骨 Mandible

17 颧弓 Zygomatic arch

18 横断面 Level of transection

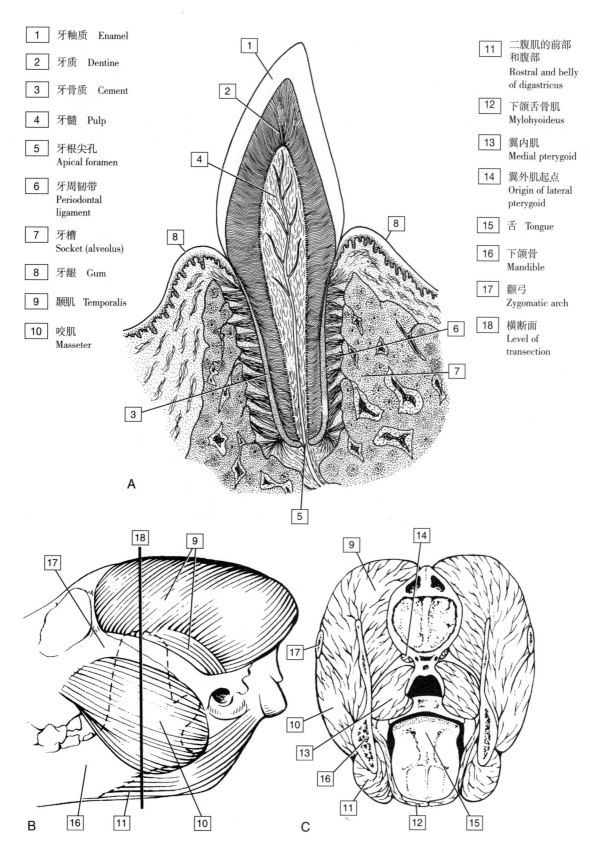

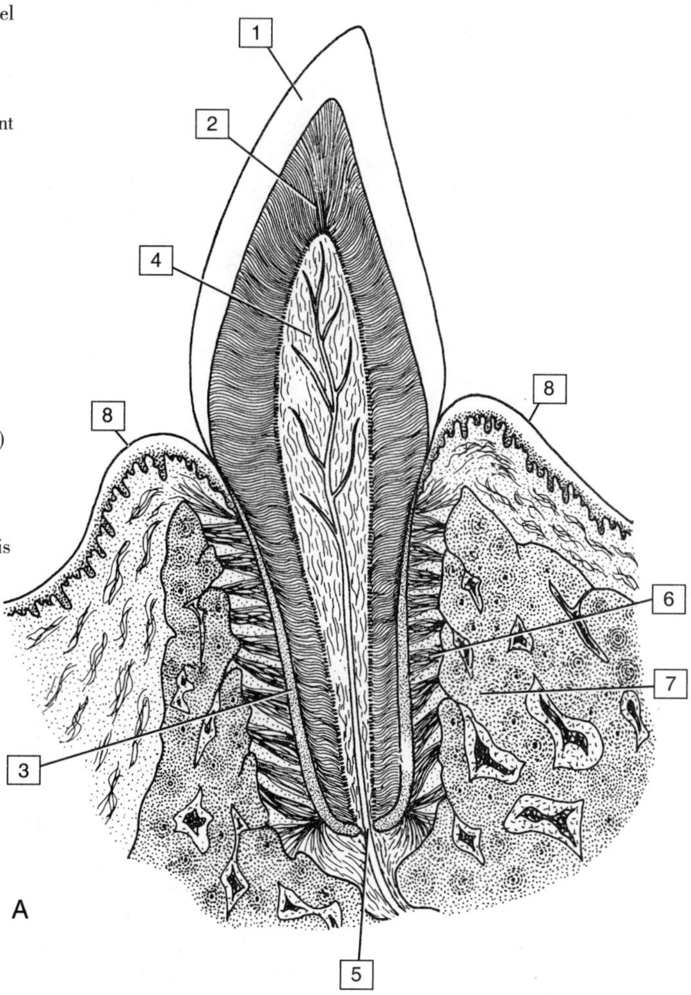

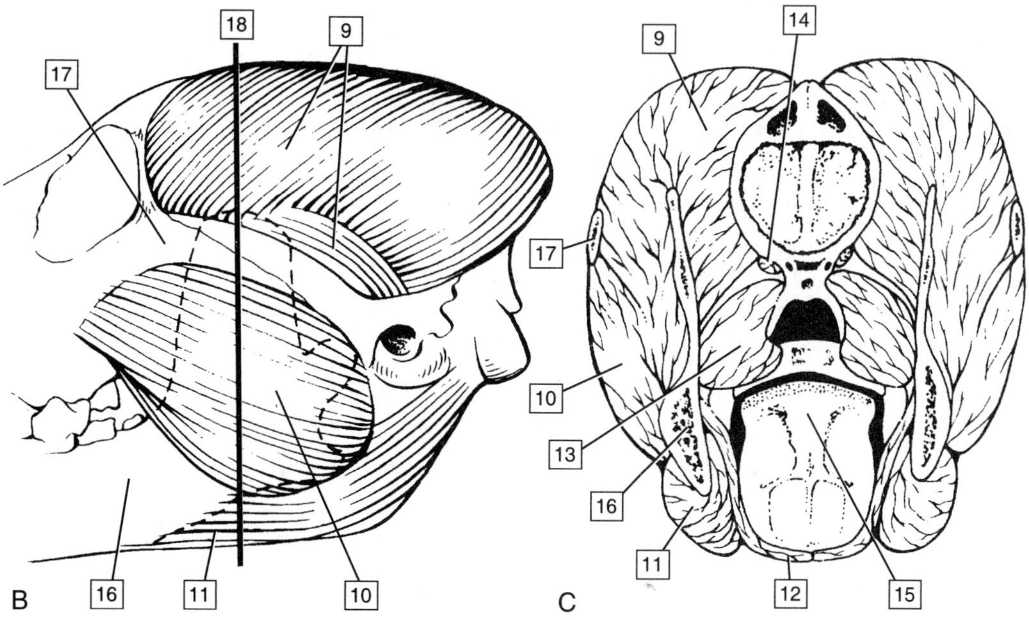

图1-8 犬单齿（A）、外侧（B）和内侧（C）咀嚼肌

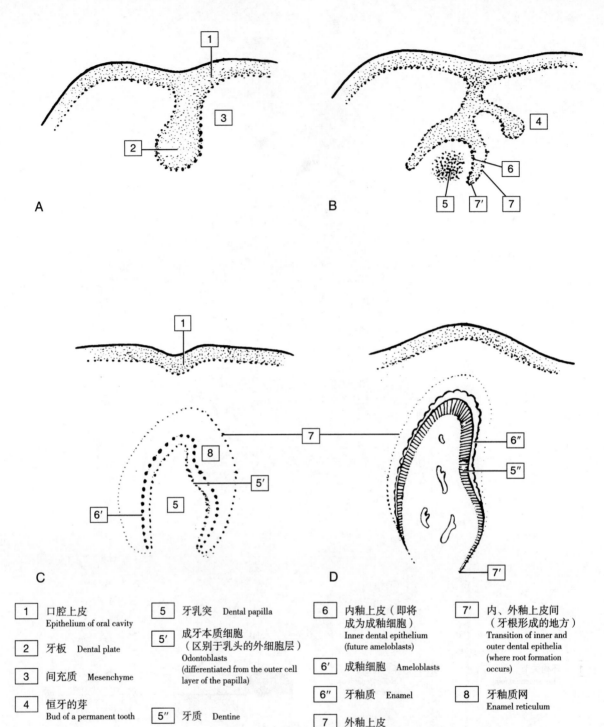

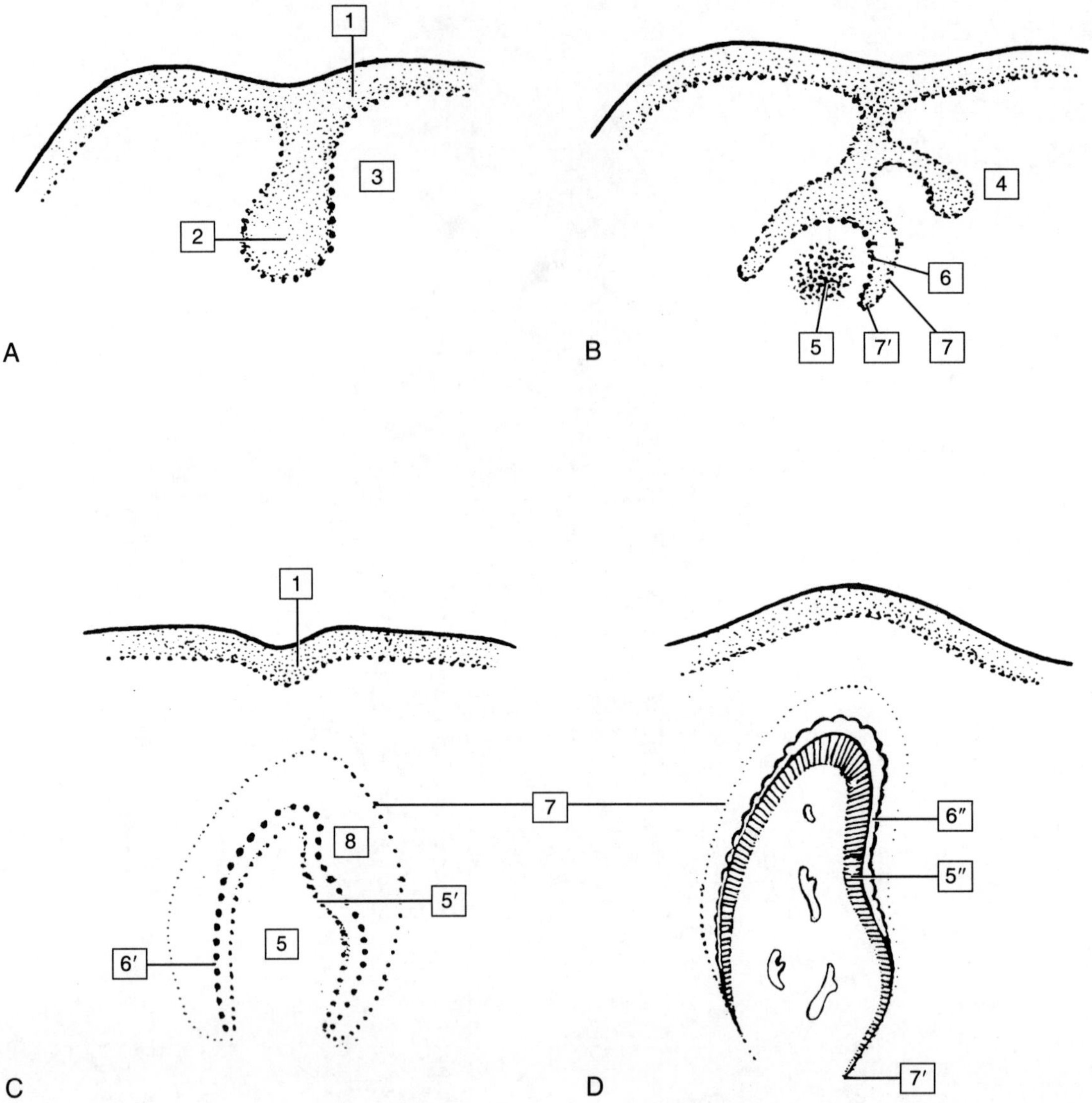

1	口腔上皮 Epithelium of oral cavity
2	牙板 Dental plate
3	间充质 Mesenchyme
4	恒牙的芽 Bud of a permanent tooth
5	牙乳突 Dental papilla
5′	成牙本质细胞（区别于乳头的外细胞层）Odontoblasts (differentiated from the outer cell layer of the papilla)
5″	牙质 Dentine
6	内釉上皮（即将成为成釉细胞）Inner dental epithelium (future ameloblasts)
6′	成釉细胞 Ameloblasts
6″	牙釉质 Enamel
7	外釉上皮 Outer dental epithelium
7′	内、外釉上皮间（牙根形成的地方）Transition of inner and outer dental epithelia (where root formation occurs)
8	牙釉质网 Enamel reticulum

图1-9 犬齿板及成釉器的发育

A. 犬齿板的发育；B、C. 成釉器的发育；D. 乳齿萌出前的发育

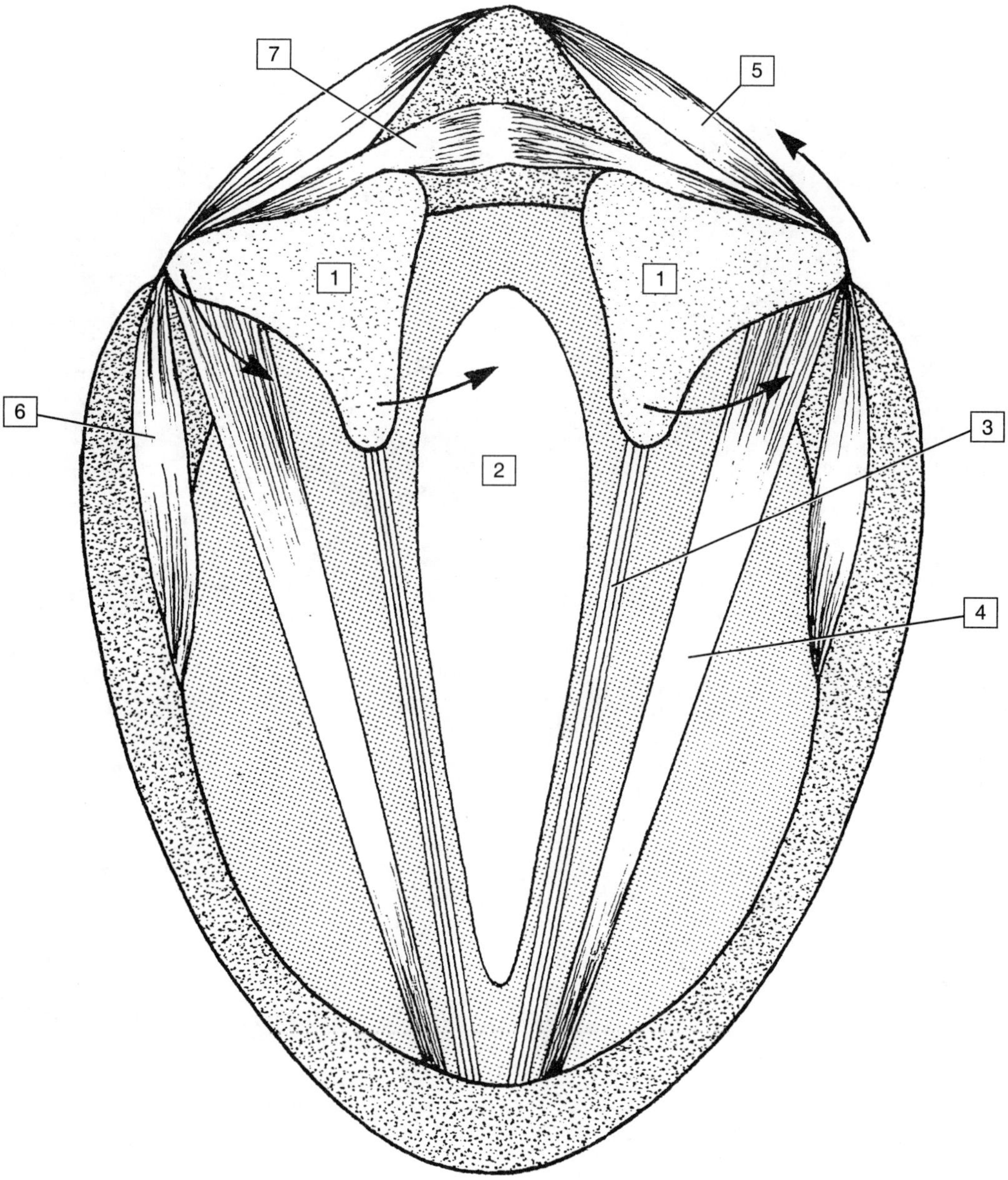

1	环杓关节 Cricoarytenoid joint		5	环杓背肌 Cricoarytenoideus dorsalis
2	声门裂 Glottic cleft		6	环杓侧肌 Cricoarytenoideus lateralis
3	声韧带 Vocal ligament in vocal fold		7	横杓肌 Arytenoideus transversus
4	甲杓肌 Thyroarytenoideus			

请写出支配 4 ~ 7 肌肉的神经。

图1-10 犬喉的横断面

左侧箭头表示环杓侧肌对杓状软骨的作用；右侧箭头表示环杓背肌对杓状软骨的作用

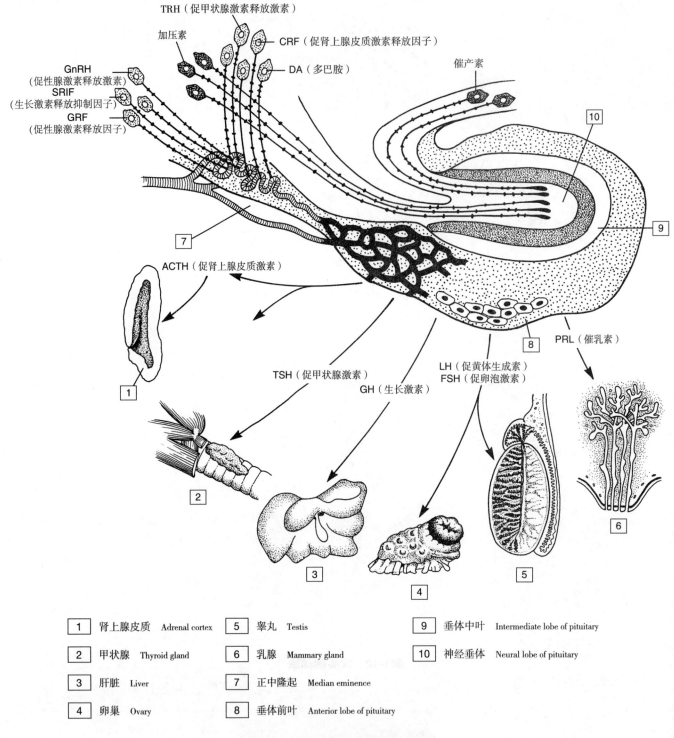

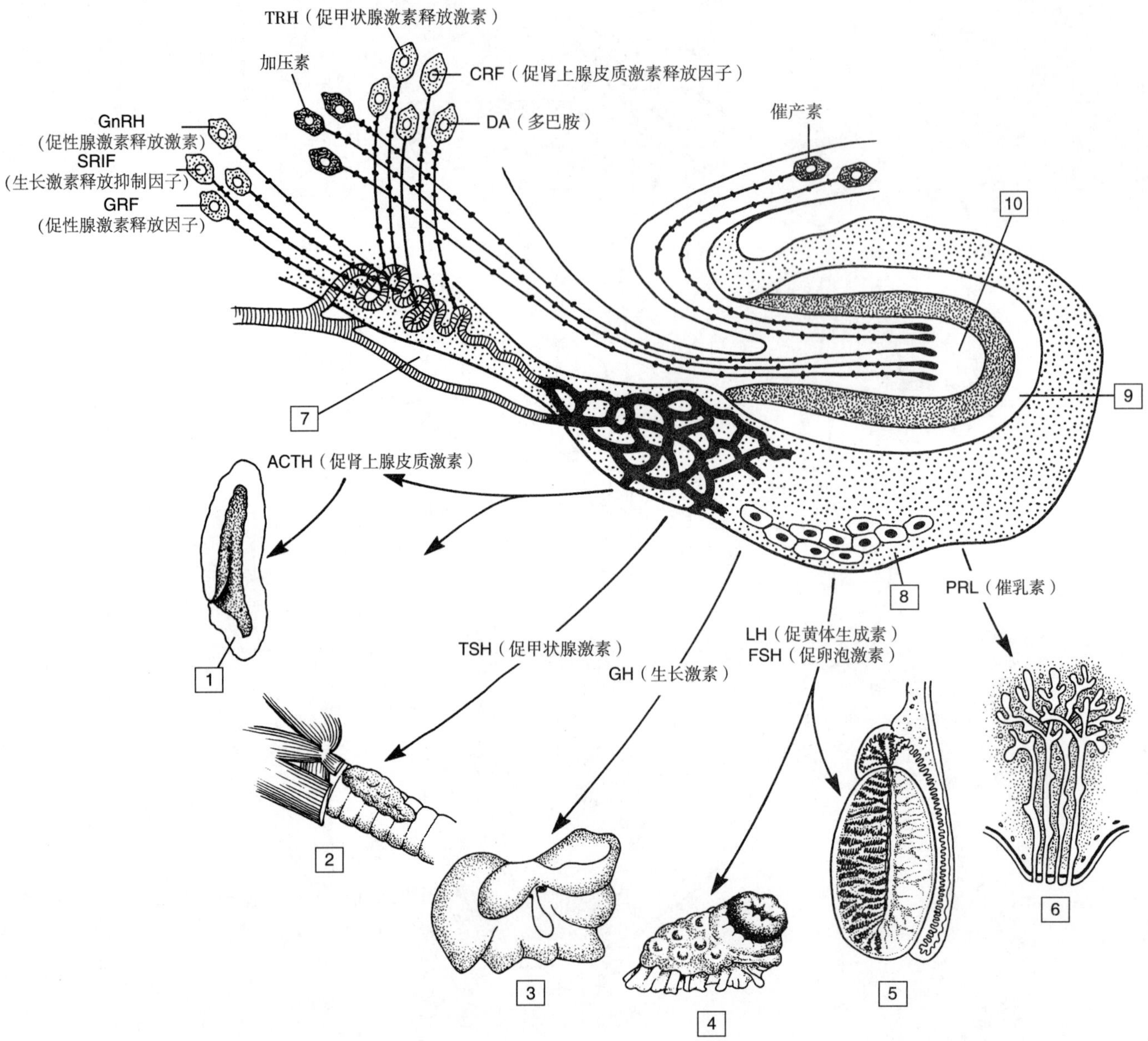

1	肾上腺皮质 Adrenal cortex	5	睾丸 Testis	9	垂体中叶 Intermediate lobe of pituitary
2	甲状腺 Thyroid gland	6	乳腺 Mammary gland	10	神经垂体 Neural lobe of pituitary
3	肝脏 Liver	7	正中隆起 Median eminence		
4	卵巢 Ovary	8	垂体前叶 Anterior lobe of pituitary		

图1-11 大脑-垂体-外周器官轴

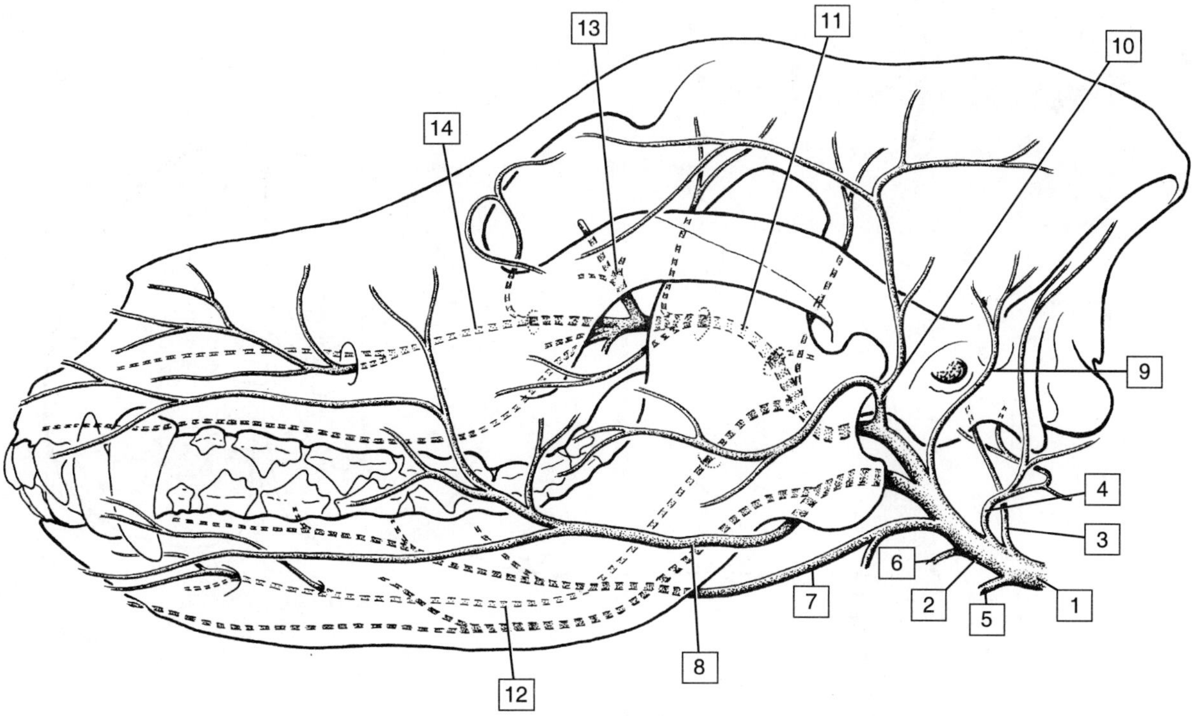

1 颈总动脉 Common carotid artery

2 颈外动脉 External carotid artery

3 颈内动脉 Internal carotid artery

4 枕动脉 Occipital artery

5 喉前动脉 Cranial laryngeal artery

6 咽升动脉 Ascending pharyngeal artery

7 舌动脉 Lingual artery

8 面动脉 Facial artery

9 耳后动脉 Caudal auricular artery

10 颞浅动脉 Superficial temporal artery

11 上颌动脉 Maxillary artery

12 下齿槽动脉 Inferior alveolar artery

13 外眼动脉 External ophthalmic artery

14 眶下动脉 Infraorbital artery

图1-12 犬头部动脉

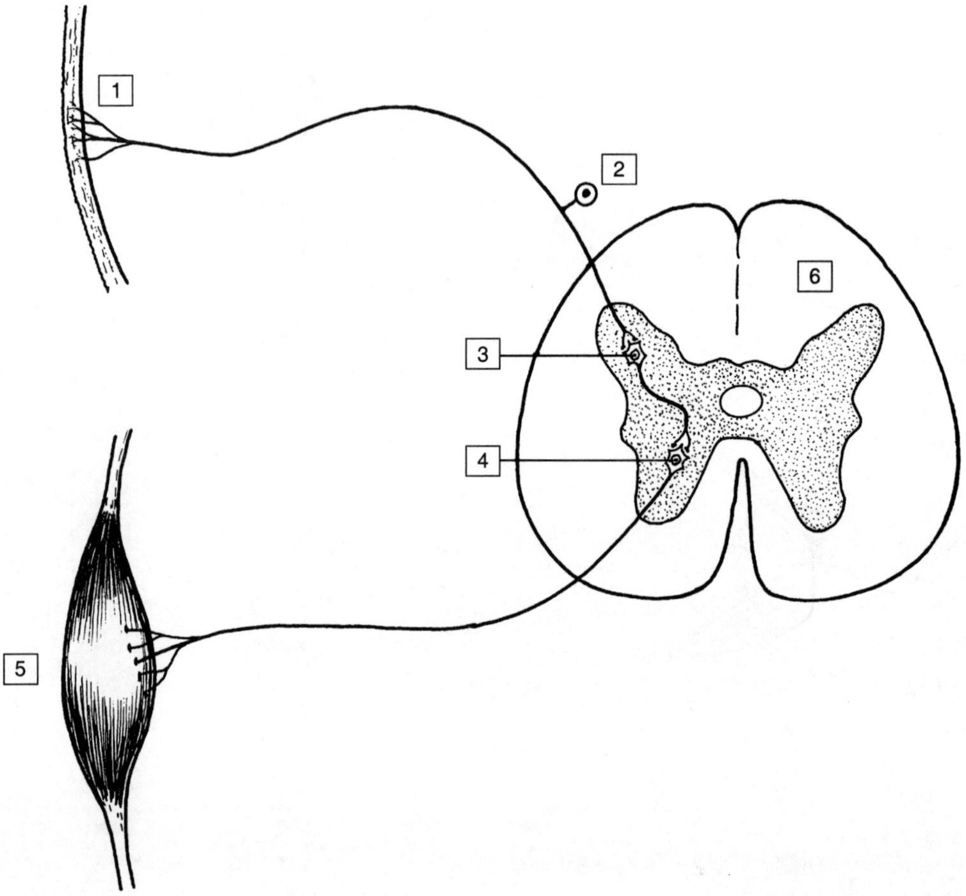

1 皮肤感受器 Skin receptor	4 传出神经元突触 Synapse at efferent neuron
2 传入神经元 Afferent neuron	5 肌肉 Muscle
3 中间神经元突触 Synapse at interneuron	6 脊髓 Spinal cord

图1-13 犬中间神经元反射弧

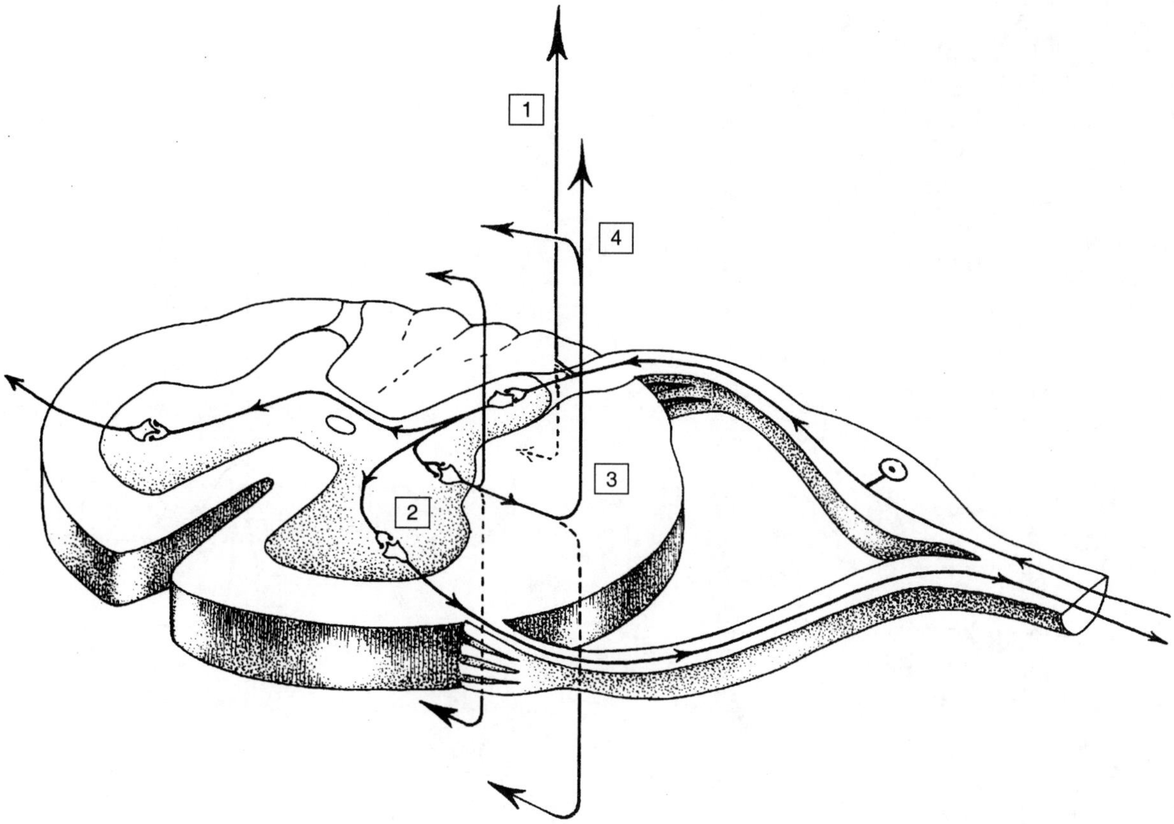

1 背索中的传入纤维向大脑投射（另一支终止于背角的神经元间）
Afferent fibers in the dorsal funiculus traveling toward the brain (others end on interneurons in the dorsal horn)

2 直接传递给传出神经元的脉冲
Impulses transmitted directly to efferent neurons

3 传递给其他中间神经元的脉冲，在脊髓前部或后部传递
Impulses transmitted to other interneurons transmitting impulses caudally or cranially within the spinal cord

4 投射到大脑的脉冲 Impulses extending to the brain

图1-14 犬脊髓内纤维传导路径

大脑半球

小脑

延髓

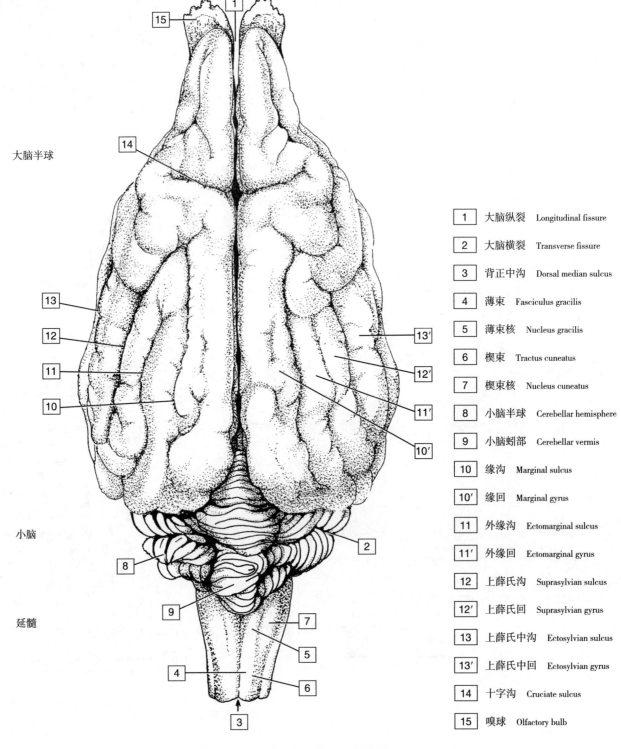

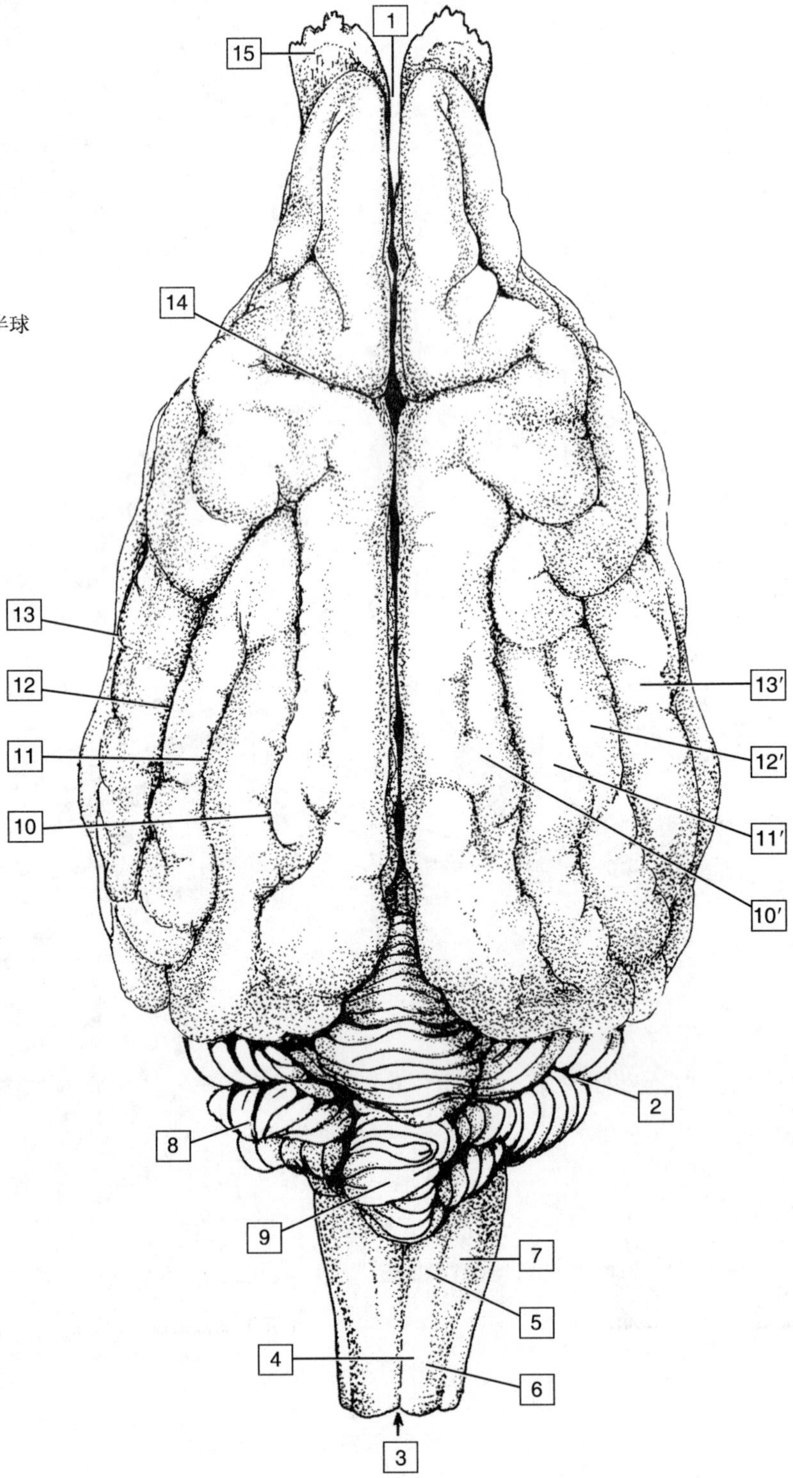

1	大脑纵裂	Longitudinal fissure
2	大脑横裂	Transverse fissure
3	背正中沟	Dorsal median sulcus
4	薄束	Fasciculus gracilis
5	薄束核	Nucleus gracilis
6	楔束	Tractus cuneatus
7	楔束核	Nucleus cuneatus
8	小脑半球	Cerebellar hemisphere
9	小脑蚓部	Cerebellar vermis
10	缘沟	Marginal sulcus
10′	缘回	Marginal gyrus
11	外缘沟	Ectomarginal sulcus
11′	外缘回	Ectomarginal gyrus
12	上薛氏沟	Suprasylvian sulcus
12′	上薛氏回	Suprasylvian gyrus
13	上薛氏中沟	Ectosylvian sulcus
13′	上薛氏中回	Ectosylvian gyrus
14	十字沟	Cruciate sulcus
15	嗅球	Olfactory bulb

图1-15　大脑，背侧观

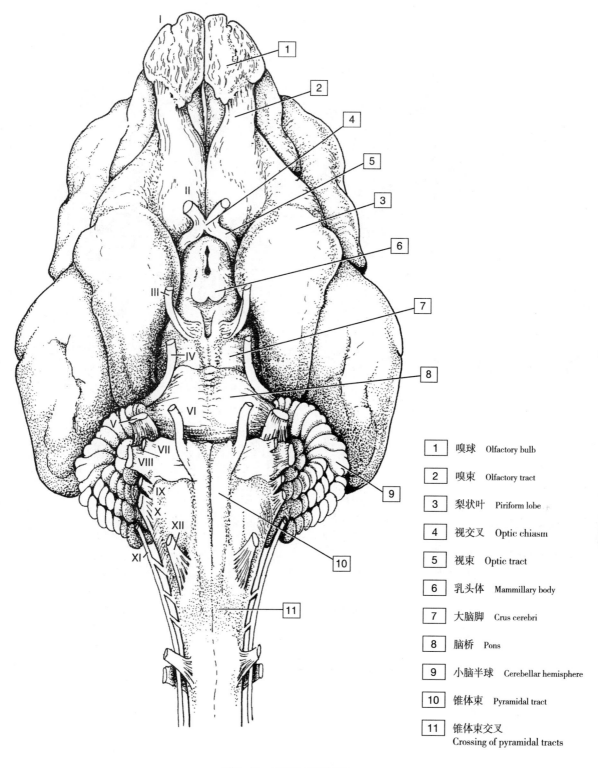

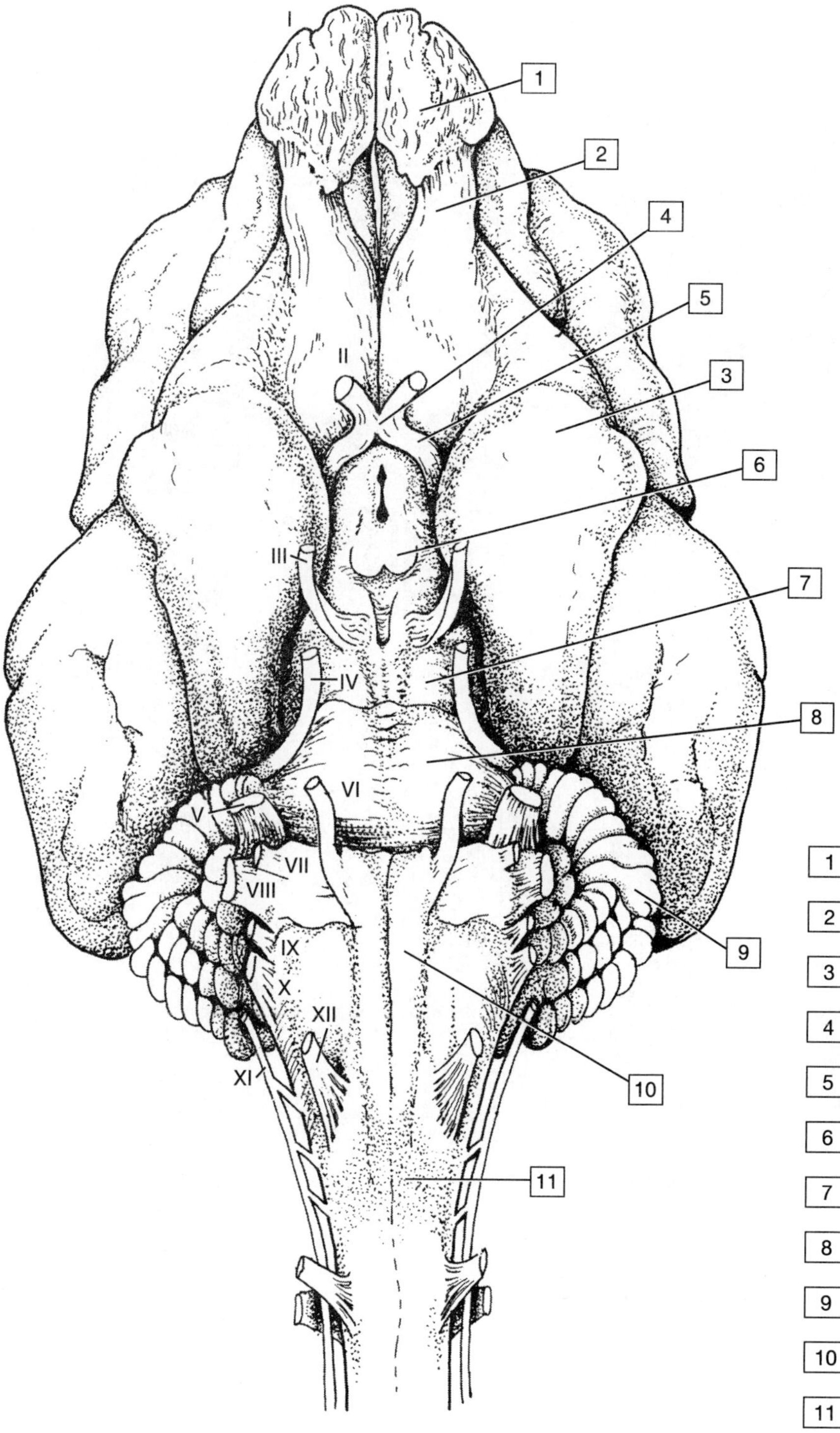

1	嗅球	Olfactory bulb
2	嗅束	Olfactory tract
3	梨状叶	Piriform lobe
4	视交叉	Optic chiasm
5	视束	Optic tract
6	乳头体	Mammillary body
7	大脑脚	Crus cerebri
8	脑桥	Pons
9	小脑半球	Cerebellar hemisphere
10	锥体束	Pyramidal tract
11	锥体束交叉	Crossing of pyramidal tracts

图1-16　大脑，腹侧观

Ⅰ~Ⅻ指相应的脑神经

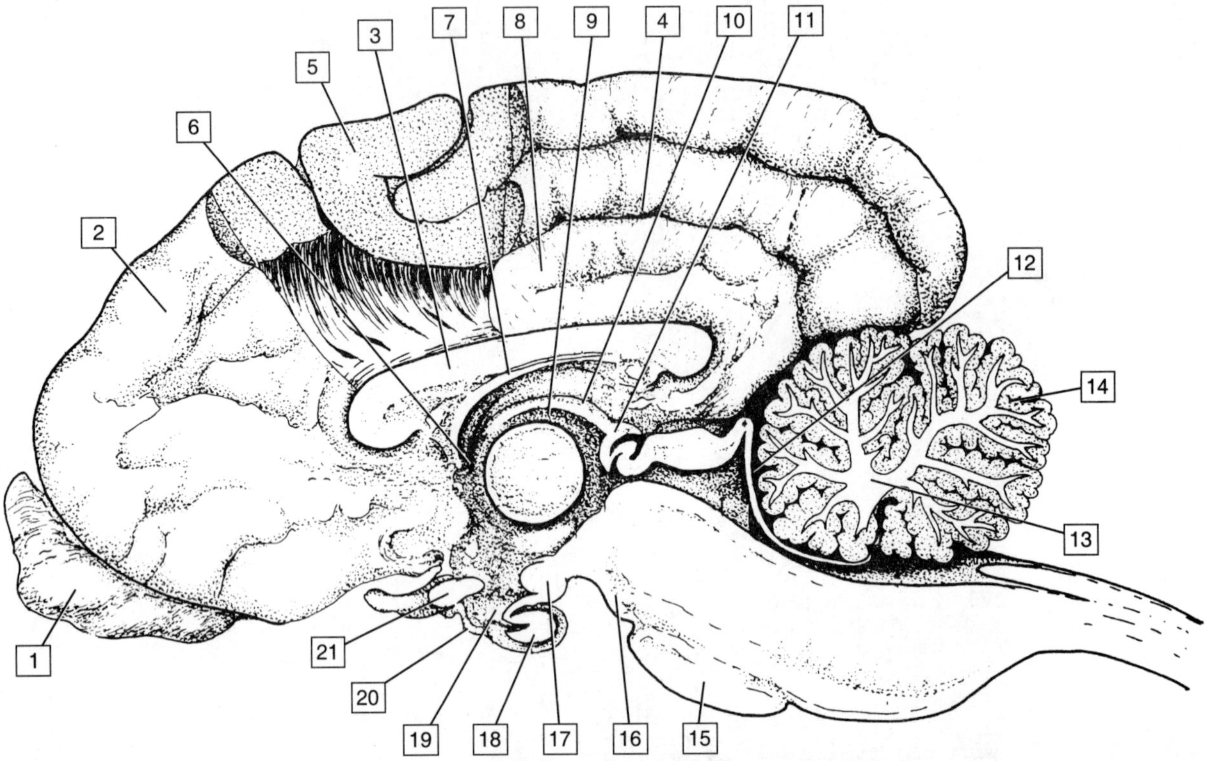

1 嗅球 Olfactory bulb
2 大脑半球 Hemisphere
3 胼胝体 Corpus callosum
4 胼胝体沟 Splenial sulcus
5 大脑皮层 Cerebral cortex
6 室间孔 Interventricular foramen
7 穹隆 Fornix
8 扣带回 Cingulate gyrus
9 丘脑 Thalamus
10 上丘脑 Epithalamus
11 松果体 Epiphysis
12 前髓帆 Rostral medullary velum
13 髓质 Corpus medullare
14 小脑皮层 Cerebellar cortex
15 脑桥 Pons
16 大脑脚 Crus cerebri
17 乳头体 Mammillary body
18 垂体 Hypophysis
19 漏斗 Infundibulum
20 灰结节 tuber cinereum
21 视交叉 Optic chiasm

图1-17 大脑，正中矢状面

大脑半球内侧壁的一部分已被切除

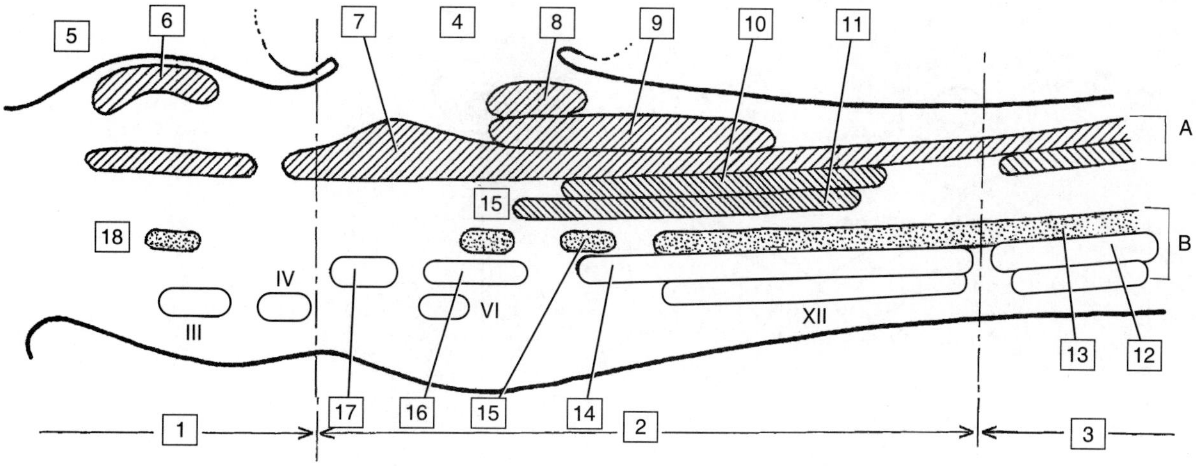

1 中脑 Mesencephalon

2 菱脑 Rhombencephalon

3 脊髓 Spinal cord

4 小脑 Cerebellum

5 中脑顶盖 Tectum mesencephali

6 前丘 Rostral colliculus (SSA)

7 三叉神经核 Trigeminal nuclei (SA)

8 耳蜗核 Cochlear nuclei (SSA)

9 前庭核 Vestibular nuclei (SSA)

10 第Ⅶ、Ⅸ和Ⅹ脑神经孤束核 Solitary nucleus of VII, IX, X (VA)

11 第Ⅶ和Ⅸ脑神经味觉核 Gustatory nuclei of VII, IX (SVA)

12 第Ⅺ脑神经运动核 Motor nucleus of XI (GSE)

13 第Ⅹ脑神经运动核 Motor nucleus of X (GVE)

14 第Ⅸ和Ⅹ脑神经疑核 Nucleus ambiguus of IX, X (GSE)

15 第Ⅶ和Ⅸ脑神经唾液核 Salivatory nuclei of VII, IX (GVE)

16 第Ⅶ脑神经运动核 Motor nucleus of VII (GSE)

17 第Ⅴ脑神经运动核 Motor nucleus of V (GSE)

18 第Ⅲ脑神经副交感神经核 Parasympathetic nucleus of III (GVE)

图1-18 犬脑干，图示成年哺乳动物神经核团

罗马数字表示脑神经；A. 传入核团；B. 传出核团

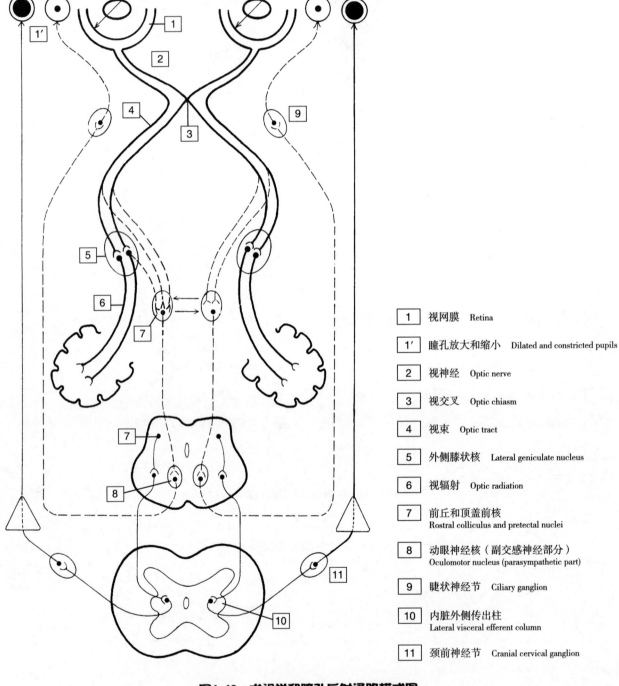

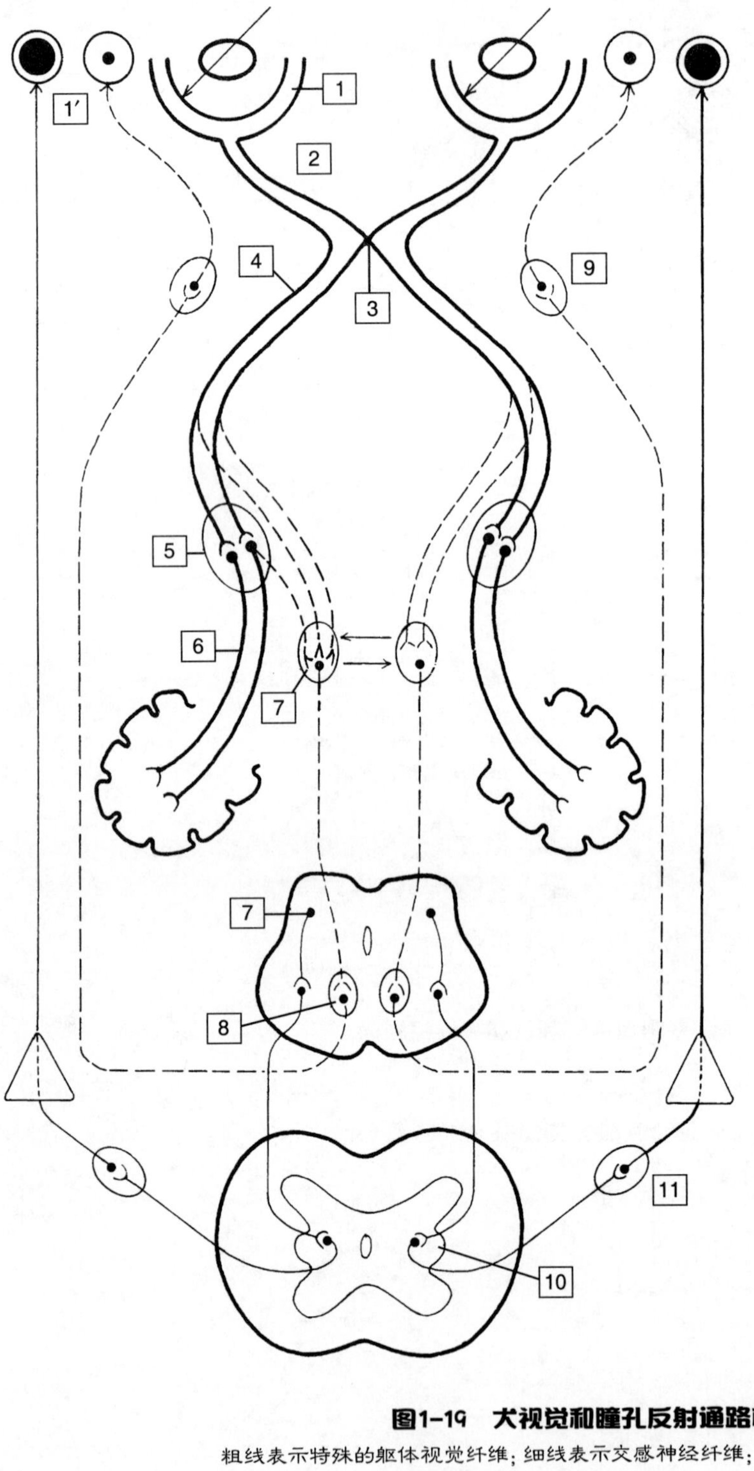

1	视网膜 Retina
1′	瞳孔放大和缩小 Dilated and constricted pupils
2	视神经 Optic nerve
3	视交叉 Optic chiasm
4	视束 Optic tract
5	外侧膝状核 Lateral geniculate nucleus
6	视辐射 Optic radiation
7	前丘和顶盖前核 Rostral colliculus and pretectal nuclei
8	动眼神经核（副交感神经部分） Oculomotor nucleus (parasympathetic part)
9	睫状神经节 Ciliary ganglion
10	内脏外侧传出柱 Lateral visceral efferent column
11	颈前神经节 Cranial cervical ganglion

图1-19 犬视觉和瞳孔反射通路模式图

粗线表示特殊的躯体视觉纤维；细线表示交感神经纤维；虚线表示副交感神经纤维

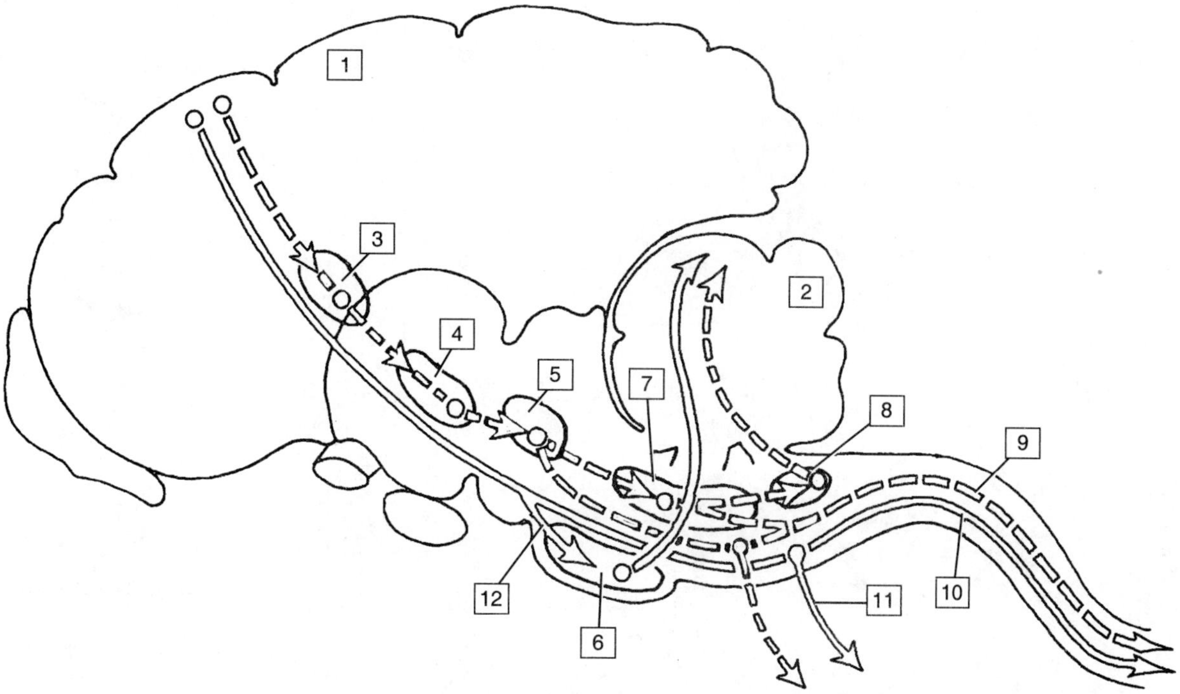

1	运动皮层 Motor cortex	7	网状结构 Reticular formation
2	小脑 Cerebellum	8	橄榄核 Olivary nucleus
3	基底核 Basal nuclei	9	红核脊髓束 Rubrospinal tract
4	黑质（中脑） Substantia nigra (mesencephalon)	10	皮质脊髓纤维 Corticospinal fibers
5	红核（中脑） Red nucleus (mesencephalon)	11	皮质延髓纤维 Corticobulbar fibers
6	脑桥核（后脑） Pontine nuclei (metencephalon)	12	皮质脑桥纤维 Corticopontine fibers

图1-20 椎体系（实线）和椎体外系（虚线）连接图

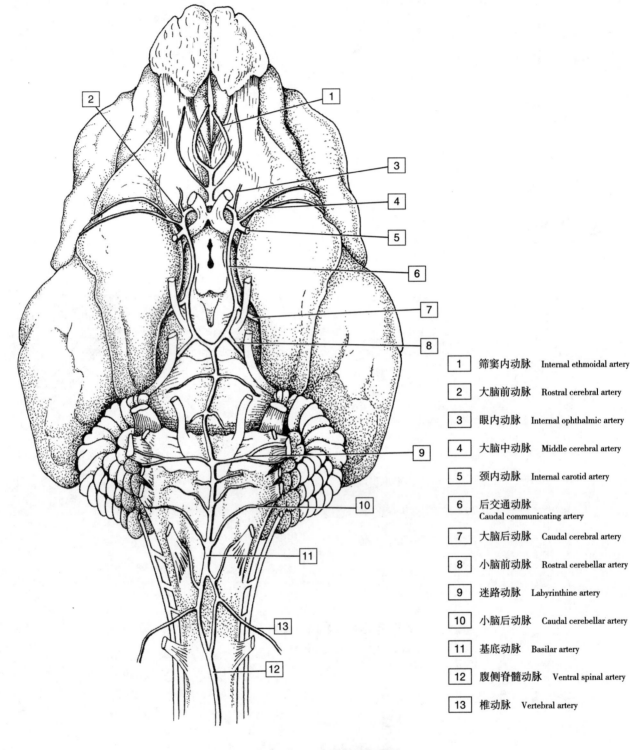

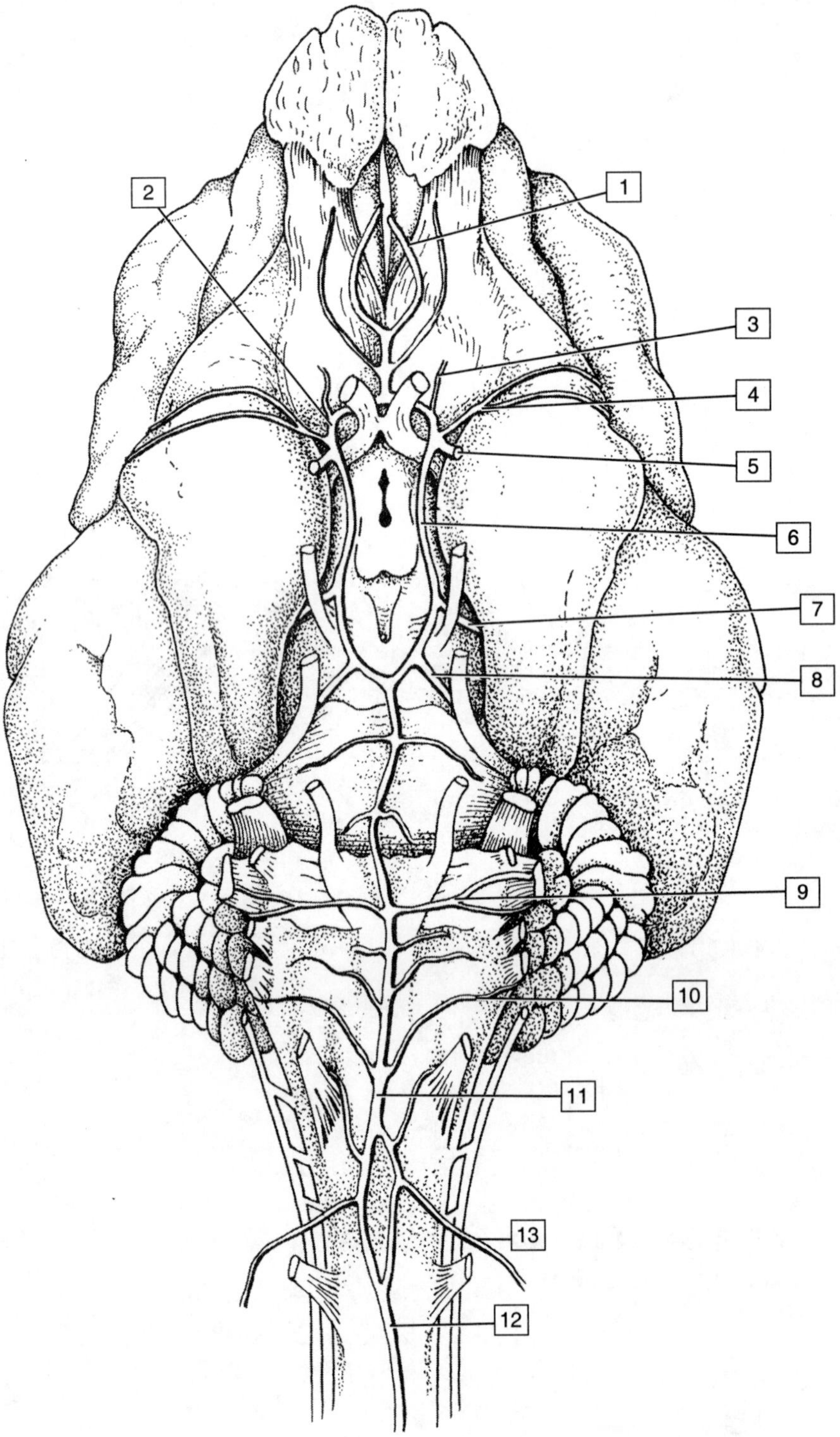

1 筛窦内动脉 Internal ethmoidal artery

2 大脑前动脉 Rostral cerebral artery

3 眼内动脉 Internal ophthalmic artery

4 大脑中动脉 Middle cerebral artery

5 颈内动脉 Internal carotid artery

6 后交通动脉 Caudal communicating artery

7 大脑后动脉 Caudal cerebral artery

8 小脑前动脉 Rostral cerebellar artery

9 迷路动脉 Labyrinthine artery

10 小脑后动脉 Caudal cerebellar artery

11 基底动脉 Basilar artery

12 腹侧脊髓动脉 Ventral spinal artery

13 椎动脉 Vertebral artery

图1-21 大脑腹侧面动脉

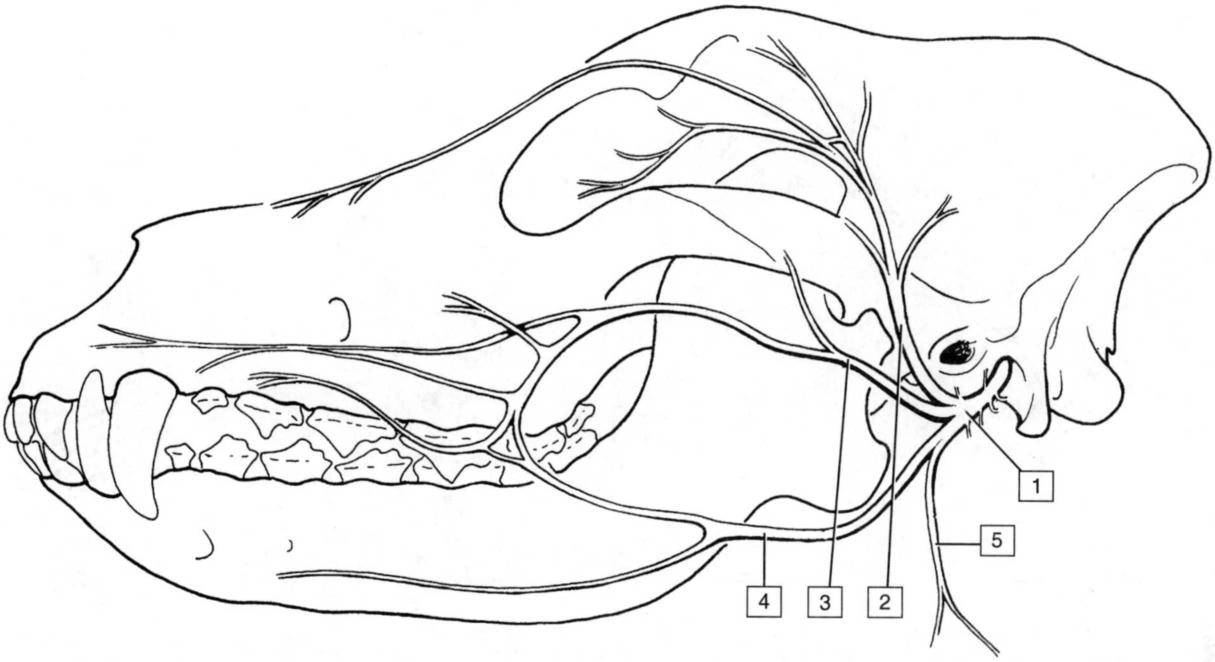

1	面神经　Facial nerve	4	腹侧颊支　Ventral buccal branch
2	耳睑神经　Auriculopalpebral nerve	5	颈支　Cervical branch
3	背侧颊支　Dorsal buccal branch		

图1-22　犬面神经分布图

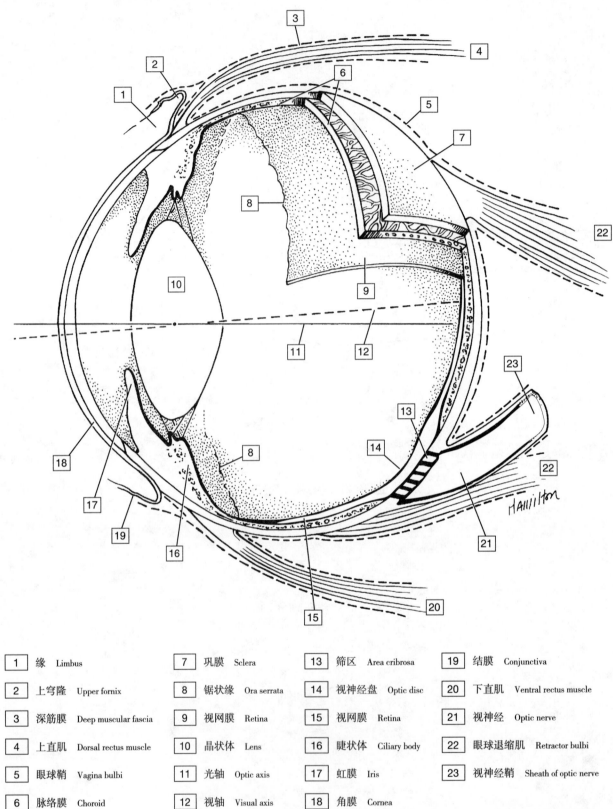

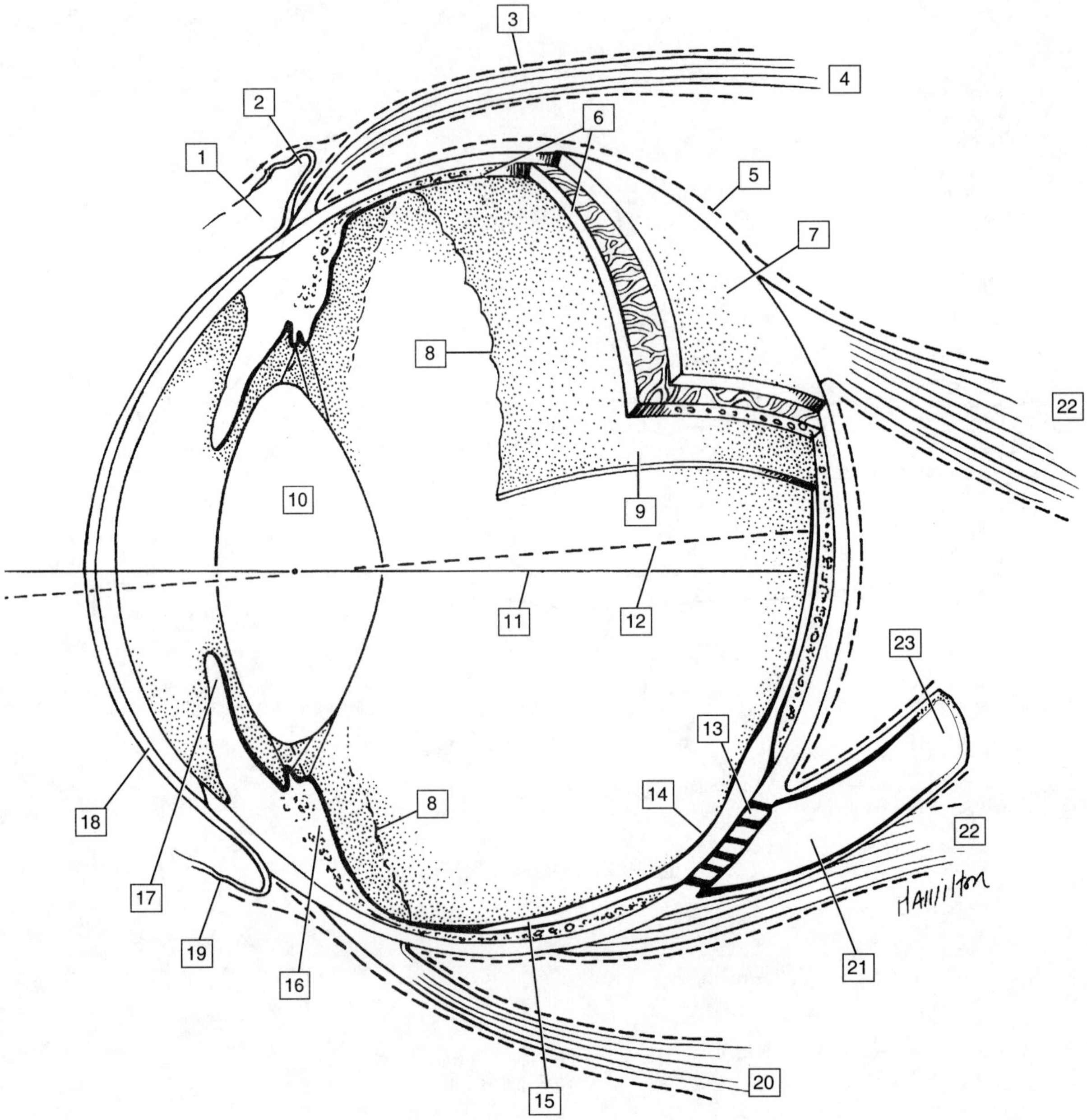

1 缘 Limbus
2 上穹隆 Upper fornix
3 深筋膜 Deep muscular fascia
4 上直肌 Dorsal rectus muscle
5 眼球鞘 Vagina bulbi
6 脉络膜 Choroid
7 巩膜 Sclera
8 锯状缘 Ora serrata
9 视网膜 Retina
10 晶状体 Lens
11 光轴 Optic axis
12 视轴 Visual axis
13 筛区 Area cribrosa
14 视神经盘 Optic disc
15 视网膜 Retina
16 睫状体 Ciliary body
17 虹膜 Iris
18 角膜 Cornea
19 结膜 Conjunctiva
20 下直肌 Ventral rectus muscle
21 视神经 Optic nerve
22 眼球退缩肌 Retractor bulbi
23 视神经鞘 Sheath of optic nerve

图1-23 剖开犬眼球，显示眼球壁三层结构

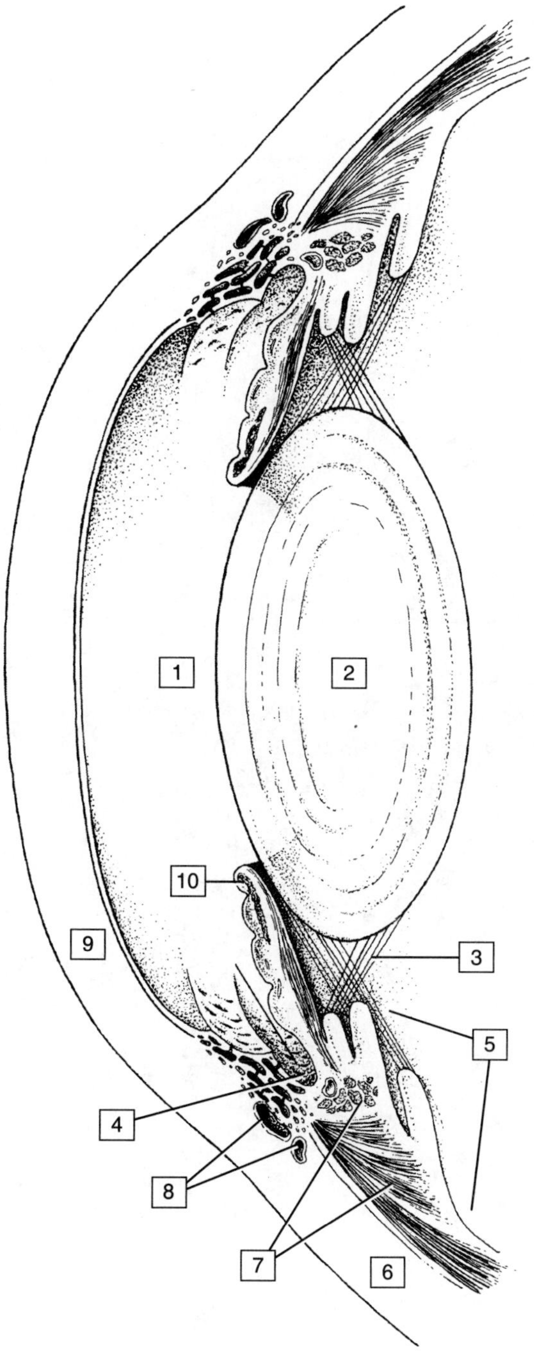

1 眼前房 Anterior chamber	6 巩膜 Sclera
2 晶状体 Lens	7 睫状肌 Ciliary muscles
3 小带纤维 Zonular fibers	8 巩膜静脉丛 Venous plexus of sclera
4 虹膜角 Iridocorneal angle	9 角膜 Cornea
5 睫状体 Ciliary body	10 虹膜括约肌和扩张肌 Iris with the sphincter and dilator muscles shown

图1-24 犬眼前部断面

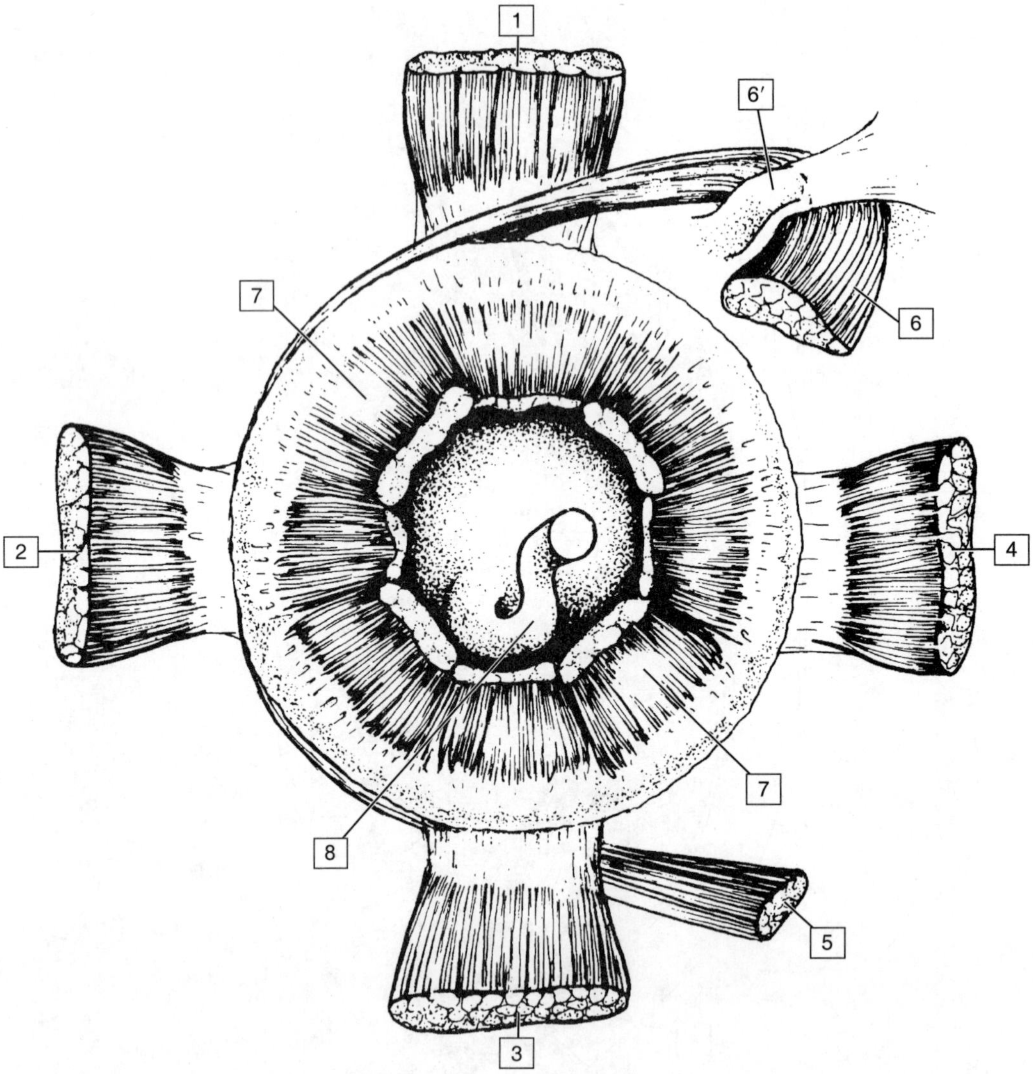

1	上直肌 Dorsal rectus muscle	6	上斜肌 Dorsal oblique muscle
2	外直肌 Lateral rectus muscle	6′	滑车 Trochlea
3	下直肌 Ventral rectus muscle	7	眼球退缩肌 Retractor bulbi
4	内直肌 Medial rectus muscle	8	视神经 Optic nerve
5	下斜肌 Ventral oblique muscle		

图1-25 犬左眼后方的肌肉

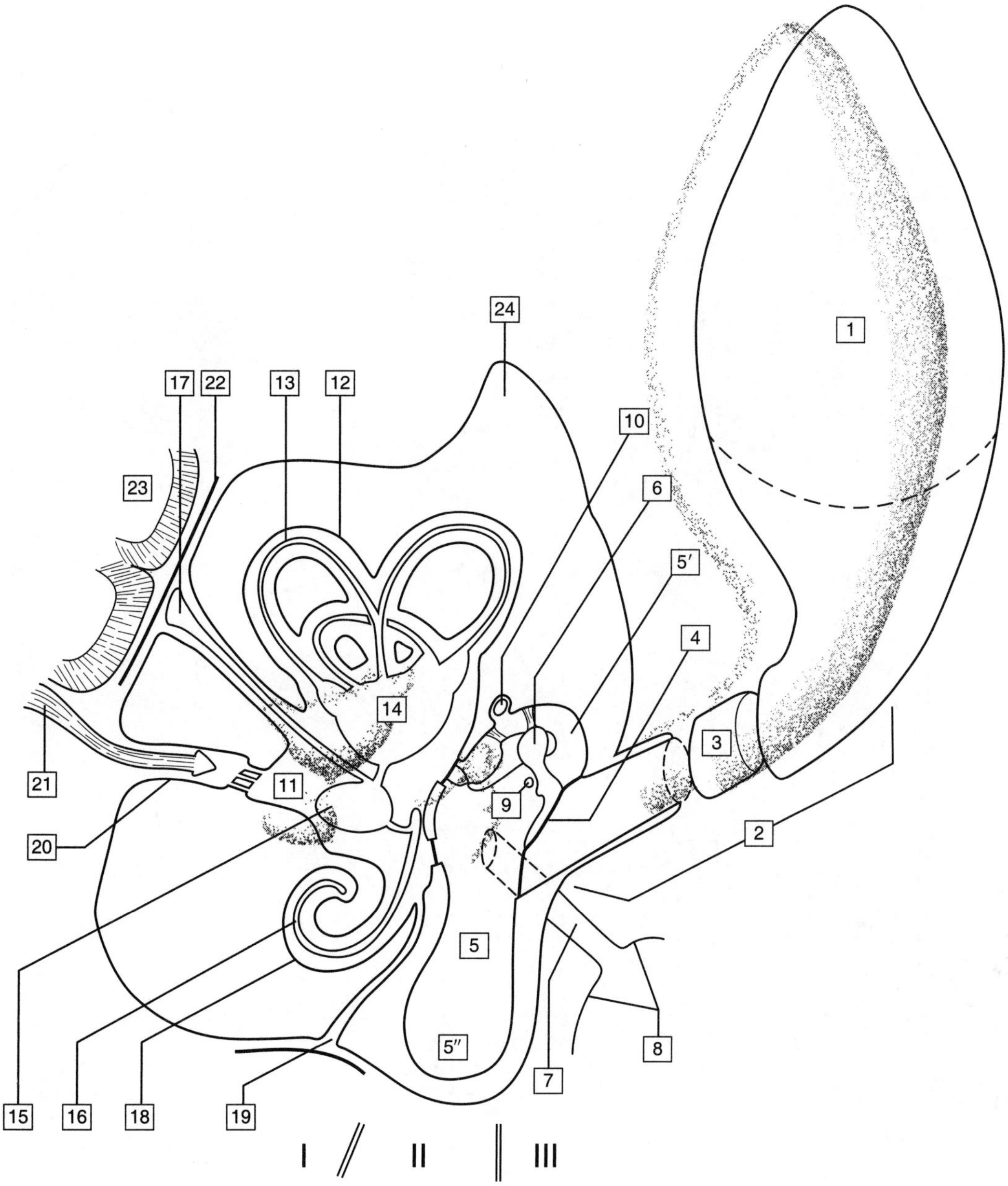

- 1 耳郭 Auricle
- 2 外耳道 External acoustic meatus
- 3 环状软骨 Annular cartilage
- 4 鼓膜 Tympanic membrane
- 5 鼓室 Tympanic cavity
- 5′ 鼓室上隐窝 Epitympanic recess
- 5″ 鼓泡 Tympanic bulla
- 6 听小骨 Auditory ossicles
- 7 咽鼓管 Auditory tube
- 8 鼻咽 Nasopharynx
- 9 鼓索 Chorda tympani
- 10 面神经 Facial nerve
- 11 前庭 Vestibule
- 12 骨半规管 Semicircular canals
- 13 膜半规管 Semicircular ducts
- 14 椭圆囊 Utriculus
- 15 球囊 Sacculus
- 16 耳蜗管 Cochlear duct
- 17 内淋巴管 Endolymphatic duct
- 18 耳蜗 Cochlea
- 19 外淋巴管 Perilymphatic duct
- 20 内耳道 Internal acoustic meatus
- 21 内耳道中的前庭耳蜗神经 Vestibulocochlear nerve in internal acoustic meatus
- 22 脑膜 Meninges
- 23 脑 Brain
- 24 颞骨岩部 Petrous temporal bone

图1-26 犬右耳，后面观模式图

Ⅰ. 内耳；Ⅱ. 中耳；Ⅲ. 外耳

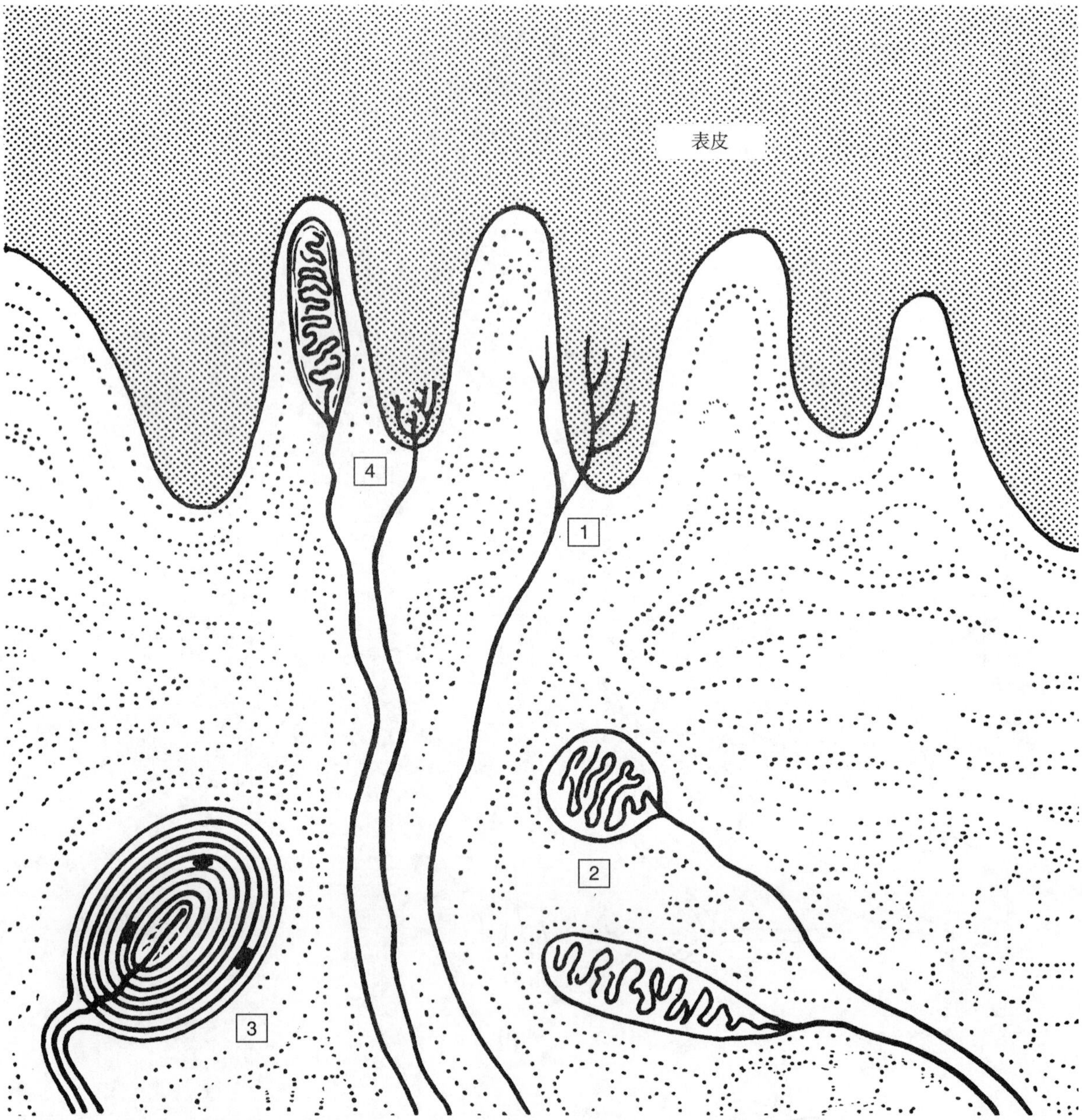

1 游离神经末梢（痛觉） Free nerve endings (pain)	3 环层小体（振动） Lamellar corpuscles (vibration)
2 球状小体（热觉和冷觉） Bulbous corpuscles (heat or cold)	4 半月板状神经末梢（触觉） Meniscoid nerve endings (touch)

图1-27　犬皮肤感觉神经末梢

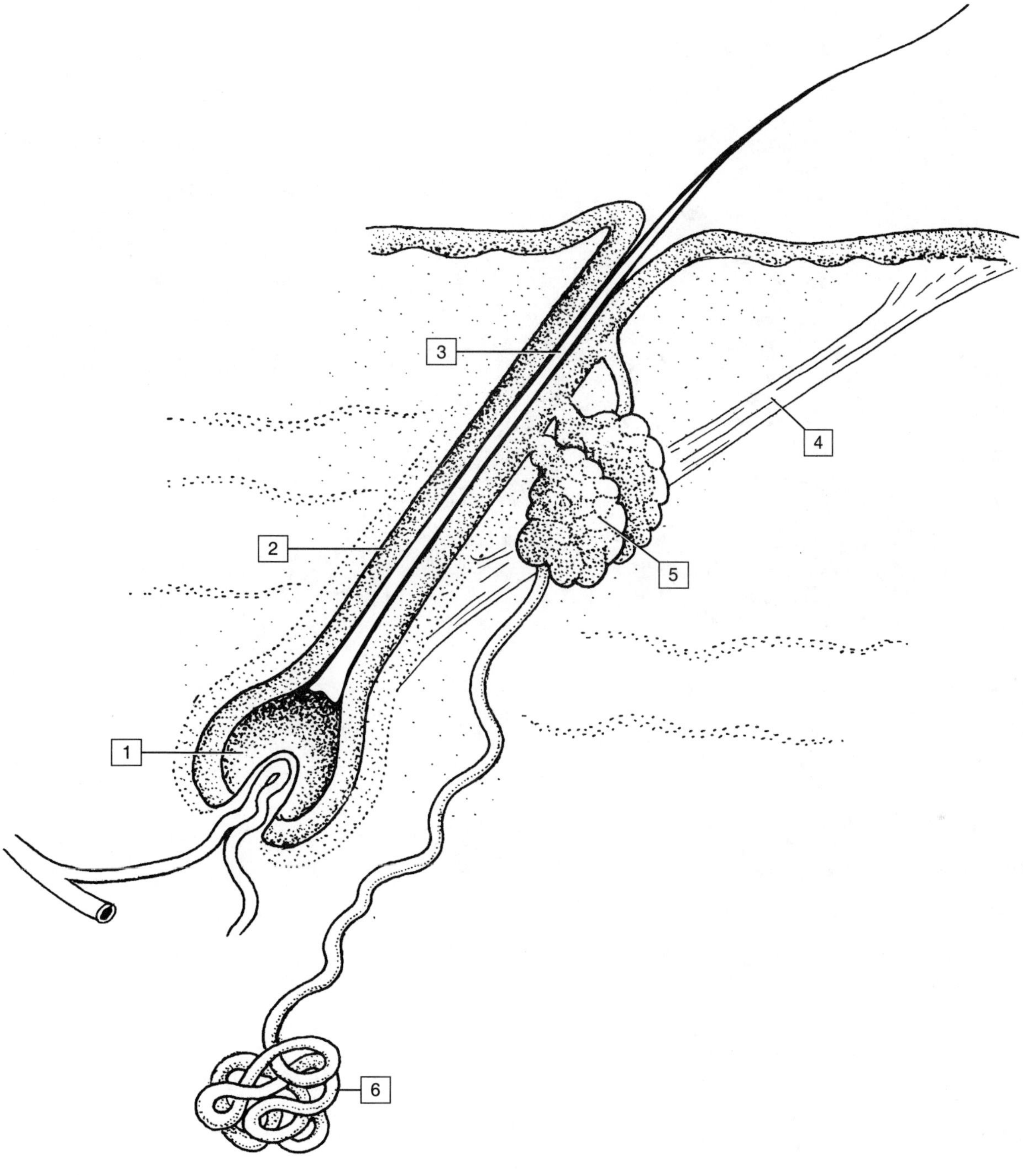

1	毛球	Bulb (hair matrix) of hair
2	毛囊	Hair follicle
3	毛根	Root of hair
4	立毛肌	Arrector pili muscle
5	皮脂腺	Sebaceous gland
6	汗腺，成年犬的汗腺独立开口，不进入毛囊	Sweat gland. In the adult, many glands open independently, not into hair follicles

图1-28 犬毛囊附属结构

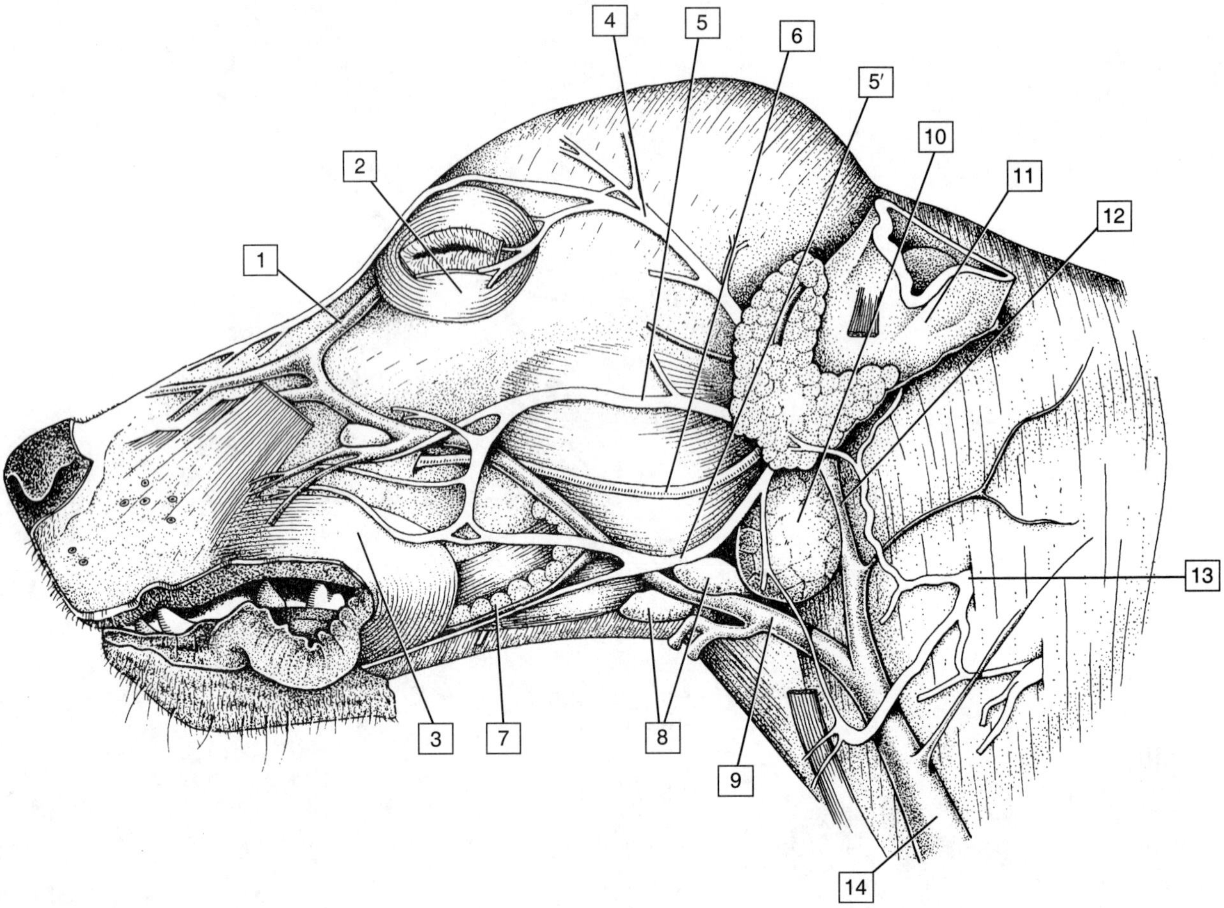

1 眼角静脉 Angularis oculi vein

2 眼轮匝肌 Orbicularis oculi

3 口轮匝肌 Orbicularis oris

4 耳睑神经 Auriculopalpebral nerve

5 面神经背侧颊支 Dorsal buccal branch of facial nerve

5′ 面神经腹侧颊支 Ventral buccal branch of facial nerve

6 腮腺管 Parotid duct

7 颊腺 Buccal salivary glands

8 下颌淋巴结 Mandibular lymph nodes

9 舌面静脉 Linguofacial vein

10 颌下腺 Mandibular gland

11 耳根 Base of ear

12 上颌静脉 Maxillary vein

13 第2颈神经 Second cervical nerve

14 颈外静脉 External jugular vein

图1-29 犬头部浅层解剖图

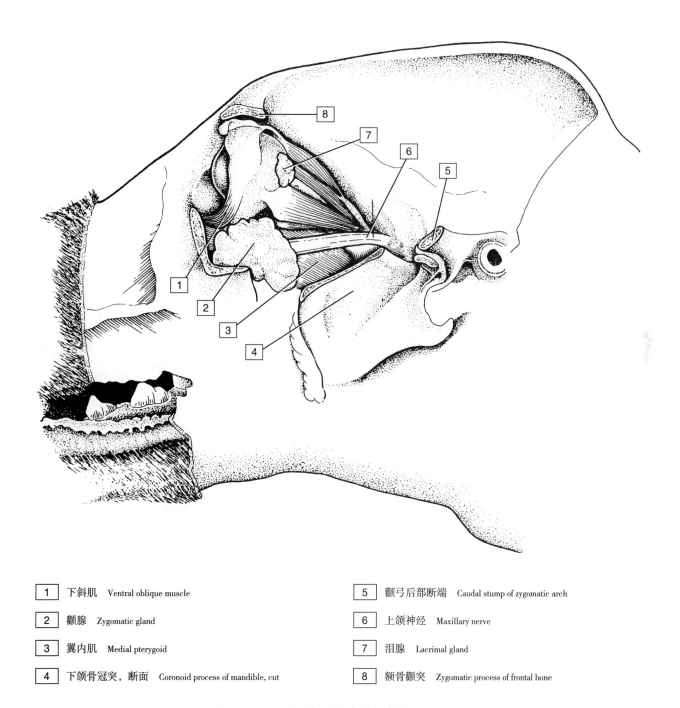

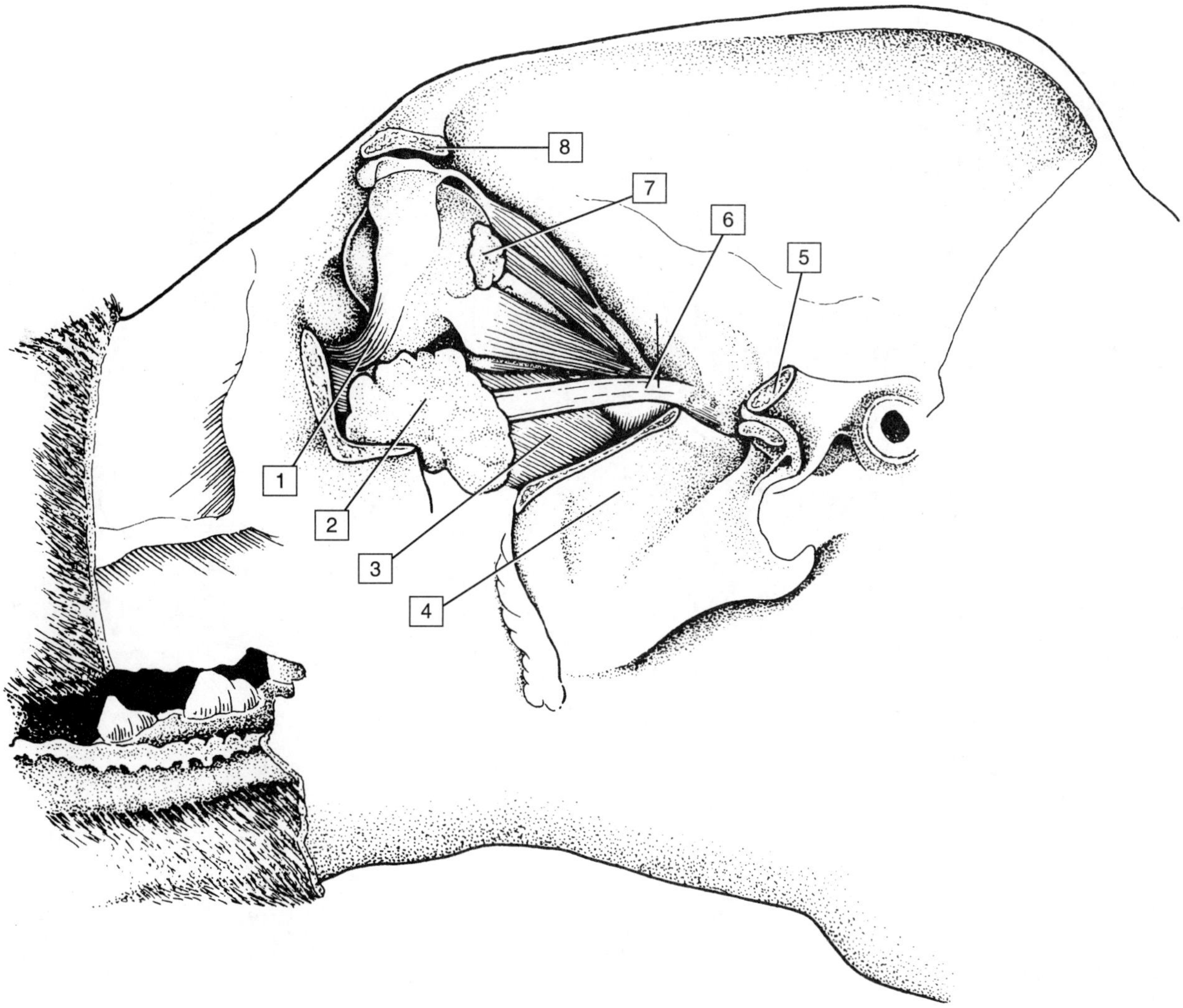

1	下斜肌 Ventral oblique muscle	5	颧弓后部断端 Caudal stump of zygomatic arch
2	颧腺 Zygomatic gland	6	上颌神经 Maxillary nerve
3	翼内肌 Medial pterygoid	7	泪腺 Lacrimal gland
4	下颌骨冠突，断面 Coronoid process of mandible, cut	8	额骨颧突 Zygomatic process of frontal bone

图1-30 犬眼眶和翼腭窝的解剖图，外侧观

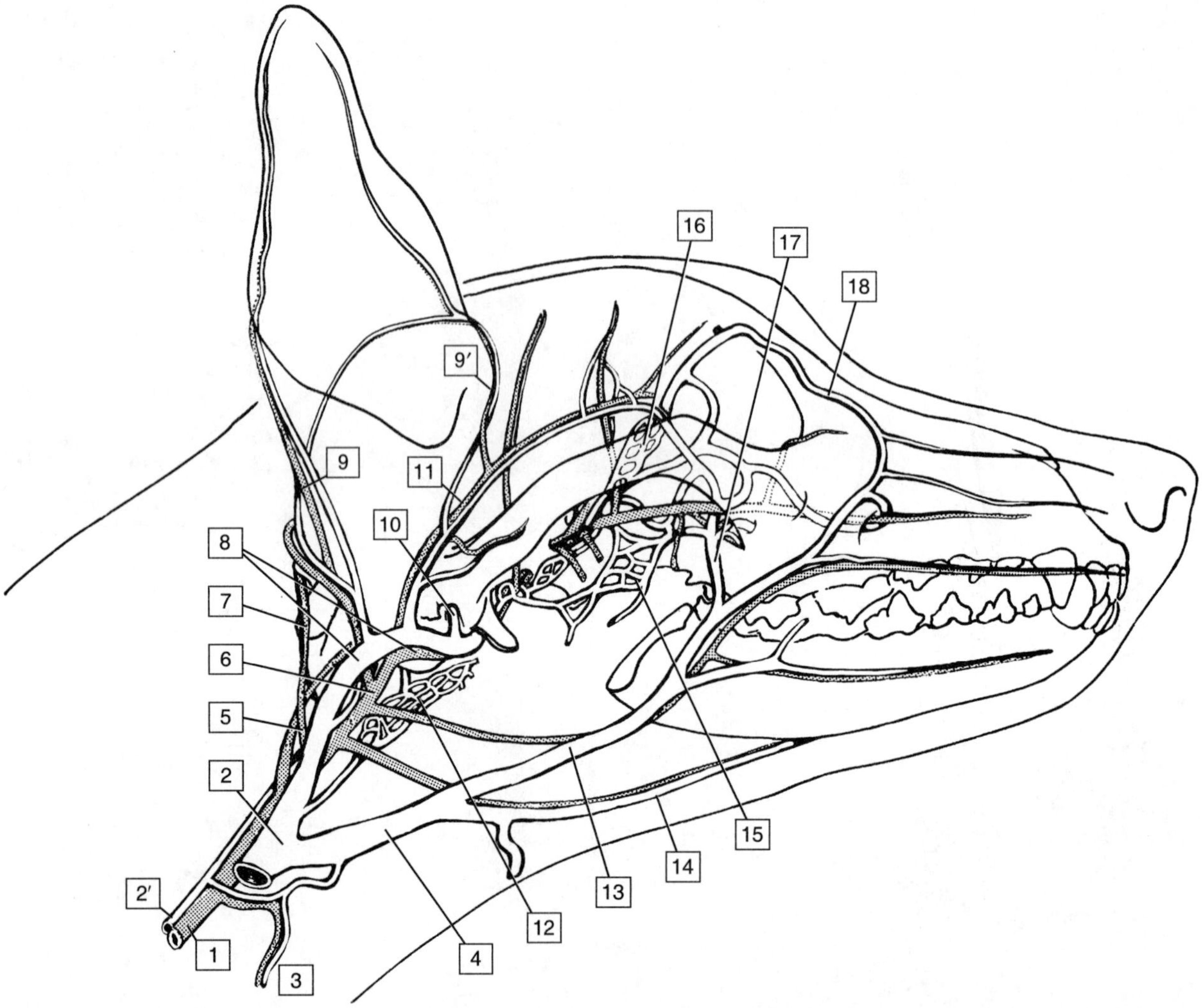

1	颈总动脉 Common carotid artery	9′	耳前动脉 Rostral auricular artery
2	颈外静脉 External jugular vein	10	背侧导管 Dorsal emissary
2′	颈内静脉 Internal jugular vein	11	颞浅动脉 Superficial temporal artery
3	甲状腺前动、静脉 Cranial thyroid vessels	12	腹侧导管和咽静脉丛 Ventral emissary and pharyngeal plexus
4	舌面静脉 Linguofacial vein	13	面静脉 Facial vein
5	颈内动脉 Internal carotid artery	14	舌静脉 Lingual vein
6	颈外动脉 External carotid artery	15	翼静脉丛 Pterygoid plexus
7	枕动脉 Occipital artery	16	眼静脉丛 Ophthalmic plexus
8	上颌动、静脉 Maxillary vessels	17	面深静脉 Deep facial vein
9	耳后动脉 Caudal auricular artery	18	眼角静脉 Angularis oculi vein

图1-31 犬头部的主要动脉（灰色）和静脉（黑色）

下颌支已被移除

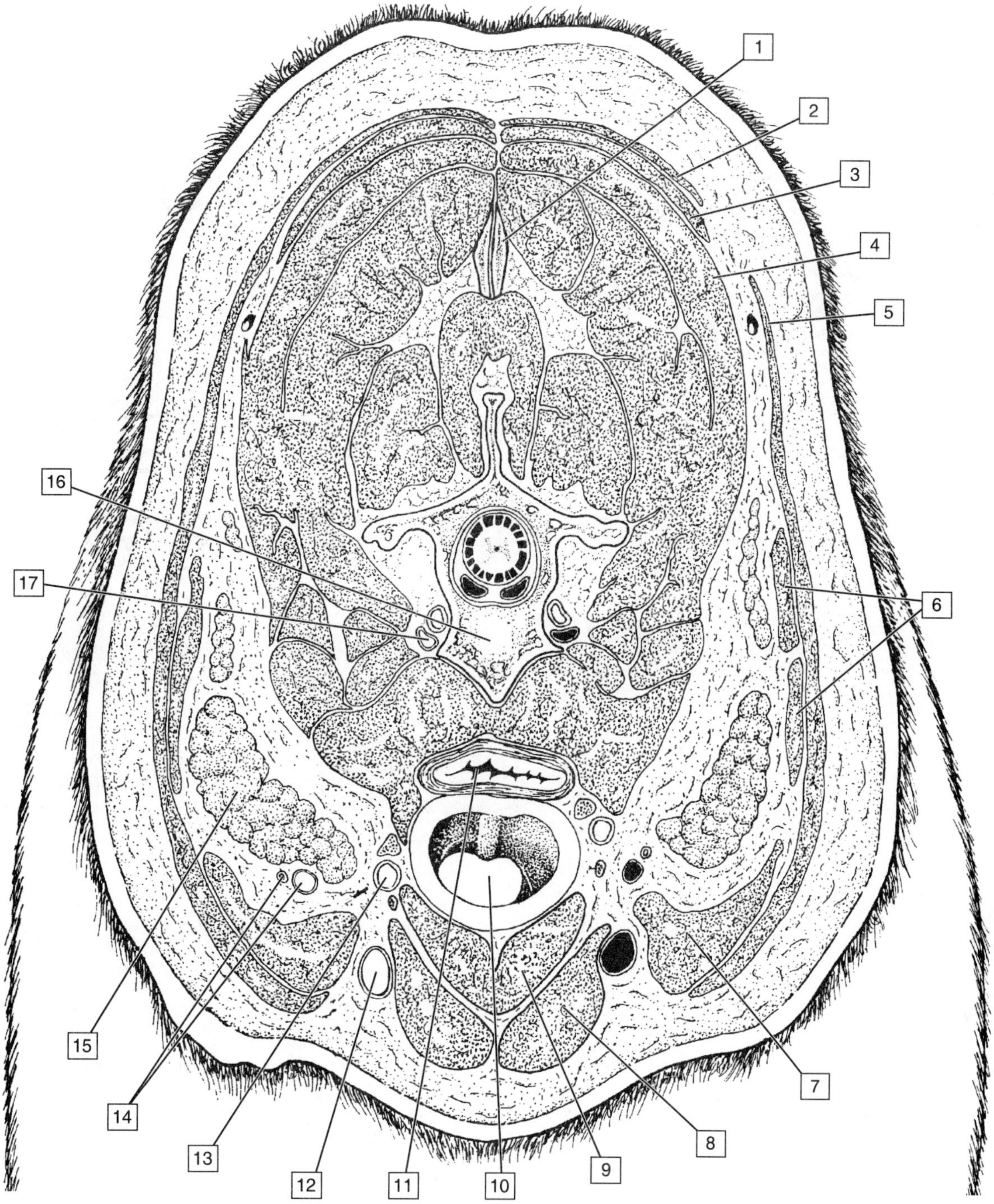

1 项韧带 Nuchal ligament

2 斜方肌 Trapezius

3 菱形肌 Rhomboideus

4 夹肌 Splenius

5 锁颈肌 Cleidocervicalis

6 肩胛横突肌 Omotransversarius

7 锁乳突肌 Cleidomastoideus

8 胸头肌 Sternocephalicus

9 胸骨甲状舌骨肌 Sternothyrohyoideus

10 气管 Trachea

11 食管 Esophagus

12 颈外静脉 External jugular vein

13 颈总动脉，迷走交感干，喉返神经 Common carotid artery, vagosympathetic trunk, and recurrent laryngeal nerve

14 颈浅动、静脉 Superficial cervical vessels

15 颈浅淋巴结 Superficial cervical lymph nodes

16 第5颈椎 Fifth cervical vertebra

17 椎动、静脉 Vertebral vessels

图1-32 犬颈部的第5颈椎横断面

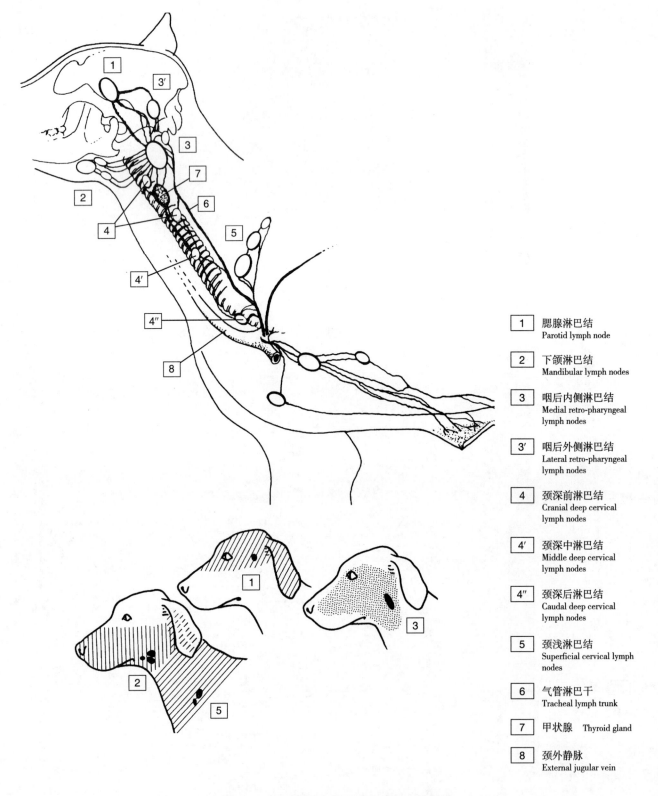

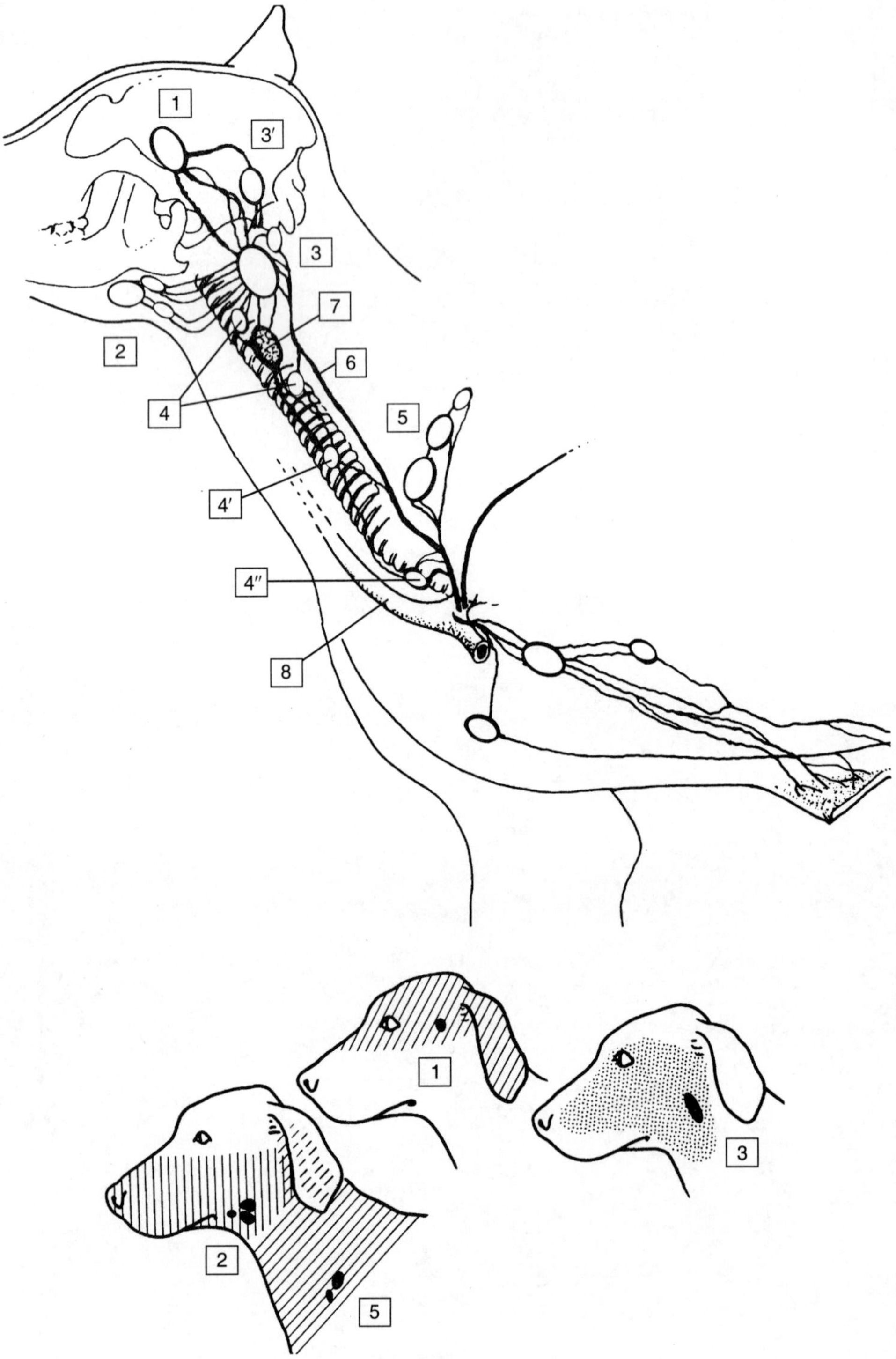

1 腮腺淋巴结 Parotid lymph node

2 下颌淋巴结 Mandibular lymph nodes

3 咽后内侧淋巴结 Medial retro-pharyngeal lymph nodes

3′ 咽后外侧淋巴结 Lateral retro-pharyngeal lymph nodes

4 颈深前淋巴结 Cranial deep cervical lymph nodes

4′ 颈深中淋巴结 Middle deep cervical lymph nodes

4″ 颈深后淋巴结 Caudal deep cervical lymph nodes

5 颈浅淋巴结 Superficial cervical lymph nodes

6 气管淋巴干 Tracheal lymph trunk

7 甲状腺 Thyroid gland

8 颈外静脉 External jugular vein

图1-33　犬头颈部主要淋巴结的引流范围

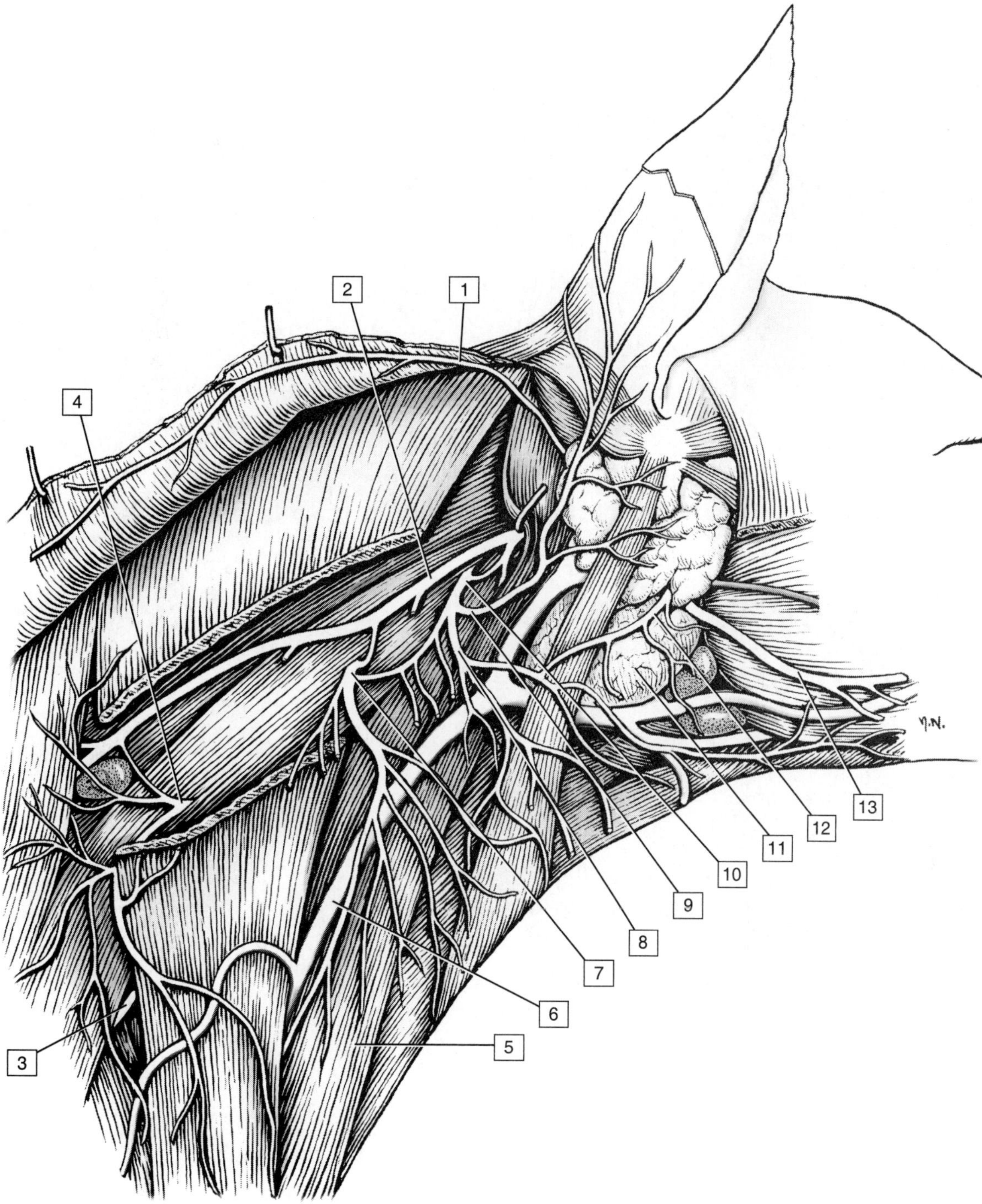

1 耳后神经（面神经）Caudal auricular nerve (Ⅶ)	5 胸头肌 Sternocephalicus	9 耳大神经 Great auricular nerve	13 面神经腹侧颊支 Ⅶ, ventral buccal branch
2 副神经 Accessory nerve	6 颈外静脉 External jugular vein	10 第2颈神经腹侧支 C2, ventral branch	
3 第5颈神经腹侧支 C5, ventral branch	7 第3颈神经腹侧支 C3, ventral branch	11 颌下腺 Mandibular gland	
4 第4颈神经腹侧支 C4, ventral branch	8 颈横神经 Transverse cervical nerve	12 面神经颈支 Ⅶ, cervical branch	

图1-34 大颈部浅层神经，外侧观

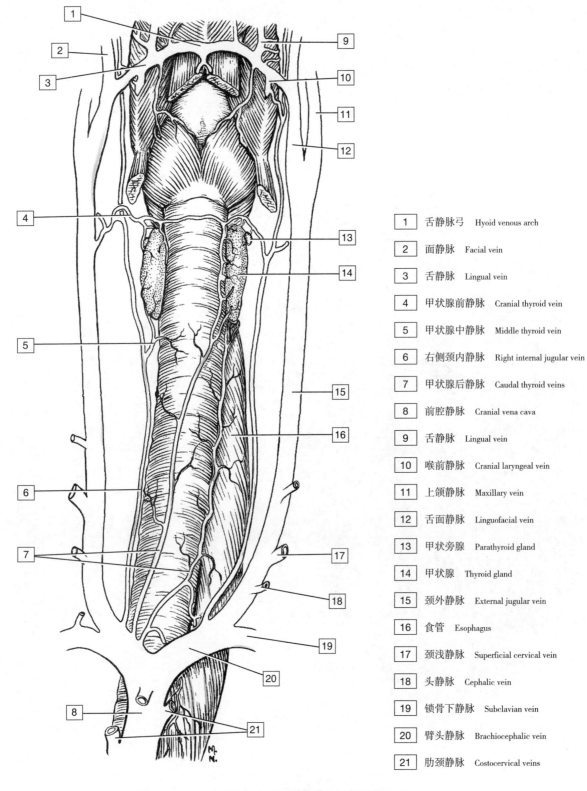

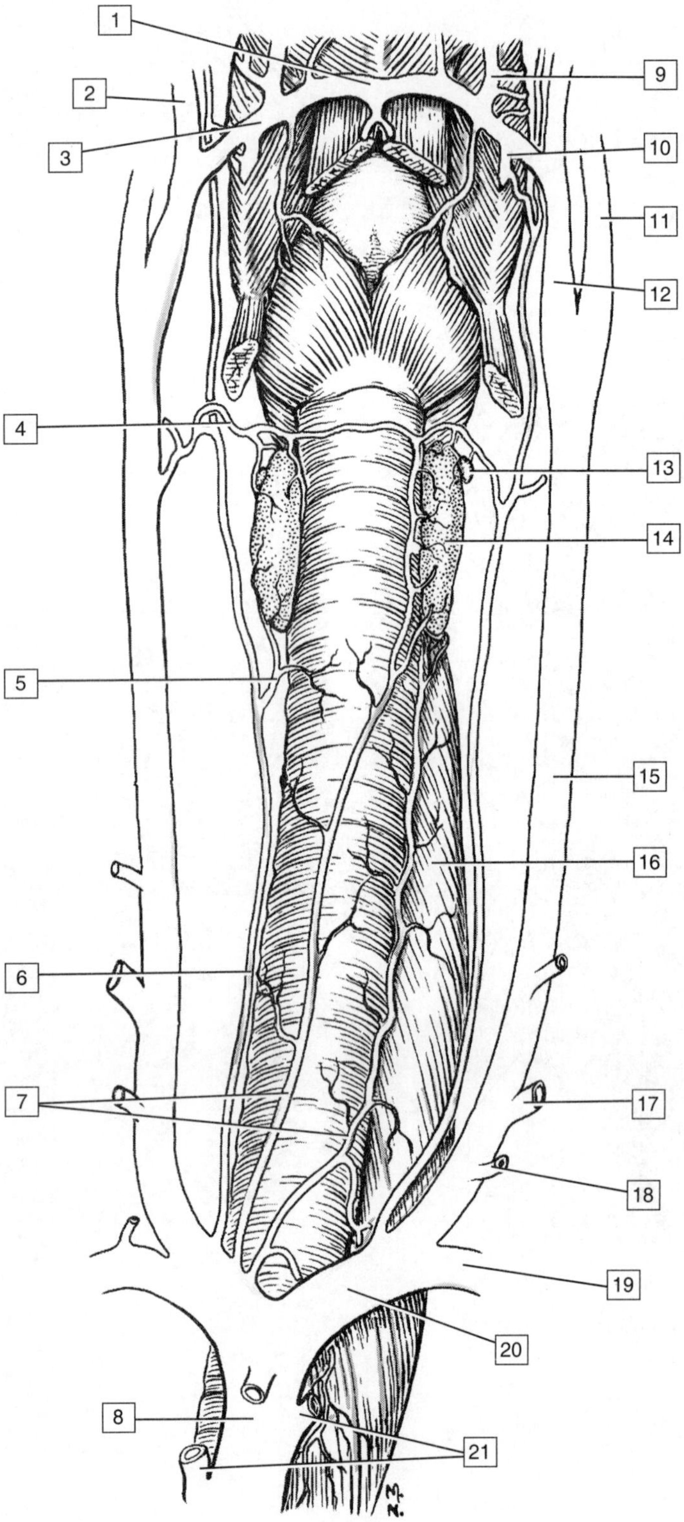

1 舌静脉弓 Hyoid venous arch
2 面静脉 Facial vein
3 舌静脉 Lingual vein
4 甲状腺前静脉 Cranial thyroid vein
5 甲状腺中静脉 Middle thyroid vein
6 右侧颈内静脉 Right internal jugular vein
7 甲状腺后静脉 Caudal thyroid veins
8 前腔静脉 Cranial vena cava
9 舌静脉 Lingual vein
10 喉前静脉 Cranial laryngeal vein
11 上颌静脉 Maxillary vein
12 舌面静脉 Linguofacial vein
13 甲状旁腺 Parathyroid gland
14 甲状腺 Thyroid gland
15 颈外静脉 External jugular vein
16 食管 Esophagus
17 颈浅静脉 Superficial cervical vein
18 头静脉 Cephalic vein
19 锁骨下静脉 Subclavian vein
20 臂头静脉 Brachiocephalic vein
21 肋颈静脉 Costocervical veins

图1-35 犬颈部静脉，腹侧观

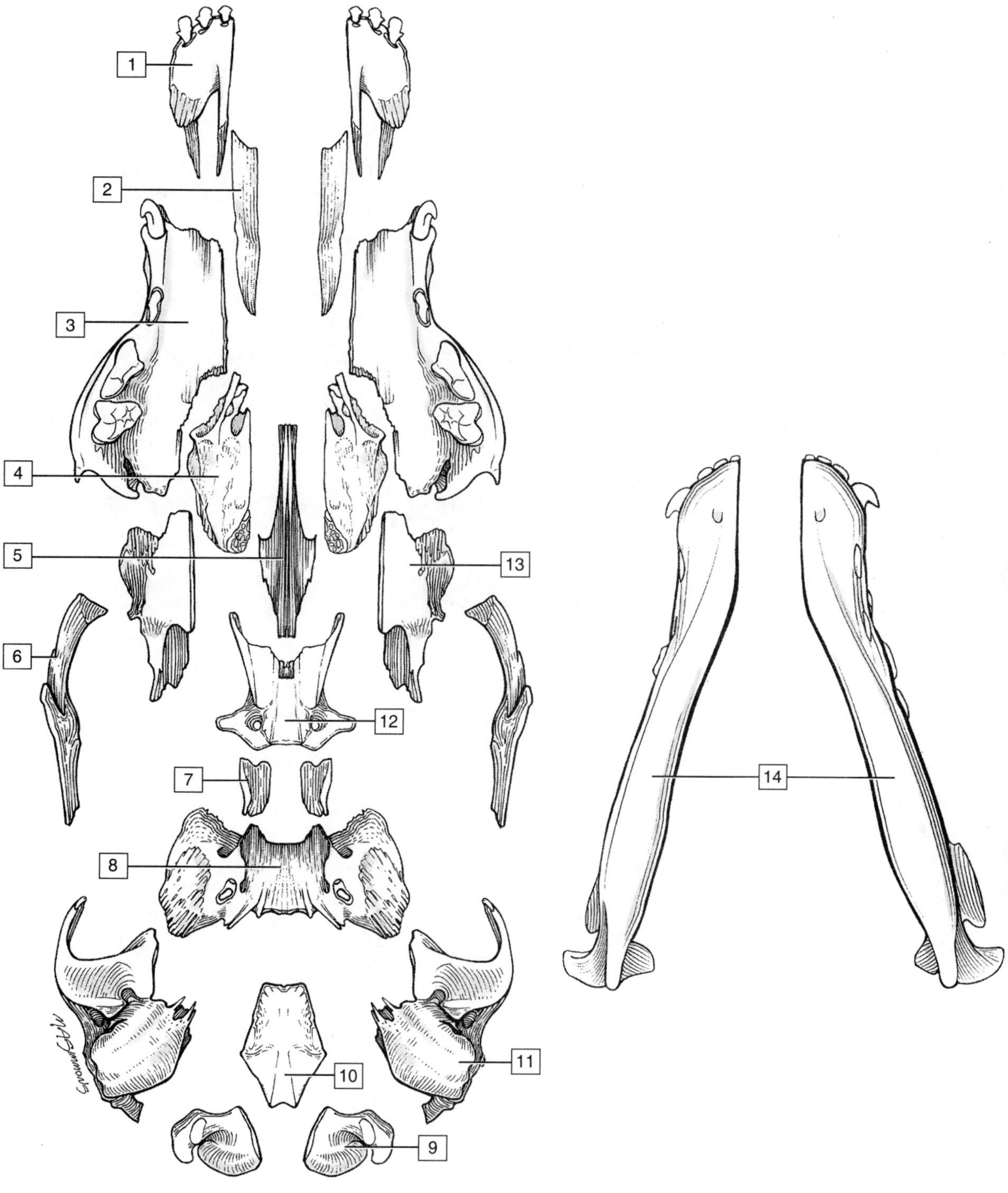

1 切齿骨 Incisive bone	5 犁骨 Vomer	9 外枕骨 Exoccipital	12 前蝶骨 Presphenoid bone
2 鼻骨 Nasal bone	6 颧骨 Zygomatic bone	10 基枕骨 Basioccipital	13 腭骨 Palatine bone
3 上颌骨 Maxilla	7 翼骨 Pterygoid bone	11 颞骨 Temporal bone	14 下颌骨 Mandible
4 筛骨 Ethmoid bone	8 基蝶骨 Basisphenoid bone		

图1-36 分离的幼犬颅骨，腹侧观

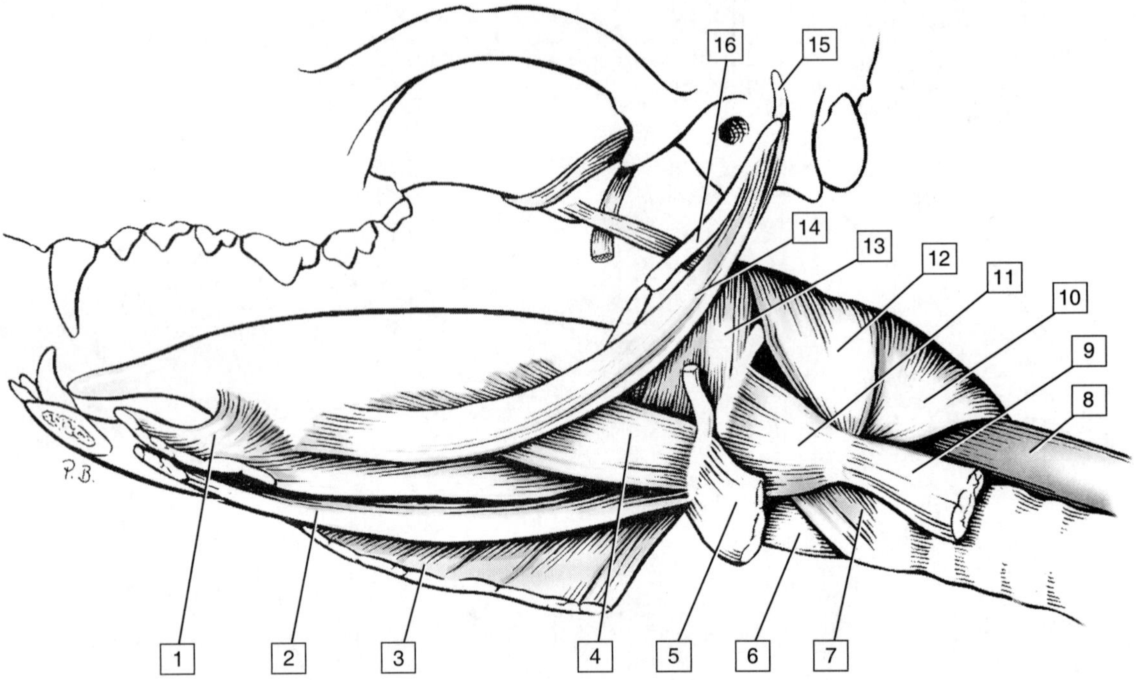

1	颏舌肌 Genioglossus	9	胸骨甲状肌 Sternothyroideus
2	颏骨肌 Geniohyoideus	10	环咽肌 Cricopharyngeus
3	下颌舌骨肌 Mylohyoideus	11	甲状舌骨肌 Thyrohyoid
4	舌肌 Hypoglossus	12	甲咽肌 Thyropharyngeus
5	胸骨舌骨肌 Sternohyoideus	13	舌骨咽肌 Hyopharyngeus
6	甲状软骨 Thyroid cartilage	14	茎突舌肌 Styloglossus
7	环甲肌 Cricothyroideus	15	鼓舌软骨 Tympanohyoid cartilage
8	食管 Esophagus	16	茎突舌骨 Stylohyoid

图1-37 犬咽部肌和舌肌，左外侧观

左侧下颌骨已切除

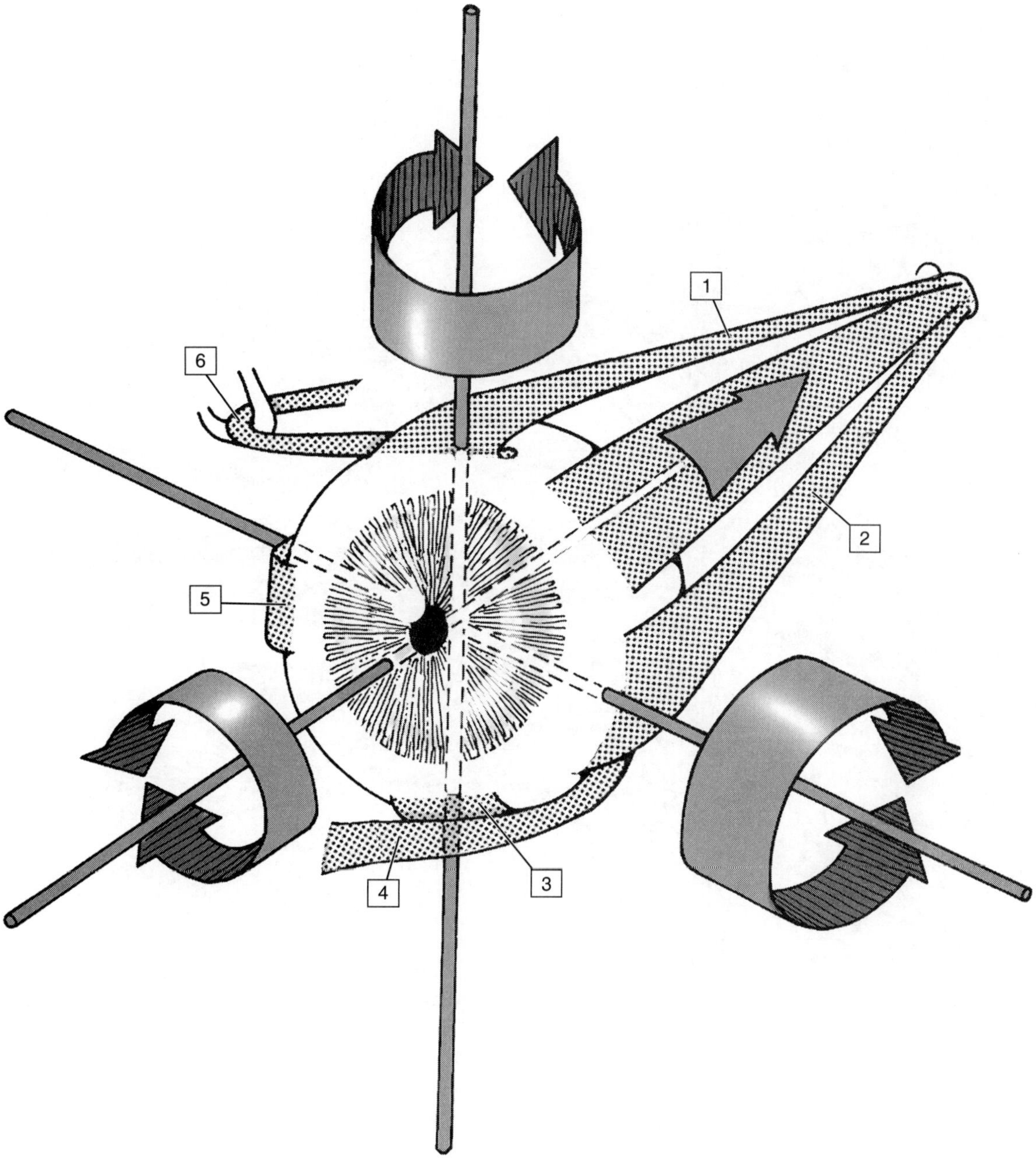

1 上直肌 Dorsal rectus muscle	4 下斜肌 Ventral oblique muscle
2 外直肌 Lateral rectus muscle	5 内直肌 Medial rectus muscle
3 下直肌 Ventral rectus muscle	6 上斜肌 Dorsal oblique muscle

图1-38　犬左眼眼外肌及其作用

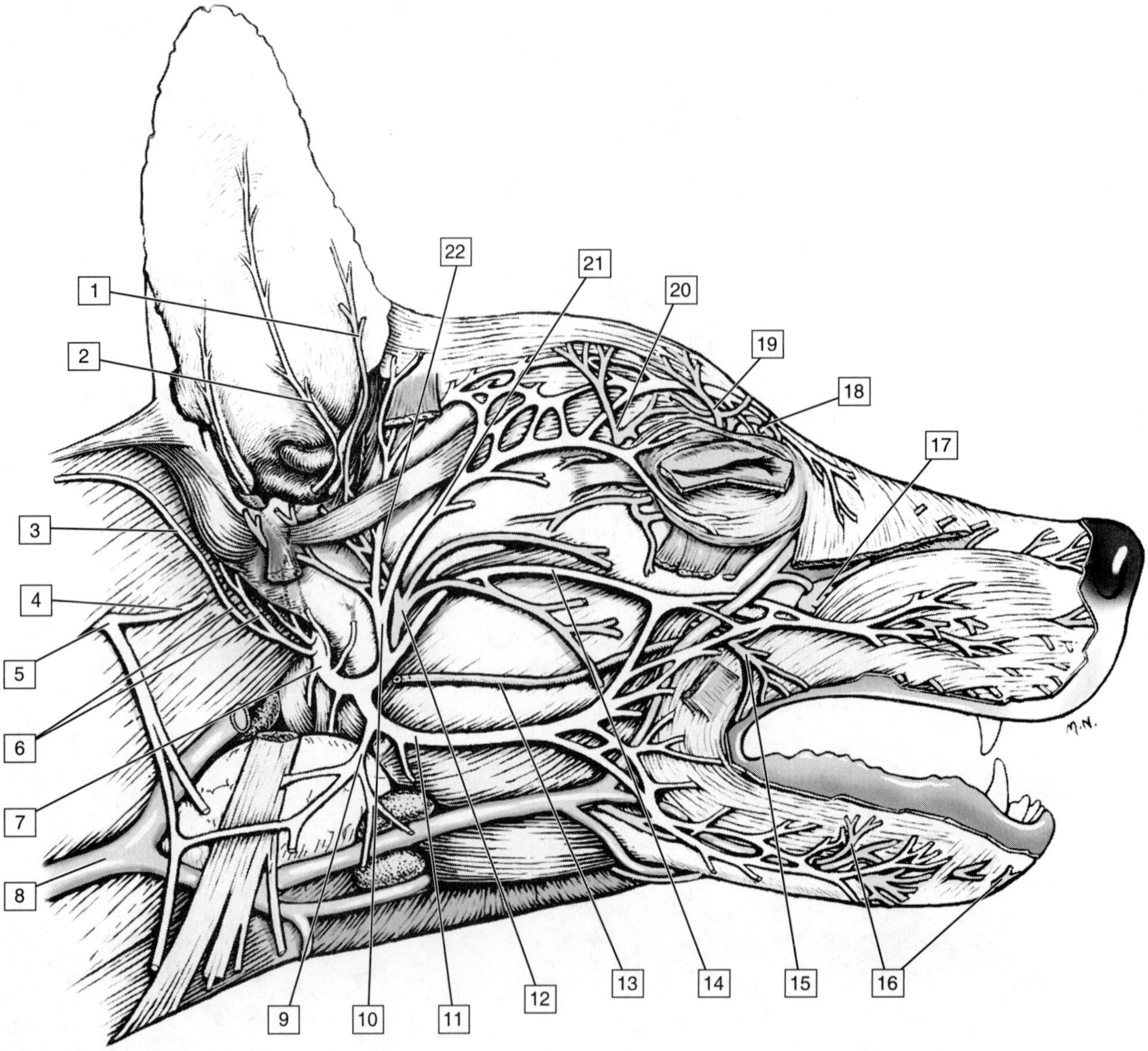

1 耳前神经 Rostral auricular nerve

2 耳内支 Internal auricular branch

3 至颈阔肌的耳后支 Caudal auricular branch to platysma

4 耳大神经 Great auricular nerve

5 第2颈神经腹侧支 C2, ventral branch

6 耳后支 Caudal auricular branches

7 面神经 Facial nerve VII

8 颈外静脉 External jugular vein

9 颈支 Cervical branch

10 耳睑神经 Auriculopalpebral nerve

11 腹侧颊支 Ventral buccal branch

12 耳颞神经V Auriculotemporal nerve V

13 腮腺管（切口） Parotid duct (cut)

14 背侧颊支 Dorsal buccal branch

15 颊神经 Buccalis nerve

16 颏神经 Mental nerves

17 眶下神经 Infraorbital nerve

18 滑车下神经 Infratrochlear nerve

19 额神经 Frontal nerve

20 颧颞神经 Zygomaticotemporal nerve

21 睑神经 Palpebral nerve

22 耳前神经 Rostral auricular nerve

图1-39 犬面神经和三叉神经的浅表分支

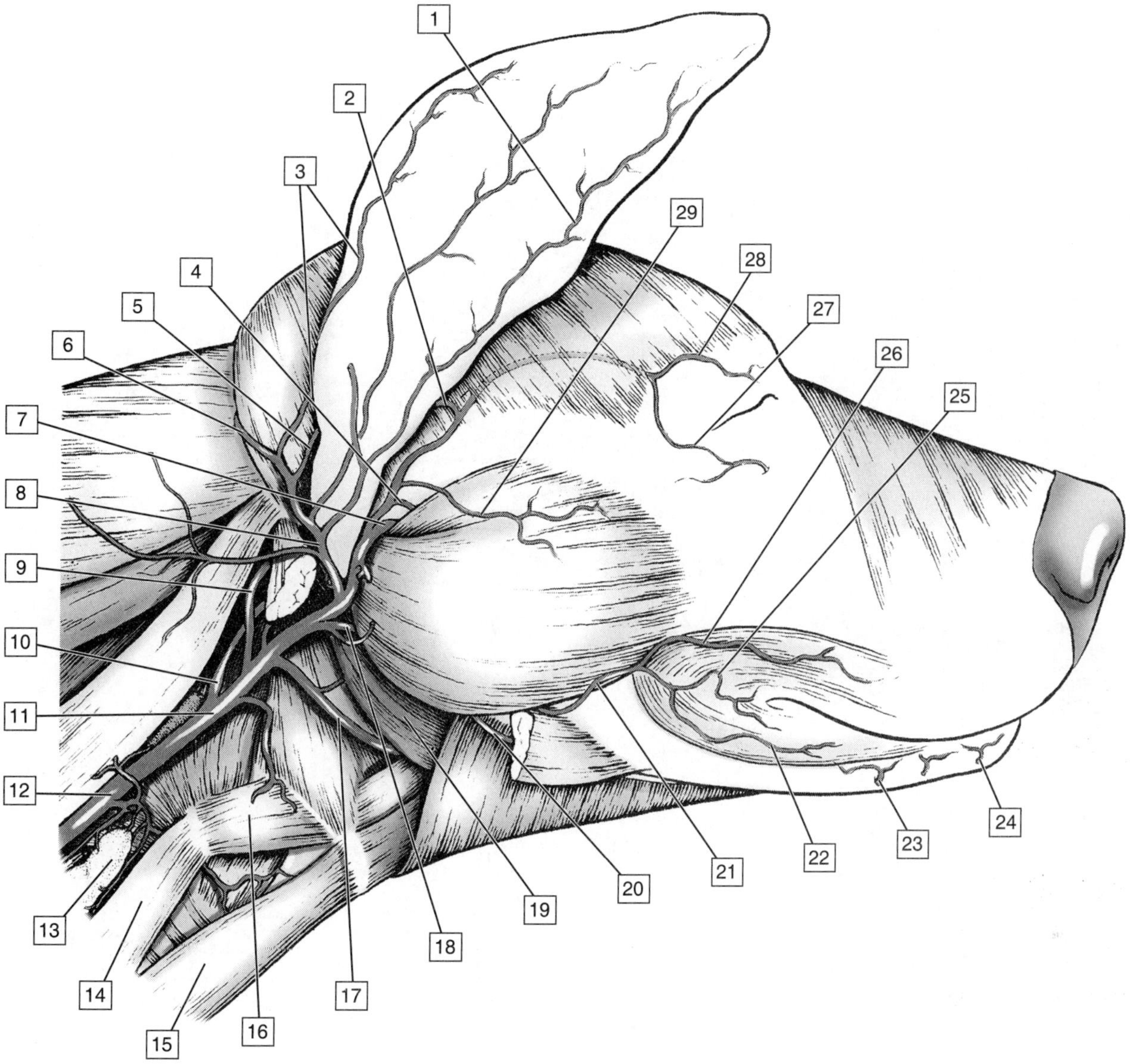

1 耳外动脉 Lateral auricular artery	8 耳后动脉 Caudal auricular artery	17 舌动脉 Lingual artery	25 口角 Angularis oris
2 耳前动脉 Rostral auricular artery	9 枕动脉 Occipital artery	18 面动脉 Facial artery	26 上唇动脉 Superior labial artery
3 耳内动脉 Medial auricular artery	10 颈内动脉 Internal carotid artery	19 茎突舌肌 Styloglossus	27 腹外侧睑动脉 Lateral ventral palpebral
4 颞浅动脉 Superficial temporal artery	11 颈外动脉 External carotid artery	20 舌下动脉 Sublingual artery	28 背外侧睑动脉 Lateral dorsal palpebral
5 耳深动脉 Deep auricular artery	12 颈总动脉 Common carotid artery	21 面动脉 Facial artery	29 面横动脉 Transverse facial artery
6 枕支 Occipital branch	13 甲状腺 Thyroid gland	22 下唇动脉 Inferior labial artery	
7 上颌动脉 Maxillary artery	14 胸骨甲状肌 Sternothyroideus	23 颏后动脉 Caudal mental artery	
	15 胸骨舌骨肌 Sternohyoideus	24 颏前动脉 Rostral mental artery	
	16 甲状舌骨肌 Thyrohyoid		

图1-40 犬颈总动脉分支,外侧观

切除部分二腹肌

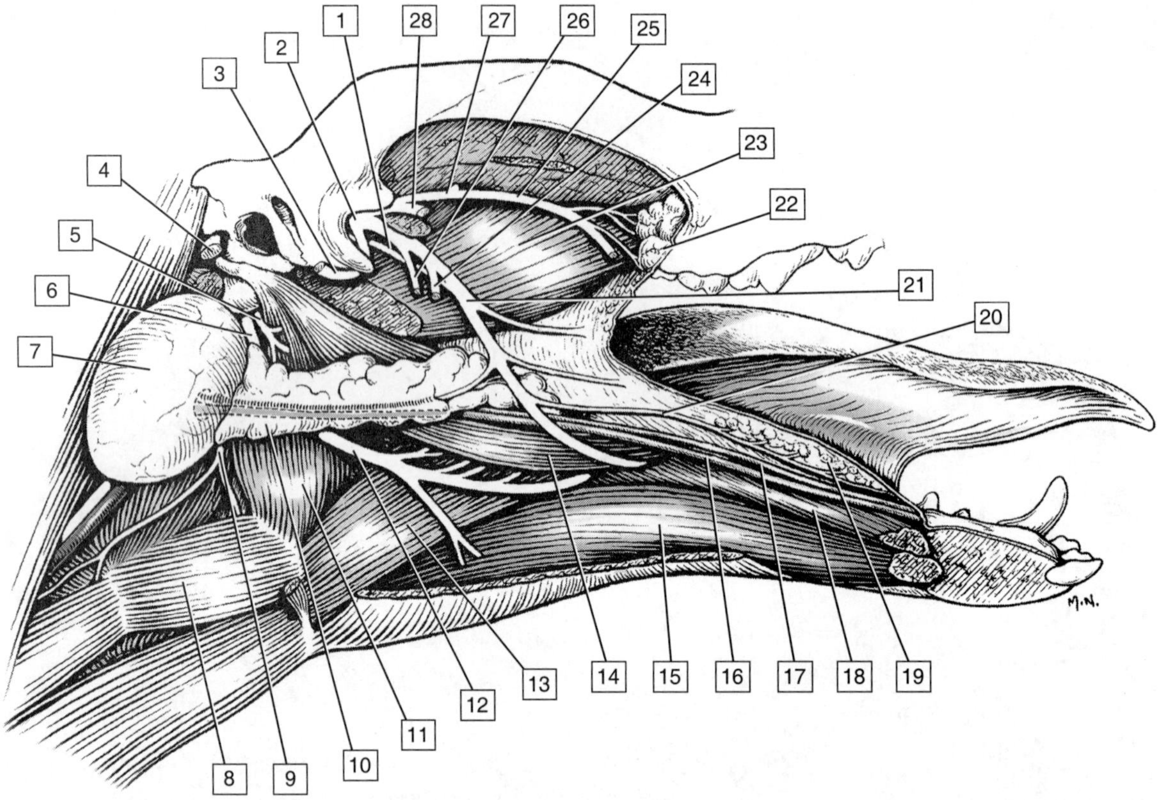

1 鼓索 Chorda tympani
2 三叉神经下颌支 Mandibular branch of V
3 耳颞神经 Auriculotemporal nerve
4 面神经 Facial nerve
5 舌咽神经 Glossopharyngeal nerve
6 舌下神经 Hypoglossal nerve
7 颌下腺 Mandibular gland
8 甲状舌骨肌 Thyrohyoid
9 喉前神经 Cranial laryngeal nerve
10 单口舌下腺 Monostomatic sublingual gland
11 舌骨咽肌 Hyopharyngeus
12 舌下神经 Hypoglossal nerve
13 舌肌 Hypoglossus
14 茎突舌肌 Styloglossus
15 颏骨肌 Geniohyoideus
16 颌下腺管 Mandibular duct
17 舌下腺管 Sublingual duct
18 颏舌肌 Genioglossus
19 多口舌下腺 Polystomatic sublingual gland
20 舌底神经 Sublingual nerve
21 舌神经 Lingual nerve
22 颧腺 Zygomatic gland
23 翼内肌 Medial pterygoid
24 下牙槽神经 Inferior alveolar nerve
25 颊神经 Buccal nerve
26 下颌舌骨肌神经 Mylohyoid nerve
27 颞深支 Deep temporal branch
28 咬肌神经 Masseteric nerve

图1-41 从中央到右侧下颌骨的犬头部肌肉、神经和唾液腺

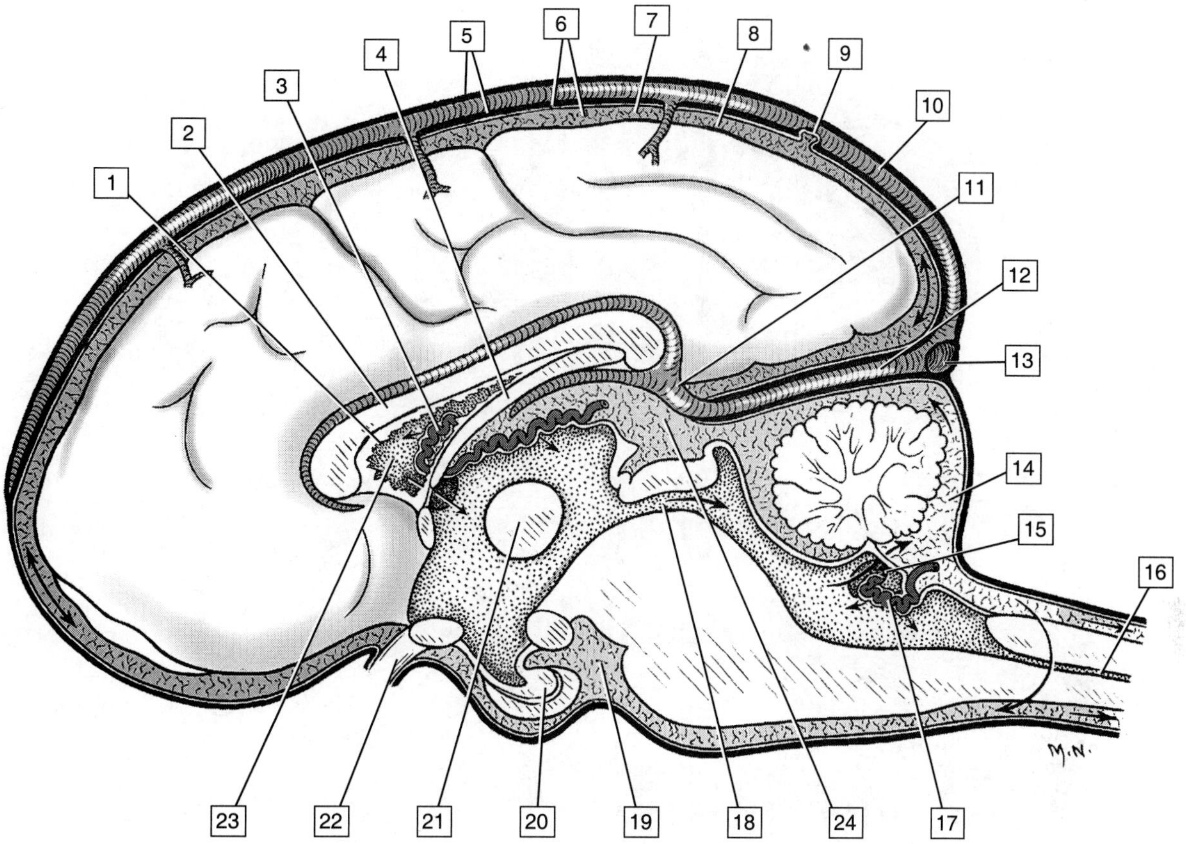

1	透明隔边缘切边 Cut edge of septum pellucidum
2	胼胝体 Corpus callosum
3	脉络丛，侧脑室 Choroid plexus, lateral ventricle
4	海马穹隆 Fornix of hippocampus
5	硬脑膜 Dura
6	蛛网膜和小梁 Arachnoid membrane and trabeculae
7	蛛网膜下腔 Subarachnoid space
8	软脑膜 Pia
9	蛛网膜绒毛 Arachnoid villus
10	背侧矢状窦 Dorsal sagittal sinus
11	大脑大静脉 Great cerebral vein
12	直窦 Straight sinus
13	横窦 Transverse sinus
14	小脑延髓池 Cerebellomedullary cistern
15	第4脑室侧孔 Lateral aperture of fourth ventricle
16	中央管 Central canal
17	脉络丛，第4脑室 Choroid plexus, fourth ventricle
18	中脑导水管 Mesencephalic aqueduct
19	脚间池 Intercrural cistern
20	神经垂体 Neurohypophysis
21	丘脑间黏合 Interthalamic adhesion
22	视神经 Optic nerve
23	尾状核上的侧脑室 Lateral ventricle over caudate nucleus
24	四叠体池 Quadrigeminal cistern

图1-42 犬正中矢状面的脑膜和脑室

箭头表示脑脊液的流动方向

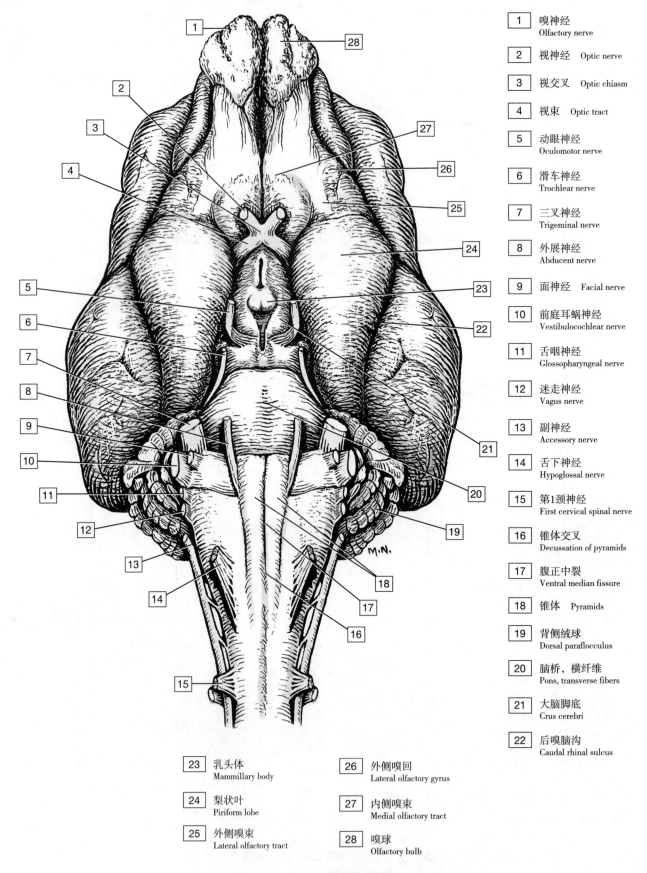

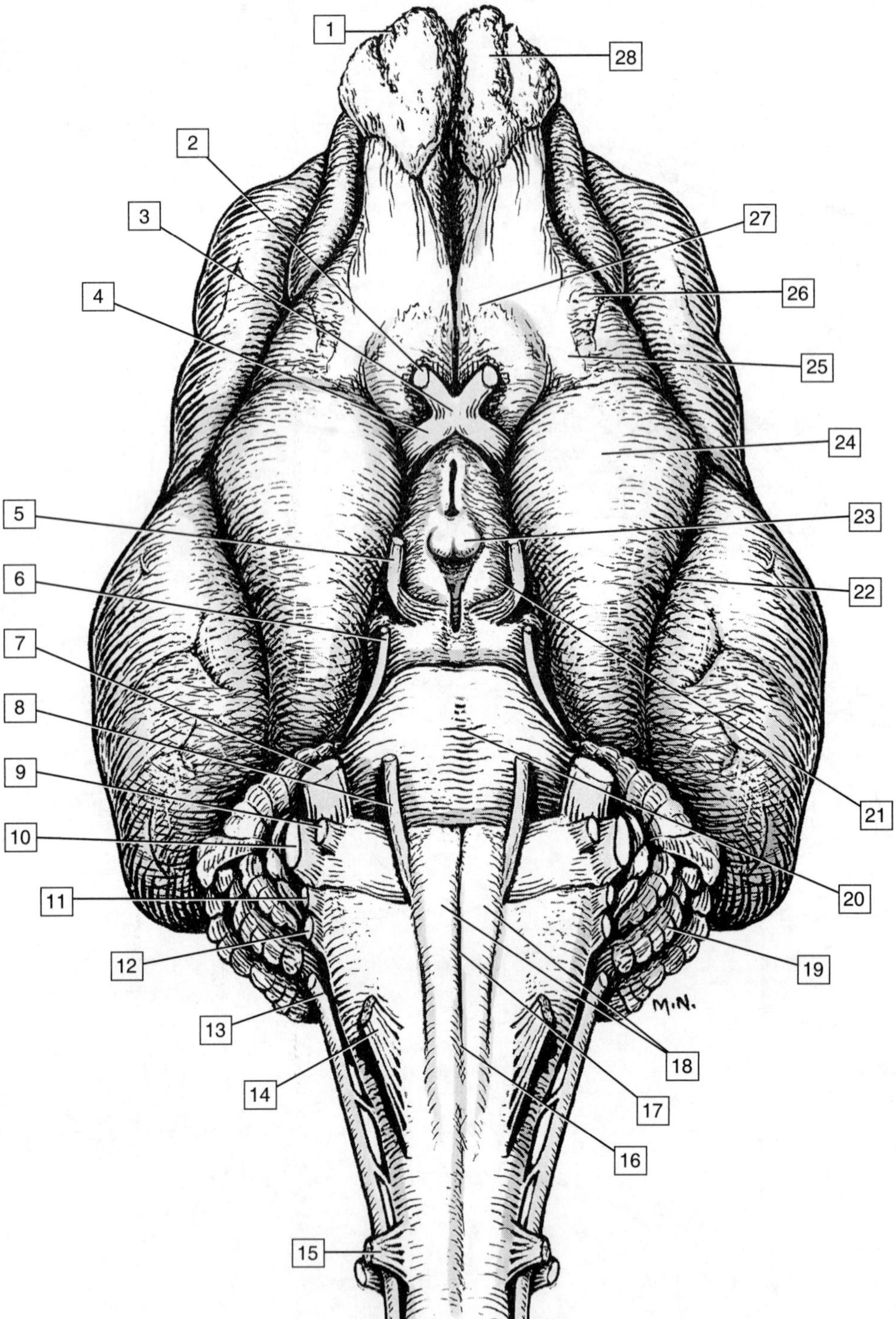

1 嗅神经 Olfactory nerve

2 视神经 Optic nerve

3 视交叉 Optic chiasm

4 视束 Optic tract

5 动眼神经 Oculomotor nerve

6 滑车神经 Trochlear nerve

7 三叉神经 Trigeminal nerve

8 外展神经 Abducent nerve

9 面神经 Facial nerve

10 前庭耳蜗神经 Vestibulocochlear nerve

11 舌咽神经 Glossopharyngeal nerve

12 迷走神经 Vagus nerve

13 副神经 Accessory nerve

14 舌下神经 Hypoglossal nerve

15 第1颈神经 First cervical spinal nerve

16 锥体交叉 Decussation of pyramids

17 腹正中裂 Ventral median fissure

18 锥体 Pyramids

19 背侧绒球 Dorsal paraflocculus

20 脑桥，横纤维 Pons, transverse fibers

21 大脑脚底 Crus cerebri

22 后嗅脑沟 Caudal rhinal sulcus

23 乳头体 Mammillary body

24 梨状叶 Piriform lobe

25 外侧嗅束 Lateral olfactory tract

26 外侧嗅回 Lateral olfactory gyrus

27 内侧嗅束 Medial olfactory tract

28 嗅球 Olfactory bulb

图1-43 犬脑、颅神经和脑干，腹侧观

猫

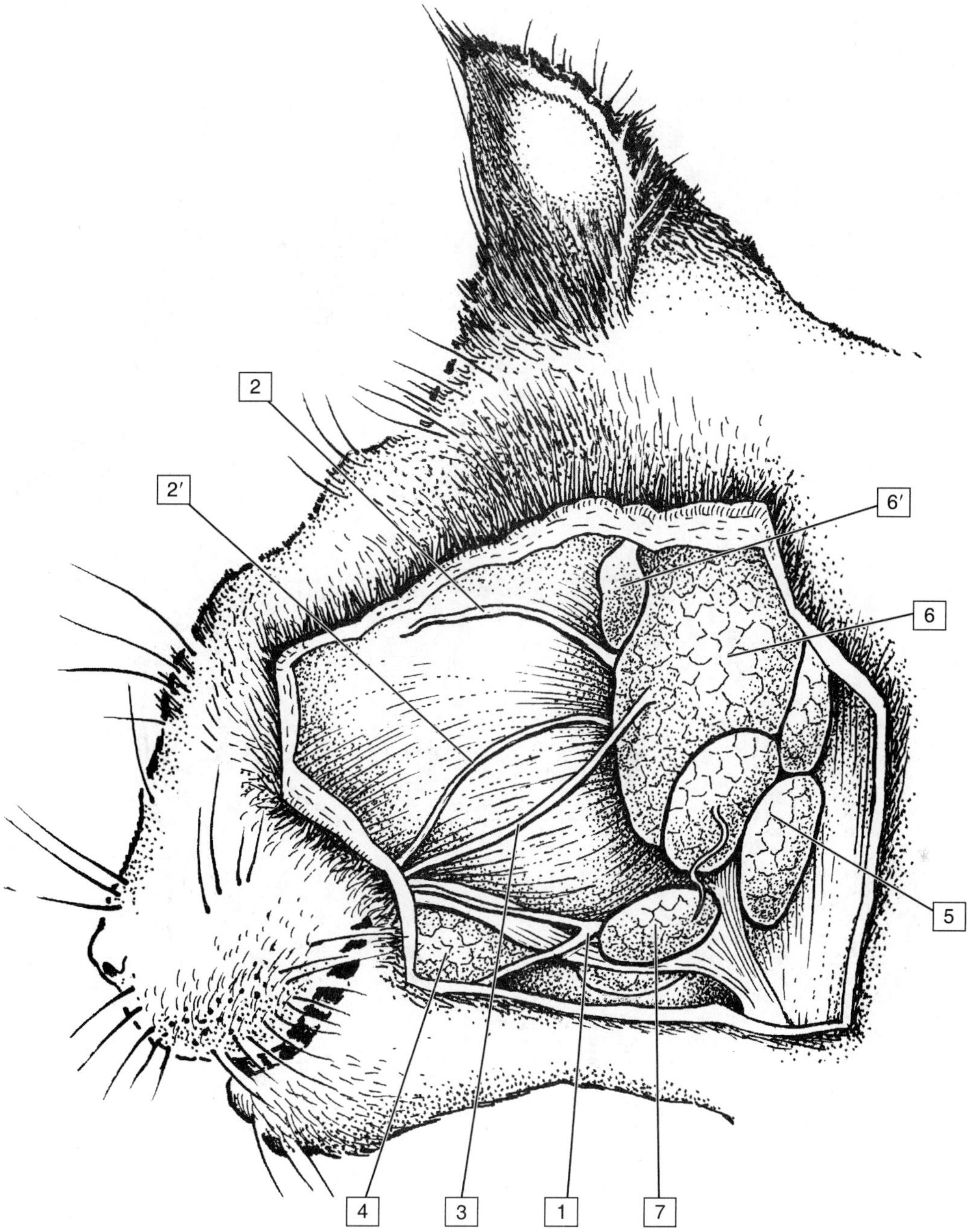

1	面静脉 Facial vein	5	颌下腺 Mandibular gland
2	面神经背侧颊支 Dorsal buccal branch of facial nerve	6	腮腺 Parotid gland
2′	面神经腹侧颊支 Ventral buccal branch of facial nerve	6′	腮腺淋巴结 Parotid lymph node
3	腮腺管 Parotid duct	7	下颌淋巴结 Mandibular lymph nodes
4	颊腺 Buccal salivary glands		

图1-44 猫头部浅层解剖图

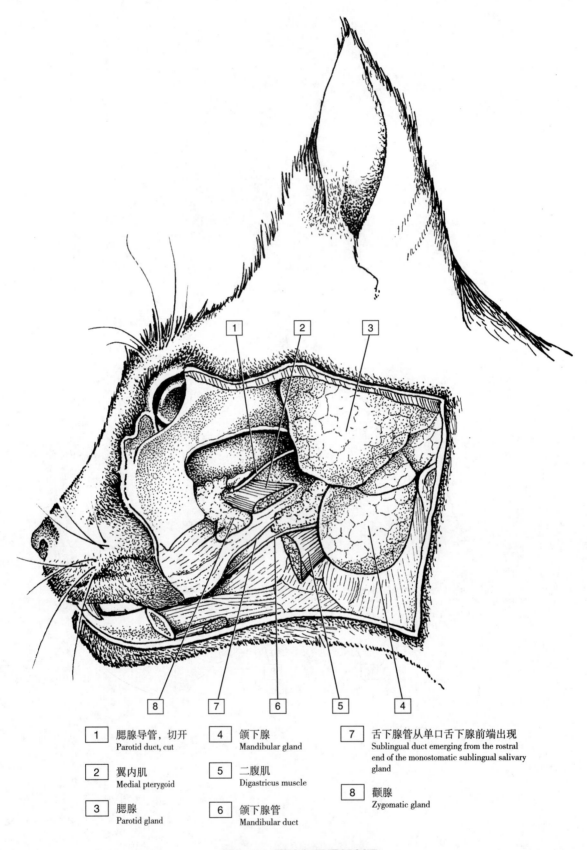

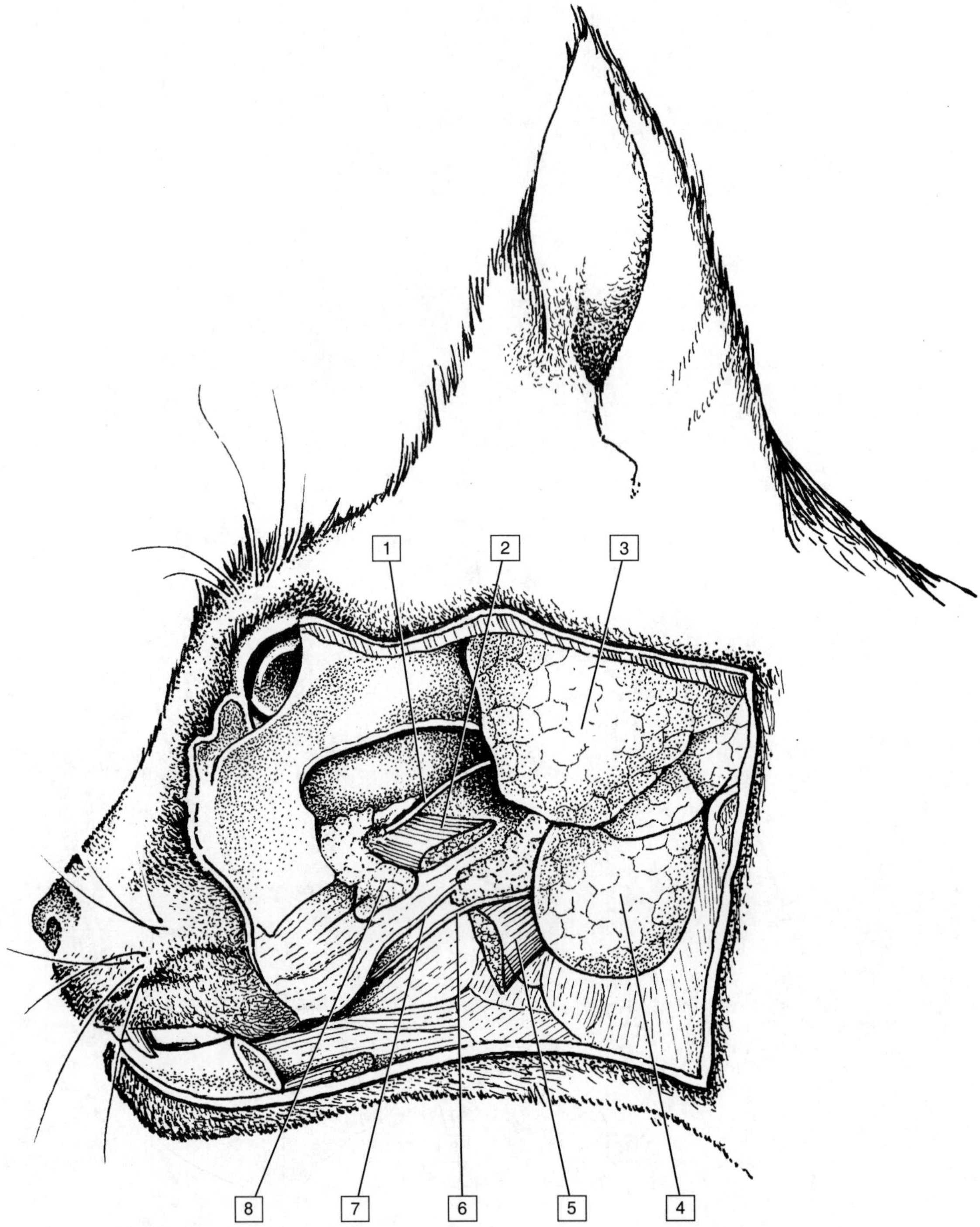

1 腮腺导管，切开 Parotid duct, cut

2 翼内肌 Medial pterygoid

3 腮腺 Parotid gland

4 颌下腺 Mandibular gland

5 二腹肌 Digastricus muscle

6 颌下腺管 Mandibular duct

7 舌下腺管从单口舌下腺前端出现 Sublingual duct emerging from the rostral end of the monostomatic sublingual salivary gland

8 颧腺 Zygomatic gland

图1-45 猫头部深层解剖图

暴露唾液腺

马

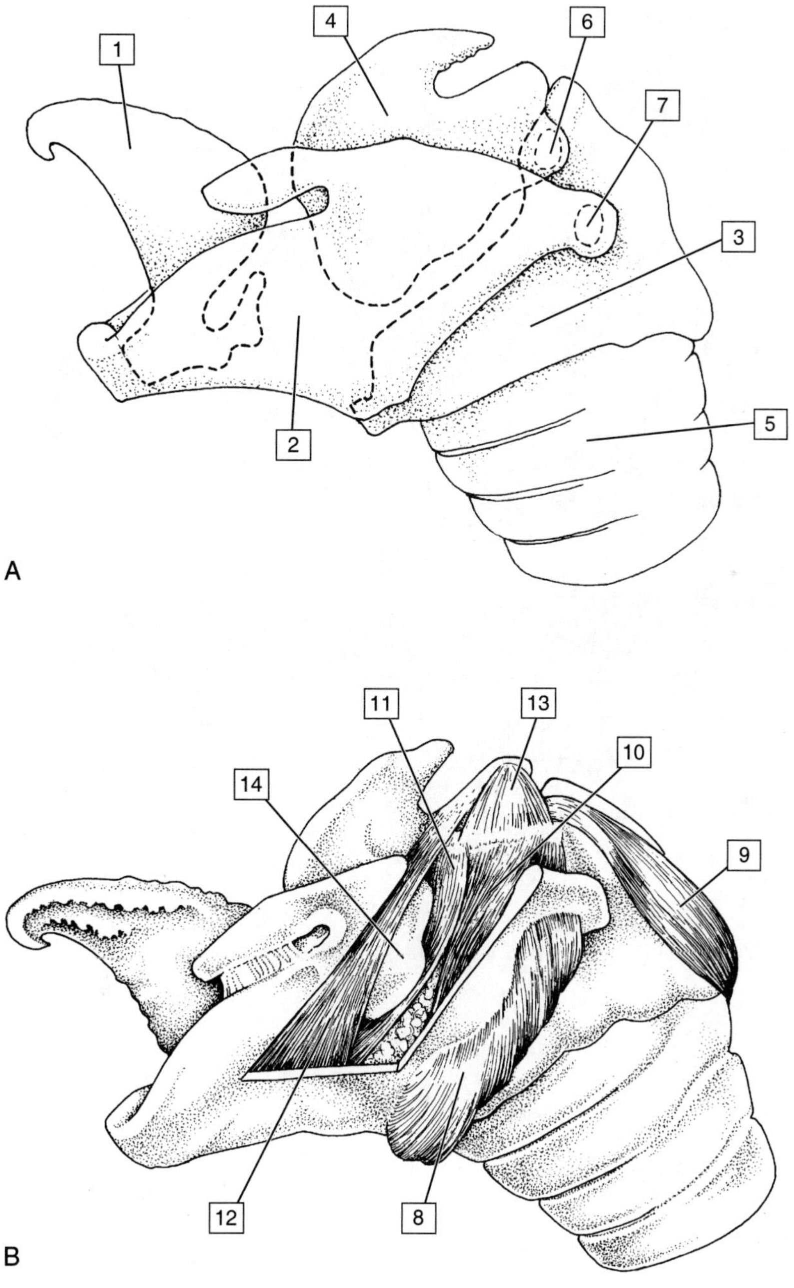

1 会厌软骨 Epiglottic cartilage

2 甲状软骨 Thyroid cartilage

3 环状软骨 Cricoid cartilage

4 杓状软骨 Arytenoid cartilage

5 气管 Trachea

6 环杓关节 Cricoarytenoid joint

7 环甲关节 Cricothyroid joint

8 环甲肌 Cricothyroideus

9 环杓背肌 Cricoarytenoideus dorsalis

10 环杓侧肌 Cricoarytenoideus lateralis

11 声带肌 Vocalis

12 喉室肌 Ventricularis

13 杓横肌 Arytenoideus transversus

14 喉室 Laryngeal ventricle

图1-46 马喉软骨外侧观（A）和马喉内部肌肉（B）

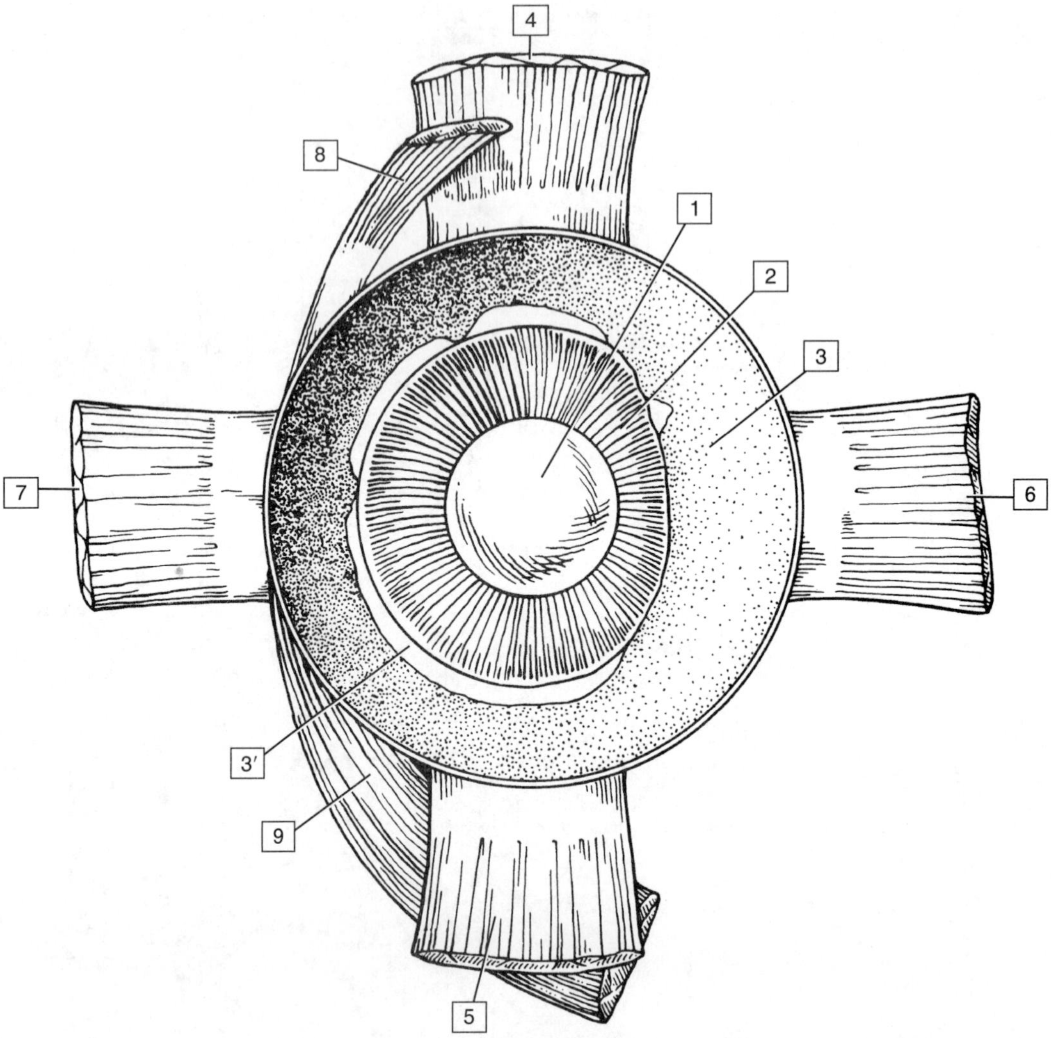

1	晶状体	Lens
2	睫状体	Ciliary body
3	被视网膜色素层覆盖的脉络膜	Choroid covered by pigmented outer layer of retina
3′	残余的视网膜神经层（已分离）	Remnants of inner nervous layer of retina, which has been removed
4	背侧肌	Dorsal muscle
5	腹侧肌	Ventral muscle
6	内直肌	Medial muscle
7	外直肌	Lateral rectus muscle
8	上斜肌	Dorsal oblique muscle
9	下斜肌	Ventral oblique muscle

图1-47　马左眼前部，从后向前观

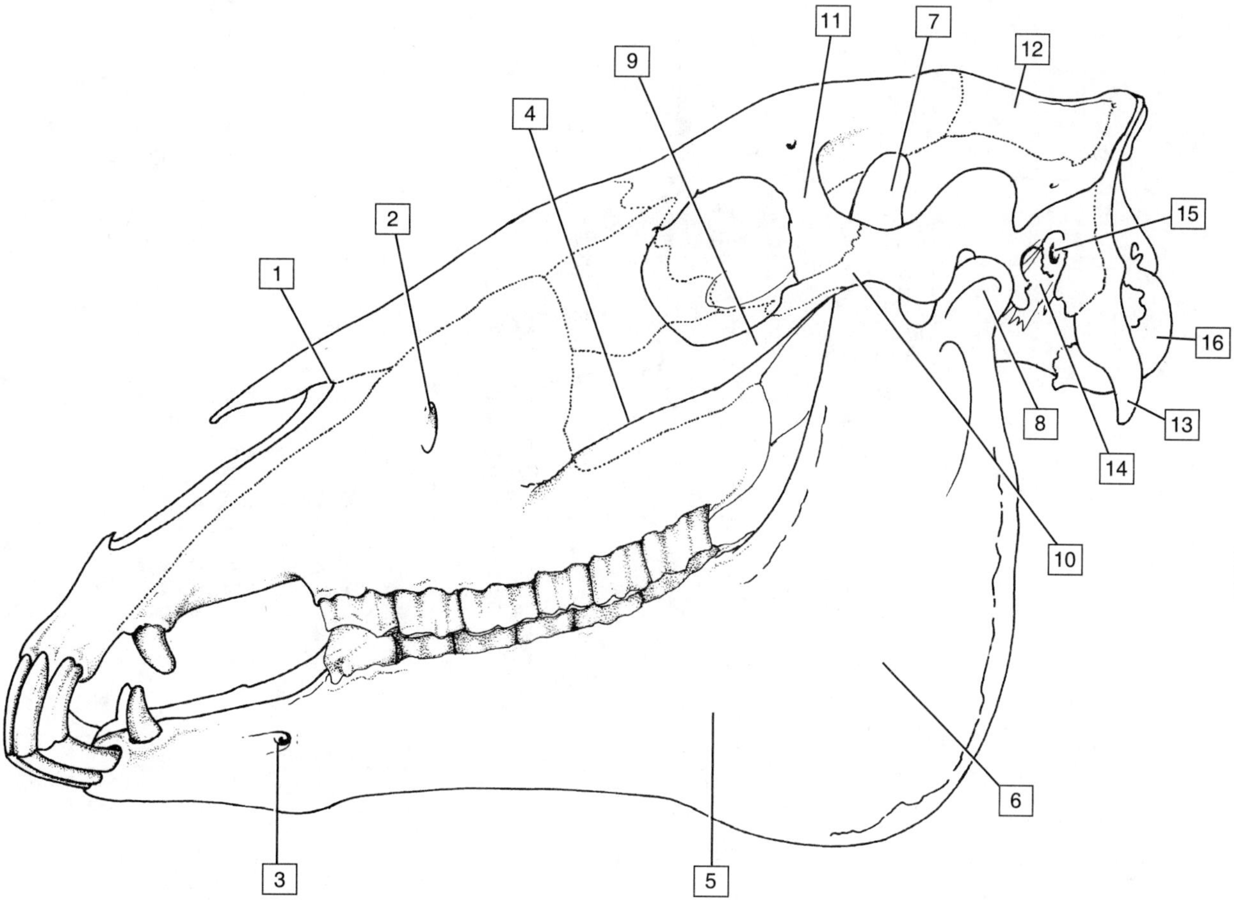

1 鼻切齿骨切迹 Nasoincisive notch

2 眶下孔 Infraorbital foramen

3 颏孔 Mental foramen

4 面嵴 Facial crest

5 下颌骨骨体 Body of mandible

6 下颌支 Ramus of mandible

7 冠突 Coronoid process

8 髁突 Condylar process

9 颧骨颞突 Temporal process of zygomatic bone

10 颞骨颧突 Zygomatic process of temporal bone

11 额骨颧突 Zygomatic process of frontal bone

12 外矢状嵴 External sagittal crest

13 髁旁突 Paracondylar process

14 茎突 Styloid process

15 外耳道 External acoustic meatus

16 枕髁 Occipital condyle

图1-48 马颅骨，外侧观

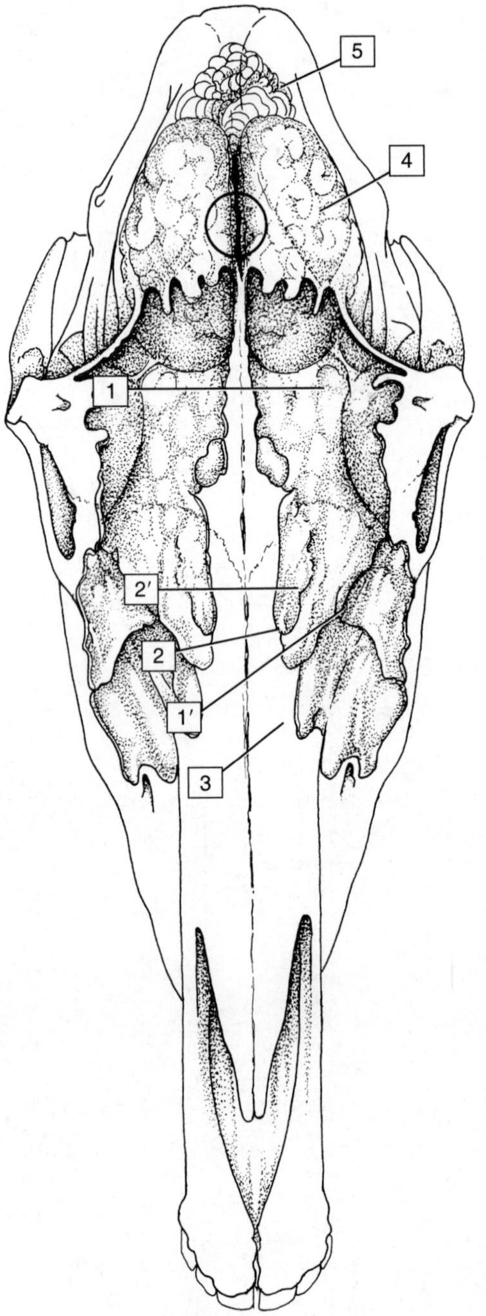

1	前额窦 Conchofrontal sinus, frontal part	2	上颌窦后部 Caudal maxillary sinus	3	上颌窦前部 Rostral maxillary sinus	5	小脑 Cerebellum
1′	鼻甲背侧前额窦 Conchofrontal sinus, dorsal conchal part	2′	额上颌缝 Frontomaxillary suture	4	大脑 Cerebrum		

图1-49　马颅骨背侧的脑、额窦、上颌窦的投影

窦内充满了填充材料。额窦向后延伸，覆盖大脑前侧，并向外侧延伸，超出眼眶的水平。圆圈表示大脑的中心

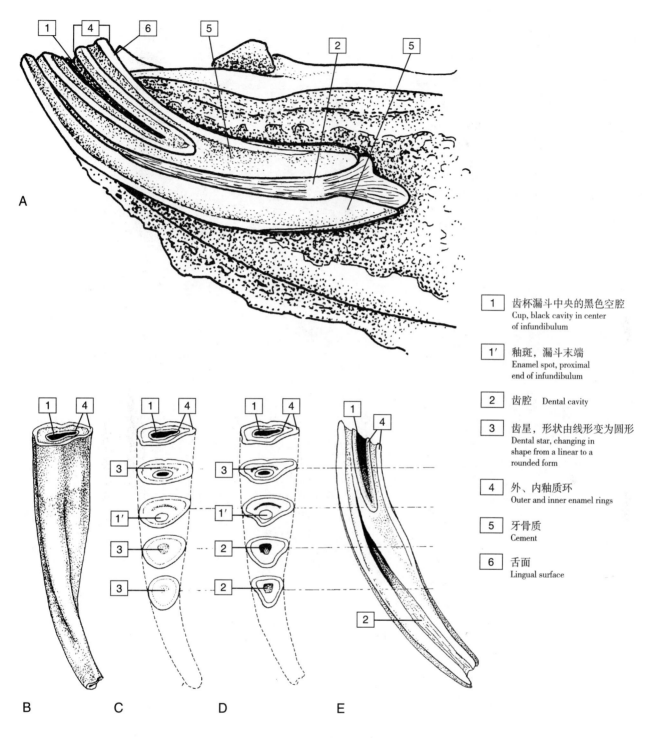

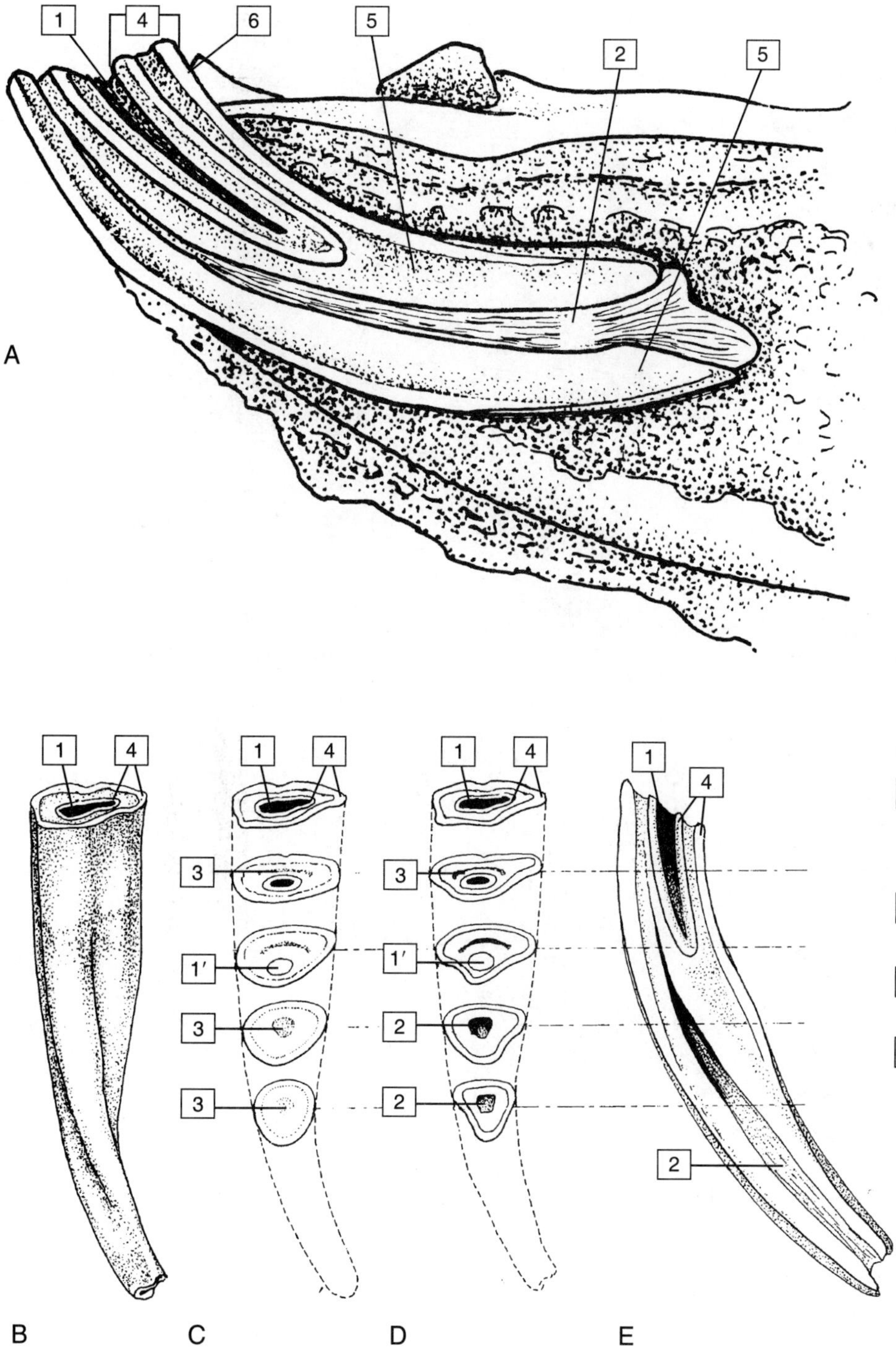

1	齿杯漏斗中央的黑色空腔 Cup, black cavity in center of infundibulum
1′	釉斑，漏斗末端 Enamel spot, proximal end of infundibulum
2	齿腔 Dental cavity
3	齿星，形状由线形变为圆形 Dental star, changing in shape from a linear to a rounded form
4	外、内釉质环 Outer and inner enamel rings
5	牙骨质 Cement
6	舌面 Lingual surface

图1-50 马下切齿的结构

A. 原位纵切面，牙冠较牙齿内嵌部分短；B. 后面观，牙冠和牙齿剩余部分之间的连接没有显示；C. 由于磨损，咬合面发生改变，齿杯变小并消失，剩余部分在一段时间后形成釉斑，齿星出现并从线状变成大的圆斑；D. 本图为年轻马牙齿的横断面；E. 切齿纵切面，显示漏斗与齿腔的关系

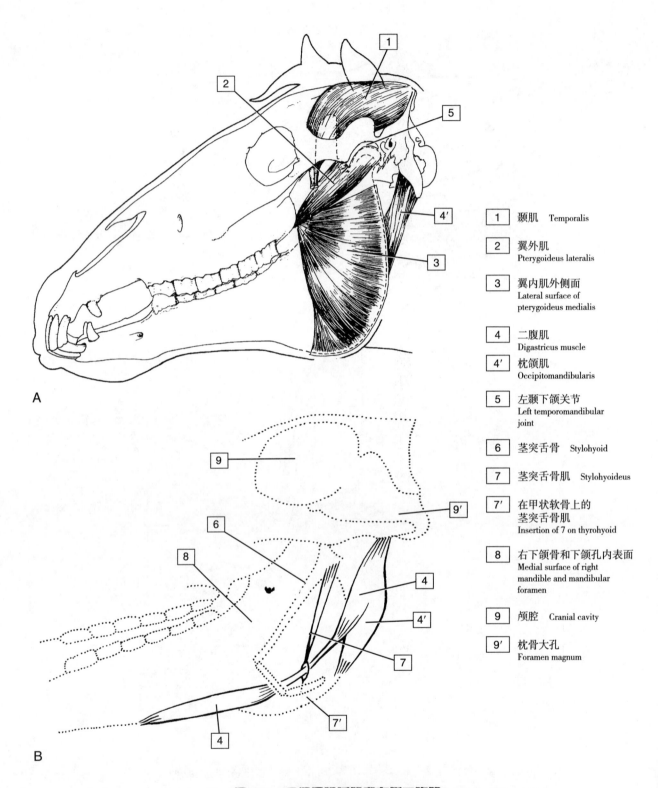

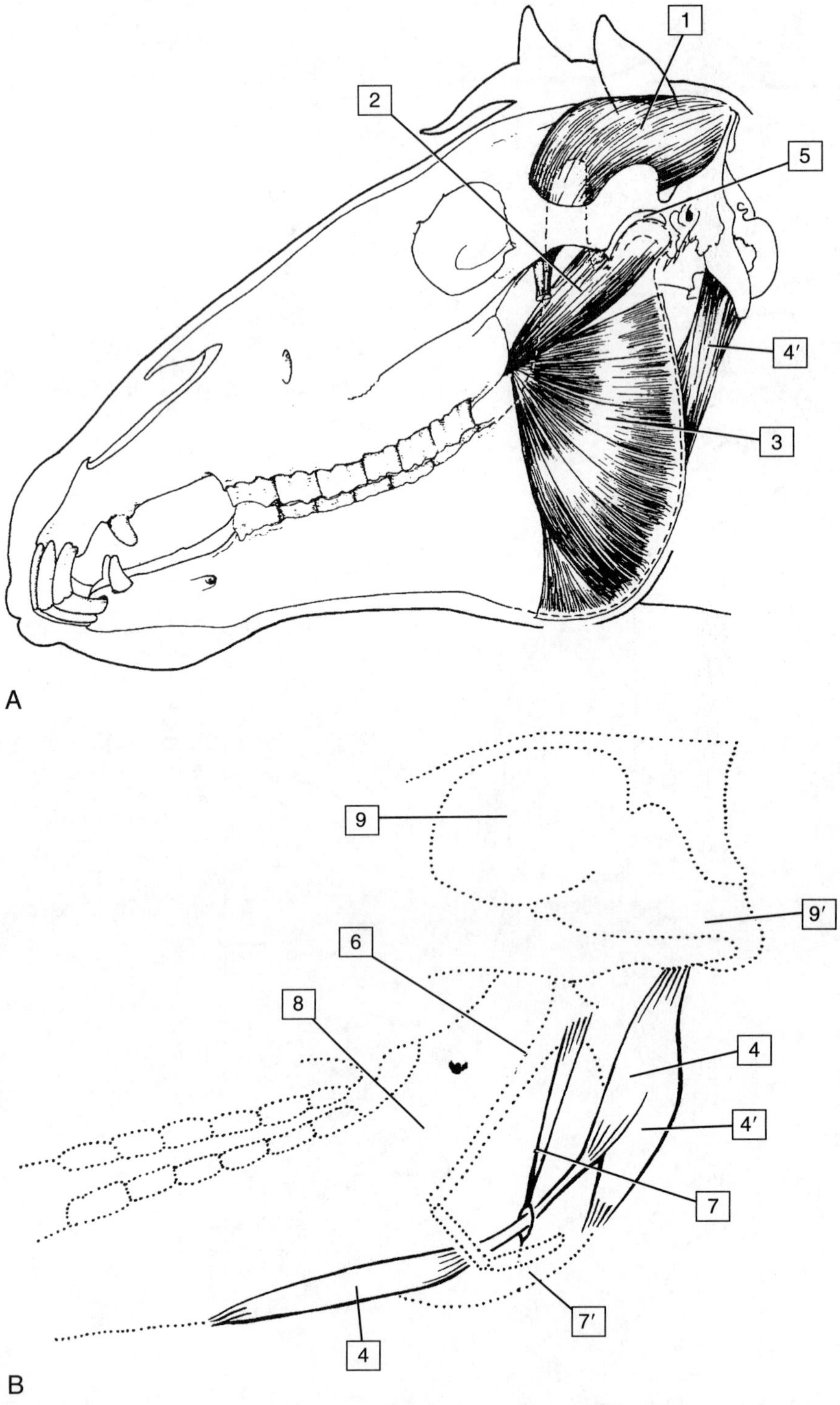

1 颞肌 Temporalis

2 翼外肌 Pterygoideus lateralis

3 翼内肌外侧面 Lateral surface of pterygoideus medialis

4 二腹肌 Digastricus muscle

4′ 枕颌肌 Occipitomandibularis

5 左颞下颌关节 Left temporomandibular joint

6 茎突舌骨 Stylohyoid

7 茎突舌骨肌 Stylohyoideus

7′ 在甲状软骨上的茎突舌骨肌 Insertion of 7 on thyrohyoid

8 右下颌骨和下颌孔内表面 Medial surface of right mandible and mandibular foramen

9 颅腔 Cranial cavity

9′ 枕骨大孔 Foramen magnum

图1-51 马深层咀嚼肌和右侧二腹肌

A. 切除左侧下颌支（虚线部分），暴露左侧深层咀嚼肌；B. 右侧二腹肌及相关结构的内侧观

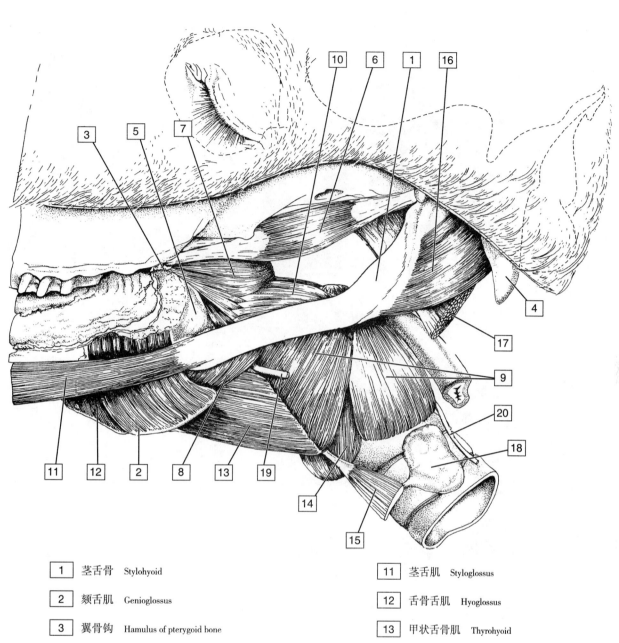

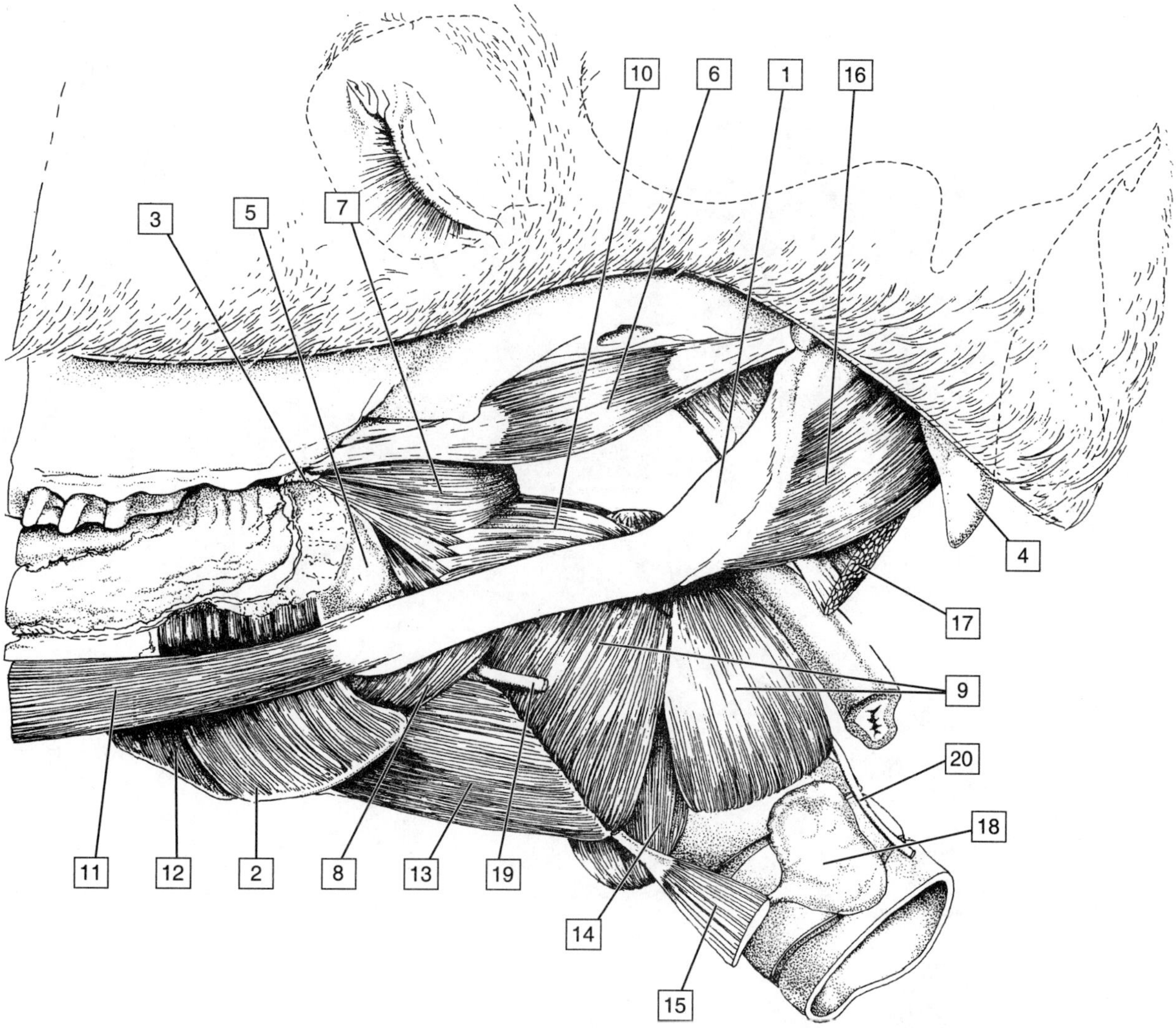

1	茎舌骨 Stylohyoid	11	茎舌肌 Styloglossus
2	颏舌肌 Genioglossus	12	舌骨舌肌 Hyoglossus
3	翼骨钩 Hamulus of pterygoid bone	13	甲状舌骨肌 Thyrohyoid
4	髁旁突 Paracondylar process	14	环甲肌 Cricothyroideus
5	颊咽筋膜 Buccopharyngeal fascia	15	胸骨甲状肌 Sternothyroideus
6	腭帆提肌 Tensor veli palatine	16	枕舌骨肌 Occipitohyoideus
7	前咽缩肌 Rostral pharyngeal constrictor	17	头长肌（断端） Longus capitis (stump)
8	中咽缩肌 Middle pharyngeal constrictor	18	甲状腺 Thyroid gland
9	后咽缩肌（甲咽肌和环咽肌） Caudal pharyngeal constrictor (thyro-and cricopharyngeus muscle)	19	喉前神经 Cranial laryngeal nerve
10	后茎突咽肌 Stylopharyngeus caudalis	20	喉返神经 Caudal (recurrent) laryngeal nerve

图1-52 马的咽、软腭和舌骨的肌肉

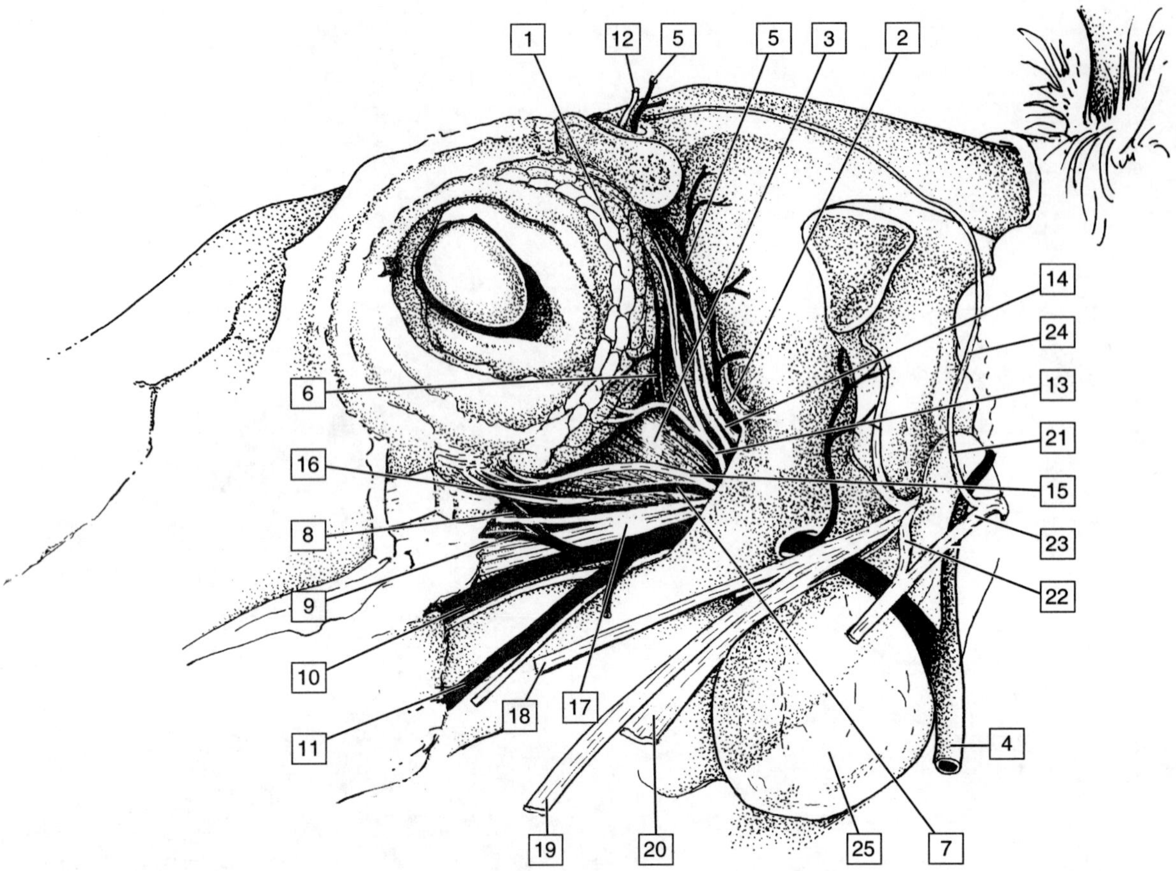

1	泪腺 Lacrimal gland	14	滑车神经 Trochlear nerve
2	眶骨膜 Periorbita	15	颧神经 Zygomatic nerve
3	外直肌 Lateral rectus muscle	16	动眼神经 Oculomotor nerve
4	上颌动脉 Maxillary artery	17	上颌神经前支 Rostral branches of maxillary nerve
5	眶上动脉 Supraorbital artery	18	颊神经 Buccal nerve
6	泪腺动脉 Lacrimal artery	19	舌神经 Lingual nerve
7	眼外动脉肌支 Muscular branch of external ophthalmic artery	20	下牙槽神经 Inferior alveolar nerve
8	颊动脉 Malar artery	21	咀嚼肌神经 Masticatory nerve
9	眶下动脉 Infraorbital artery	22	耳颞神经 Auriculotemporal nerve
10	腭大动脉 Major palatine artery	23	面神经 Facial nerve
11	颊动脉 Buccal artery	24	耳睑神经 Auriculopalpebral nerve
12	眶上神经 Supraorbital nerve	25	咽鼓管囊 Guttural pouch
13	泪腺神经 Lacrimal nerve		

图1-53 马眼眶的解剖图

颧弓和眶骨膜被切除

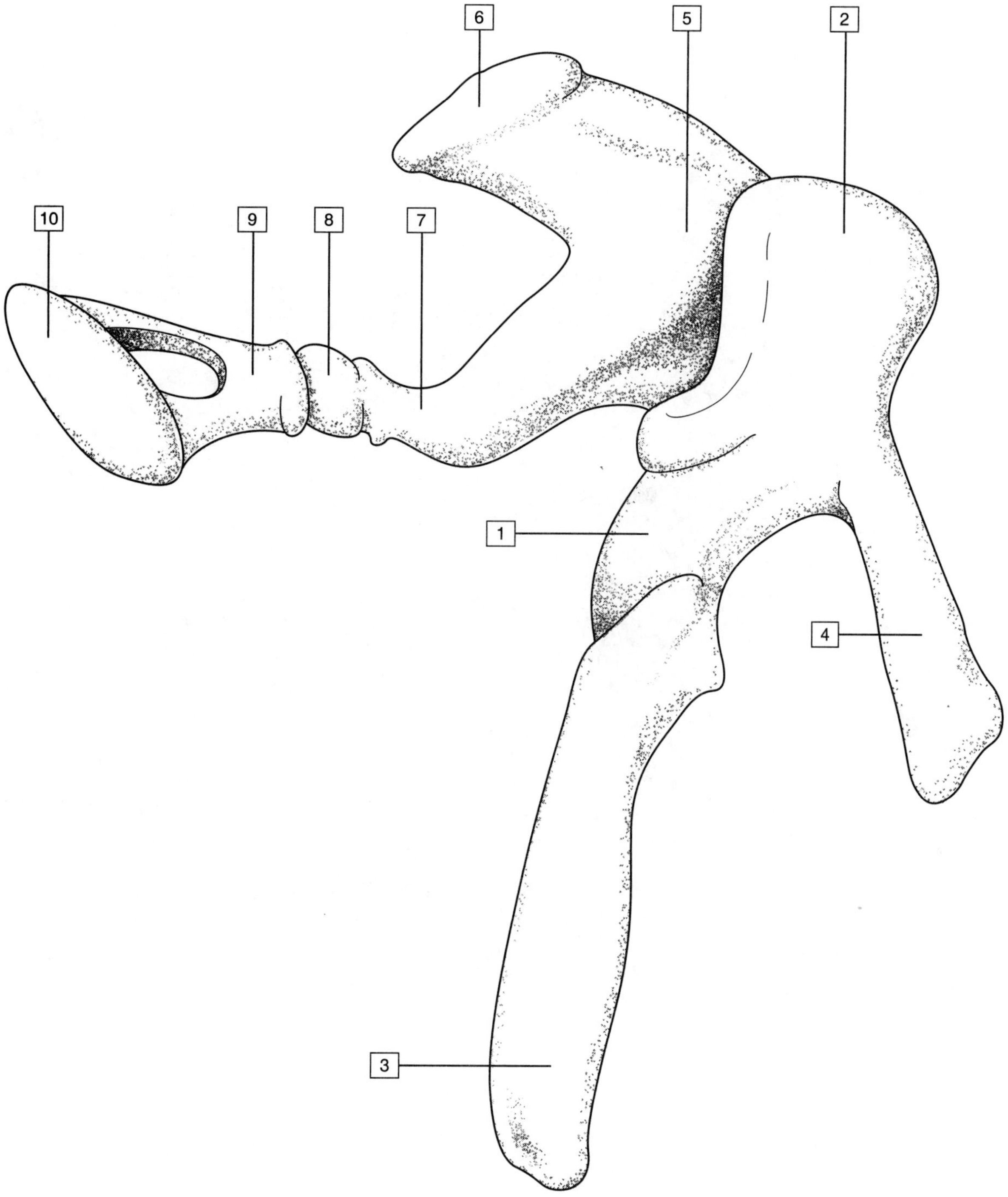

1 锤骨 Malleus	4 喙突 Rostral process	7 砧骨长脚 Long crus
2 锤骨头 Head of malleus	5 砧骨 Incus	8 豆状突 Os lenticulare
3 锤骨柄 Handle of malleus	6 砧骨短脚 Short crus	9 镫骨头 Head of stapes

10 镫骨底 Base (footplate) of stapes

图1-54 马左侧听小骨，前内侧观

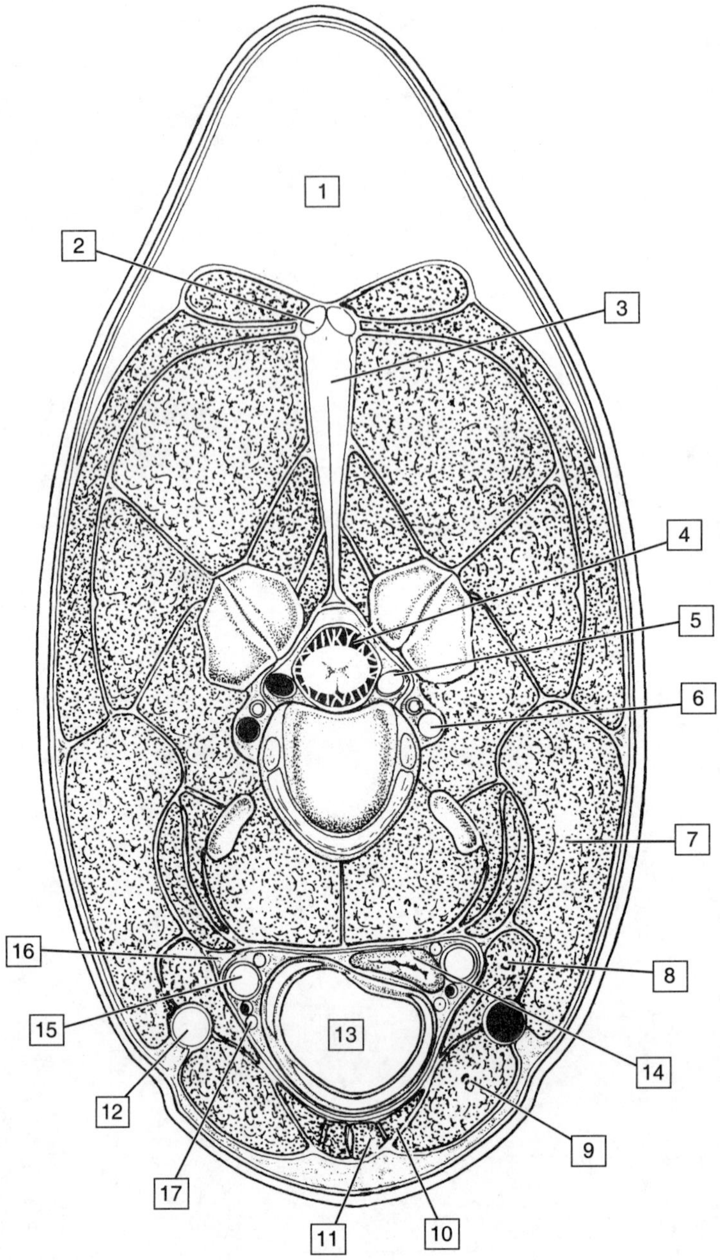

1	嵴 Crest	5	椎内静脉丛 Internal vertebral venous plexus	9	胸头肌 Sternocephalicus	14	食管 Esophagus
2	项韧带索状部 Funicular part of nuchal ligament	6	椎动脉和椎静脉 Vertebral artery and vein	10	胸骨甲状肌 Sternothyroideus	15	颈总动脉 Common carotid artery
3	项韧带板状部 Laminar part of nuchal ligament	7	臂头肌 Brachiocephalicus	11	胸骨舌骨肌 Sternohyoideus	16	迷走交感干 Vagosympathetic trunk
4	蛛网膜下腔 Subarachnoid space	8	肩胛舌骨肌 Omohyoideus	12	颈外静脉 External jugular vein	17	喉返神经 Recurrent laryngeal nerve
				13	气管 Trachea		

图1-55　马颈部第4颈椎横断面

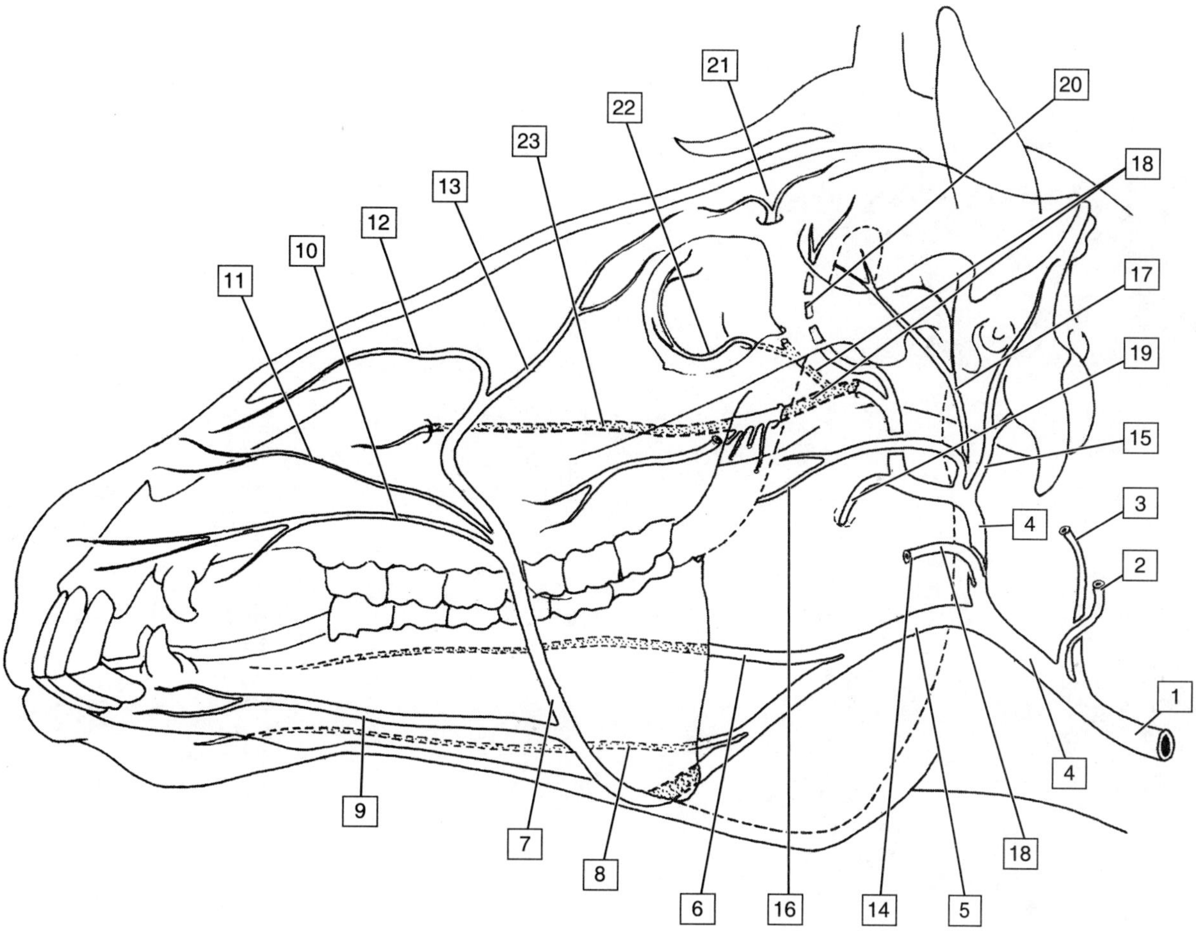

1	颈总动脉	Common carotid artery
2	枕动脉	Occipital artery
3	颈内动脉	Internal carotid artery
4	颈外动脉	External carotid artery
5	舌面动脉	Linguofacial artery
6	舌动脉	Lingual artery
7	面动脉	Facial artery
8	舌下动脉	Sublingual artery
9	下唇动脉	Inferior labial artery
10	上唇动脉	Superior labial artery
11	鼻外侧动脉	Lateral nasal artery
12	鼻背侧动脉	Dorsal nasal artery
13	眼角动脉	Angularis oculi artery
14	咬肌动脉	Masseteric artery
15	耳后动脉	Caudal auricular artery
16	为清晰可见面横动脉，已将腹侧剥离	Transverse facial artery, displaced ventrally for clarity
17	颞浅动脉	Superficial temporal artery
18	上颌动脉	Maxillary artery
19	下齿槽动脉	Inferior alveolar artery
20	颞深后动脉	Caudal deep temporal artery
21	眶上动脉	Supraorbital artery
22	颊动脉	Malar artery
23	眶下动脉	Infraorbital artery

图1-56　马头部主要动脉

牛

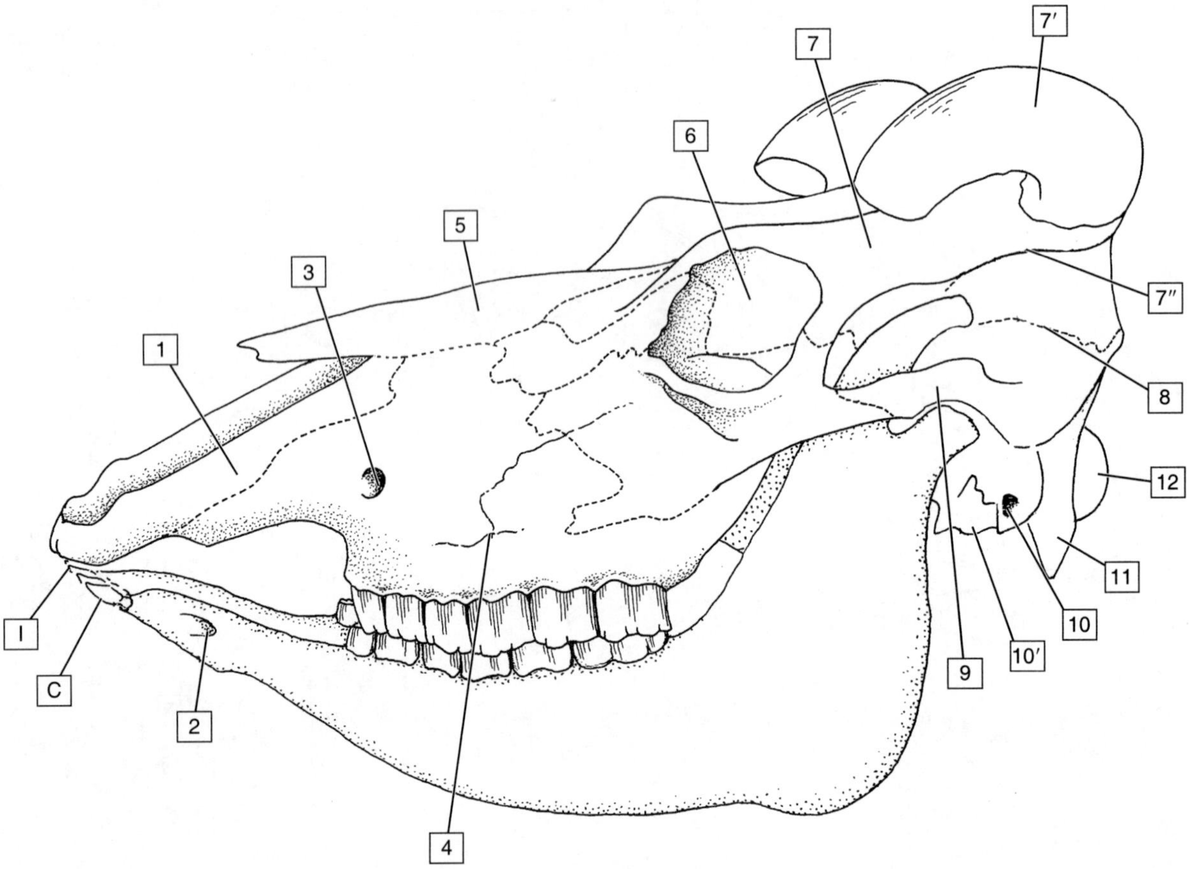

1	切齿骨	Incisive bone
2	颏孔	Mental foramen
3	眶下孔	Infraorbital foramen
4	面部结节	Facial tuberosity
5	鼻骨	Nasal bone
6	眼眶	Orbit
7	额骨	Frontal bone
7′	额骨角突	Horn surrounding cornual process of frontal bone
7″	颞线	Temporal line
8	颞窝	Temporal fossa
9	颧弓	Zygomatic arch
10	外耳道	External acoustic meatus
10′	鼓泡	Tympanic bulla
11	髁旁突	Paracondylar process
12	枕髁	Occipital condyle
C	犬齿	Canine tooth
I	切齿	Incisors

图1-57　牛颅骨和下颌骨

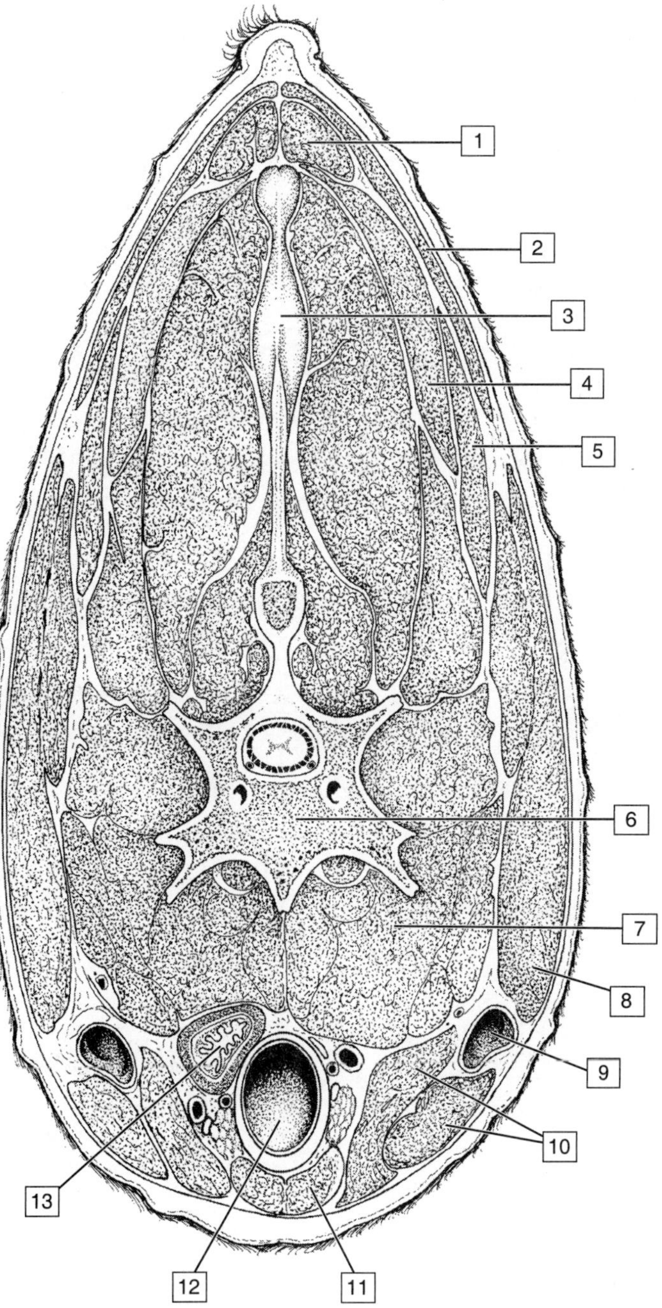

1	菱形肌	Rhomboideus
2	斜方肌	Trapezius
3	项韧带	Nuchal ligament
4	夹肌	Splenius
5	肩胛横突肌	Omotransversarius
6	椎骨	Vertebra
7	颈长肌	Longus colli
8	臂头肌	Brachiocephalicus
9	颈静脉沟内的颈外静脉	External jugular vein in jugular groove
10	胸头肌，下颌部和乳突肌	Sternocephalicus,mandibular, and mastoid parts
11	胸骨舌骨肌和胸骨甲状肌结合处	Combined sternohyoideus and sternothyroideus
12	气管	Trachea
13	食管（腹侧有神经、血管和胸腺）	Esophagus (ventral to it, nerves, blood vessels, and thymus)

图1-58 牛颈部横断面

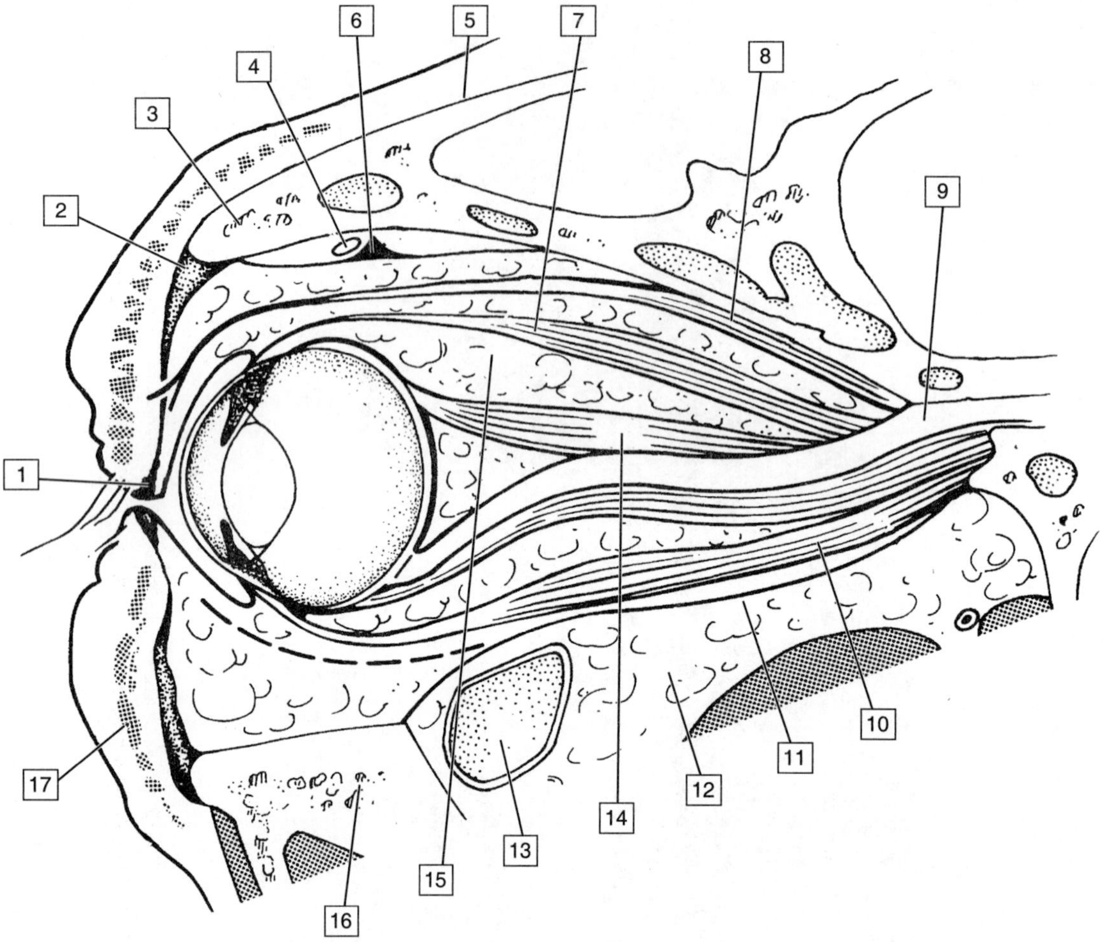

1 睑板 Tarsus	10 下直肌 Ventral rectus muscle
2 眶隔 Orbital septum	11 眶骨膜 Periorbita
3 眶缘 Orbital margin	12 眶外脂肪 Extraorbital fat
4 上斜肌 Dorsal oblique muscle	13 泪泡，上颌窦的后隐窝 Lacrimal bulla, a caudal recess of the maxillary sinus
5 面部骨膜 Periosteum of face	14 眼球退缩肌 Retractor bulbi
6 滑车 Trochlea	15 眶内脂肪 Intraperiorbital fat
7 上直肌 Dorsal rectus muscle	16 颧弓 Zygomatic arch
8 上睑提肌 Levator palpebrae superioris	17 眼轮匝肌 Orbicularis oculi
9 视神经孔内的视神经 Optic nerve in optic foramen	

图1-59 牛右眼

沿眼眶轴切开，前内侧面

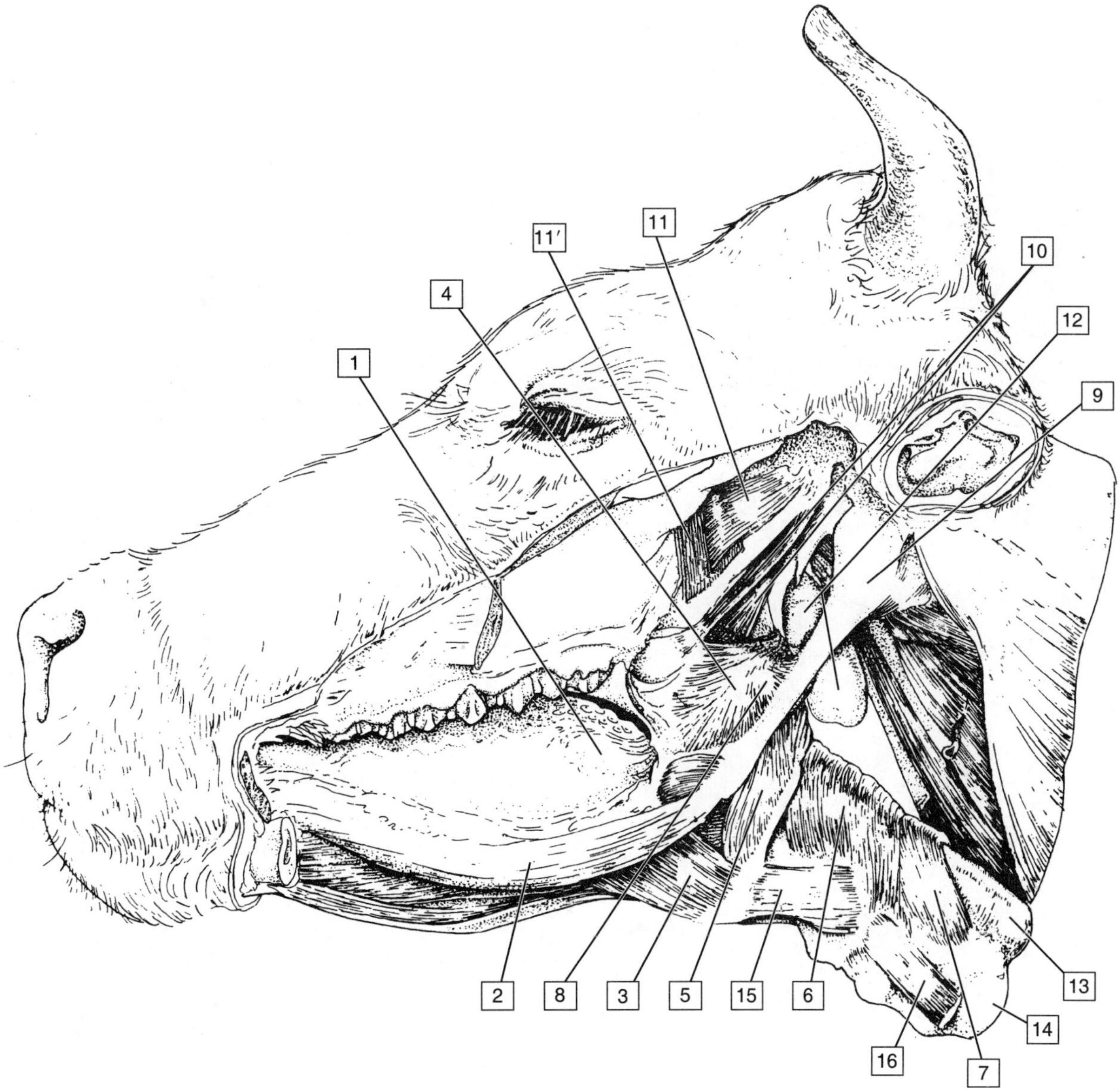

1	舌根	Root of tongue
2	茎突舌肌	Styloglossus
3	舌骨舌肌	Hyoglossus
4	前咽缩肌	Rostral pharyngeal constrictor
5	中咽缩肌（舌咽肌）	Middle pharyngeal constrictor
6	下咽缩肌（甲咽肌）	Caudal pharyngeal constrictor (thyropharyngeus muscle)
7	下咽缩肌（环咽肌）	Caudal pharyngeal constrictor (cricopharyngeus muscle)
8	后茎突咽肌	Stylopharyngeus caudalis
9	茎突舌骨	Stylohyoid
10	腭帆提肌和腭帆张肌	Tensor and levator veli palatine
11	翼外侧肌	Pterygoideus lateralis
11′	残留的翼内侧肌	Remnants of pterygoideus medialis
12	咽后内侧淋巴结	Medial retropharyngeal lymph node
13	食管	Esophagus
14	气管	Trachea
15	甲状舌骨肌	Thyrohyoid
16	胸骨甲状肌	Sternothyroideus

图1-60 牛颅骨基部和舌与咽、喉的连接

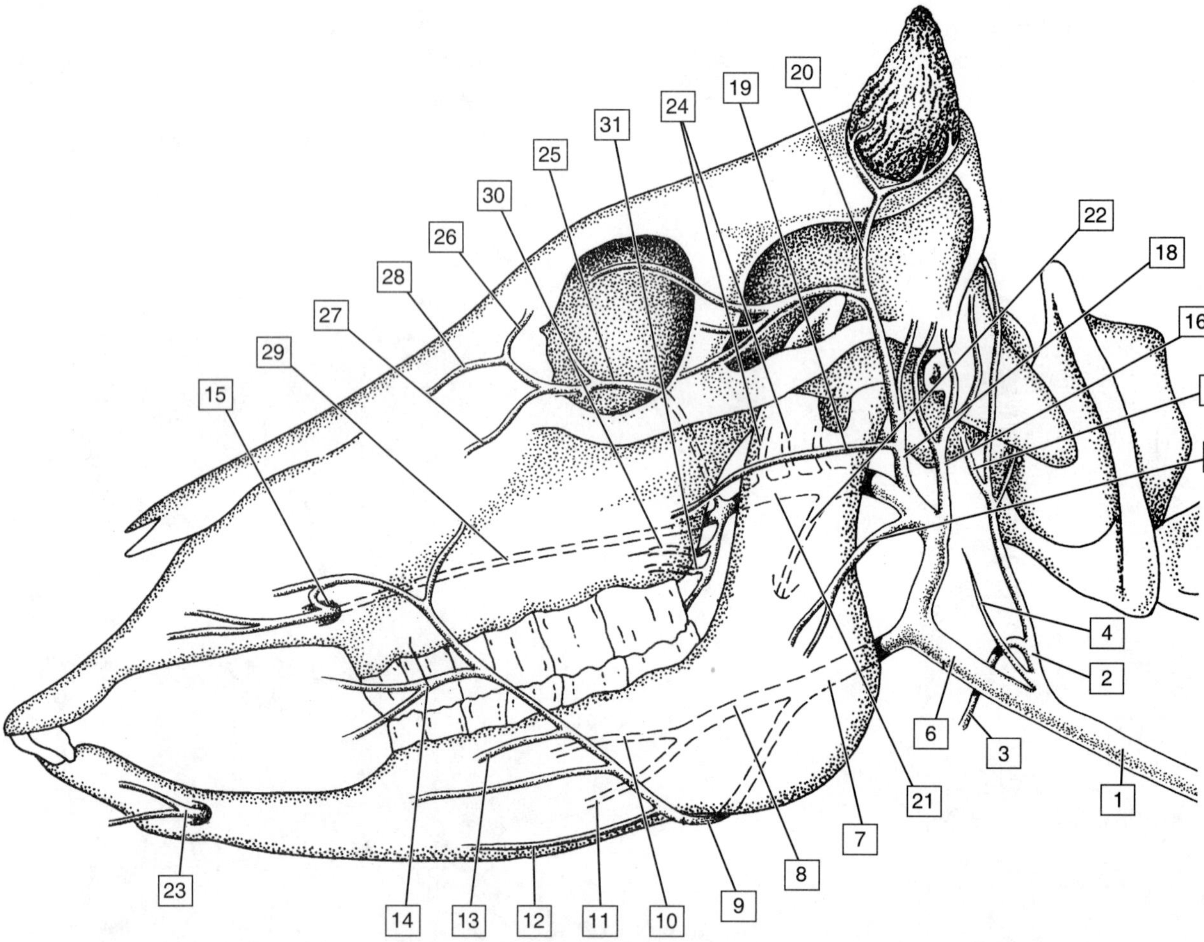

1 颈总动脉 Common carotid artery

2 枕动脉 Occipital artery

3 腭升动脉 Ascending palatine artery

4 残留的颈内动脉 Remnant of internal carotid artery

5 脑膜内动脉 Medial meningeal artery

6 颈外动脉 External carotid artery

7 舌面干 Linguofacial trunk

8 舌动脉 Lingual artery

9 面动脉 Facial artery

10 舌深动脉 Deep lingual artery

11 舌下动脉 Sublingual artery

12 颏下动脉 Submental artery

13 下唇动脉 Inferior labial artery

14 上唇动脉 Superior labial artery

15 眶下孔 Infraorbital foramen

16 耳后动脉 Caudal auricular artery

17 咬肌支 Masseteric branch

18 颞浅动脉 Superficial temporal artery

19 面横动脉 Transverse facial artery

20 角动脉 Cornual artery

21 上颌动脉 Maxillary artery

22 下齿槽动脉 Inferior alveolar artery

23 颏动脉 Mental artery

24 硬膜外异网的前后支 Rostral and caudal branches to rete mirabile

25 颊动脉 Malar artery

26 眼的角动脉 Angular a. of the eye

27 鼻后外侧动脉 Caudal lateral nasal artery

28 鼻背侧动脉 Dorsal nasal artery

29 眶下动脉 Infraorbital artery

30 蝶腭动脉 Sphenopalatine artery

31 腭大动脉和腭小动脉 Major and minor palatine artery

图1-61　牛左颈总动脉分支

猪

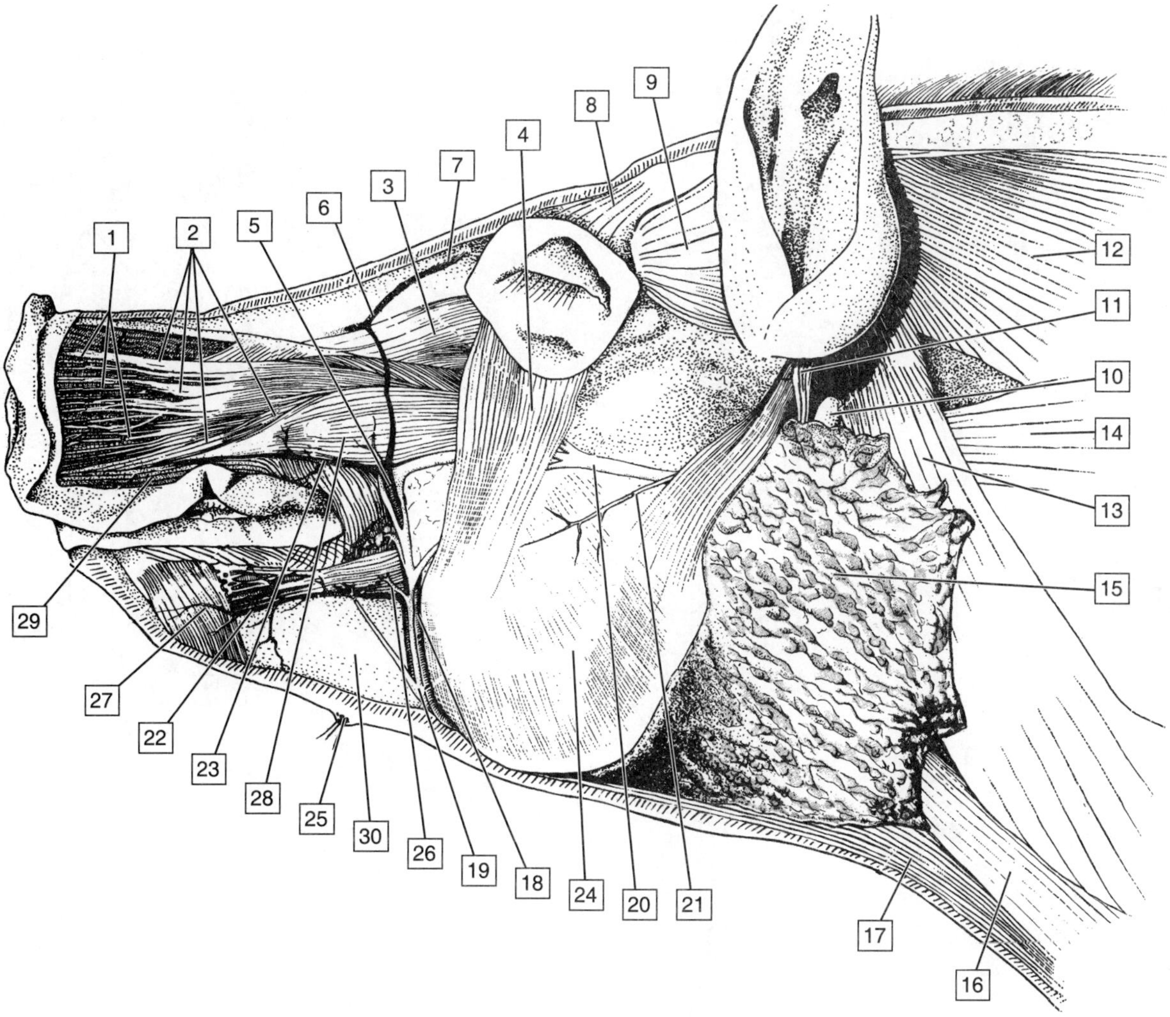

1 鼻唇提肌纤维束 Cut fasciculi of levator nasolabialis

2 犬齿肌 Caninus

3 上唇提肌 Levator labii superioris

4 颧肌 Zygomaticus

5 面静脉 Facial vein

6 鼻背侧静脉 Dorsal nasal vein

7 额叶静脉 Frontal vein

8 眼角提肌 Levator anguli oculi

9 额盾肌 Frontoscutularis

10 咽后外侧淋巴结 Lateral retropharyngeal lymph node

11 腮耳肌 Parotidoauricularis

12 斜方肌 Trapezius

13 枕乳突肌 Brachiocephalicus

14 肩胛横突肌 Omotransversarius

15 腮腺 Parotid gland

16 胸头肌 Sternocephalicus

17 胸骨舌骨肌 Sternohyoideus

18 腮腺管 Parotid duct

19, 20 面神经的腹侧颊变和背侧颊支 Ventral and dorsal buccal branches of facial nerve

21 面横神经 Transverse facial nerve

22 下唇静脉 Inferior labial vein

23 上唇静脉 Superior labial vein

24 咬肌 Masseter

25 头部毛发和腺体 Mental hairs and gland

26 下唇降肌 Depressor labii inferioris

27 颏肌 Mentalis

28 上唇降肌 Depressor labii superioris

29 口轮匝肌 Orbicularis oris

30 下颌骨 Mandible

图1-62 猪头部的浅层解剖图

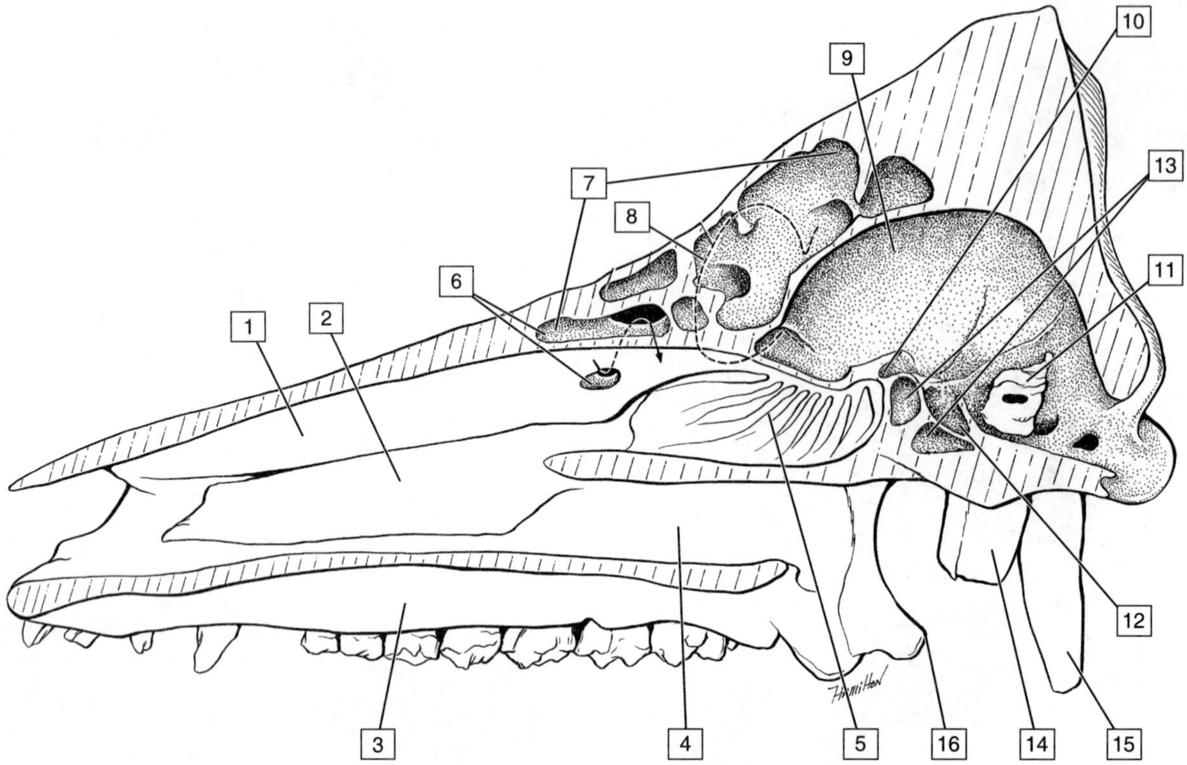

1 上鼻甲骨，6处切开示鼻窦
Dorsal turbinate bone, fenestrated at 6 to show conchal sinus

2 下鼻甲骨 Ventral turbinate bone

3 硬腭 Hard palate

4 鼻后孔 Choana

5 眼底筛骨的鼻腔部
Ethmoturbinates in fundus of nasal cavity

6 鼻窦 Conchal sinus

7 正中切开暴露额窦部分
Portion of frontal sinus exposed by paramedian saw cut

8 眼眶 orbit

9 颅腔 Cranial cavity

10 视神经管 Optic canal

11 颞骨岩部 Petrous temporal bone

12 垂体窝 Fossa for hypophysis

13 蝶窦 Sphenoid sinus

14 鼓泡 Tympanic bulla

15 髁旁突 Paracondylar process

16 翼骨钩 Hamulus of pterygoid bone

图1-63 猪颅骨旁正中面

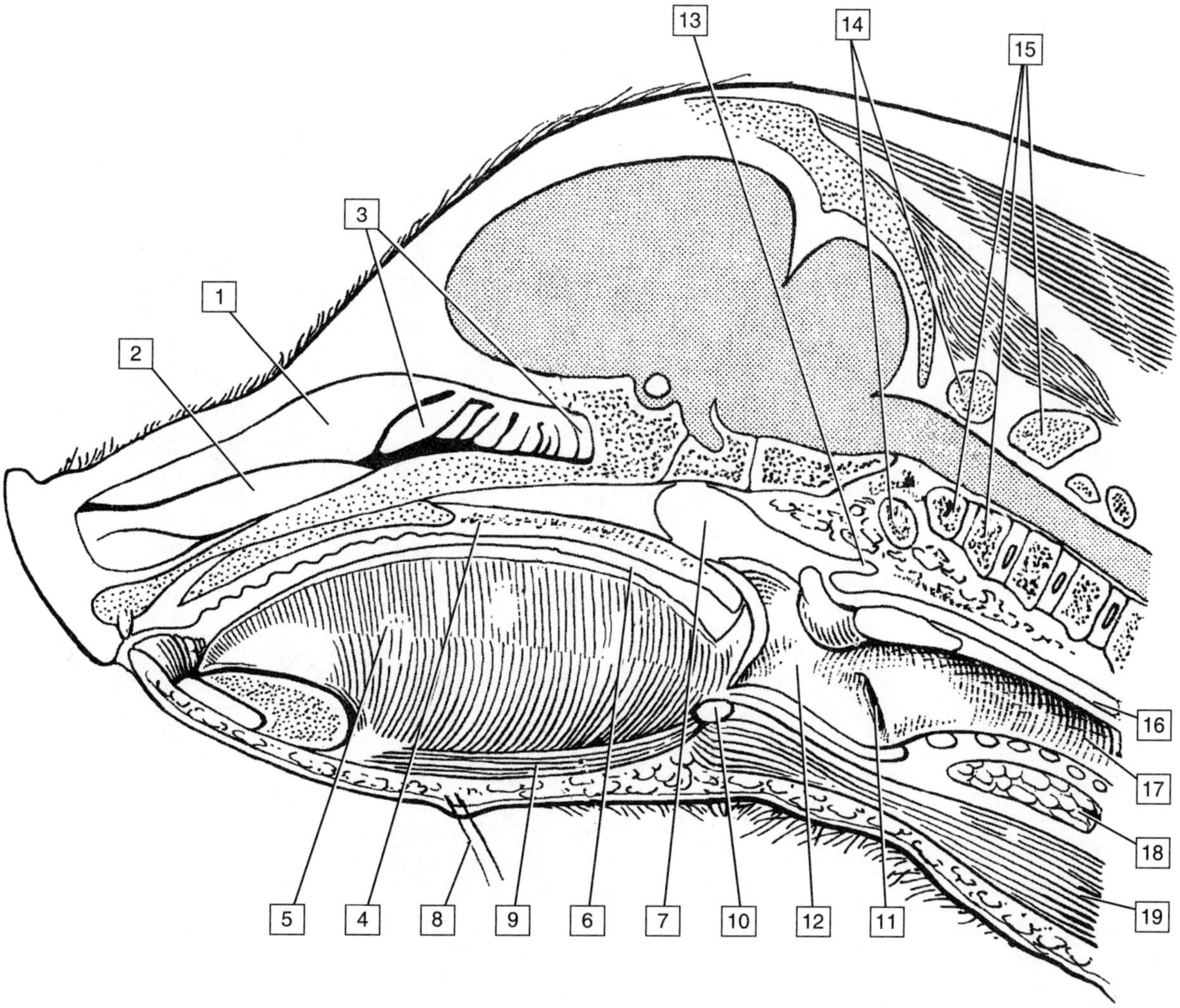

1	上鼻甲	Dorsal nasal concha
2	下鼻甲	Ventral nasal concha
3	筛鼻甲	Ethmoidal conchae
4	软腭	Soft palate
5	舌	Tongue
6	口咽	Oropharynx
7	鼻咽	Nasopharynx
8	头部毛发	Mental hairs
9	颏舌骨肌	Geniohyoideus
10	基舌骨	Basihyoid
11	喉室	Laryngeal ventricle
12	喉	Larynx
13	咽憩室	Pharyngeal diverticulum
14	寰椎	Atlas
15	枢椎	Axis
16	食管	Esophagus
17	气管	Trachea
18	甲状腺	Thyroid gland
19	胸骨舌骨肌	Sternohyoideus

图1-64 4周龄猪头部正中面

鼻中隔已被剥离

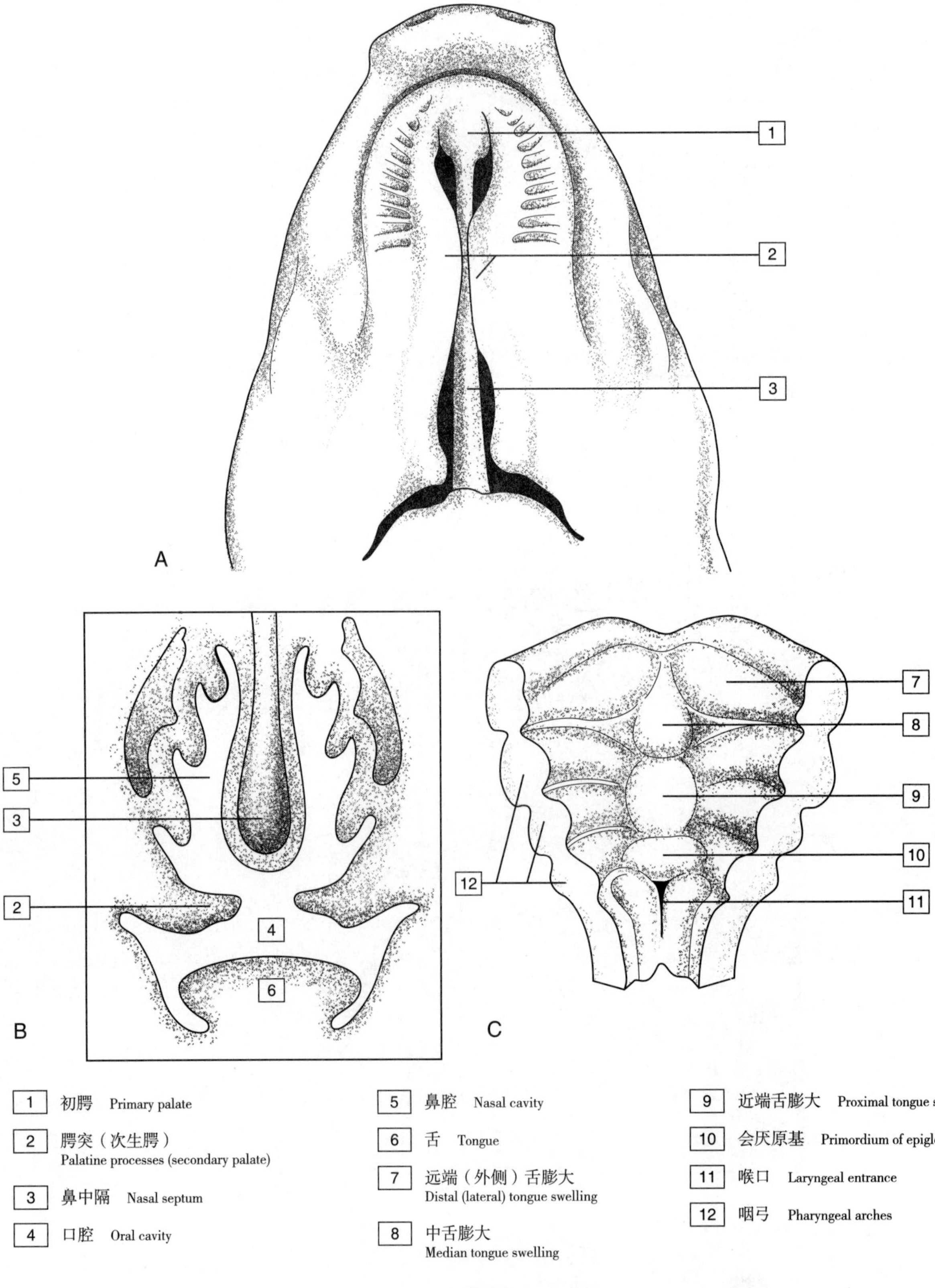

1 初腭 Primary palate	5 鼻腔 Nasal cavity	9 近端舌膨大 Proximal tongue swelling
2 腭突（次生腭） Palatine processes (secondary palate)	6 舌 Tongue	10 会厌原基 Primordium of epiglottis
3 鼻中隔 Nasal septum	7 远端（外侧）舌膨大 Distal (lateral) tongue swelling	11 喉口 Laryngeal entrance
4 口腔 Oral cavity	8 中舌膨大 Median tongue swelling	12 咽弓 Pharyngeal arches

图1-65 猪腭及舌的发育

A. 上腭的发育（腹侧观）；B. 第2腭闭合前经口腔和鼻腔横断面；C. 口腔底舌的发育

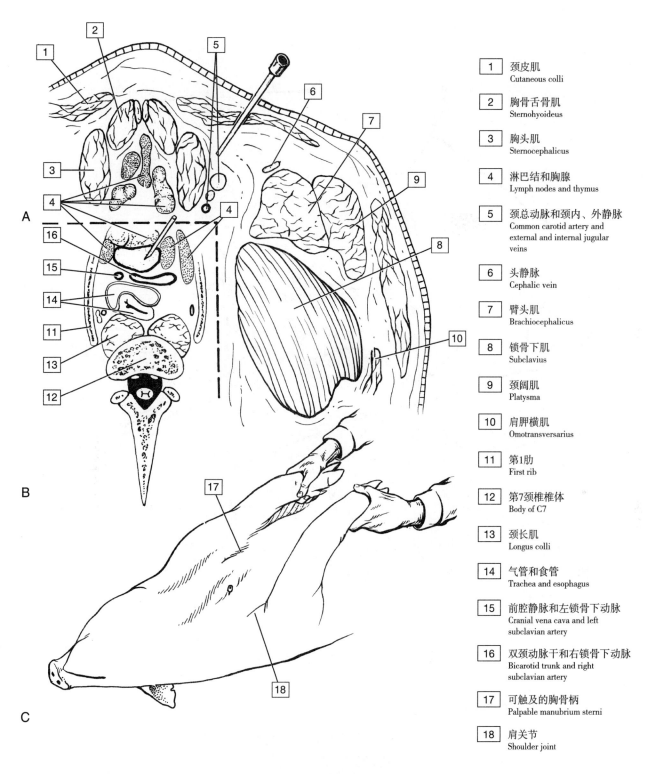

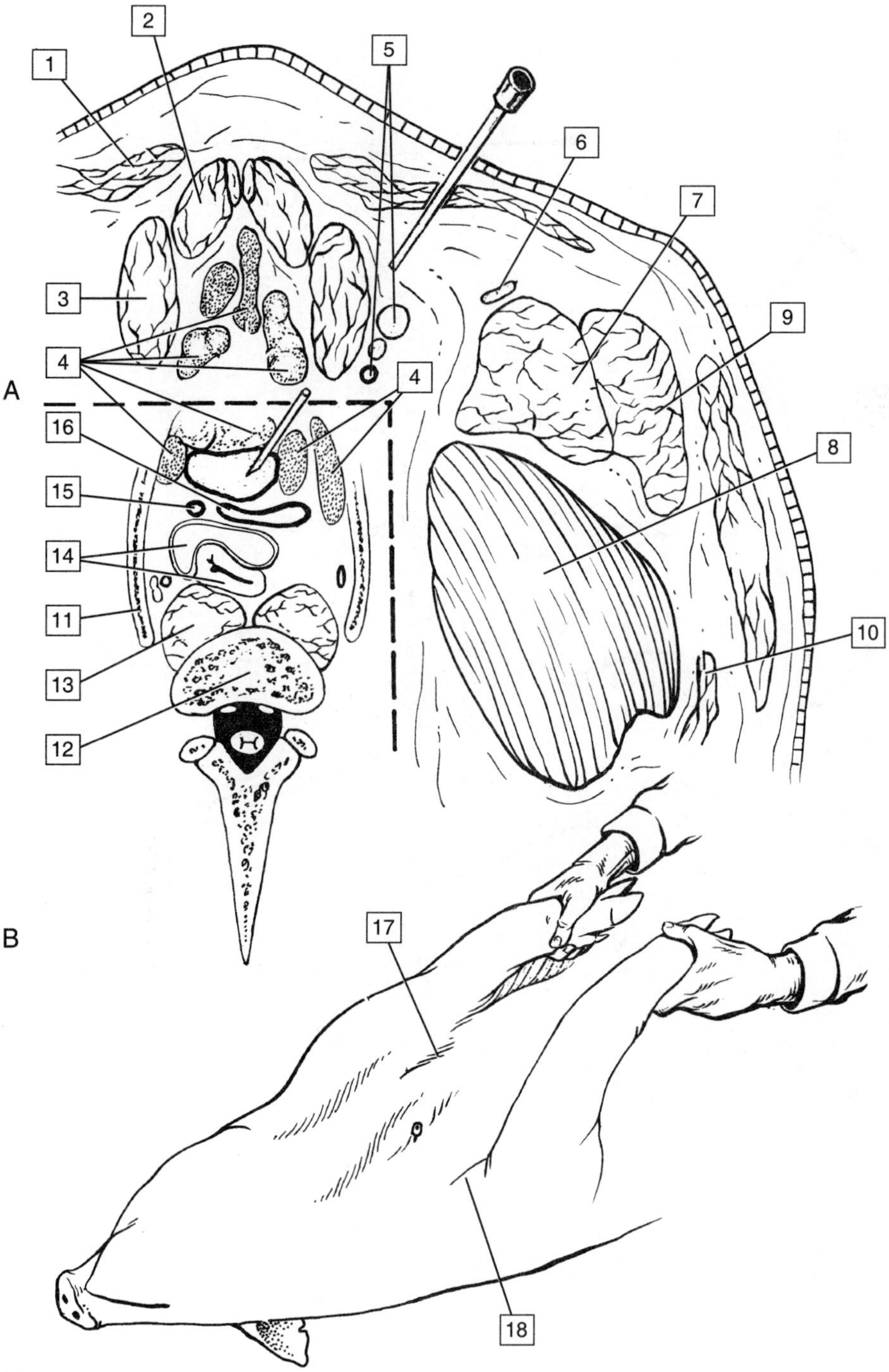

1 颈皮肌 Cutaneous colli
2 胸骨舌骨肌 Sternohyoideus
3 胸头肌 Sternocephalicus
4 淋巴结和胸腺 Lymph nodes and thymus
5 颈总动脉和颈内、外静脉 Common carotid artery and external and internal jugular veins
6 头静脉 Cephalic vein
7 臂头肌 Brachiocephalicus
8 锁骨下肌 Subclavius
9 颈阔肌 Platysma
10 肩胛横肌 Omotransversarius
11 第1肋 First rib
12 第7颈椎椎体 Body of C7
13 颈长肌 Longus colli
14 气管和食管 Trachea and esophagus
15 前腔静脉和左锁骨下动脉 Cranial vena cava and left subclavian artery
16 双颈动脉干和右锁骨下动脉 Bicarotid trunk and right subclavian artery
17 可触及的胸骨柄 Palpable manubrium sterni
18 肩关节 Shoulder joint

图1-66　猪颈腹侧横断面

A. 颈腹侧横断面，头部至胸骨柄；B. 虚线所示区域为第1肋后的局部图像；
C. 猪背部接触地面进行前腔静脉穿刺，见针的位置

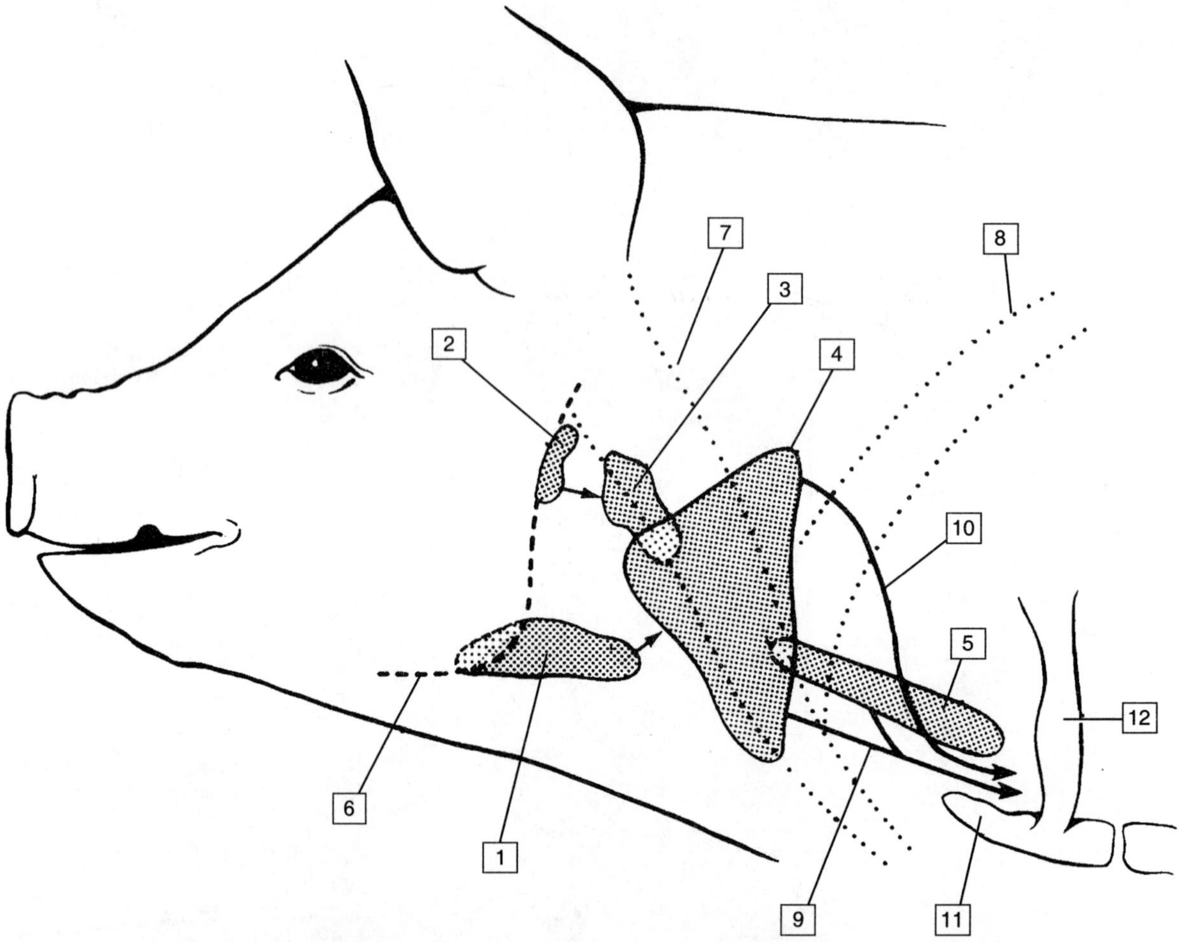

1 下颌淋巴中心 Mandibular lymph center	7 臂头肌 Brachiocephalicus
2 腮淋巴中心 Parotid lymph center	8 锁骨下肌 Subclavius
3 咽后淋巴中心 Retropharyngeal lymph center	9 气管淋巴干 Tracheal lymph trunk
4 颈浅淋巴中心 Superficial cervical lymph center	10 从背侧颈浅淋巴结流出的淋巴 Lymph from dorsal superficial cervical nodes
5 颈深淋巴中心 Deep cervical lymph center	11 胸骨柄 Manubrium sterni
6 下颌骨 Mandible	12 第1肋 First rib

图1-67 猪头部和颈部的淋巴中心

箭头表示淋巴流向

鸟类

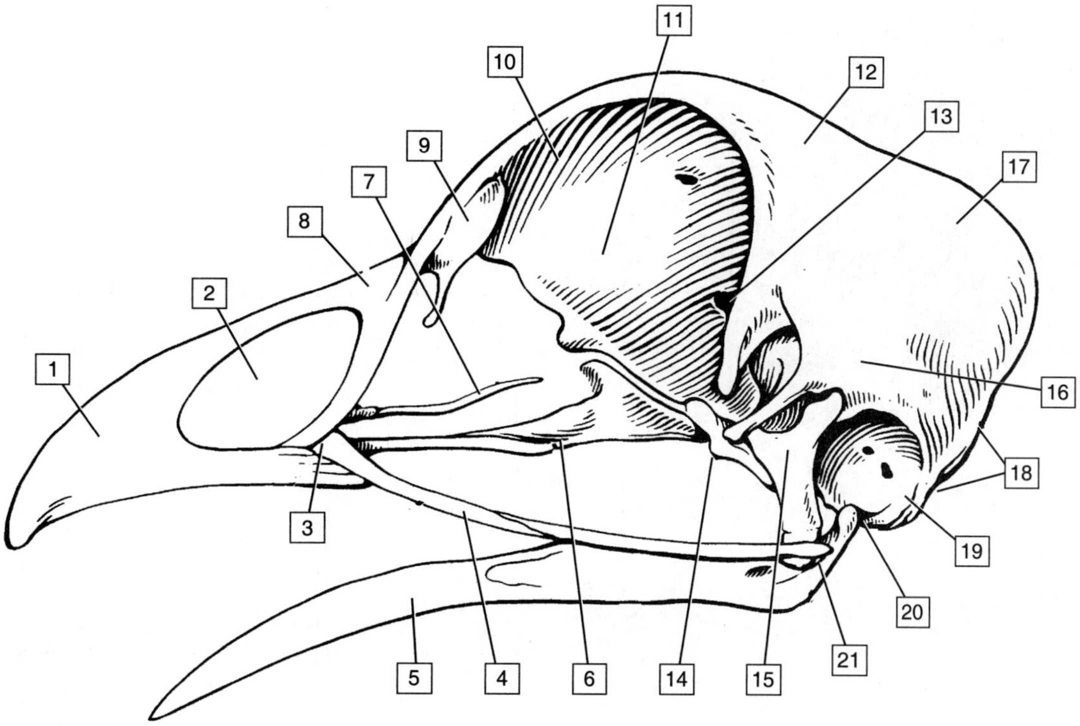

1	前颌骨	Premaxilla
2	鼻孔	Nasal aperture
3	上颌骨	Maxilla
4	颧弓	Jugal arch
5	下颌骨	Mandible
6	腭骨	Palatine bone
7	犁骨	Vomer
8	鼻骨	Nasal bone
9	泪骨	Lacrimal bone
10	眼眶	Orbit
11	眶间隔	Interorbital septum
12	额骨	Frontal bone
13	视神经孔	Optic foramen
14	翼骨	Pterygoid bone
15	方骨	Quadrate bone
16	颞骨	Temporal bone
17	顶骨	Parietal bone
18	枕骨	Occipital bone
19	带耳蜗和前庭窗的鼓室	Tympanic cavity with cochlear and vestibular windows
20	蝶骨	Sphenoid bone
21	关节骨	Articular bone

图1-68　鸟类的头骨

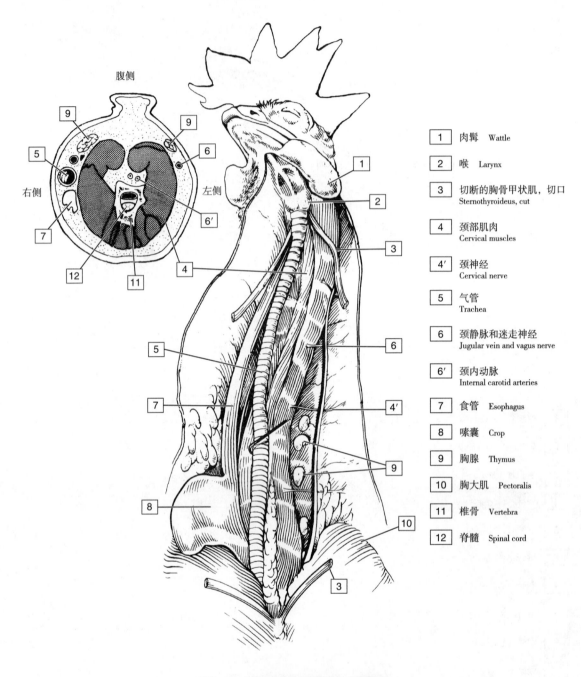

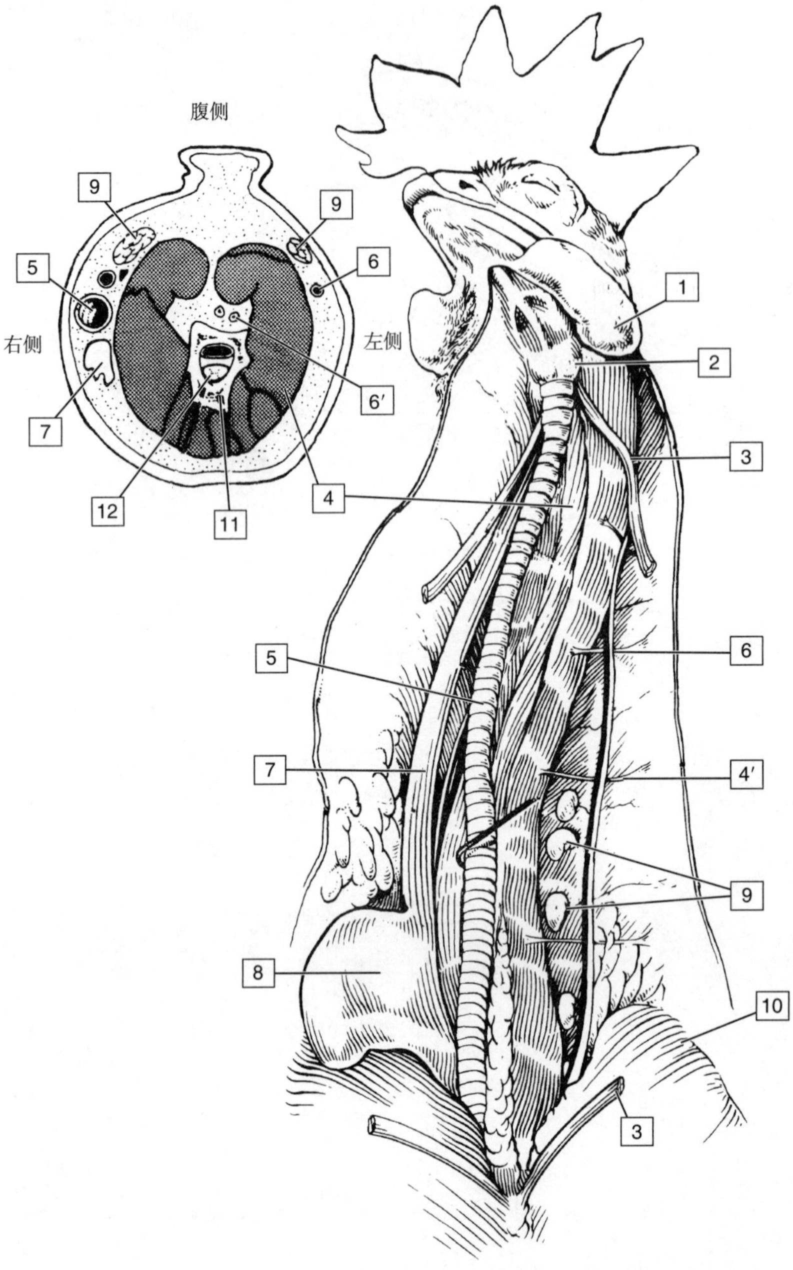

1	肉髯 Wattle
2	喉 Larynx
3	切断的胸骨甲状肌，切口 Sternothyroideus, cut
4	颈部肌肉 Cervical muscles
4′	颈神经 Cervical nerve
5	气管 Trachea
6	颈静脉和迷走神经 Jugular vein and vagus nerve
6′	颈内动脉 Internal carotid arteries
7	食管 Esophagus
8	嗉囊 Crop
9	胸腺 Thymus
10	胸大肌 Pectoralis
11	椎骨 Vertebra
12	脊髓 Spinal cord

图1-69　家禽颈部解剖图，腹侧观

小图显示颈中部的横断面

不同动物的解剖结构比较

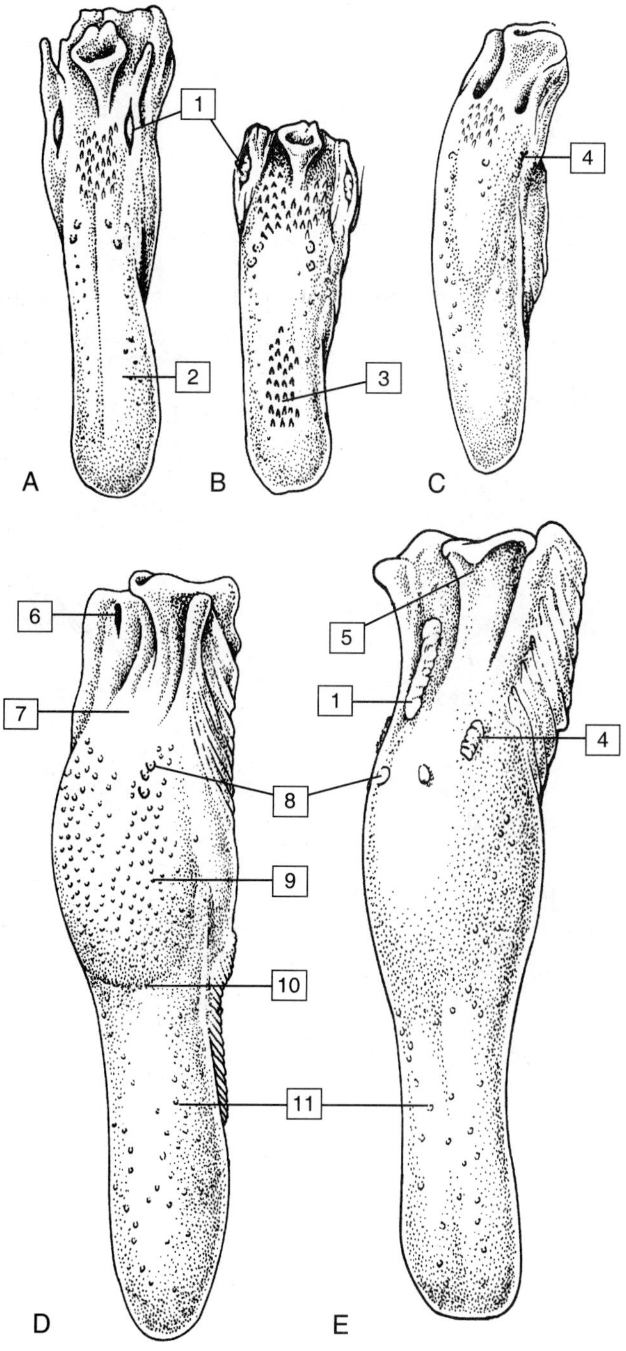

1	腭扁桃体 Palatine tonsil	5	会厌 Epiglottis	9	舌圆枕 Torus linguae
2	舌正中沟 Median groove	6	扁桃体窦 Tonsillar sinus	10	舌隐窝 Fossa linguae
3	丝状乳头 Filiform papillae	7	舌根 Root of tongue	11	菌状乳头 Fungiform papillae
4	叶状乳头 Foliate papillae	8	轮廓乳头 Vallate papillae		

图1-70 不同动物的舌和会厌，背侧观

A. 犬；B. 猫；C. 猪；D. 牛；E. 马

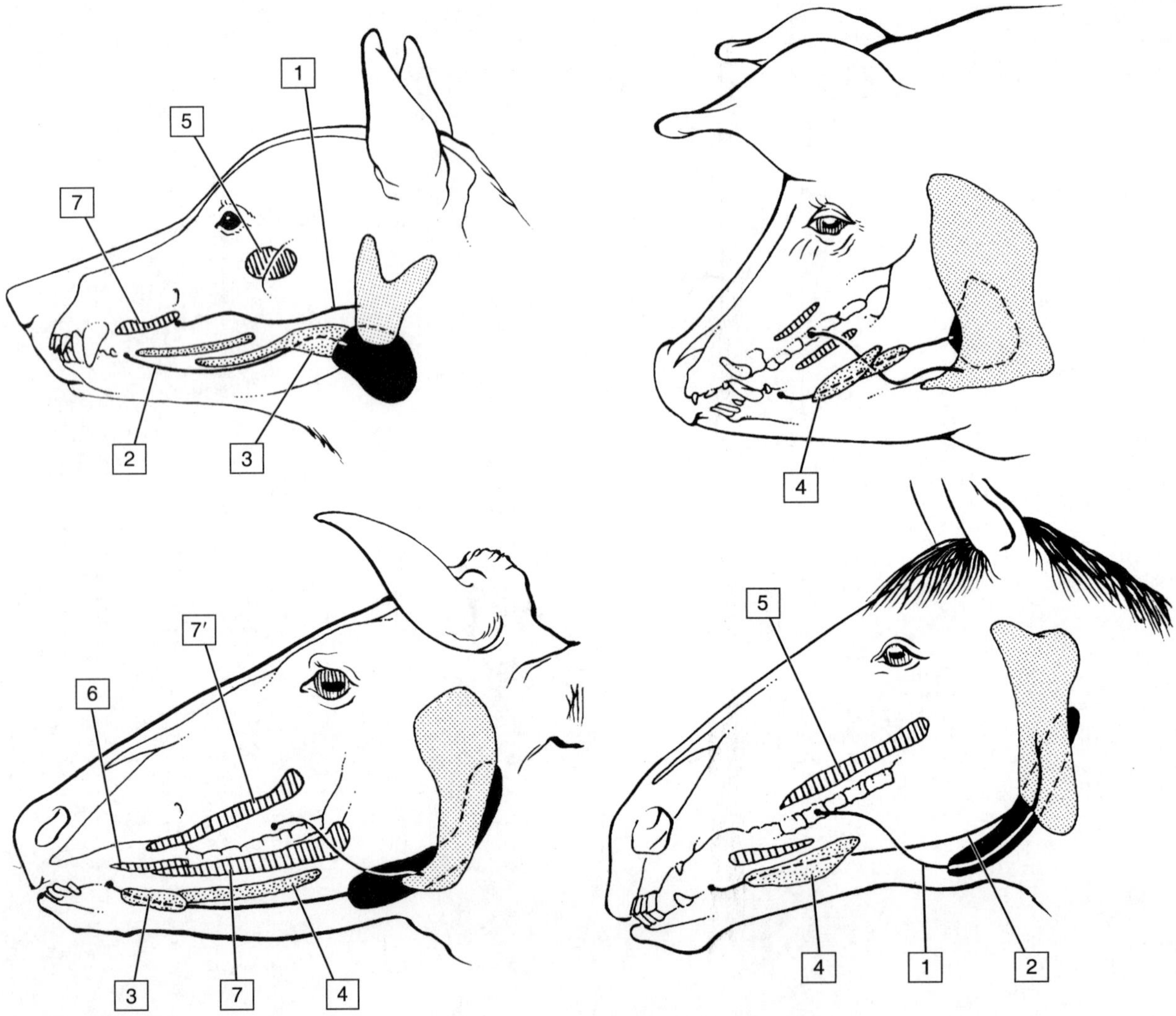

1 腮腺管 Parotid duct

2 颌下腺管 Mandibular duct

3 单口舌下腺
Compact (monostomatic) part of sublingual gland

4 多口舌下腺
Diffuse (polystomatic) part of sublingual gland

5 背侧颊腺（犬颧腺）
Dorsal buccal glands (zygomatic gland in the dog)

6 内侧颊腺 Middle buccal glands

7 腹侧颊腺 Ventral buccal glands

7′ 背侧颊腺 Dorsal buccal gland

图1-71 犬、猪、牛和马的大唾液腺

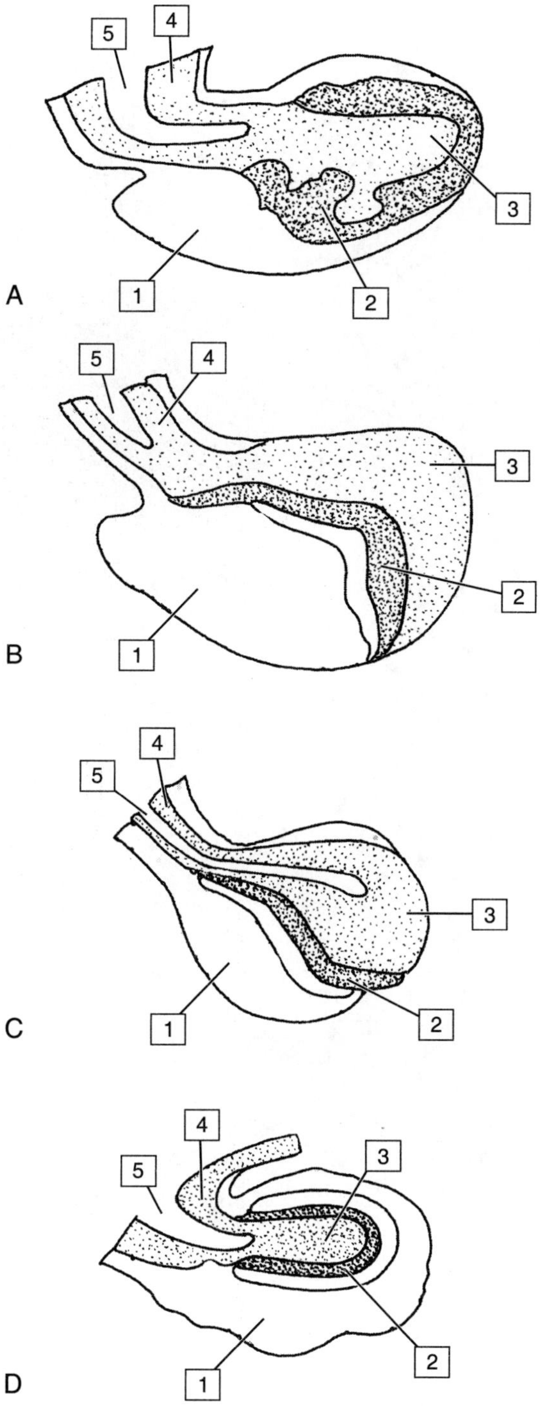

1	腺垂体	Adenohypophysis	4	垂体柄	Hypophysial stalk
2	中间部	Intermediate part	5	第3脑室隐窝	Recess of third ventricle
3	神经垂体	Neurohypophysis			

图1-72 不同动物的垂体的正中切面，垂体的前末端位于左侧

A. 马；B. 牛；C. 猪；D. 犬

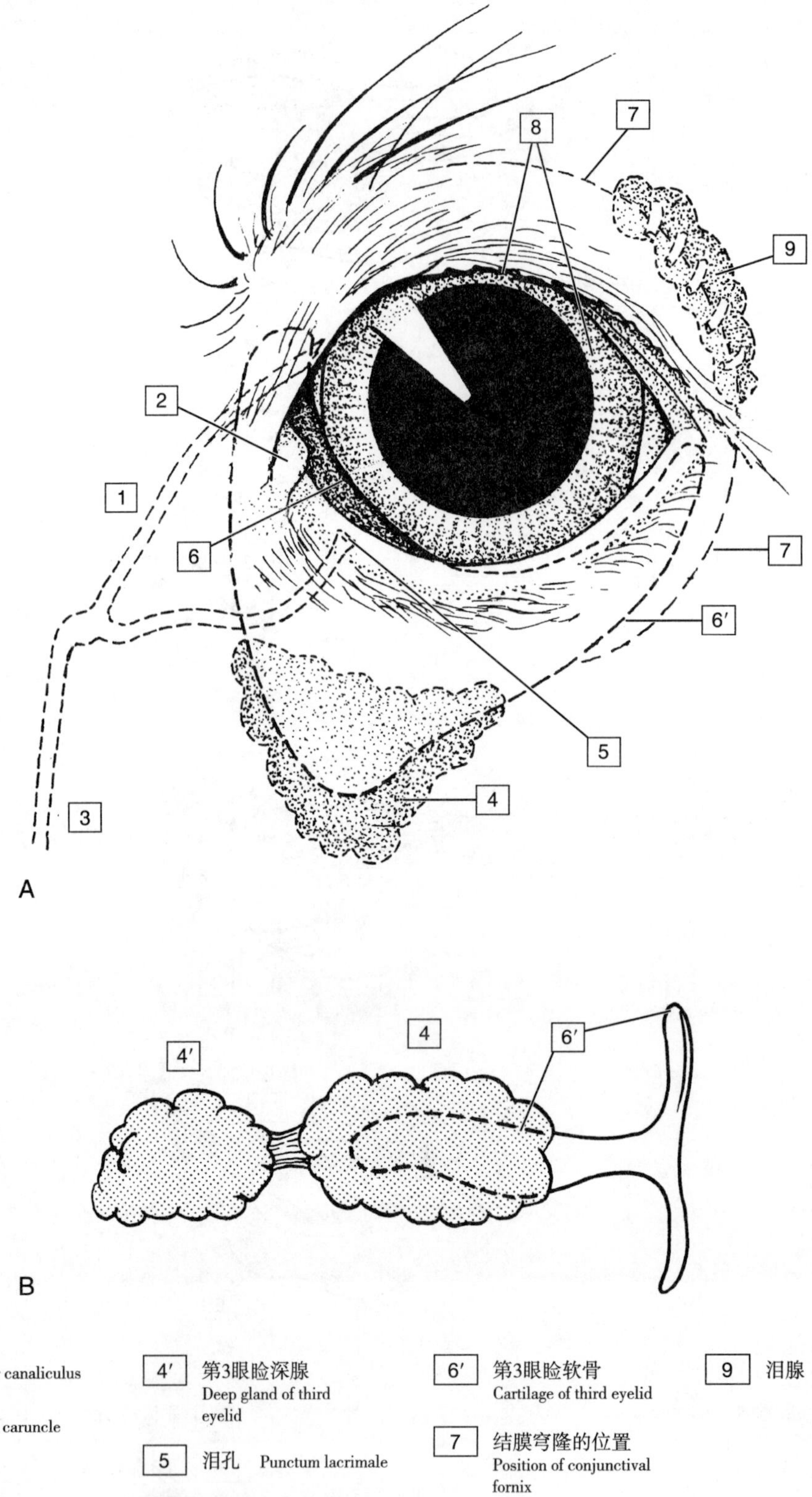

1 上泪管 Upper canaliculus

2 泪阜 Lacrimal caruncle

3 鼻泪管 Nasolacrimal duct

4 第3眼睑腺 Gland of third eyelid

4′ 第3眼睑深腺 Deep gland of third eyelid

5 泪孔 Punctum lacrimale

6 第3眼睑 Third eyelid

6′ 第3眼睑软骨 Cartilage of third eyelid

7 结膜穹隆的位置 Position of conjunctival fornix

8 瞳孔 Pupil

9 泪腺 Lacrimal gland

图1-73 犬和猪的第3眼睑

A. 犬左眼的第3眼睑和泪腺；B. 猪的第3眼睑软骨与相关腺体

2

颈部、背部和脊柱

犬

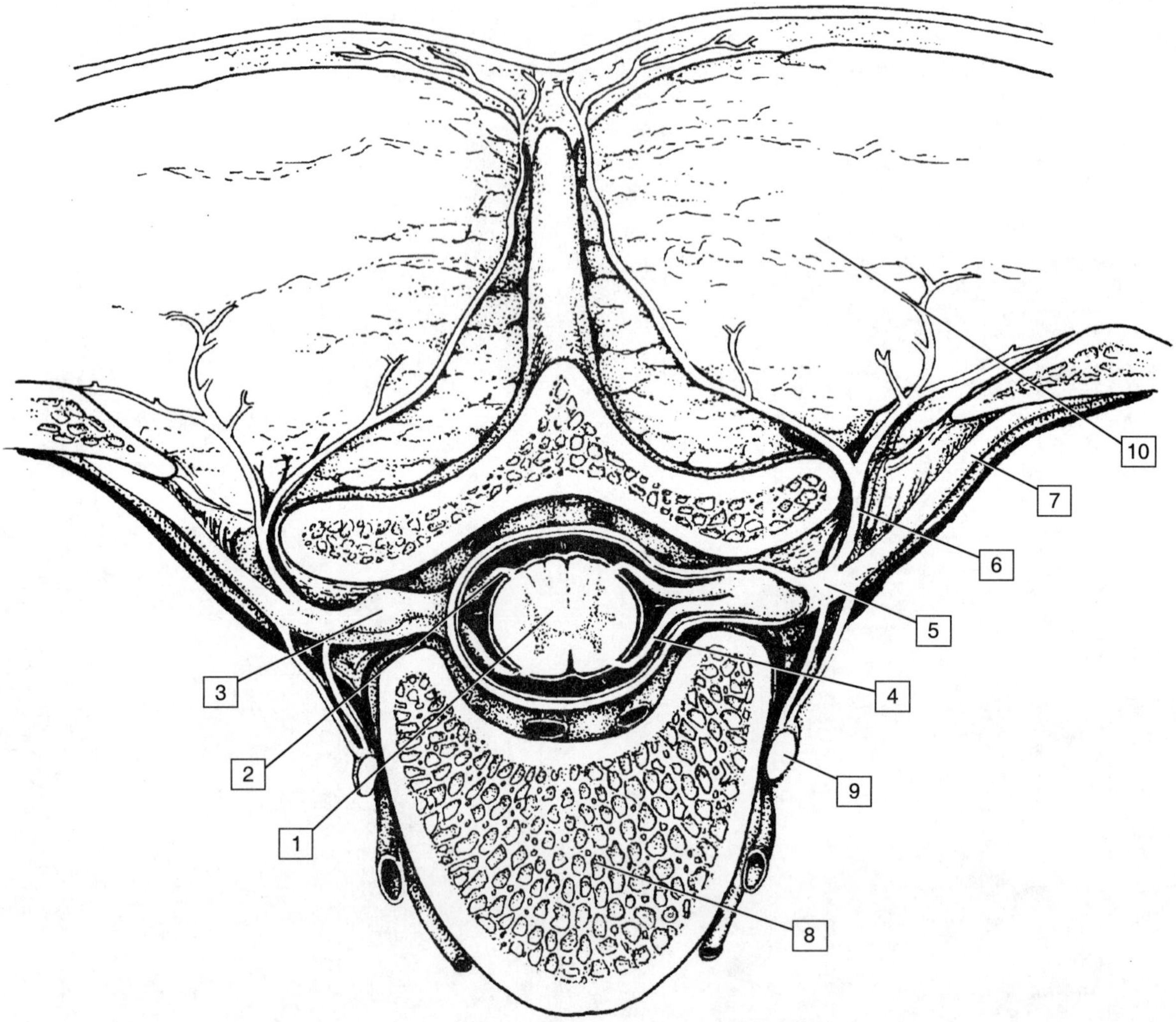

1	脊髓	Spinal cord	6	脊神经背侧支	Dorsal branch of spinal nerve
2	背根	Dorsal root	7	脊神经腹侧支	Ventral branch of spinal nerve
3	脊神经节	Spinal ganglion	8	椎体	Body of vertebra
4	腹根	Ventral root	9	交感干	Sympathetic trunk
5	脊神经	Spinal nerve	10	轴上肌	Epaxial muscles

图2-1　犬脊柱的横断面

图中显示脊神经的分支

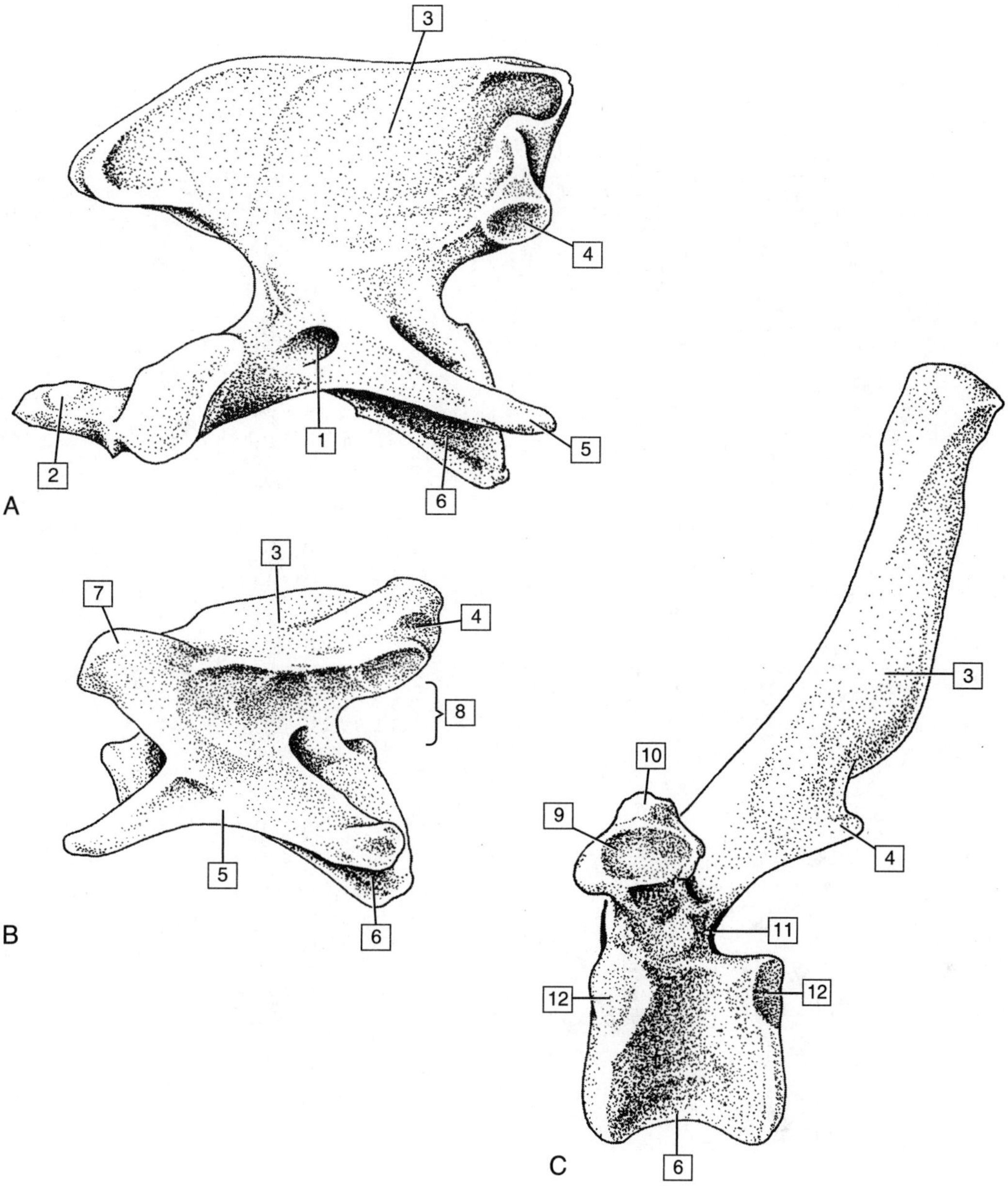

1 横突孔 Transverse foramen	7 前关节突 Cranial articular process
2 齿突 Dens	8 椎孔的位置 Position of vertebral foramen
3 棘突 Spinous process	9 横突肋凹 Transverse process with costal fovea
4 后关节突 Caudal articular process	10 乳突 Mammillary process
5 横突 Transverse process	11 椎后切迹 Caudal vertebral notch
6 椎体 Body of vertebra	12 肋凹 Costal foveae

图2-2 犬的颈椎和胸椎

A. 枢椎，外侧观；B. 第5颈椎，外侧观；C. 胸椎，左外侧观

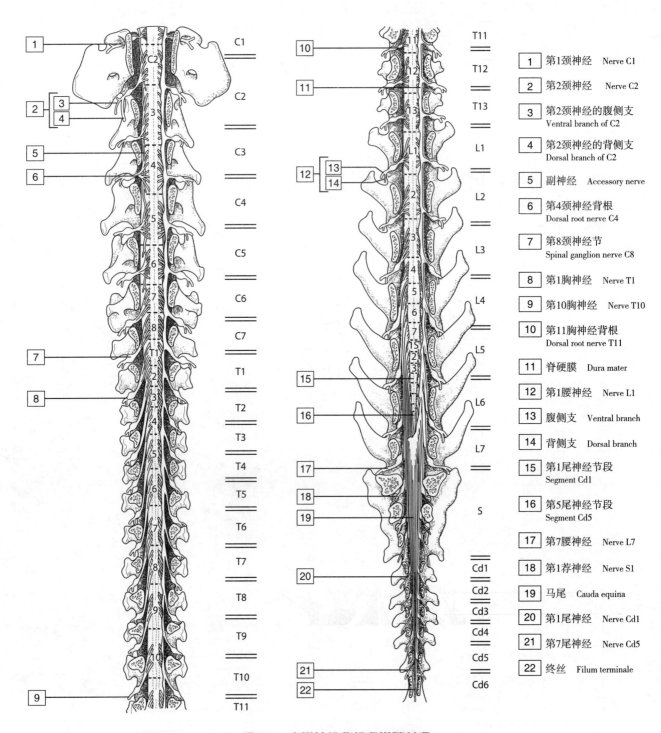

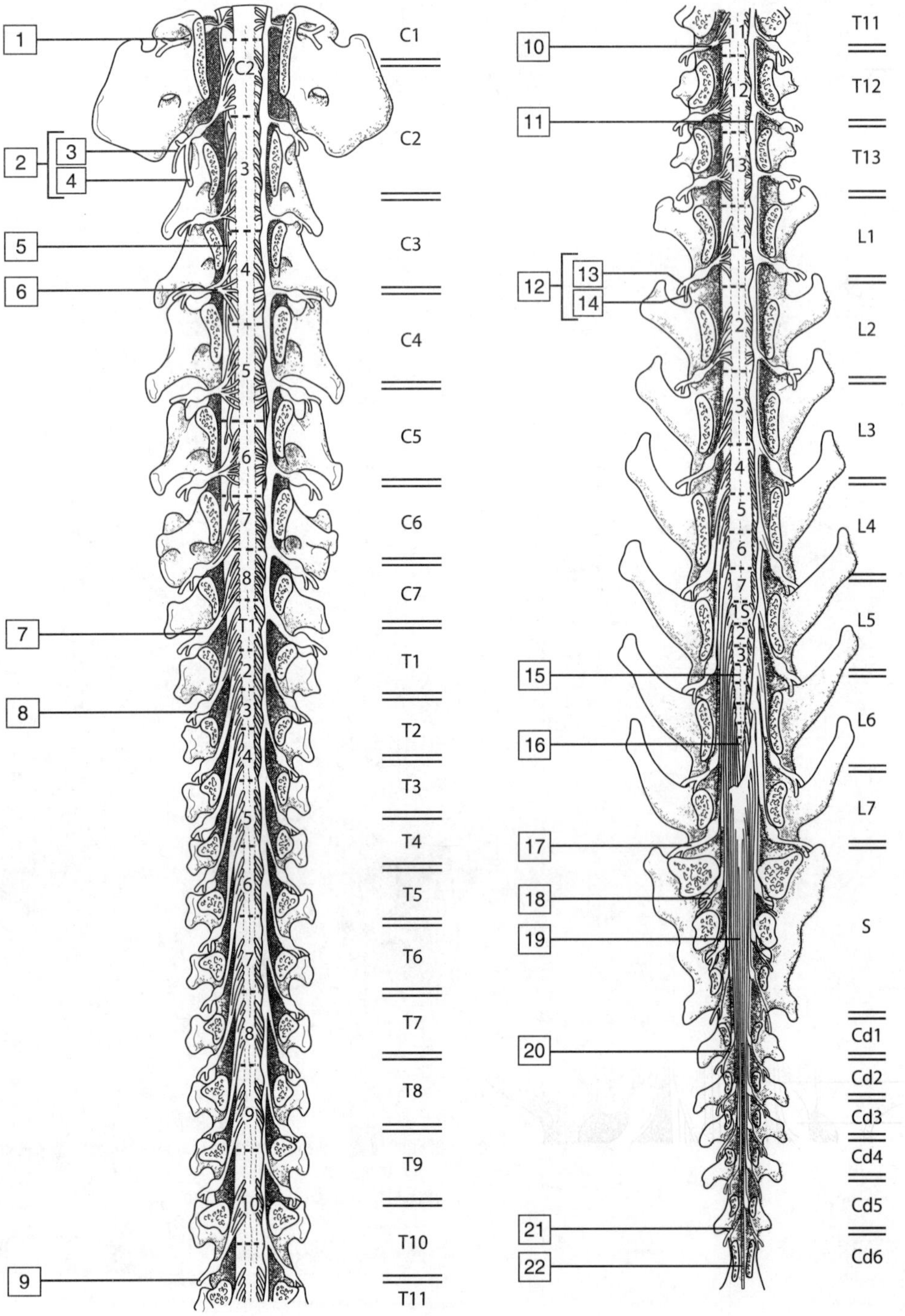

1 第1颈神经 Nerve C1

2 第2颈神经 Nerve C2

3 第2颈神经的腹侧支 Ventral branch of C2

4 第2颈神经的背侧支 Dorsal branch of C2

5 副神经 Accessory nerve

6 第4颈神经背根 Dorsal root nerve C4

7 第8颈神经节 Spinal ganglion nerve C8

8 第1胸神经 Nerve T1

9 第10胸神经 Nerve T10

10 第11胸神经背根 Dorsal root nerve T11

11 脊硬膜 Dura mater

12 第1腰神经 Nerve L1

13 腹侧支 Ventral branch

14 背侧支 Dorsal branch

15 第1尾神经节段 Segment Cd1

16 第5尾神经节段 Segment Cd5

17 第7腰神经 Nerve L7

18 第1荐神经 Nerve S1

19 马尾 Cauda equina

20 第1尾神经 Nerve Cd1

21 第7尾神经 Nerve Cd5

22 终丝 Filum terminale

图2-3 犬脊神经背根和脊髓节段

背侧观，已剥离椎弓。两张图中左侧的脊硬膜均已剥离

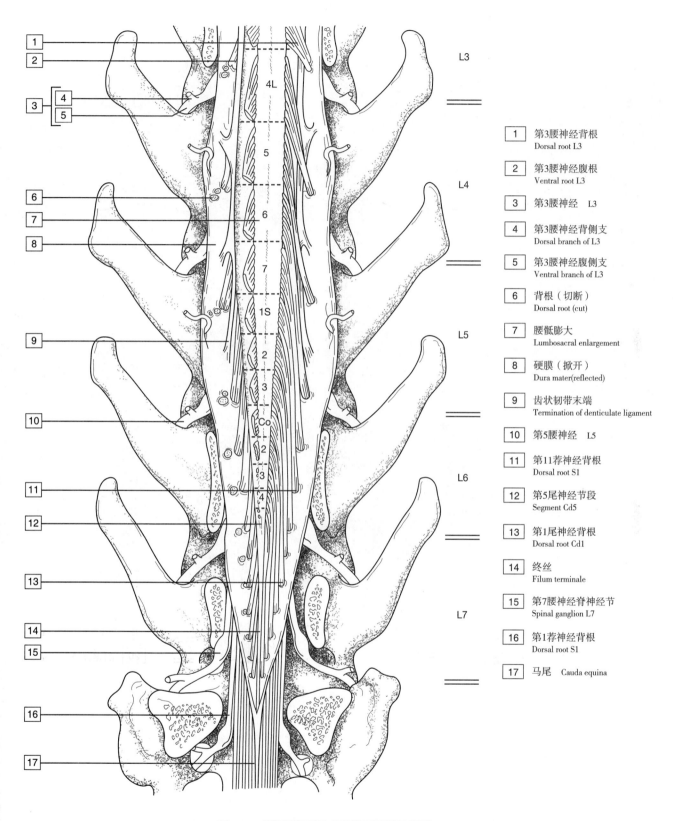

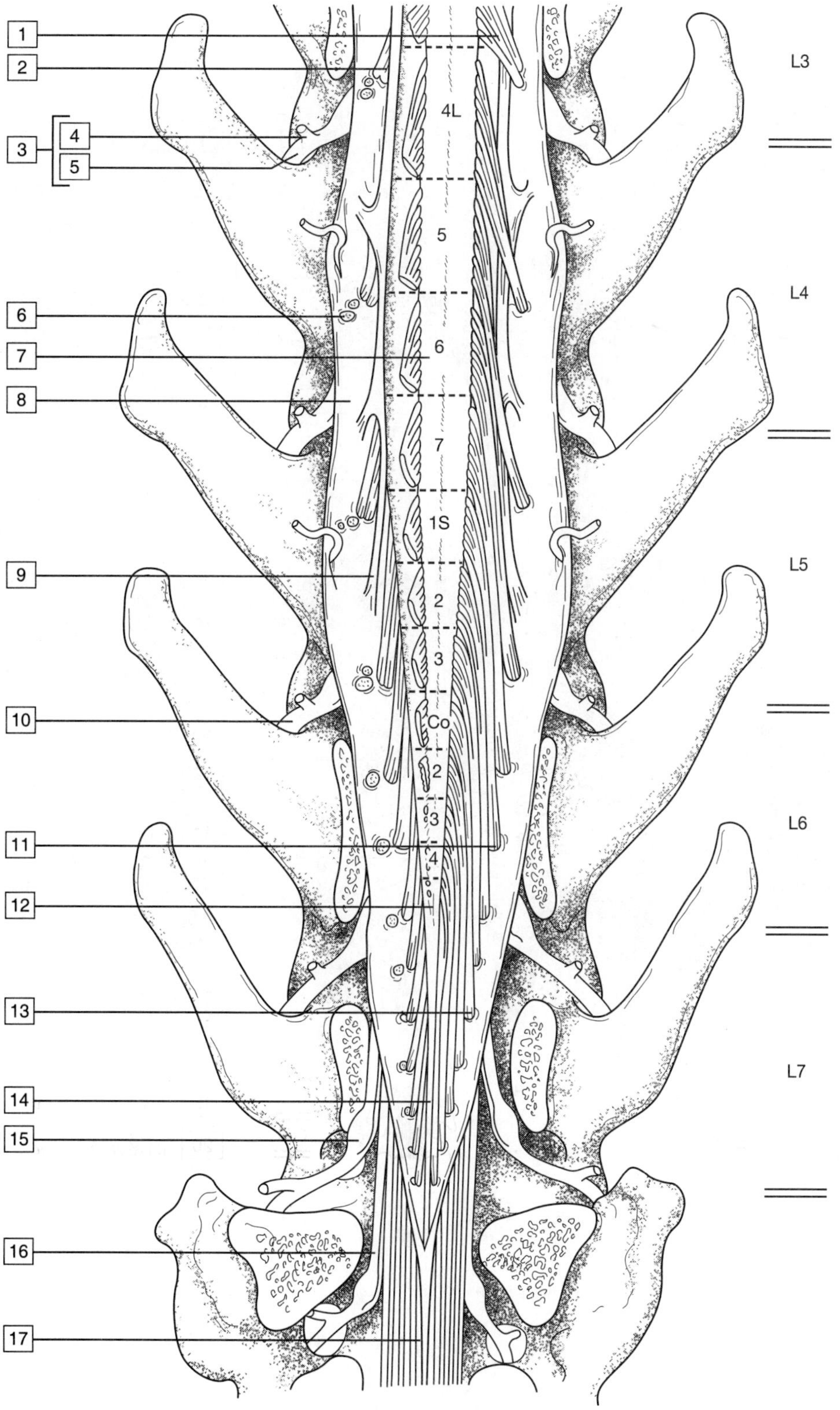

1 第3腰神经背根 Dorsal root L3

2 第3腰神经腹根 Ventral root L3

3 第3腰神经 L3

4 第3腰神经背侧支 Dorsal branch of L3

5 第3腰神经腹侧支 Ventral branch of L3

6 背根（切断） Dorsal root (cut)

7 腰骶膨大 Lumbosacral enlargement

8 硬膜（掀开） Dura mater(reflected)

9 齿状韧带末端 Termination of denticulate ligament

10 第5腰神经 L5

11 第11荐神经背根 Dorsal root S1

12 第5尾神经节段 Segment Cd5

13 第1尾神经背根 Dorsal root Cd1

14 终丝 Filum terminale

15 第7腰神经脊神经节 Spinal ganglion L7

16 第1荐神经背根 Dorsal root S1

17 马尾 Cauda equina

图2-4 硬膜掀开的犬脊髓末端示意图

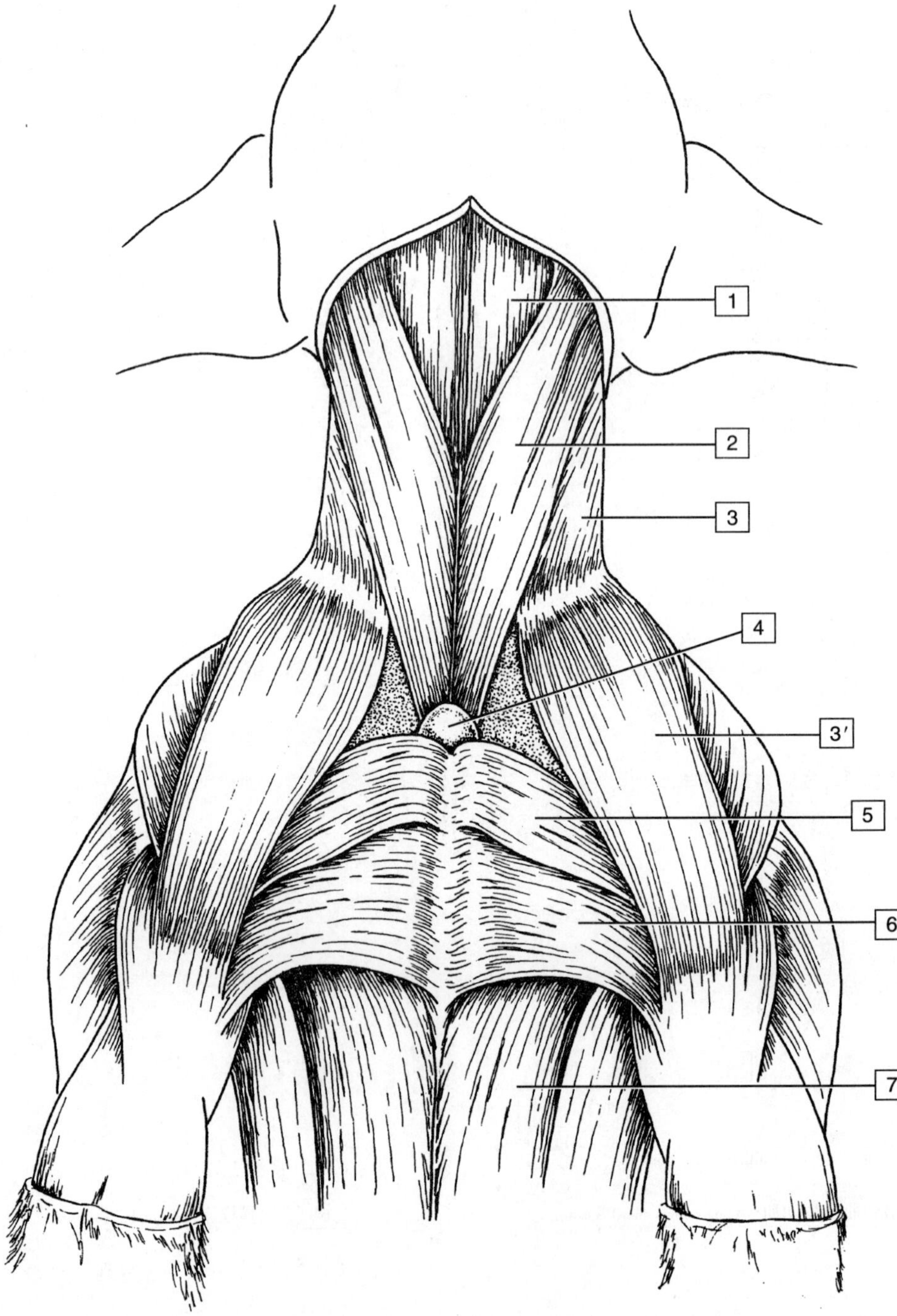

1	胸骨舌骨肌和胸骨甲状肌合并	Combined sternohyoideus and sternothyroideus	4	胸骨柄	Manubrium sterni
2	胸头肌	Sternocephalicus	5	胸降肌	Pectoralis descendens
3	臂二头肌：锁颈肌	Brachiocephalicus: cleidocervicalis	6	胸横肌	Pectoralis transversus
3′	臂二头肌：锁臂肌	Brachiocephalicus: cleidobrachialis	7	胸升肌	Pectoralis profundus

图2-5　犬颈部和胸部的腹侧肌

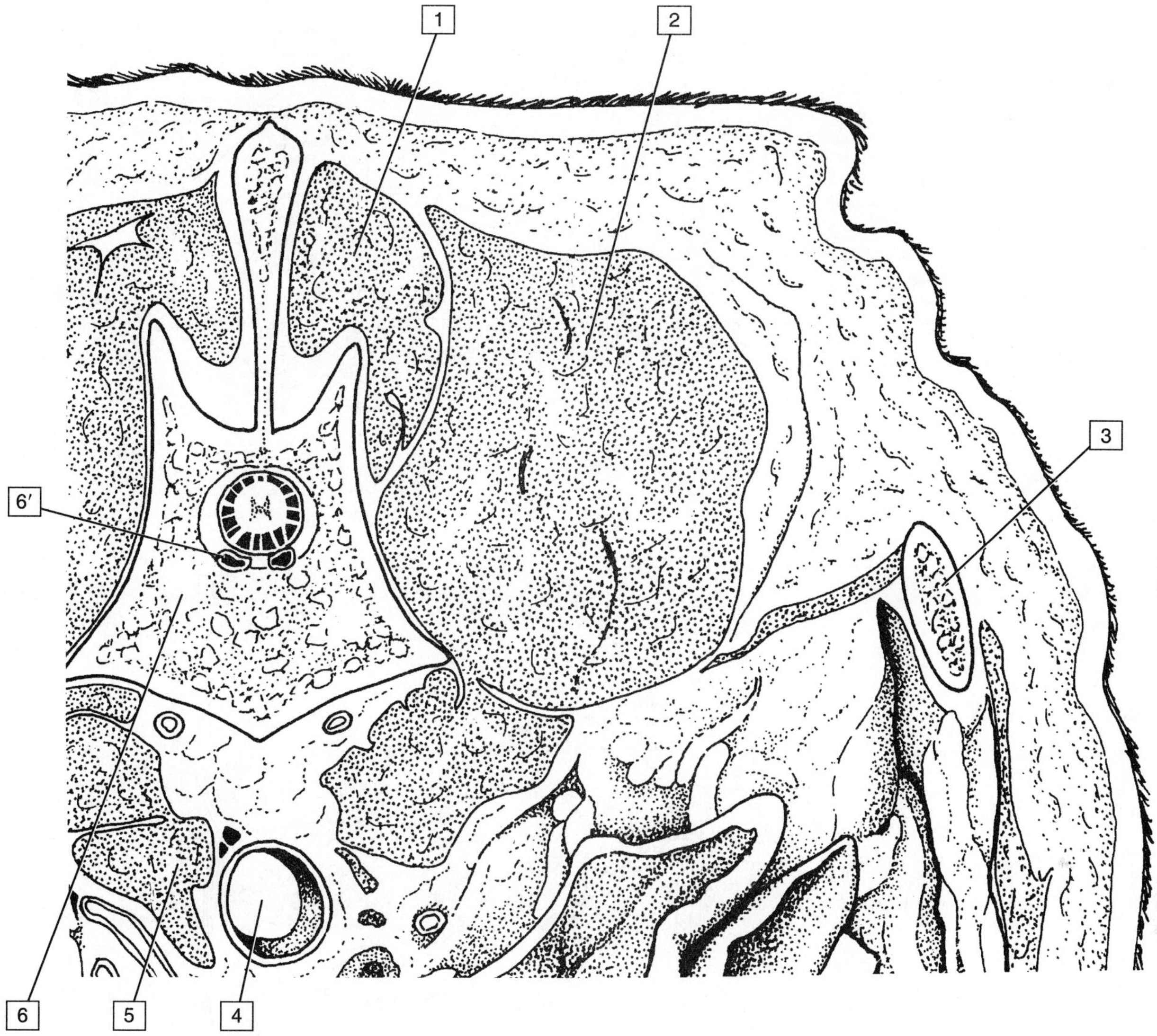

1 多裂肌和棘肌 Multifidus and spinalis	5 右膈脚 Right crus of diaphragm
2 背最长肌和髂肋肌 Longissimus and iliocostalis	6 第1腰椎 First lumbar vertebra
3 最后肋 Last rib	6′ 椎内静脉丛 Internal vertebral venous plexus
4 主动脉 Aorta	

图2-6 犬背部在第1腰椎的横断面

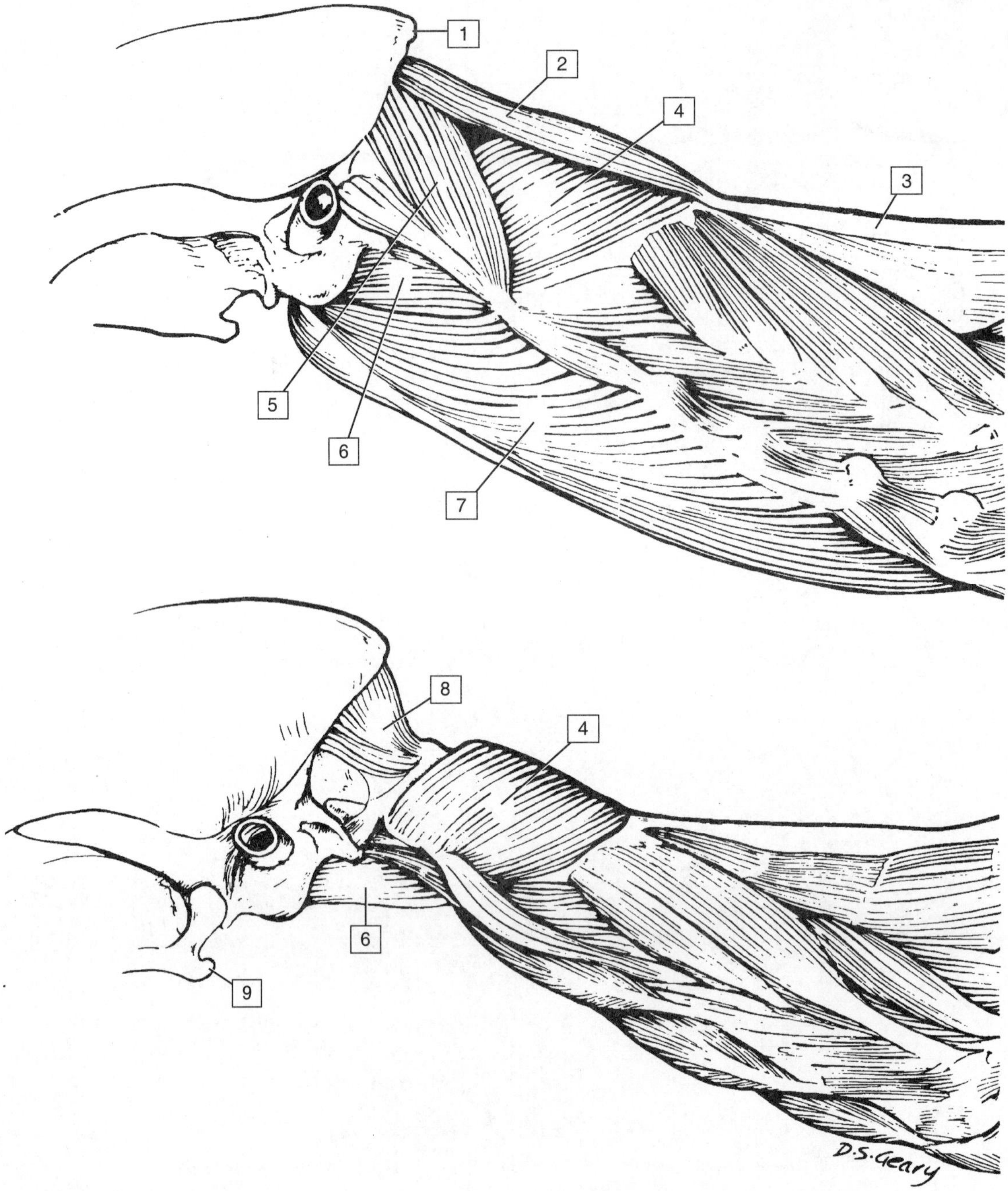

1	枕外隆凸	External occipital protuberance	6	头腹侧直肌	Rectus capitis ventralis
2	头背侧大直肌	Rectus capitis dorsalis major	7	头最长肌	Longissimus capitis
3	项韧带	Nuchal ligament	8	头背侧小直肌	Rectus capitis dorsalis minor
4	头后斜肌	Obliquus capitis caudalis	9	下颌角突	Angular process of mandible
5	头前斜肌	Obliquus capitis cranialis			

图2-7　大寰枕关节和寰枢关节相关的肌肉，外侧观

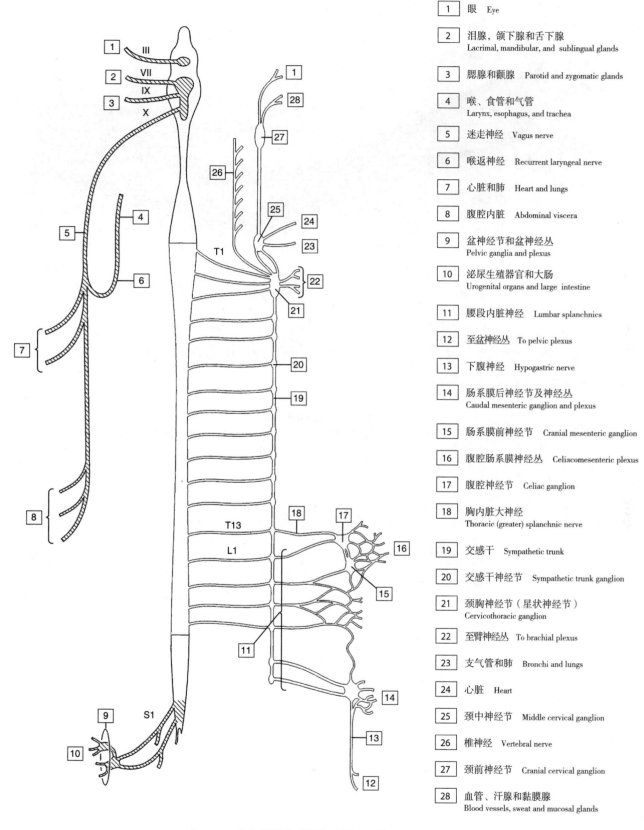

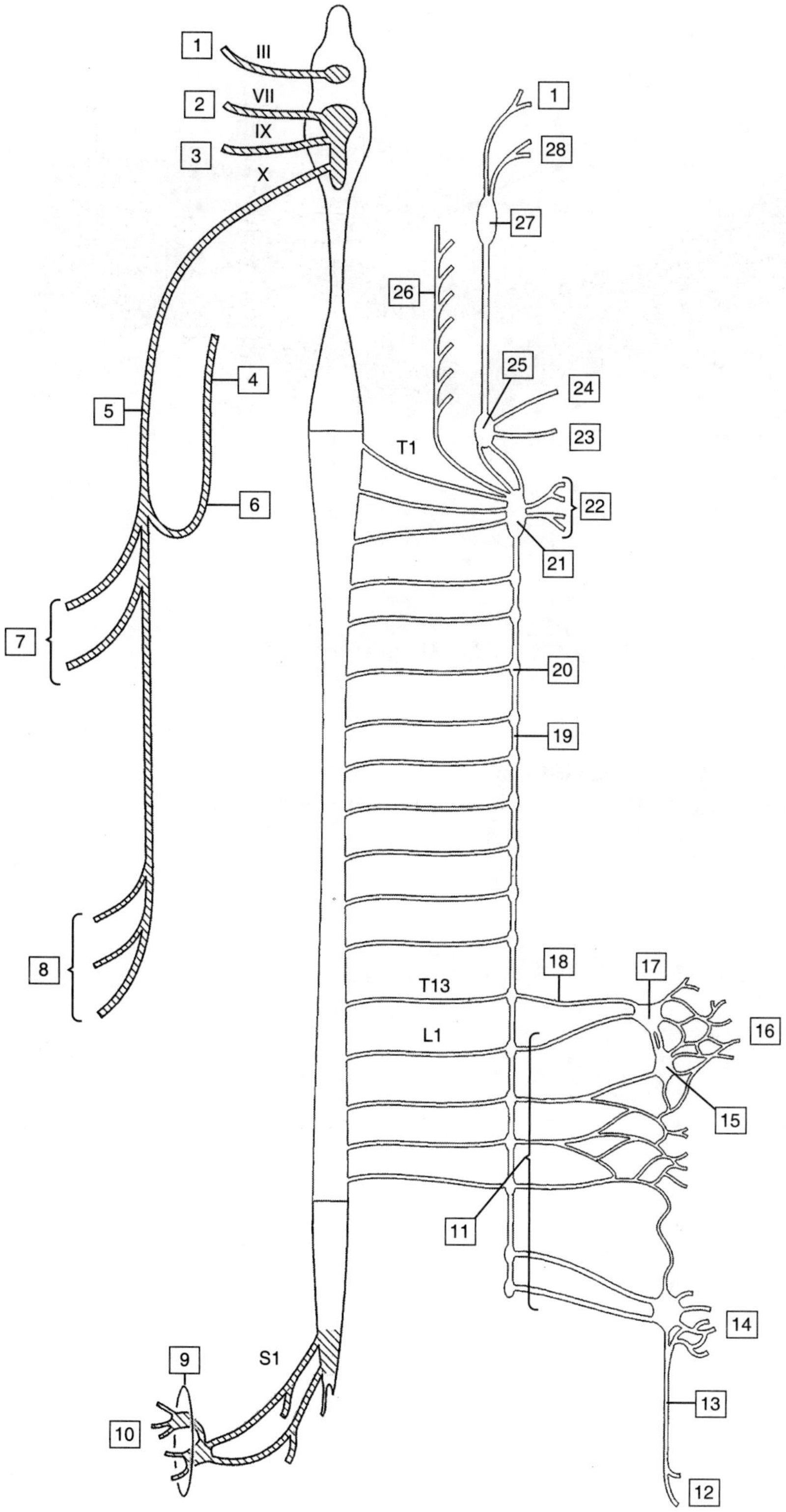

1 眼 Eye

2 泪腺，颌下腺和舌下腺 Lacrimal, mandibular, and sublingual glands

3 腮腺和颧腺 Parotid and zygomatic glands

4 喉、食管和气管 Larynx, esophagus, and trachea

5 迷走神经 Vagus nerve

6 喉返神经 Recurrent laryngeal nerve

7 心脏和肺 Heart and lungs

8 腹腔内脏 Abdominal viscera

9 盆神经节和盆神经丛 Pelvic ganglia and plexus

10 泌尿生殖器官和大肠 Urogenital organs and large intestine

11 腰段内脏神经 Lumbar splanchnics

12 至盆神经丛 To pelvic plexus

13 下腹神经 Hypogastric nerve

14 肠系膜后神经节及神经丛 Caudal mesenteric ganglion and plexus

15 肠系膜前神经节 Cranial mesenteric ganglion

16 腹腔肠系膜神经丛 Celiacomesenteric plexus

17 腹腔神经节 Celiac ganglion

18 胸内脏大神经 Thoracic (greater) splanchnic nerve

19 交感干 Sympathetic trunk

20 交感干神经节 Sympathetic trunk ganglion

21 颈胸神经节（星状神经节） Cervicothoracic ganglion

22 至臂神经丛 To brachial plexus

23 支气管和肺 Bronchi and lungs

24 心脏 Heart

25 颈中神经节 Middle cervical ganglion

26 椎神经 Vertebral nerve

27 颈前神经节 Cranial cervical ganglion

28 血管、汗腺和黏膜腺 Blood vessels, sweat and mucosal glands

图2-8 犬交感神经和副交感神经的外周分布

图中Ⅲ为动眼神经，Ⅶ为面神经，Ⅸ为舌咽神经，Ⅹ为迷走神经

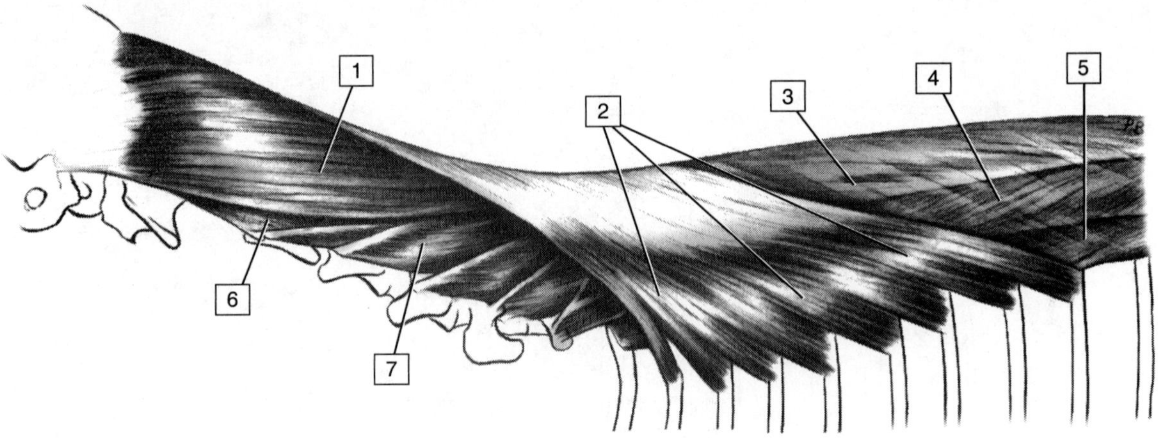

1 夹肌 Splenius	5 髂肋肌 Iliocostalis
2 前背侧锯肌 Serratus dorsalis cranialis	6 头最长肌 Longissimus capitis
3 棘肌和半棘肌 Spinalis at semispinalis	7 颈最长肌 Longissimus cervicis
4 背最长肌 Longissimus	

图2-9 犬的轴上肌

图中每块肌肉都可以跨越其他椎骨，因此每块肌肉可相互重叠，肌肉界限模糊不清

马

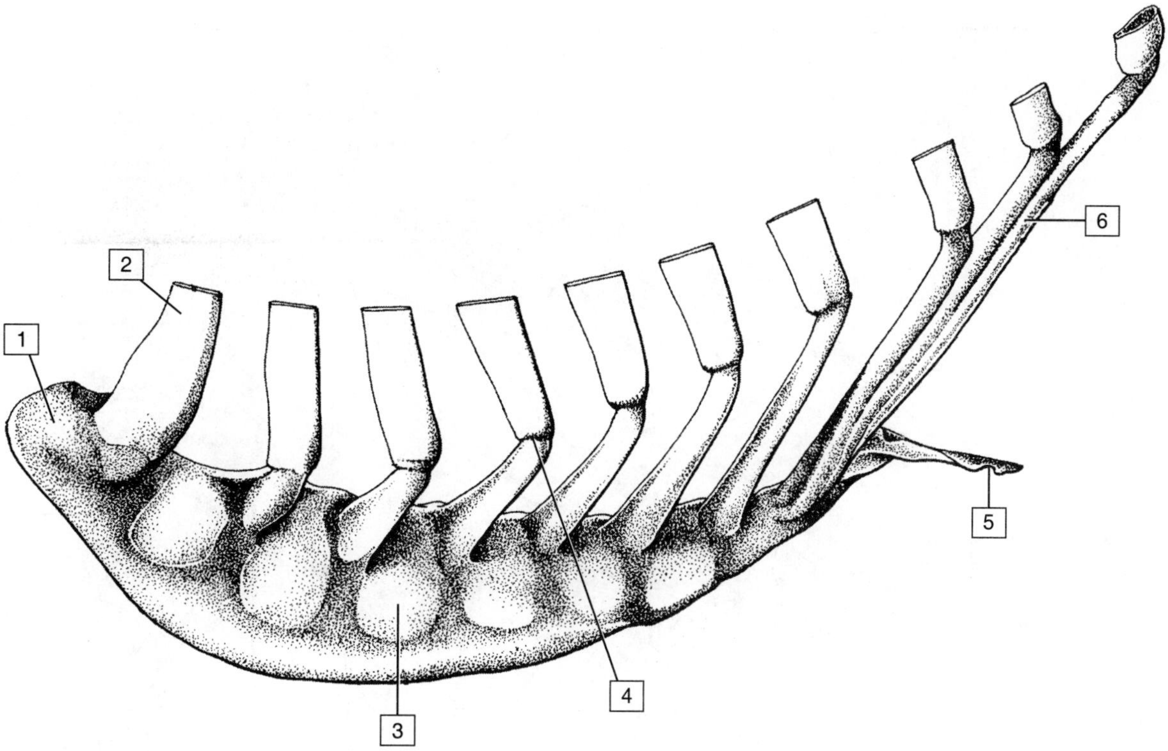

1	胸骨柄 Manubrium sterni	4	肋软骨接合处 Costochondral junction
2	第1肋 First rib	5	剑状软骨 Xiphoid cartilage
3	胸骨节 Sternebra	6	肋弓 Costal arch

图2-10 马胸骨和肋软骨，外侧观

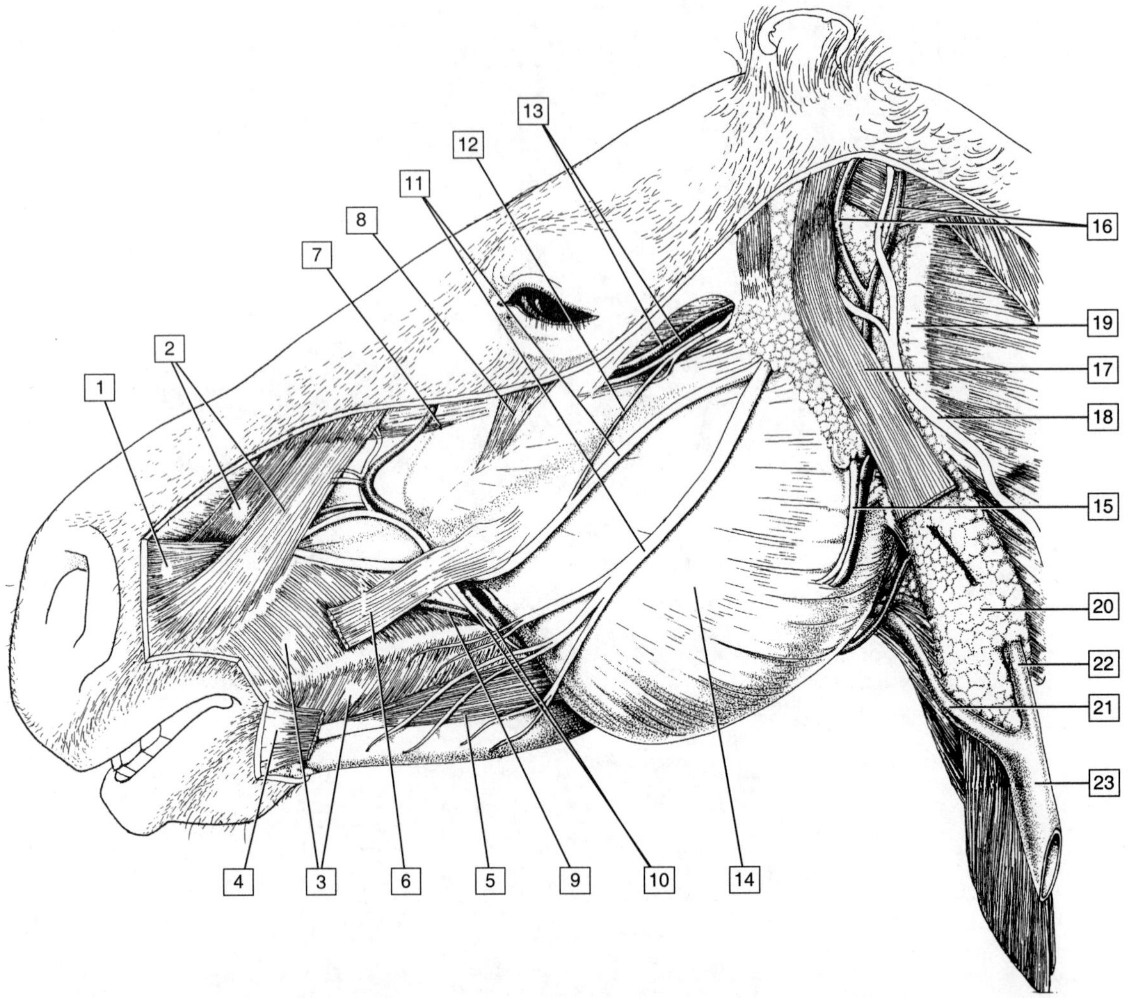

1 犬齿肌 Caninus

2 鼻唇提肌 Levator nasolabialis

3 颊肌 Buccinator

4 皮肌与口轮匝肌的连接处 Stump of cutaneous muscle joining orbicularis oris

5 下唇降肌 Depressor labii inferioris

6 颧肌 Zygomaticus

7 上唇提肌 Levator labii superioris

8 颧肌 Malaris

9 腮腺管 Parotid duct

10 面动脉和面静脉 Facial artery and vein

11 面神经颊支 Buccal branches of facial nerve

12 耳颞神经前交通支 Rostral communicating branch of auriculotemporal nerve

13 面横动脉、面横静脉，颞耳神经的面横支 Transverse facial artery and vein and transverse facial branch of auriculotemporal nerve

14 咬肌 Masseter

15 咬肌动脉和咬肌静脉 Masseteric artery and vein

16 耳静脉 Auricular veins

17 腮耳肌 Parotid auricularis

18 耳大神经（第2神经） Great auricular nerve (C2)

19 寰椎翼 Wing of atlas

20 腮腺 Parotid gland

21 舌面静脉 Linguofacial vein

22 上颌静脉 Maxillary vein

23 颈外静脉 External jugular vein

图2-11 马头部浅层解剖图

牛

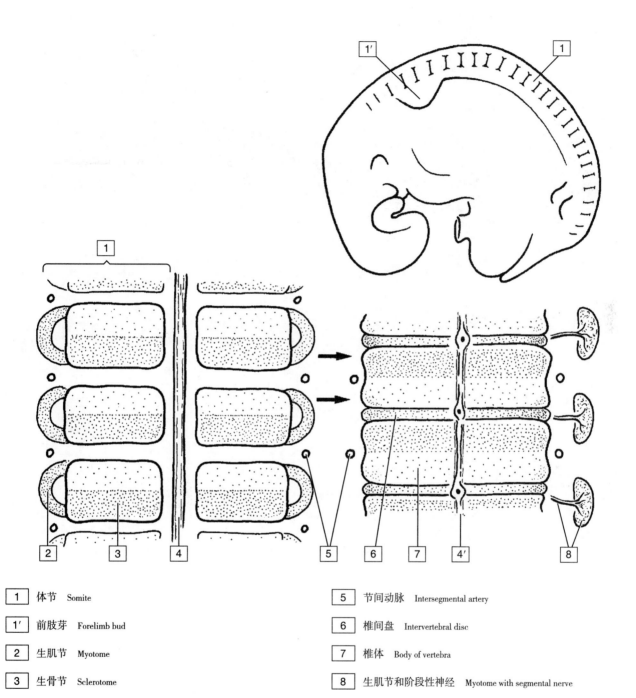

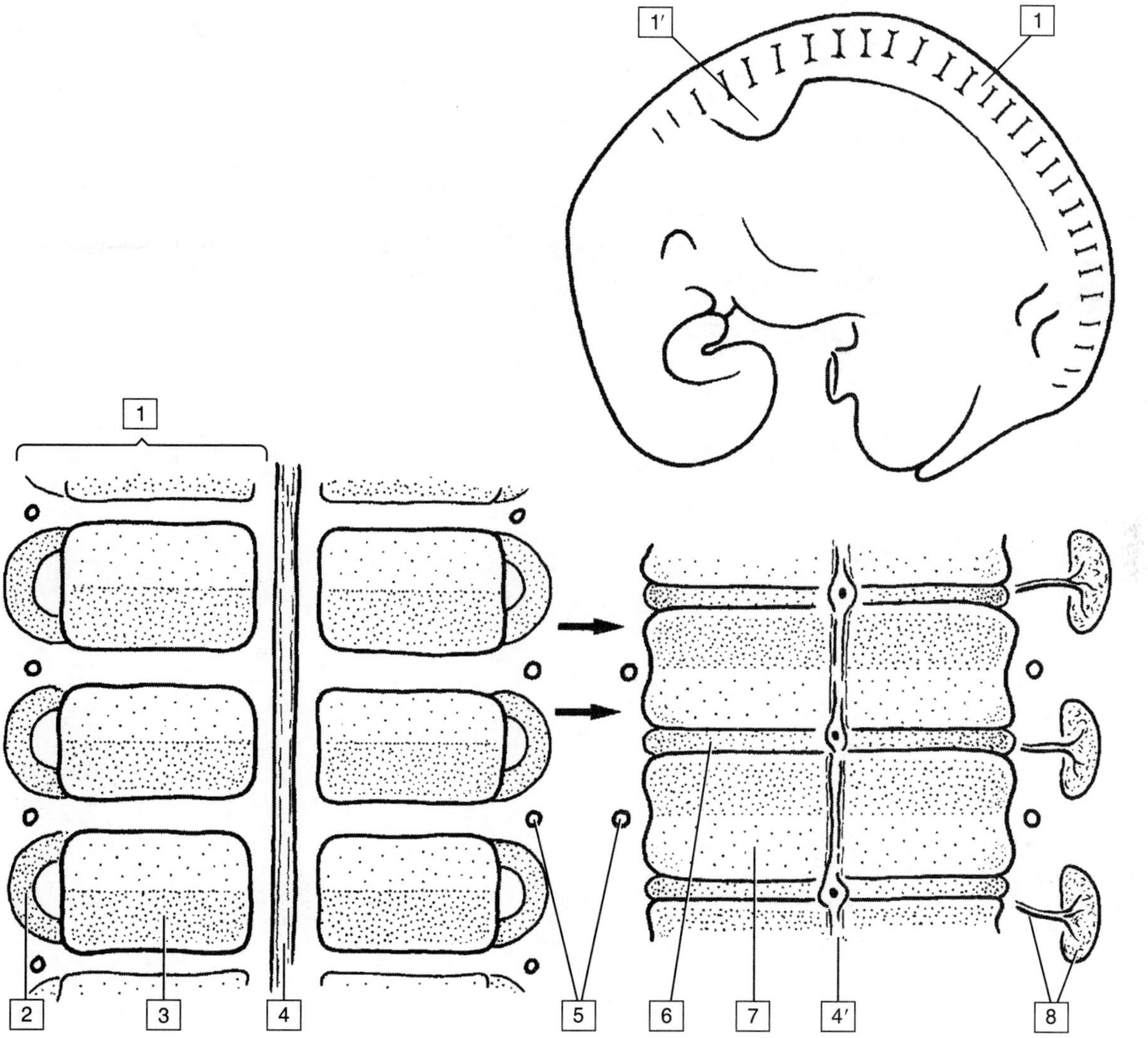

1 体节 Somite	5 节间动脉 Intersegmental artery
1′ 前肢芽 Forelimb bud	6 椎间盘 Intervertebral disc
2 生肌节 Myotome	7 椎体 Body of vertebra
3 生骨节 Sclerotome	8 生肌节和阶段性神经 Myotome with segmental nerve
4 脊索 Notochord	
4′ 脊髓在椎间盘中央生成髓核 Notochord giving rise to the nucleus pulposus in the center of the intervertebral disc (6)	

图2-12 牛近轴侧中胚层的分割

10毫米的牛胚胎（上图），椎骨、相关血管和神经发育的两个阶段（下图），箭头表示每个椎骨由相邻的两对体节形成

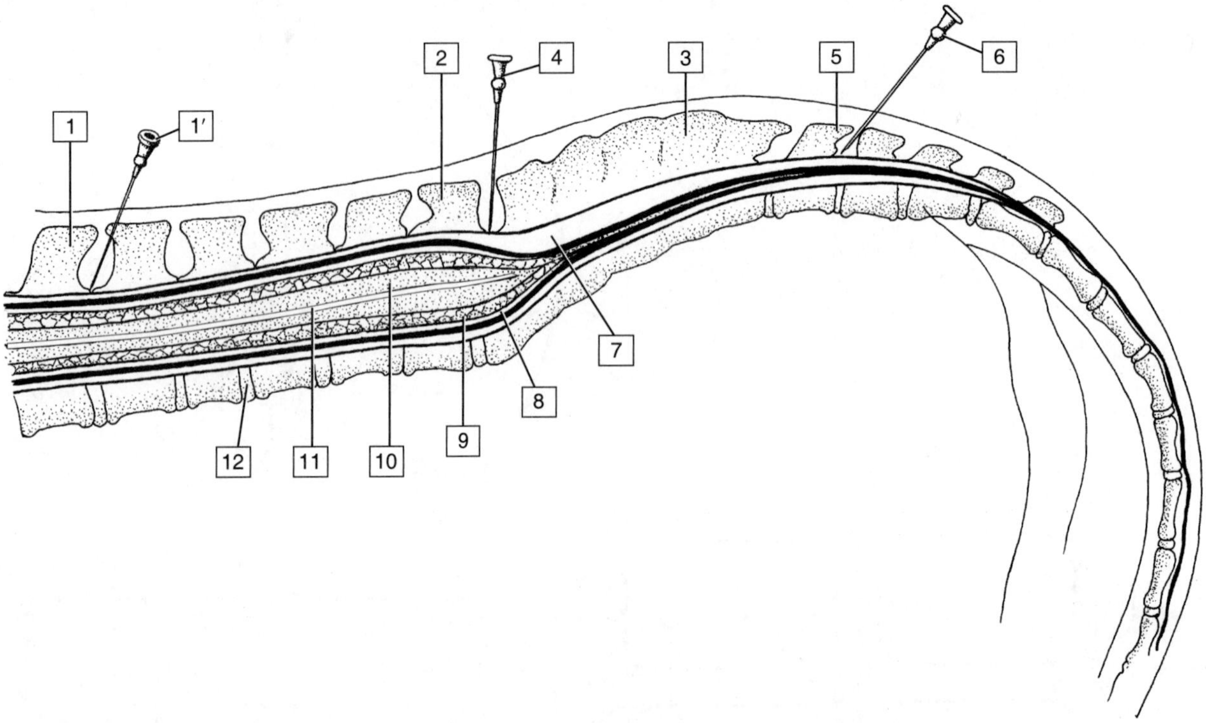

1 第1腰椎 First lumbar vertebra	7 硬膜外腔 Epidural space
1′ 针头位于侧面麻醉部位 Needle in position for flank anesthesia	8 硬膜 Dura mater
2 最后一节腰椎（第6腰椎） Last lumbar vertebra (L6)	9 蛛网膜下腔 Subarachnoid space
3 荐骨 Sacrum	10 脊髓 Spinal cord
4 针头位于腰荐间隙 Needle in lumbosacral space	11 中央管 Central canal
5 第1尾椎 First caudal vertebra	12 椎间盘 Intervertebral disc
6 在第1和第2尾椎之间的针头 Needle between first and second caudal vertebrae (tail block)	

图2-13　牛尾部椎管及其内容物

针头所示为硬膜外注射部位

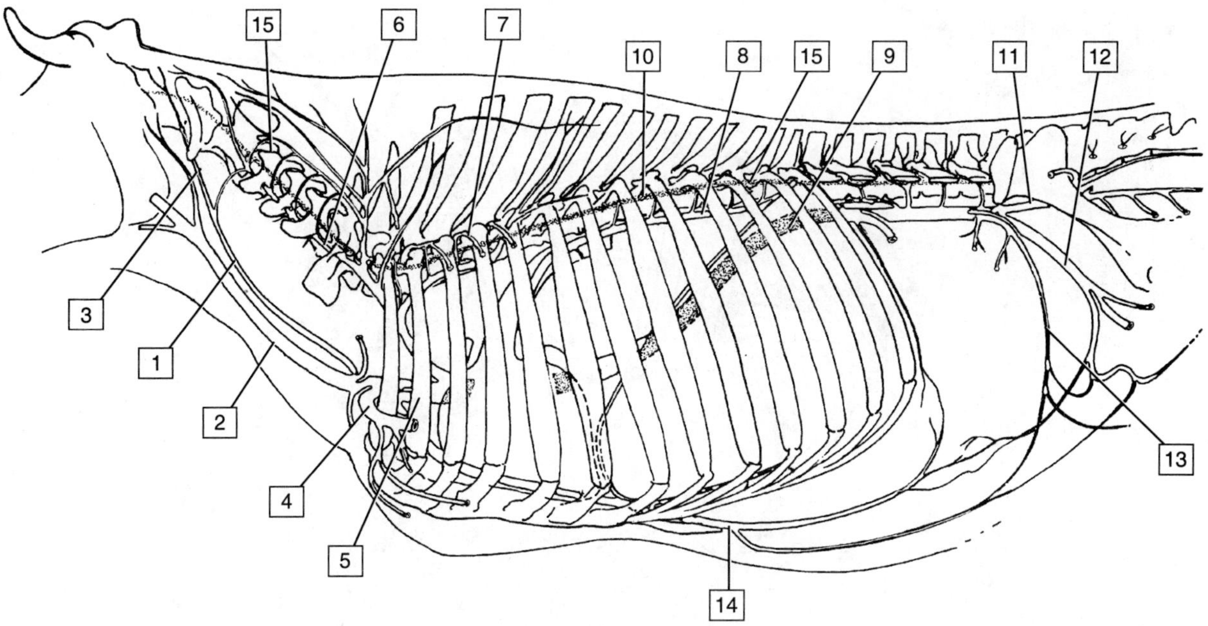

1 颈内静脉 Internal jugular vein

2 颈外静脉 External jugular vein

3 枕静脉 Occipital vein

4 腋静脉 Axillary vein

5 第2肋 Second rib

6 椎静脉 Vertebral vein

7 最上肋间静脉 Supreme intercostal vein

8 左奇静脉 Left azygous vein

9 后腔静脉 Caudal vena cava

10 肋间静脉 Intercostal vein

11 髂内静脉 Internal iliac vein

12 髂外静脉 External iliac vein

13 旋髂深静脉 Deep circumflex iliac vein

14 腹壁前静脉 Cranial epigastric vein

15 在椎管内的椎内神经丛
Internal vertebral plexus, stippled in the vertebral canal

图 2-14 牛椎丛-奇静脉系统与主要静脉的连接

注意椎内丛和肋间静脉之间的连接，以及神经丛和椎静脉分支之间的连接

鸟类

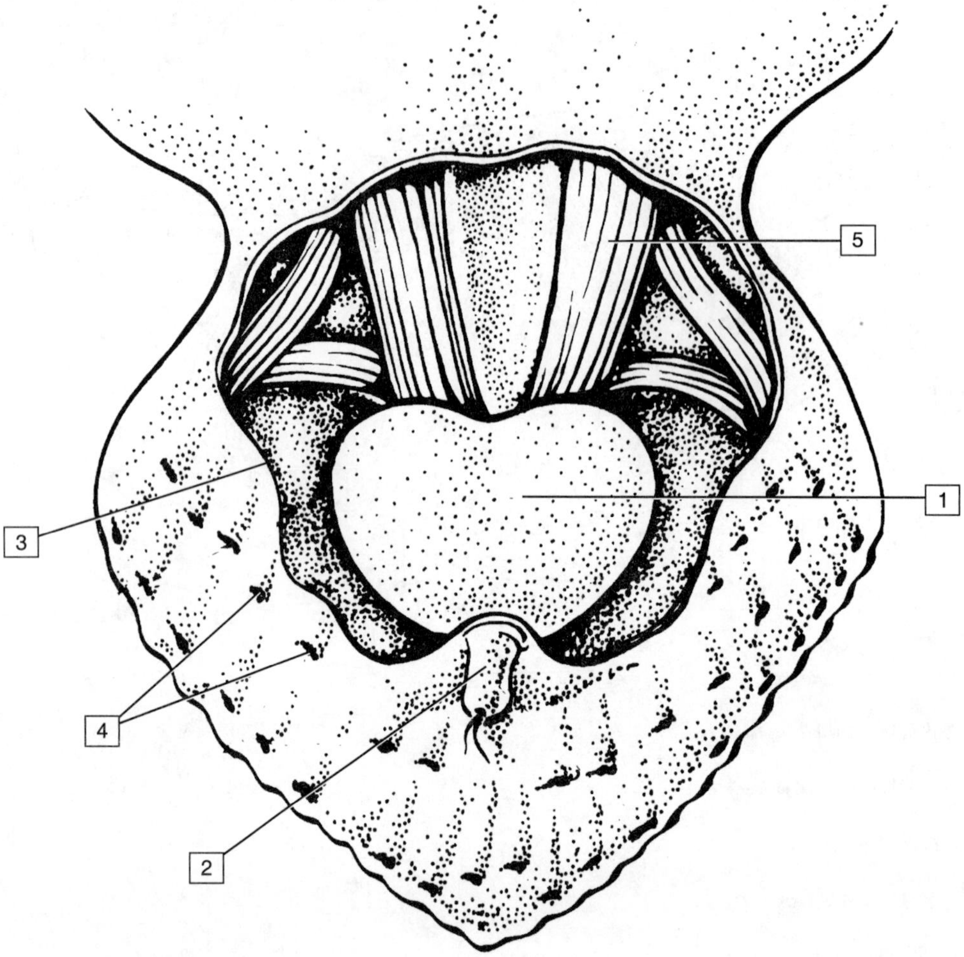

1 尾脂腺 Uropygial gland	4 羽毛毛囊 Feather follicles
2 尾脂腺乳头，分泌物通过该结构被排出 Papilla of uropygial gland through which the secretion is extruded	5 尾椎及相关肌肉 Caudal vertebrae and associated muscles
3 皮肤的断端 Cut edge of skin	

图2-15 鸟类尾脂腺，背侧观

不同动物的解剖结构比较

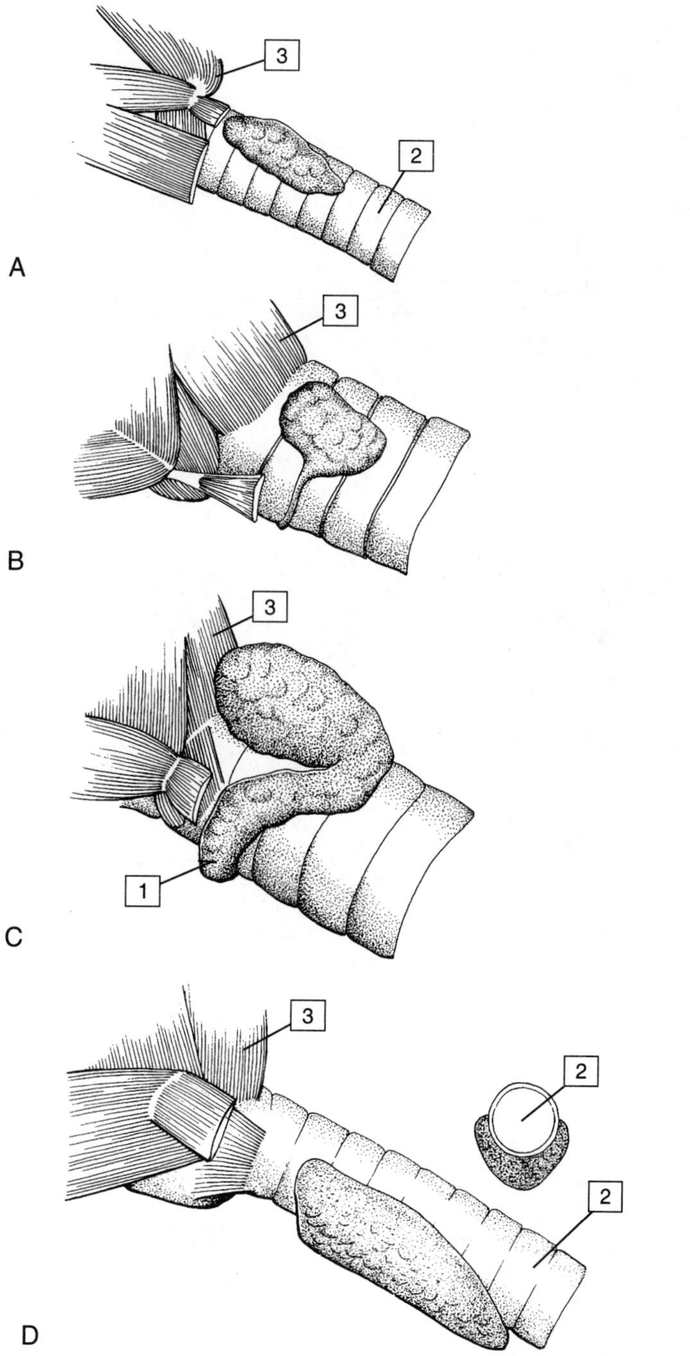

1	峡部 Isthmus	3	环咽肌 Cricopharyngeus
2	气管 Trachea		

图2-16 犬(A)、马(B)、牛(C)、猪(D)的甲状腺

图D显示了猪气管下横断面的连接

3

胸　部

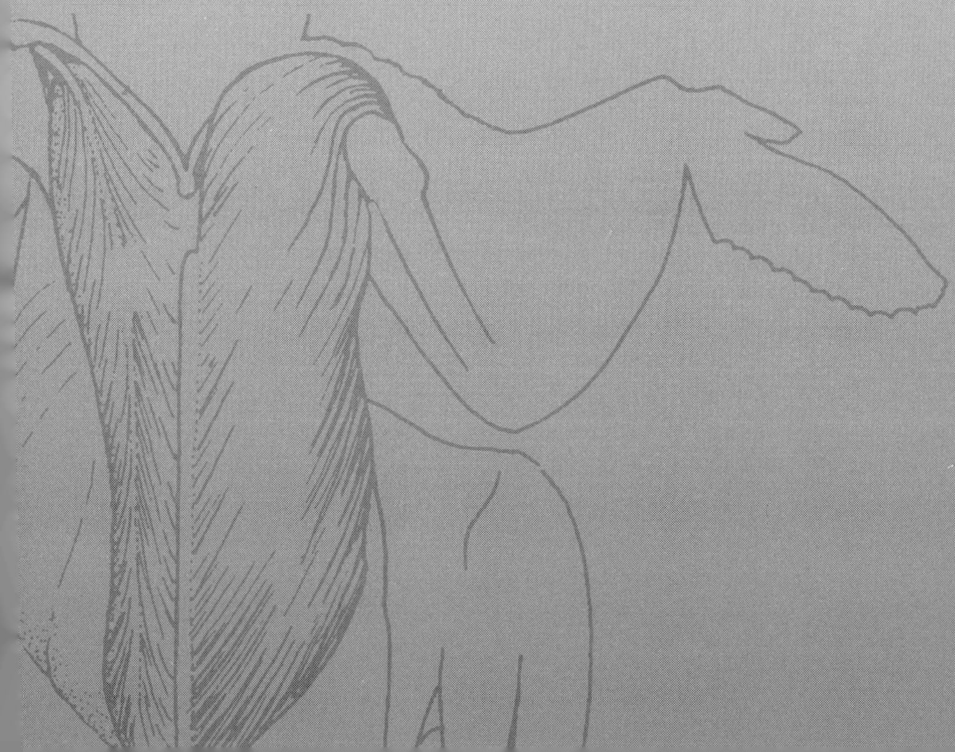

犬

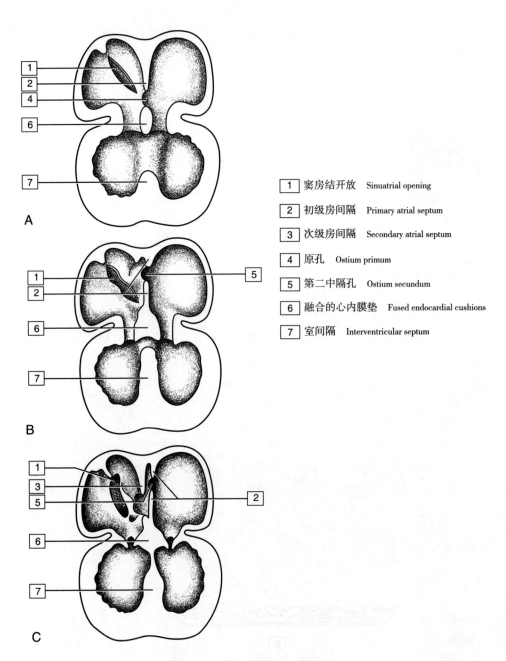

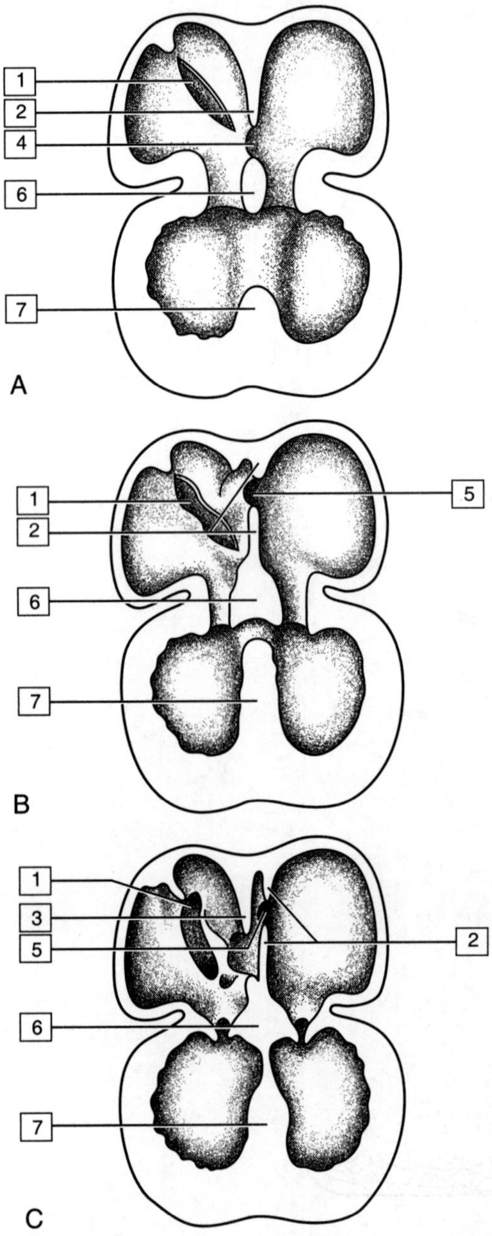

1 窦房结开放 Sinuatrial opening

2 初级房间隔 Primary atrial septum

3 次级房间隔 Secondary atrial septum

4 原孔 Ostium primum

5 第二中隔孔 Ostium secundum

6 融合的心内膜垫 Fused endocardial cushions

7 室间隔 Interventricular septum

图3-1 心房和心室分区示意图

A. 初级房间隔已形成，室间隔开始发育；B. 初级房间隔与心内膜垫融合，并形成二级孔（5）；C. 次级房间隔已形成，在初级和次级房间隔之间的通道（卵圆孔）连接了左右心房。

注意室间隔和心内膜垫的融合。注意A、B、C图是心脏的常规表示，与物种无关

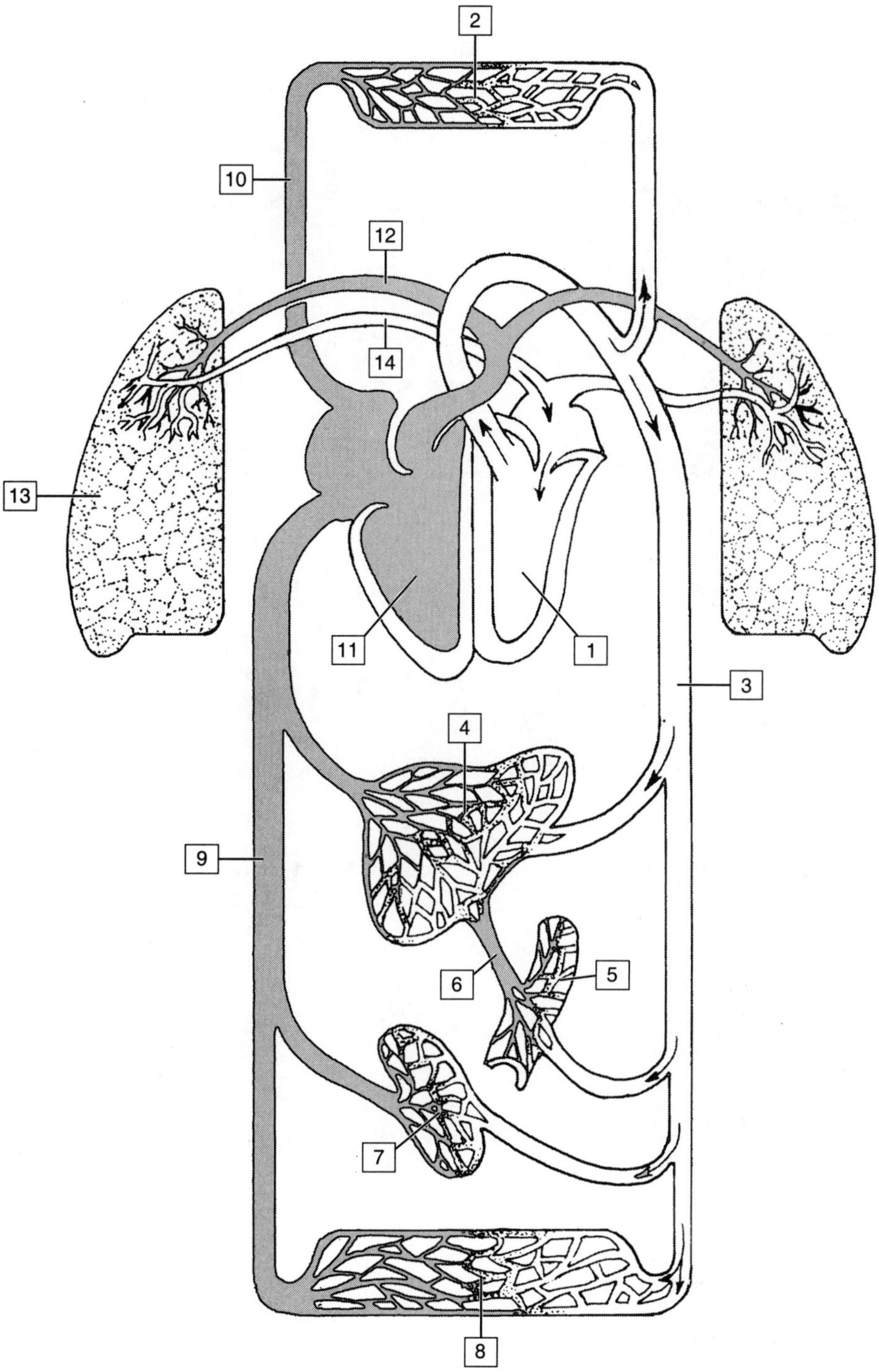

体循环 Systemic circulation

1 心脏左侧 Left side of the heart
2 头部的血管 Vessels in the cranial part of the body
3 主动脉 Aorta
4 肝脏 Liver
5 肠 Intestines
6 门静脉 Portal vein
7 肾 Kidneys
8 身体其他部位的血管 Vessels in the caudal part of the body
9 后腔静脉 Caudal vena cava
10 前腔静脉 Cranial vena cava

肺循环 Pulmonary circulation

11 心脏右侧 Right side of the heart
12 肺动脉 Pulmonary artery
13 肺 Lung
14 肺静脉 Pulmonary vein

图3-2 犬的心血管循环模式图

携带含氧血液的血管以白色显示，缺氧血液的血管以黑色显示

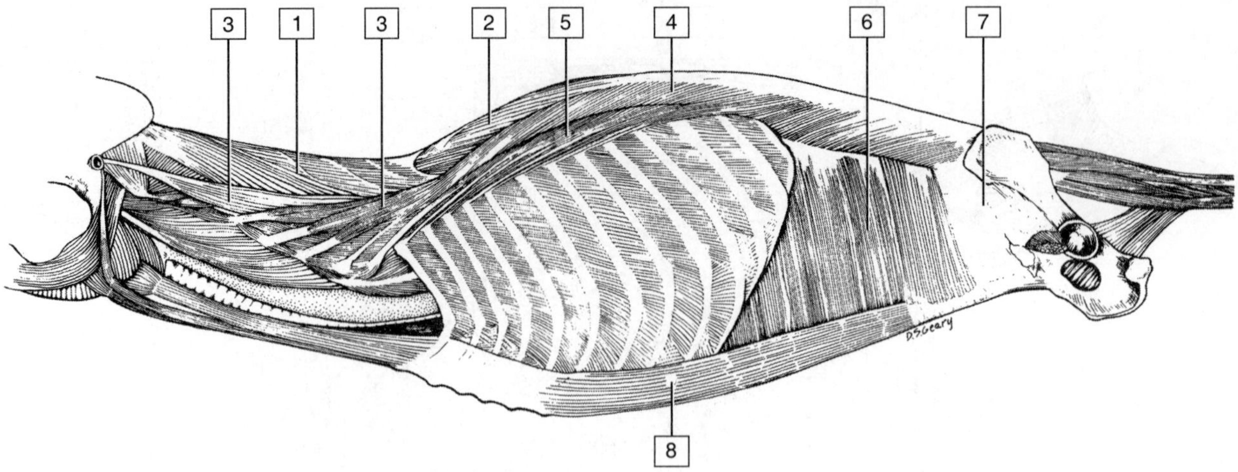

1	头半棘肌 Semispinalis capitis	5	髂肋肌 Iliocostalis
2	棘肌和半棘肌 Spinalis and semispinalis	6	腹横肌 Transversus abdominis
3	头最长肌和颈最长肌 Longissimus capitis and cervicis	7	腹横筋膜 Transverse fascia
4	胸最长肌 Longissimus thoracis	8	腹直肌 Rectus abdominis

图3-3 犬躯干深层肌

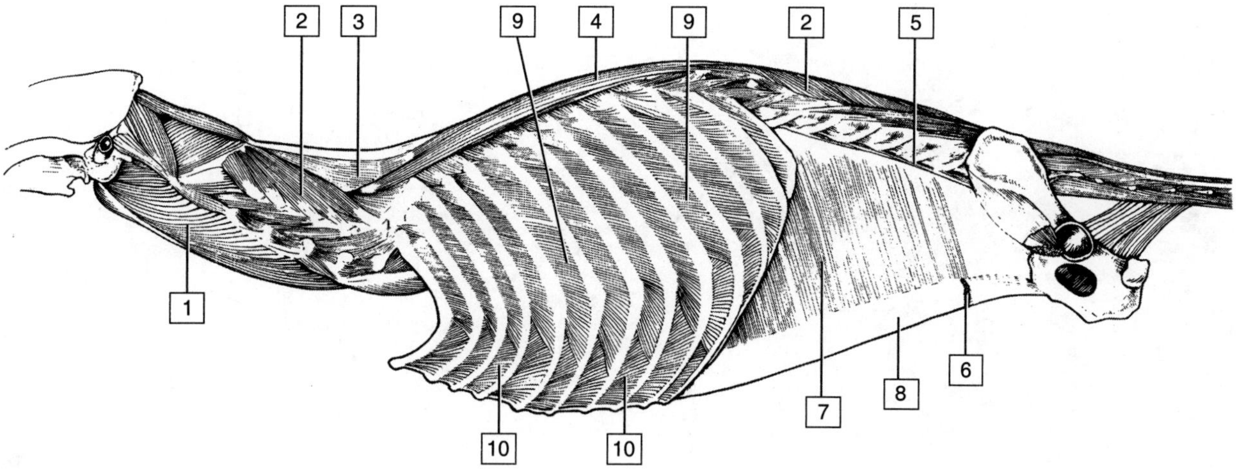

1	头长肌 Longus capitis	6	腹直肌 Rectus abdominis
2	多裂肌 Multifidus	7	腹横肌 Transversus abdominis
3	颈棘肌 Spinalis cervicis	8	腹横肌腱膜 Aponeurosis of transversus abdominis
4	棘肌和半棘肌 Spinalis and semispinalis	9	肋间外肌 External intercostal muscles
5	腰方肌 Quadratus lumborum	10	肋间内肌 Internal intercostal muscles

图3-4 犬躯干最深层肌

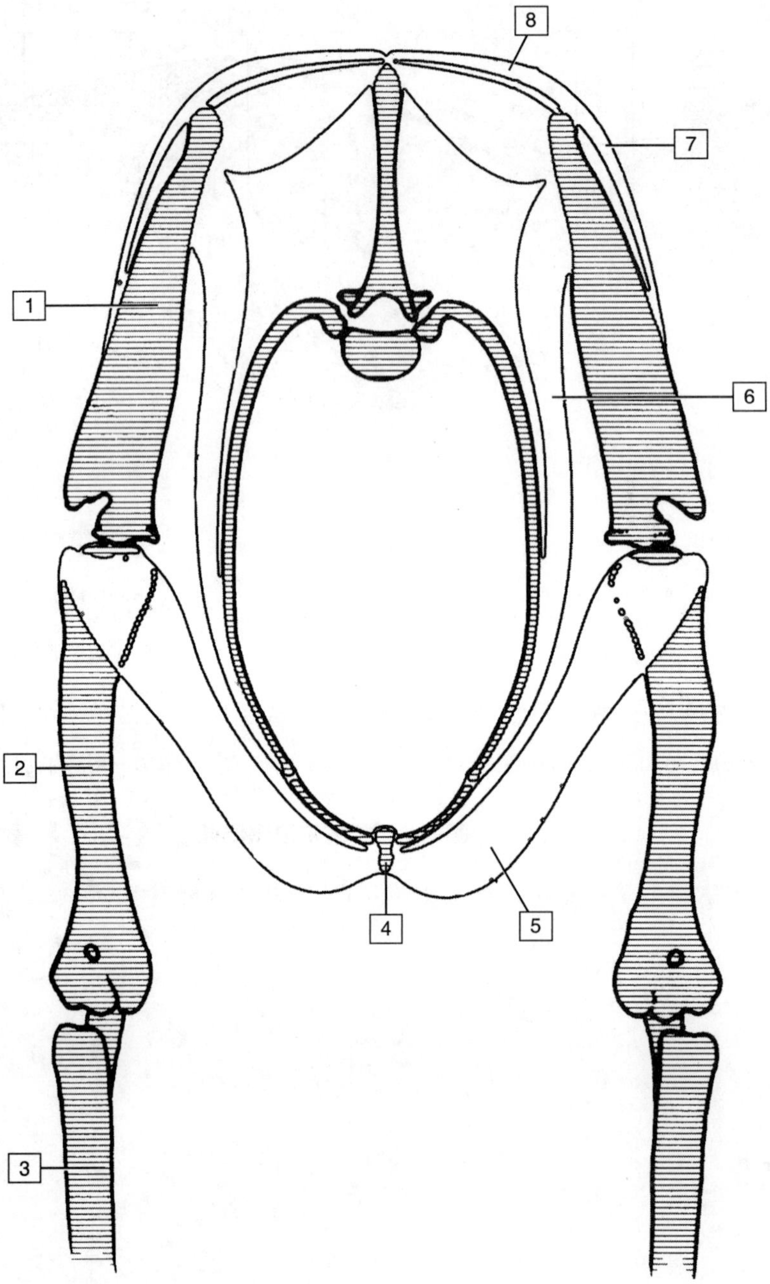

1	肩胛骨 Scapula	5	胸深肌（胸升肌） Pectoralis profundus (ascendens)
2	肱骨 Humerus	6	腹侧锯肌 Serratus ventralis
3	桡骨和尺骨 Radius and ulna	7	斜方肌 Trapezius
4	胸骨 Sternum	8	菱形肌 Rhomboideus

图3-5 犬胸部和前肢间的肌肉悬吊

哪些神经支配肌肉 6 和 8 ？

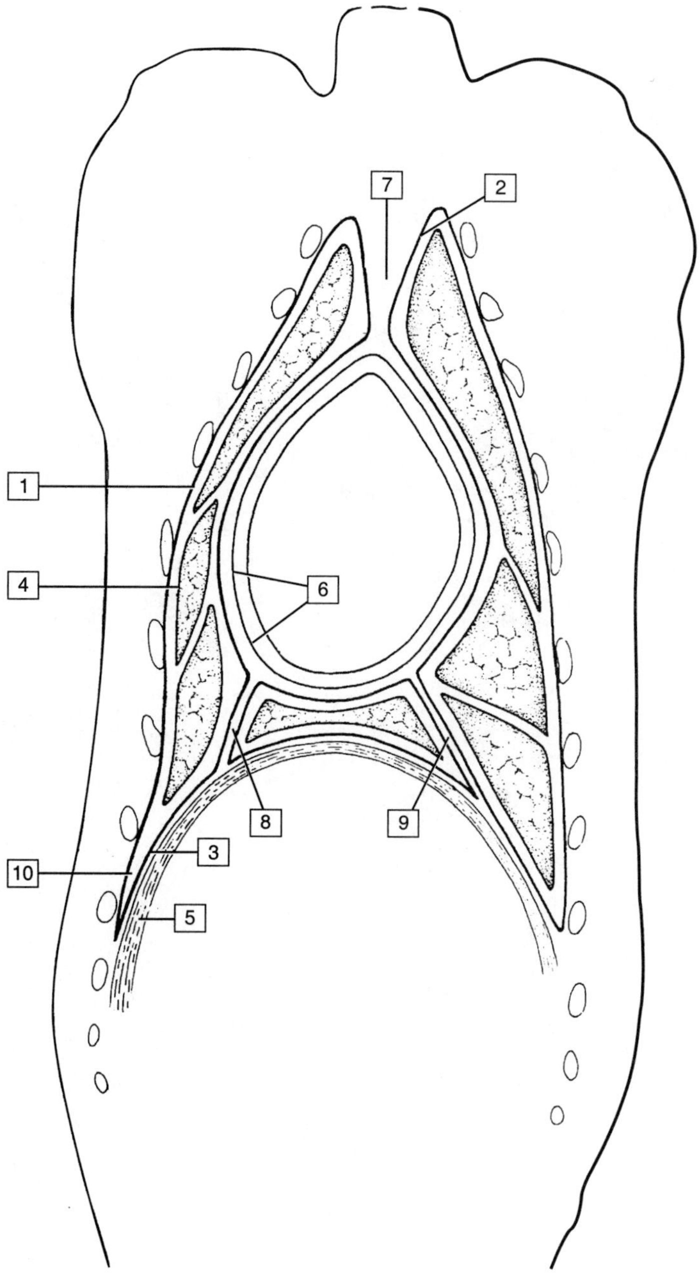

1	肋胸膜 Costal pleura	7	前纵隔 Cranial mediastinum
2	纵隔胸膜 Mediastinal pleura	8	后纵隔 Caudal mediastinum
3	膈胸膜 Diaphragmatic pleura	9	腔静脉皱襞 Plica venae cavae
4	胸膜脏层 Visceral pleura	10	肋膈隐窝 Costodiaphragmatic recess
5	膈 Diaphragm		
6	心包膜壁层：它的外层纤维层与内部的浆膜层紧密相连 Parietal pericardium; its outer fibrous layer tightly adheres to its inner serous layer		

图3-6 犬胸膜和心包膜的分布

加粗线表示胸膜

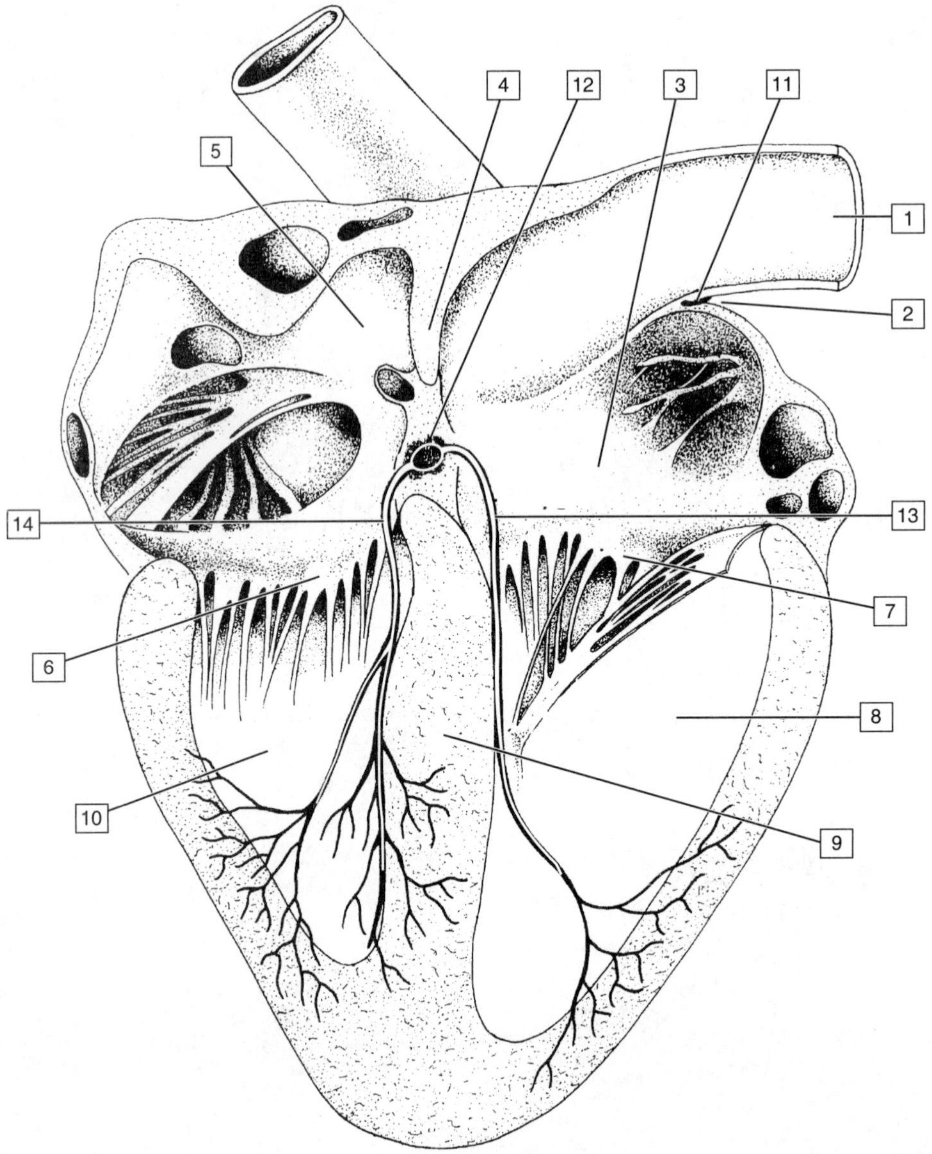

1	前腔静脉	Cranial vena cava	8	右心室	Right ventricle
2	终嵴	Terminal sulcus	9	室间隔	Interventricular septum
3	右心房	Right atrium	10	左心室	Left ventricle
4	房间隔	Interatrial septum	11	窦房结	Sinoatrial node
5	左心房	Left atrium	12	房室结	Atrioventricular node
6	左房室瓣	Left atrioventricular valve	13	房室束的左右束支	Right and left limbs of atrioventricular bundle
7	右房室瓣	Right atrioventricular valve	14	房室束的左右束支	Right and left limbs of atrioventricular bundle

图3-7 犬心腔纵切面

请在这张图上画出心脏传导系统。

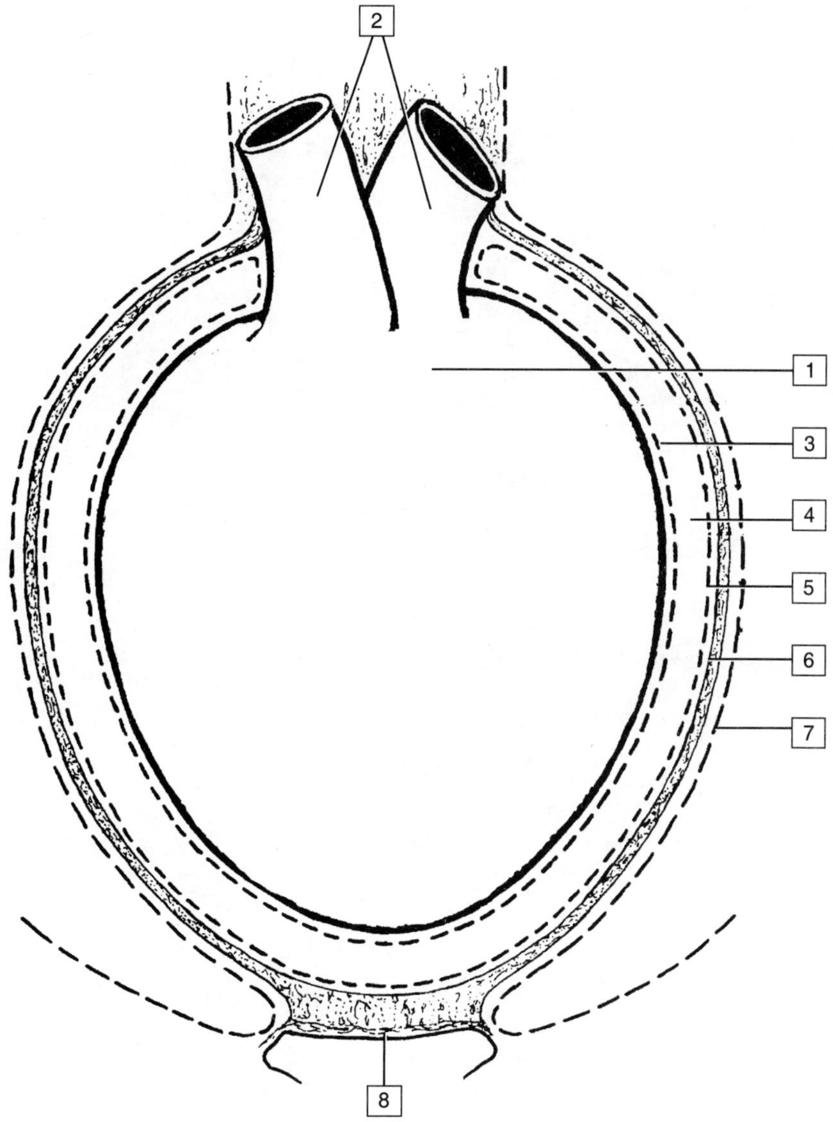

1 心脏 Heart	5 心包膜壁层 Parietal pericardium
2 大血管 Great vessels	6 心包膜壁层的结缔组织层 Connective tissue layer of the parietal pericardium
3 心包膜（心外膜） Visceral pericardium (epicardium)	7 纵隔胸膜 Mediastinal pleura
4 心包腔（放大的尺寸） Pericardial cavity (exaggerated in size)	8 胸骨心包韧带 Sternopericardial ligament

图3-8 犬心包膜

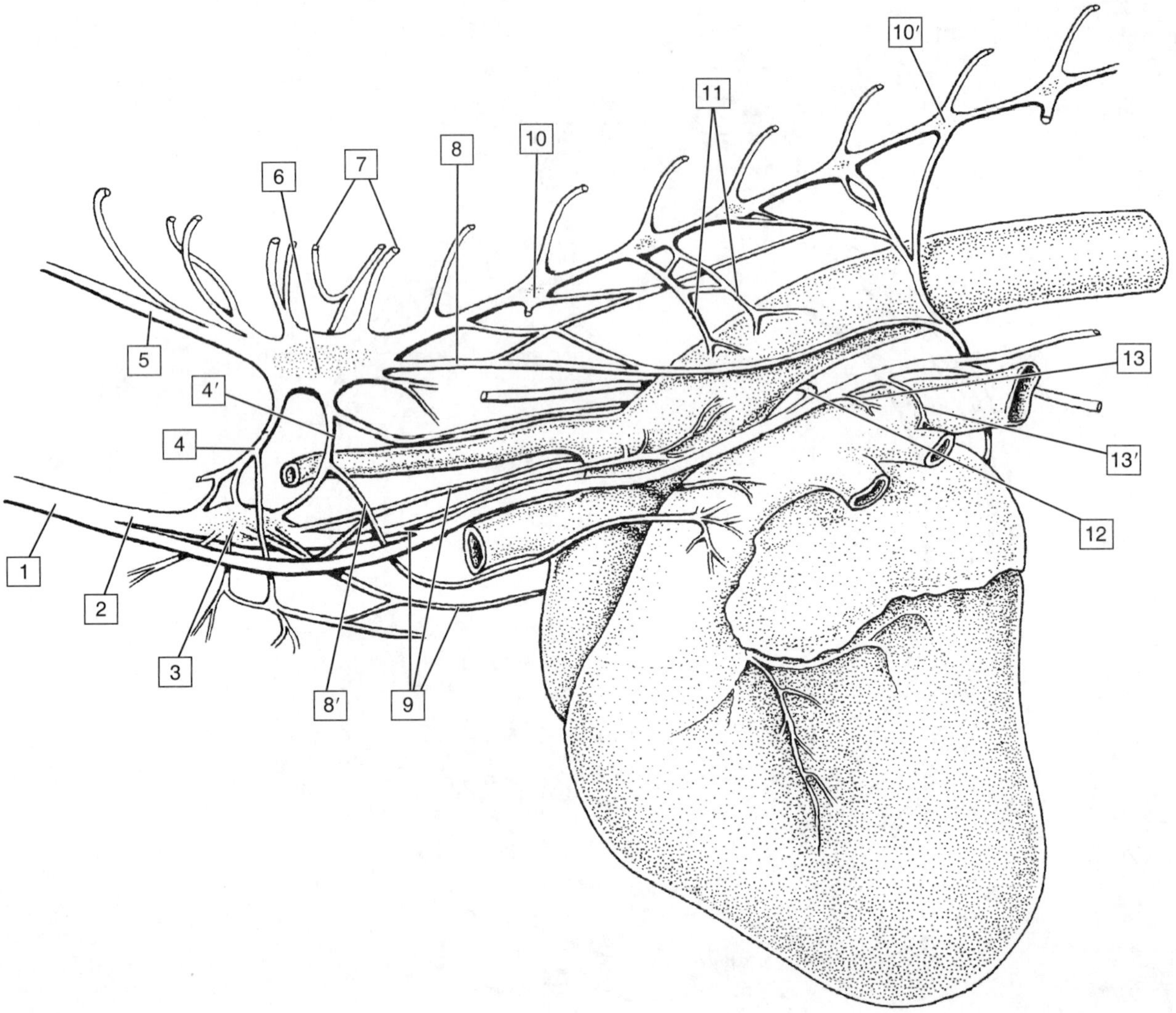

1	迷走神经干 Vagosympathetic trunk	8′	颈胸后腹侧心神经 Caudoventral cervicothoracic cardiac nerve
2	交感干 Sympathetic trunk	9	椎心神经 Vertebral cardiac nerve
3	颈中神经节 Middle cervical ganglion	10	第3胸神经节 Third thoracic ganglion
4	锁骨下袢前支 Cranial limb of ansa subclavia	10′	第7胸神经节 Seventh thoracic ganglion
4′	锁骨下袢后支 Caudal limb of ansa subclavia	11	胸心神经 Thoracic cardiac nerve
5	椎神经 Vertebral nerve	12	左喉返神经 Left recurrent laryngeal nerve
6	颈胸神经节 Cervicothoracic ganglion	13	心迷走神经前支 Cranial vagal cardiac nerve
7	交通支 Communicating branches	13′	心迷走神经后支 Caudal vagal cardiac nerve
8	颈胸后背侧心神经 Caudodorsal cervicothoracic cardiac nerve		

图3-9 犬心脏神经和相关神经节，左外侧观

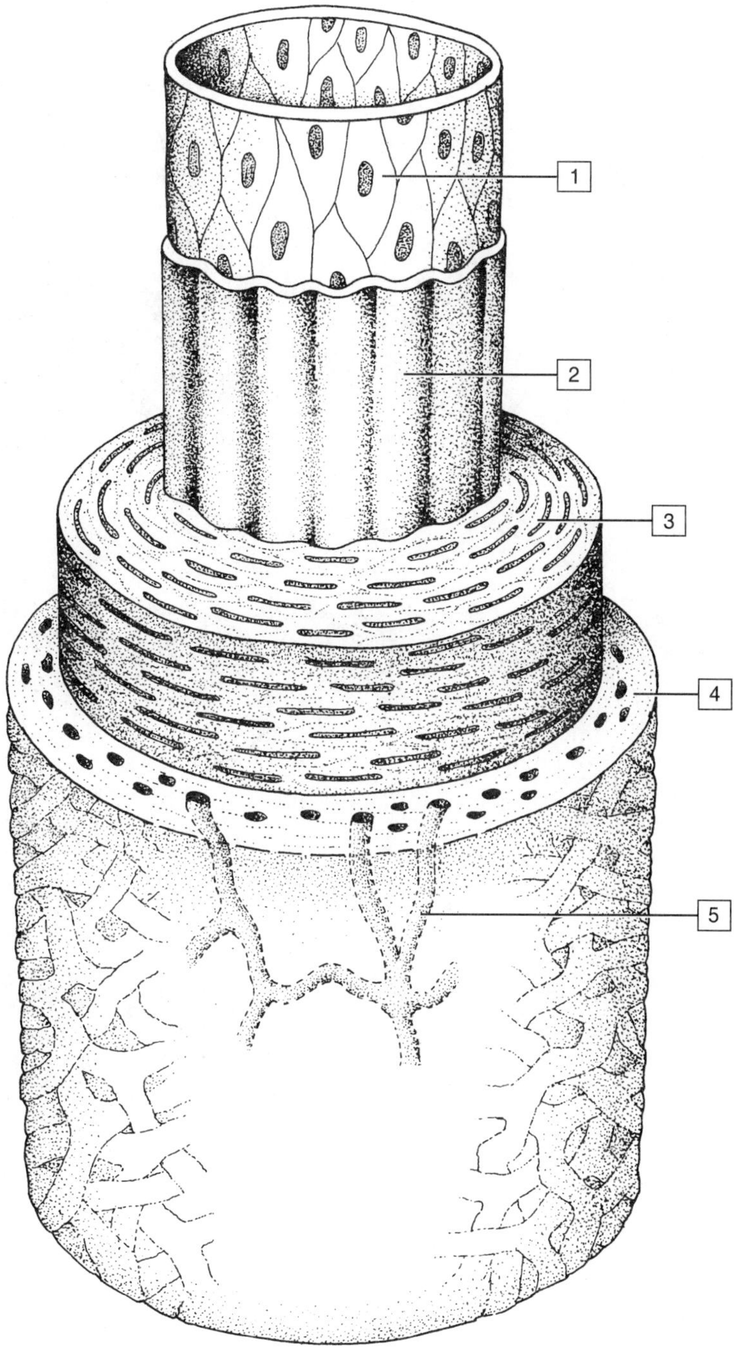

1	内皮 Endothelium	4	外膜 Tunica adventitia
2	内弹性膜 Inner elastic membrane	5	血管滋养管 Vasa vasorum
3	中膜 Tunica media		

图3-10 大动脉壁的组成

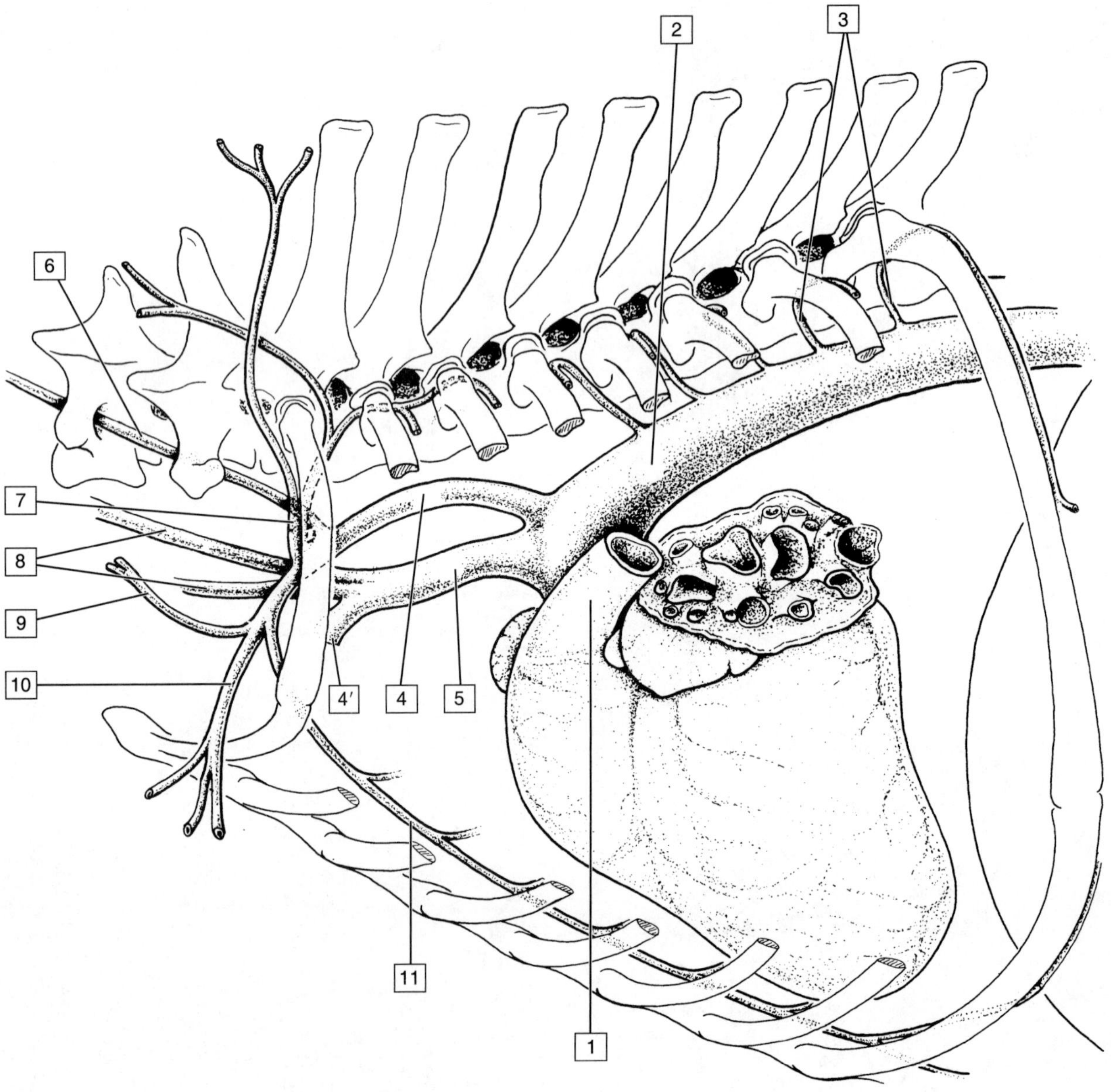

1	肺动脉干 Pulmonary trunk	6	椎动脉 Vertebral artery
2	主动脉 Aorta	7	肋颈干 Costocervical trunk
3	肋间动脉 Intercostal artery	8	左、右颈总动脉 Left and right common carotid artery
4	右锁骨下动脉 Right subclavian artery	9	颈浅动脉 Superficial cervical artery
4′	左锁骨下动脉 Left subclavian artery	10	腋动脉 Axillary artery
5	头臂干 Brachiocephalic trunk	11	胸廓内动脉 Internal thoracic artery

图3-11 犬主动脉弓的分支

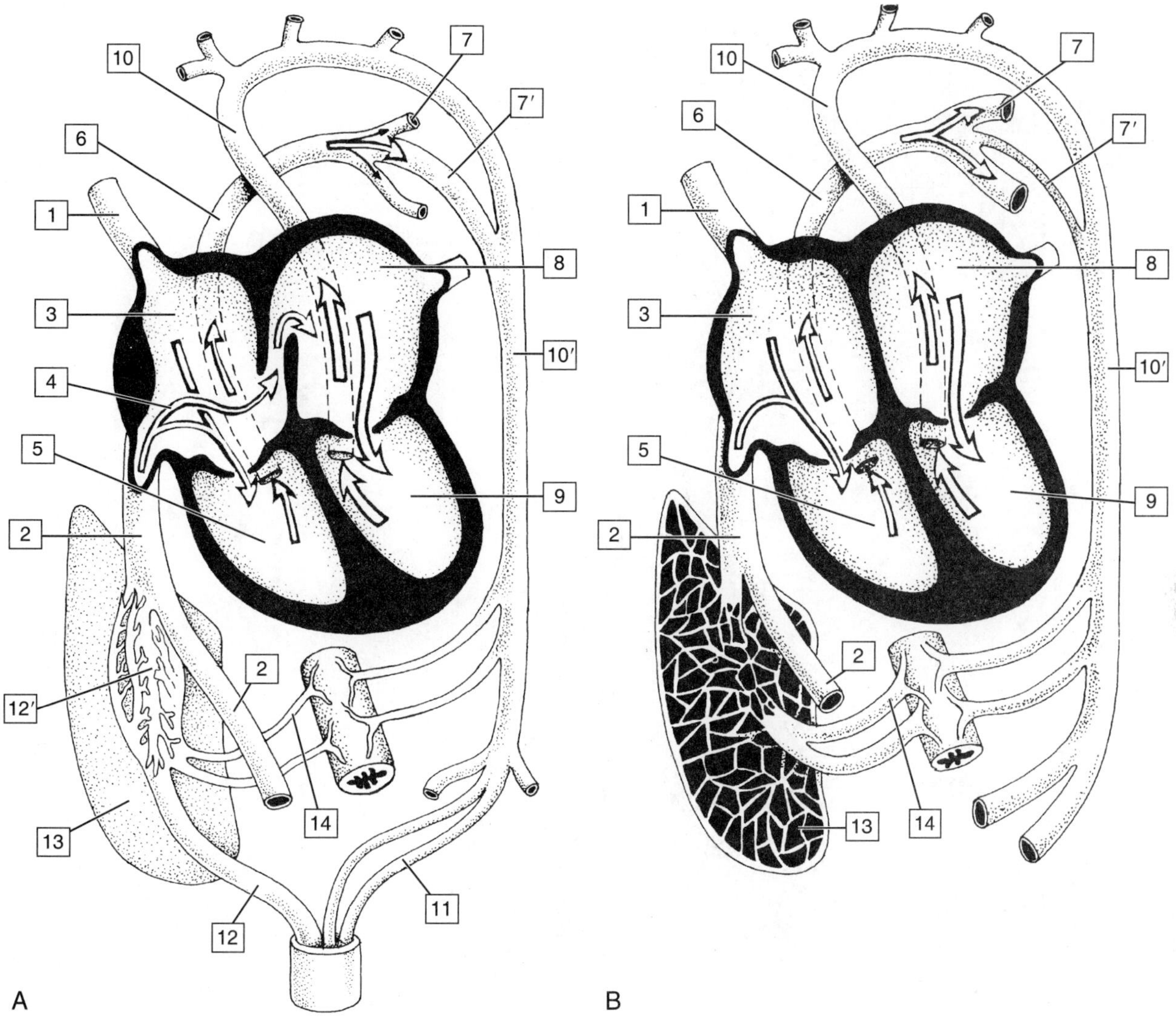

1	前腔静脉 Cranial vena cava	9	左心室 Left ventricle
2	后腔静脉 Caudal vena cava	10	主动脉弓 Aortic arch
3	右心房 Right atrium	10′	降主动脉 Descending aorta
4	血液通过卵圆孔进入 Arrow entering oval foramen	11	脐动脉 Umbilical artery
5	右心室 Right ventricle	12	脐静脉 Umbilical vein
6	肺动脉干 Pulmonary trunk	12′	静脉导管 Ductus venosus
7	肺动脉 Pulmonary artery	13	肝脏 Liver
7′	动脉导管（图B中为退化后遗迹） Ductus arteriosus (in B, vestige)	14	门静脉 Portal vein
8	左心房 Left atrium		

图3-12　犬胎儿（A）及胎儿出生后（B）的循环系统

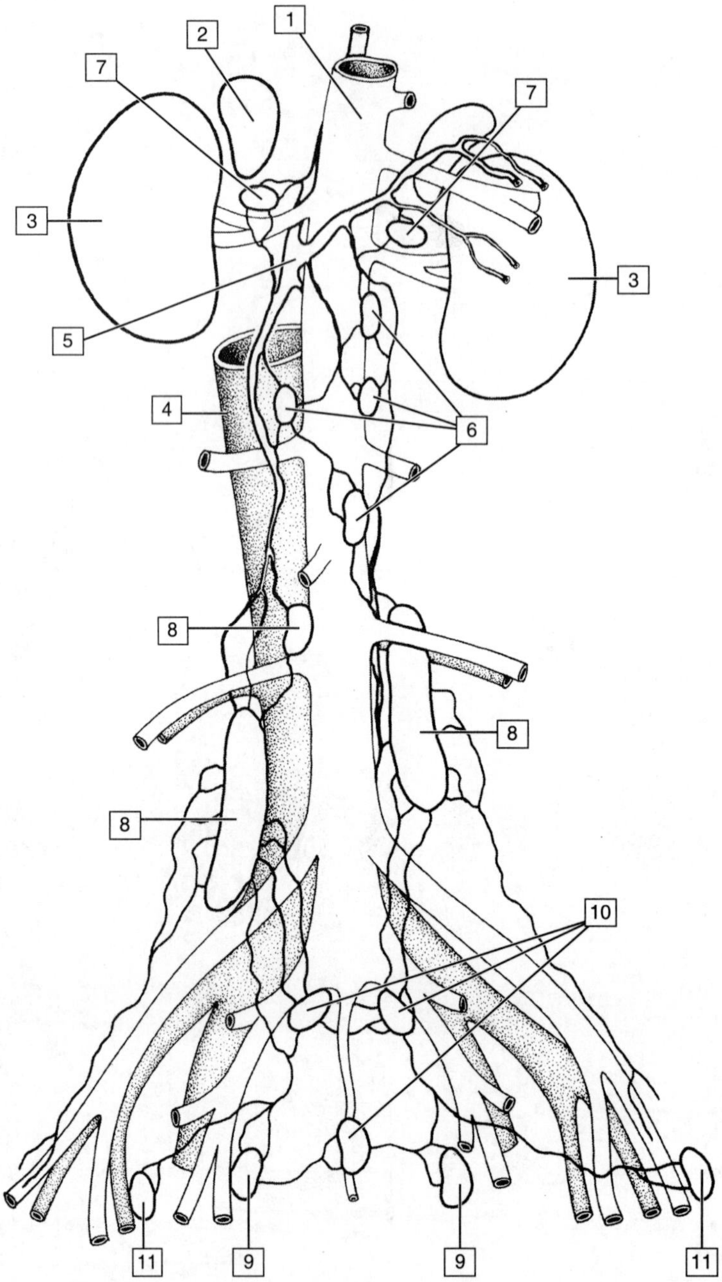

1	主动脉 Aorta	7	肾淋巴结 Renal nodes
2	肾上腺 Adrenals	8	髂内侧淋巴结 Medial iliac nodes
3	肾 Kidneys	9	下腹淋巴结 Hypogastric nodes
4	后腔静脉 Caudal vena cava	10	荐淋巴结 Sacral nodes
5	乳糜池 Cisterna chyli	11	腹股沟深淋巴结（髂股淋巴结） Deep inguinal (iliofemoral) nodes
6	腰主动脉淋巴结 Lumbar aortic nodes		

图3-13 犬腰荐部淋巴引流，腹侧观

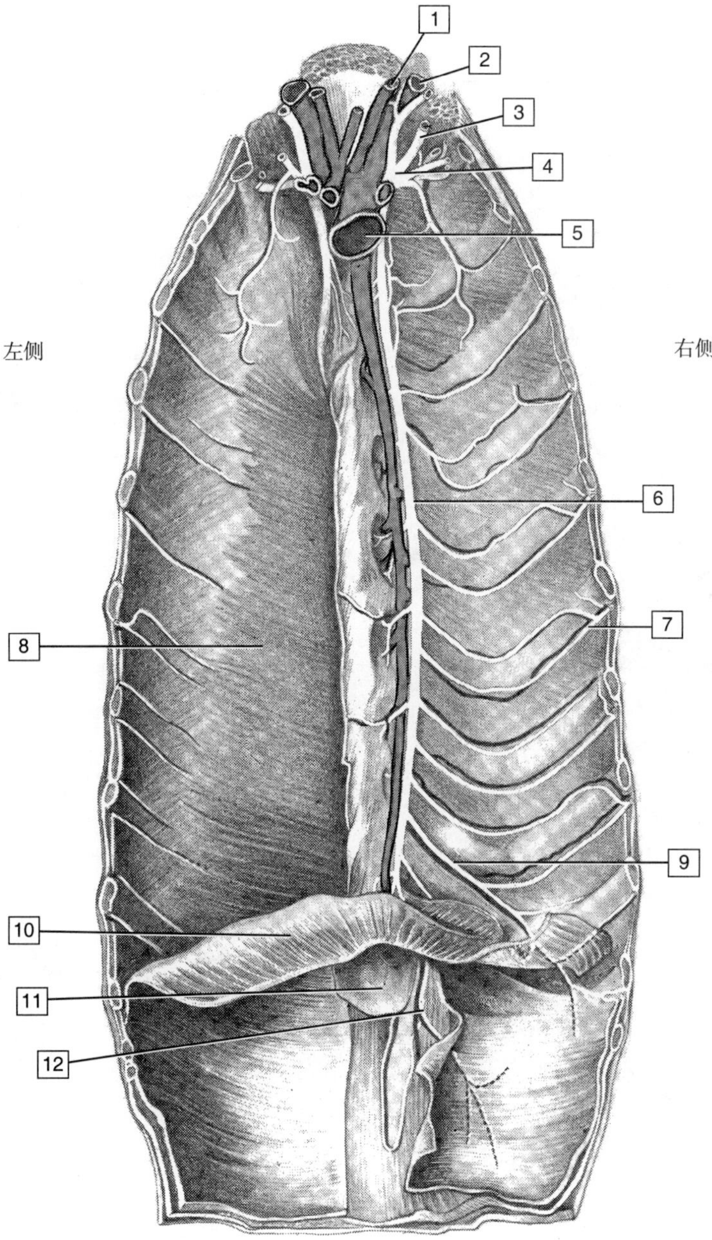

1	颈内静脉 Internal jugular vein	7	肋间动脉 Intercostal artery
2	颈外静脉 External jugular vein	8	胸横肌 Transversus thoracis
3	椎动脉 Vertebral artery	9	肌膈动脉 Musculophrenic artery
4	右锁骨下动脉 Right subclavian artery	10	膈 Diaphragm
5	前腔静脉 Cranial vena cava	11	剑状软骨 Xiphoid cartilage
6	胸廓内动脉 Internal thoracic artery	12	腹壁前动脉 Cranial epigastric artery

图3-14 犬胸腔底壁血管

右侧胸横肌已被剥离

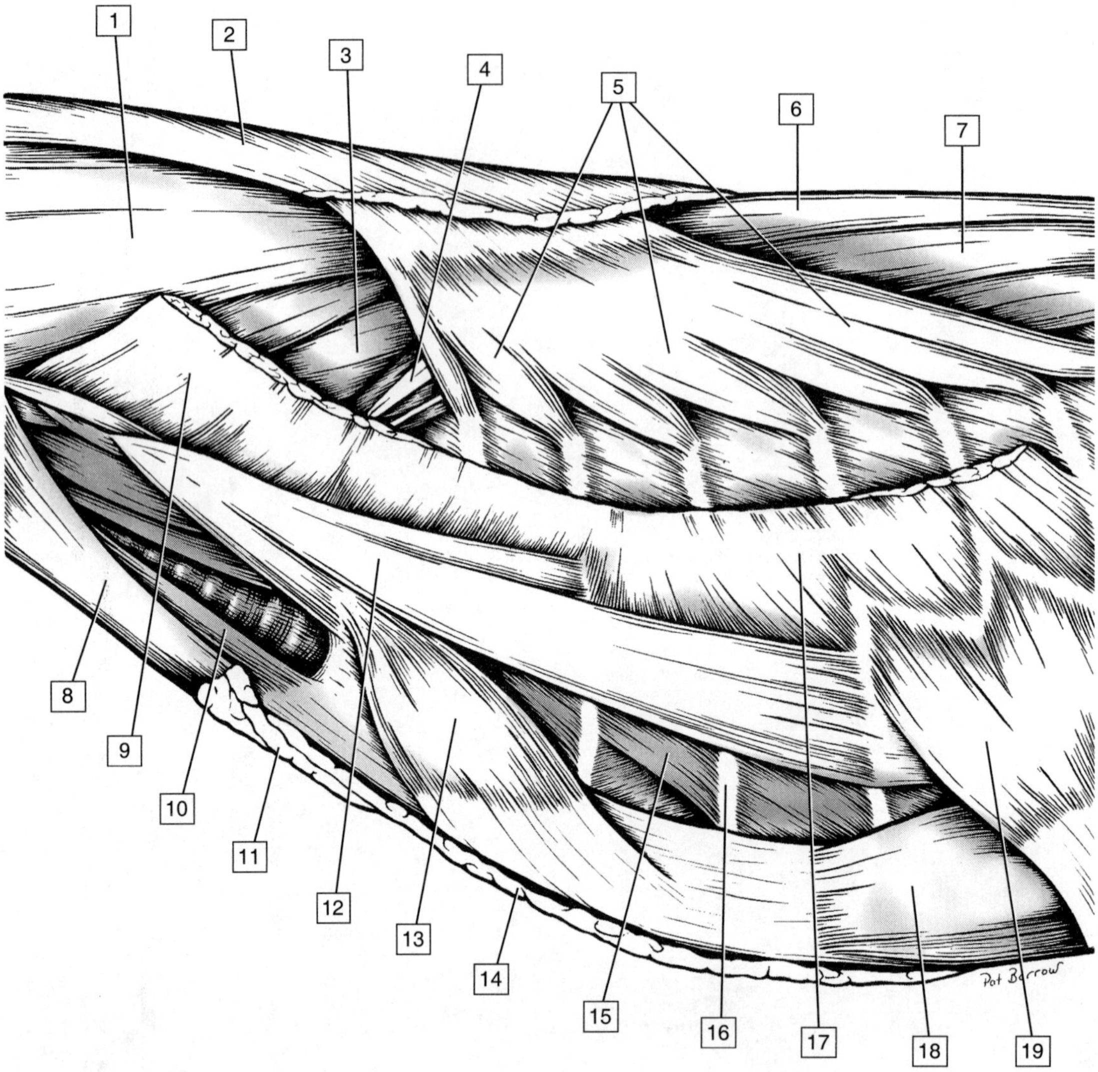

1	夹肌 Splenius	11	胸浅肌 Superficial pectoral
2	菱形肌 Rhomboideus	12	斜角肌 Scalenus
3	颈最长肌 Longissimus cervicis	13	胸直肌 Rectus thoracis
4	胸最长肌 Longissimus thoracis	14	胸深肌 Deep pectoral
5	背侧锯肌 Serratus dorsalis cranialist	15	肋间外肌 External intercostal muscle
6	棘肌和胸半棘肌 Spinalis and semispinalis thoracis	16	第4肋 Fourth rib
7	胸最长肌 Longissimus thoracis	17	腹侧锯肌（胸部） Serratus ventralis (thoracis)
8	胸头肌 Sternocephalicus	18	腹直肌 Rectus abdominis
9	腹侧锯肌（颈部） Serratus ventralis (cervicis)	19	腹外斜肌 External abdominal oblique
10	胸骨甲状肌 Sternothyroideus		

图3-15 犬颈部和胸部肌肉，外侧观

猫

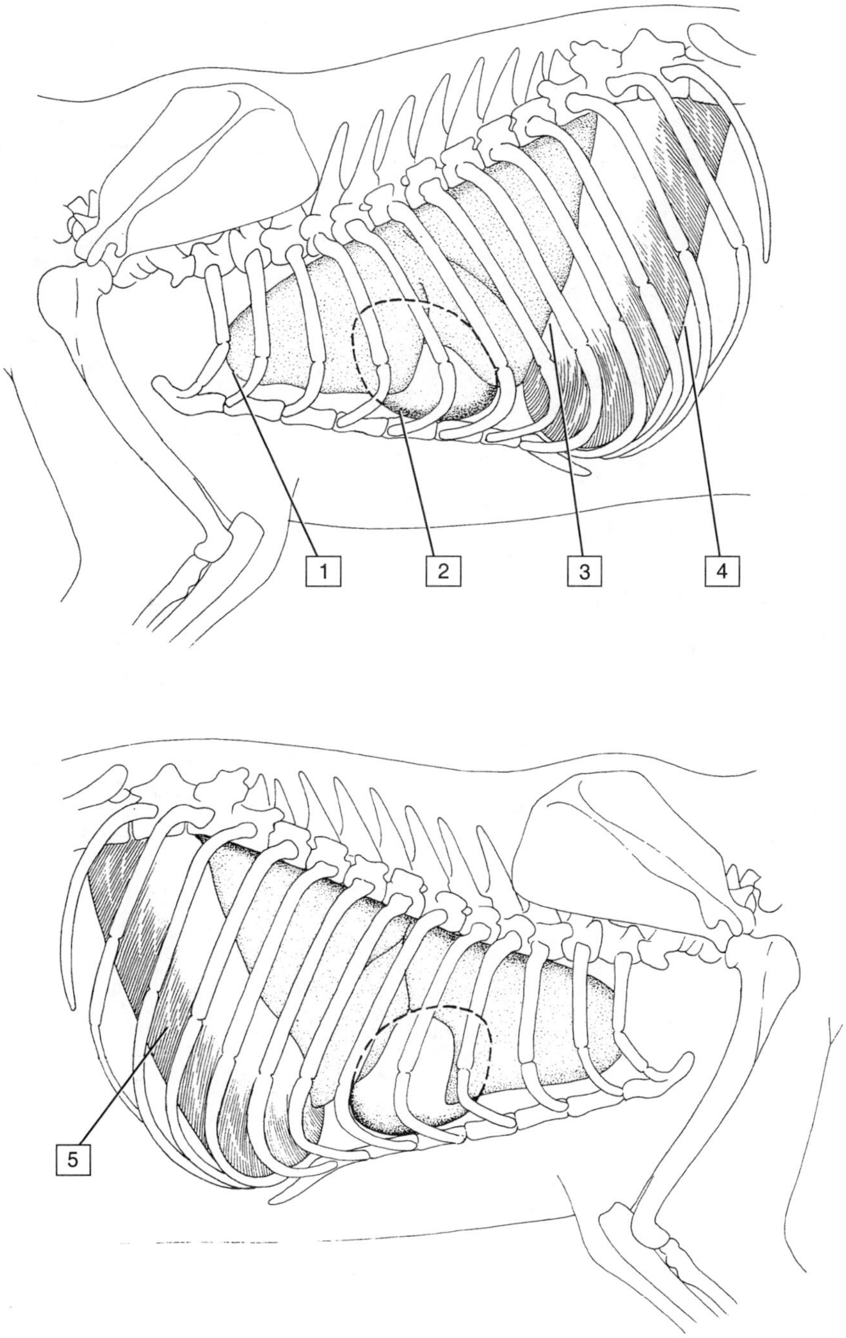

1 左肺尖 Apex of left lung

2 心脏 Heart

3 肺底缘 Basal border of lung

4 胸膜投影线 Line of pleural reflection

5 膈 Diaphragm

图3-16 猫心脏和肺的左、右侧体表投影

请在图中画出肺动脉瓣和左、右房室瓣的体表投影。

马

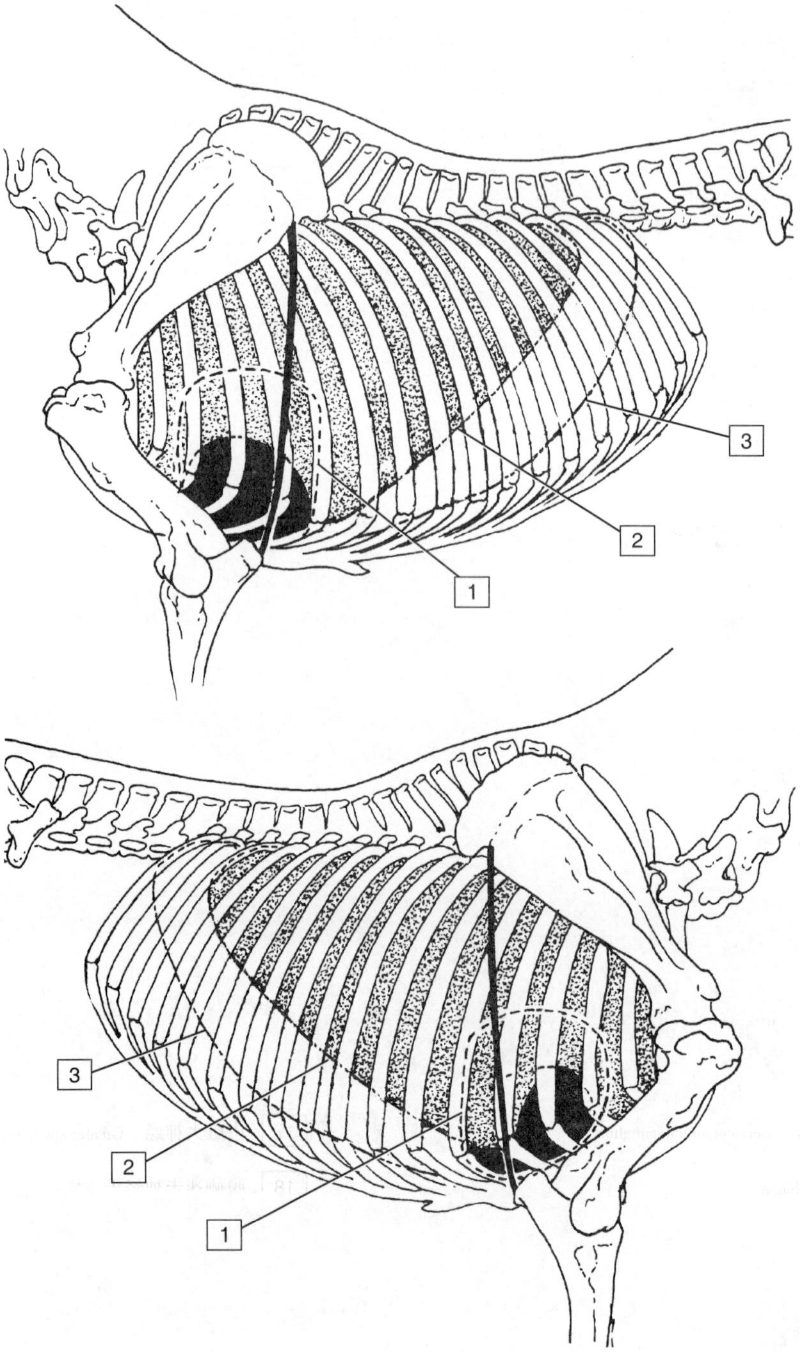

1 心脏轮廓 Outline of heart

2 肺底缘 Basal border of lung

3 胸膜投影线 Line of pleural reflection

图3-17 马心脏和肺在左、右胸壁的投影

粗线为臂三头肌的后缘

牛

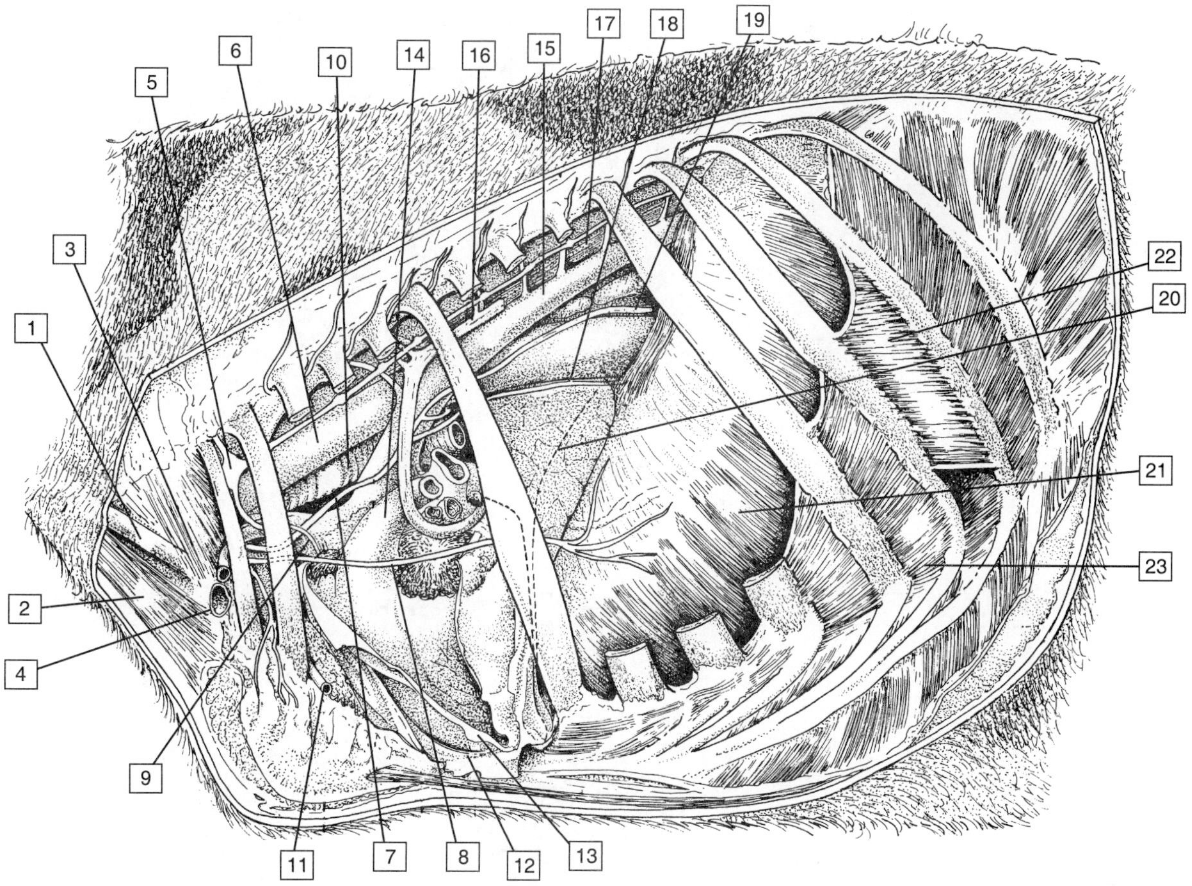

1	颈外静脉 External jugular vein	13	心包膜，已掀开 Pericardium, reflected
2	胸头肌 Sternocephalicus	14	肺动脉干 Pulmonary trunk
3	腋动脉 Axillary artery	15	主动脉 Aorta
4	腋静脉 Axillary vein	16	左奇静脉 Left azygous vein
5	颈胸神经节 Cervicothoracic ganglion	17	内脏大神经 Greater splanchnic nerve
6	食管 Esophagus	18	腹侧迷走神经干 Ventral vagal trunk
7	迷走神经 Vagus nerve	19	背侧迷走神经干 Dorsal vagal trunk
8	膈神经 Phrenic nerve	20	膈前界 Cranial extent of diaphragm
9	一条心脏神经 One of the cardiac nerves	21	膈 Diaphragm
10	气管 Trachea	22	肋间内肌 Internal intercostal muscle
11	胸廓内动脉 Internal thoracic artery	23	肋间外肌 External intercostal muscle
12	纵隔胸膜 Mediastinal pleura		

图3-18 牛胸腔，左外侧观

左肺和部分胸膜已被剥离

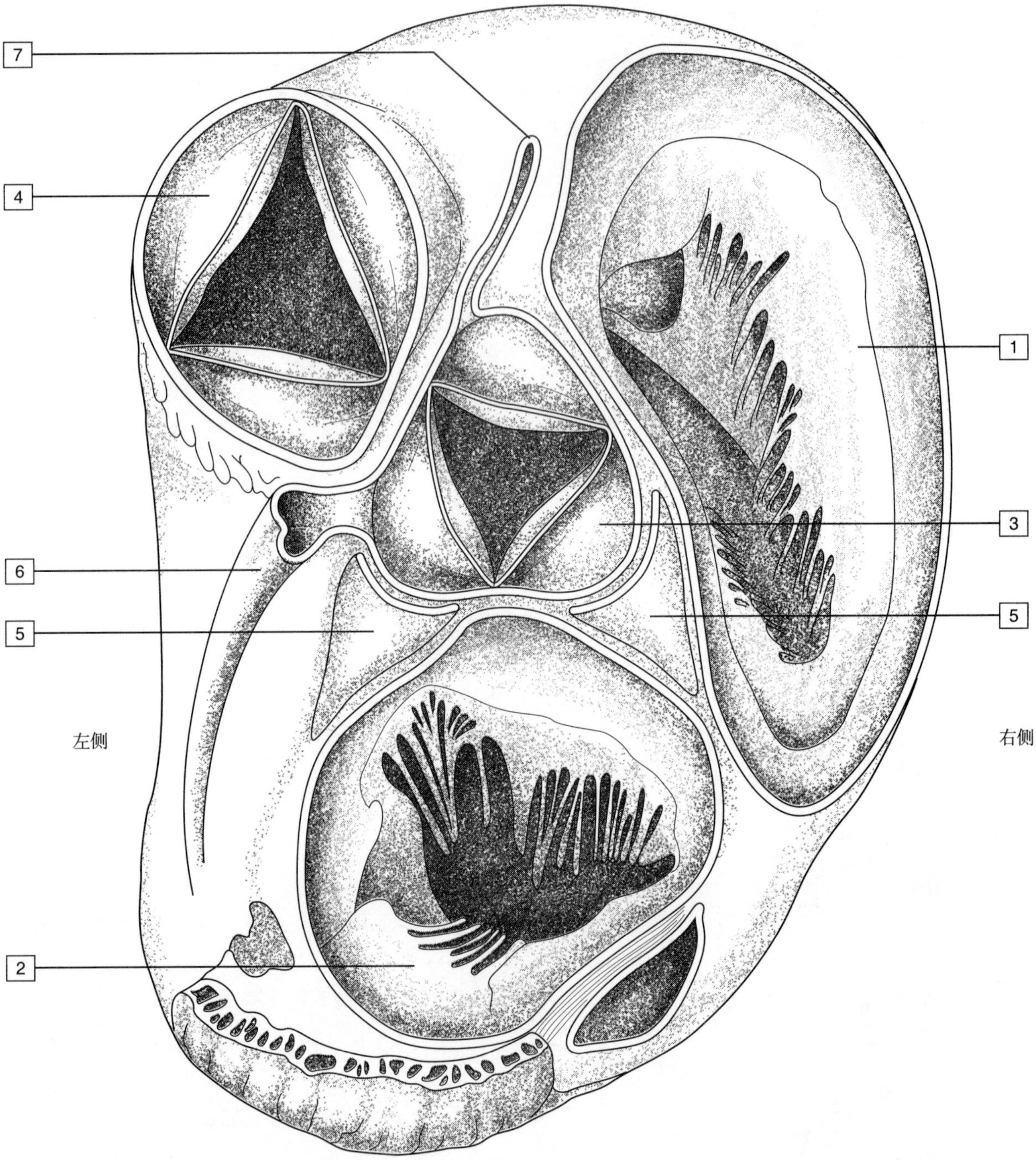

1	右房室瓣 Right atrioventricular valve	5	心骨 Ossa cordis
2	左房室瓣 Left atrioventricular valve	6	左冠状动脉 Left coronary artery
3	主动脉瓣 Aortic valve	7	右冠状动脉 Right coronary artery
4	肺动脉瓣 Pulmonary valve		

图3-19 牛心脏的基部，背侧观

已移除心房

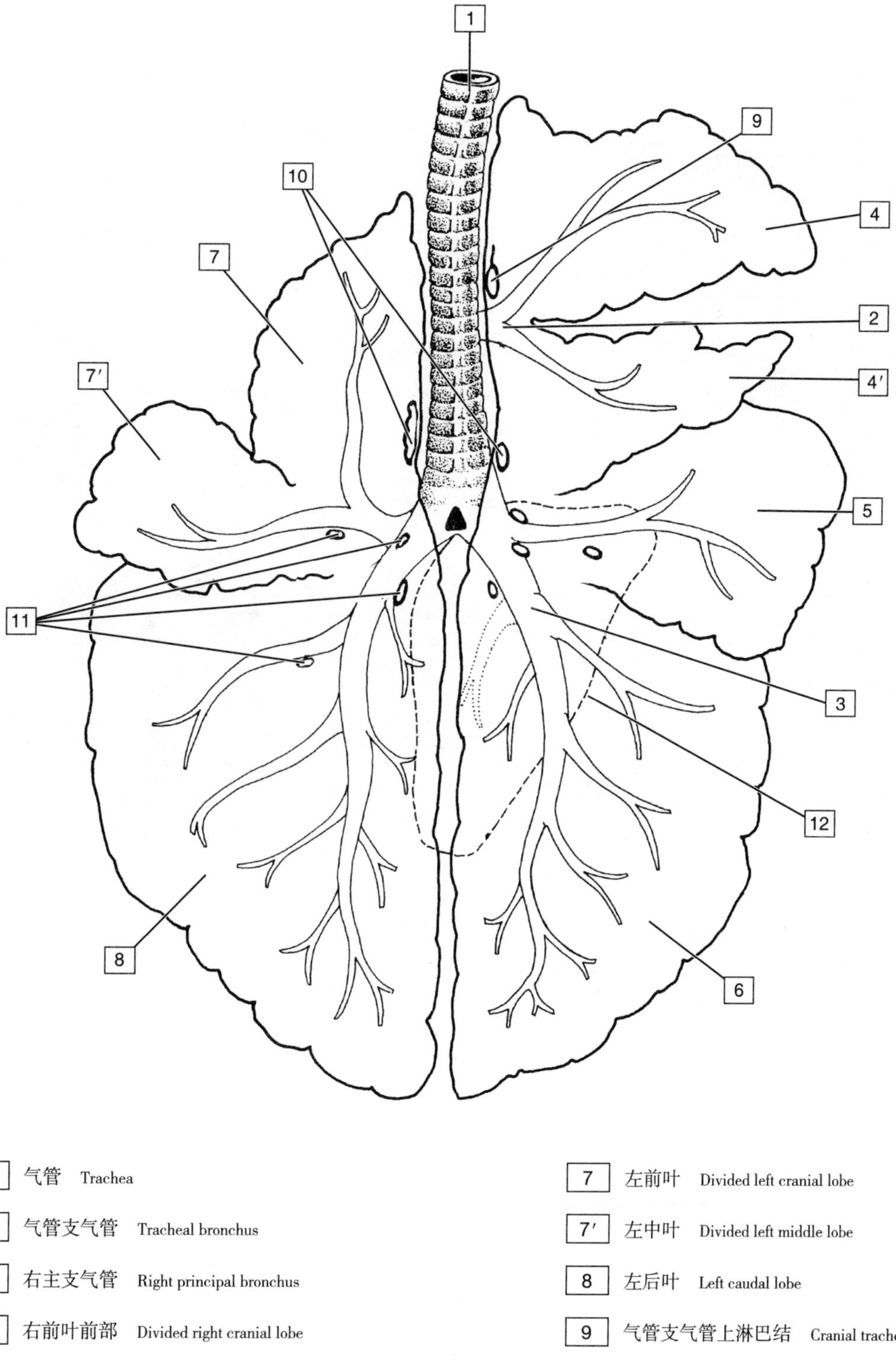

1	气管 Trachea	7	左前叶 Divided left cranial lobe
2	气管支气管 Tracheal bronchus	7′	左中叶 Divided left middle lobe
3	右主支气管 Right principal bronchus	8	左后叶 Left caudal lobe
4	右前叶前部 Divided right cranial lobe	9	气管支气管上淋巴结 Cranial tracheobronchial lymph node
4′	右前叶前部 Divided right cranial lobe	10	气管支气管淋巴结 Tracheobronchial lymph nodes
5	中叶 Middle lobe	11	肺淋巴结 Pulmonary lymph nodes
6	右后叶 Right caudal lobe	12	右肺副叶轮廓 Outline of accessory lobe of right lung

图3-20 牛肺叶及支气管树模式图，背侧观

猪

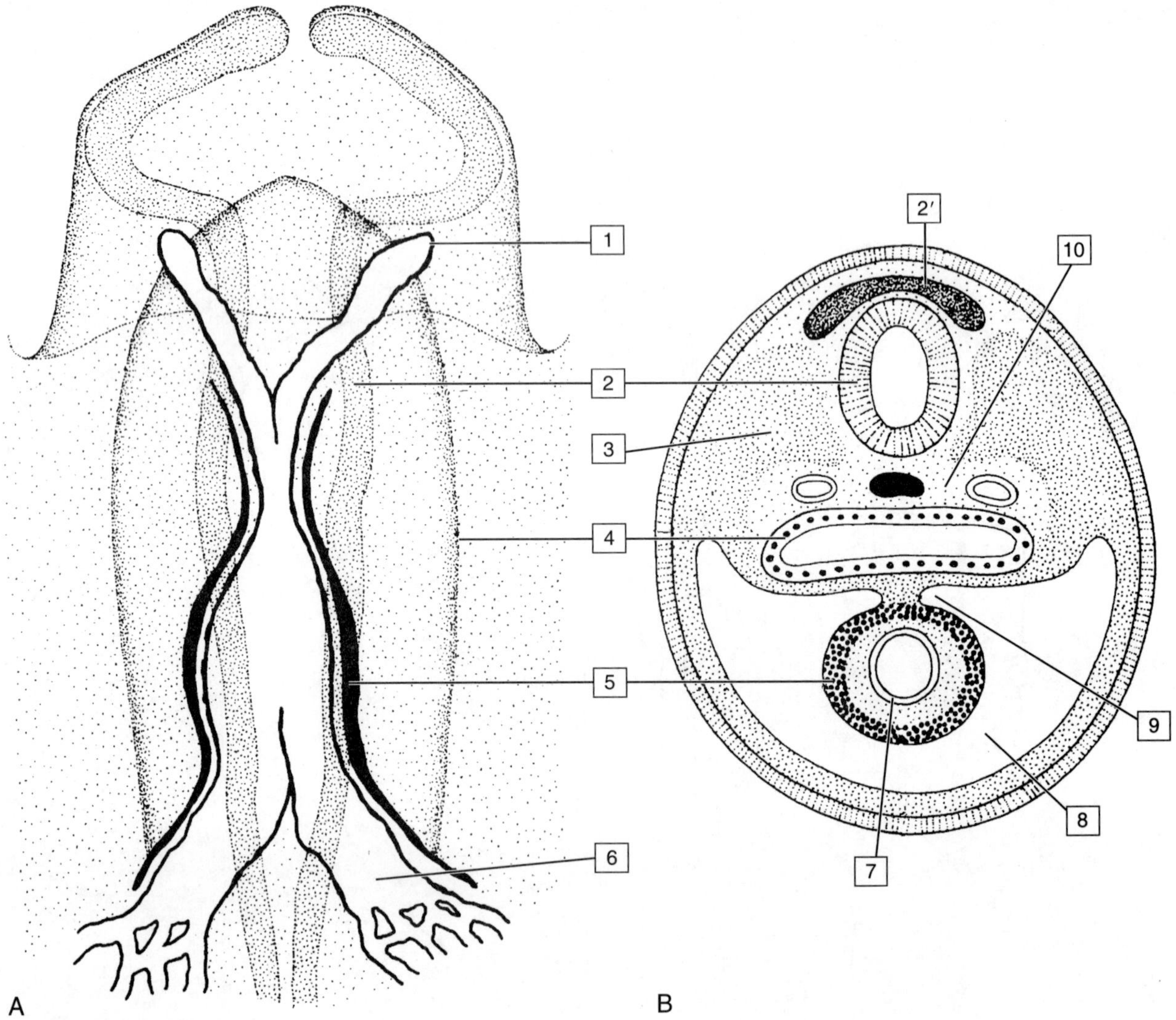

1	第1主动脉弓 First aortic arch	6	卵黄静脉 Vitelline vein
2	神经管 Neural tube	7	心内膜管 Endocardial tube
2′	神经嵴 Neural crest	8	心包腔 Pericardial cavity
3	体节 Somite	9	背侧心肌膜 Dorsal mesocardium
4	前肠 Foregut	10	脊索和背侧主动脉 Notochord and dorsal aortae
5	融合的心内膜管的心外膜壁 Epimyocardial wall of the fused endocardial tubes		

图3-21 心内膜管融合后的猪胚胎

A. 心内膜管融合后15日龄的猪胚胎前部（腹侧观）；B. 在图A中5水平上的7~8体节猪胚胎横断面

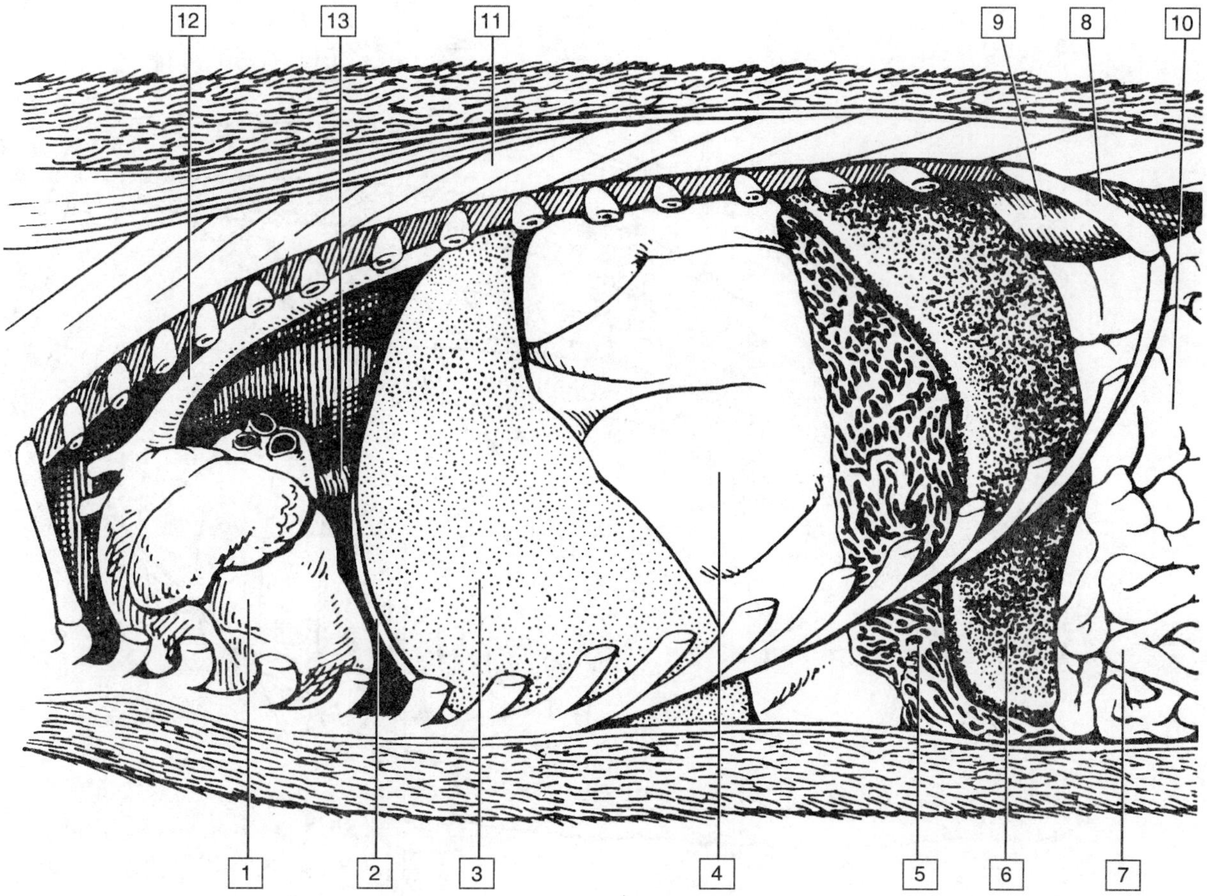

1 心脏 Heart	8 最后肋 Last rib
2 膈 Diaphragm	9 左肾 Left kidney
3 肝左叶 Left lobe of liver	10 升结肠 Ascending colon
4 膨大的胃 Stomach, greatly dilated	11 背部肌肉 Back muscles
5 大网膜（胃脾韧带） Greater omentum, gastrosplenic ligament	12 主动脉 Aorta
6 脾 Spleen	13 后腔静脉 Caudal vena cava
7 空肠 Jejunum	

图3-22 猪心脏原位图

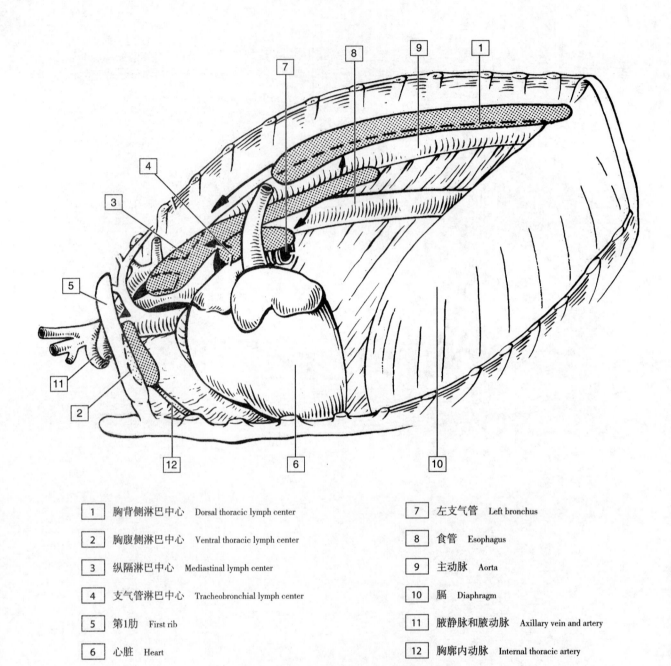

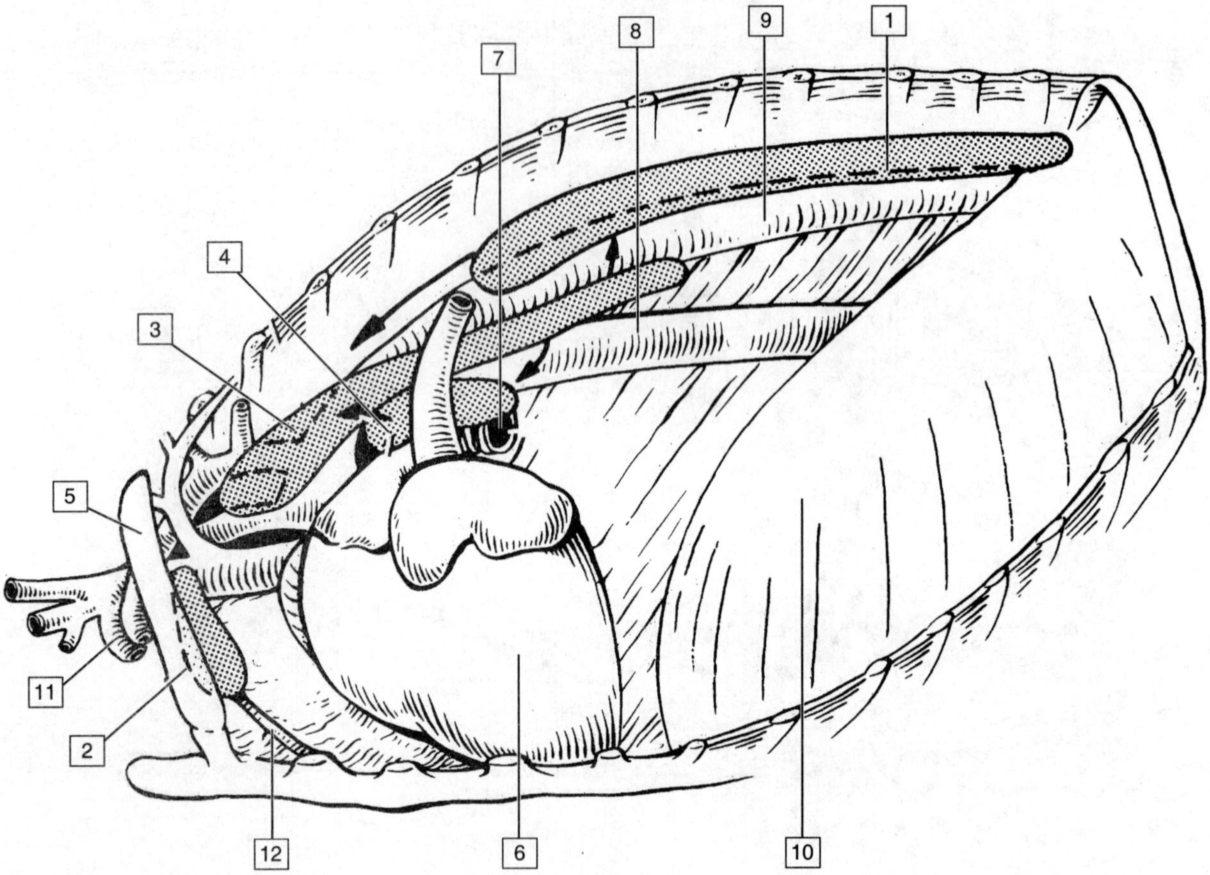

1	胸背侧淋巴中心 Dorsal thoracic lymph center	7	左支气管 Left bronchus
2	胸腹侧淋巴中心 Ventral thoracic lymph center	8	食管 Esophagus
3	纵隔淋巴中心 Mediastinal lymph center	9	主动脉 Aorta
4	支气管淋巴中心 Tracheobronchial lymph center	10	膈 Diaphragm
5	第1肋 First rib	11	腋静脉和腋动脉 Axillary vein and artery
6	心脏 Heart	12	胸廓内动脉 Internal thoracic artery

图3-23 猪胸腔淋巴中心，左外侧观

鸟类

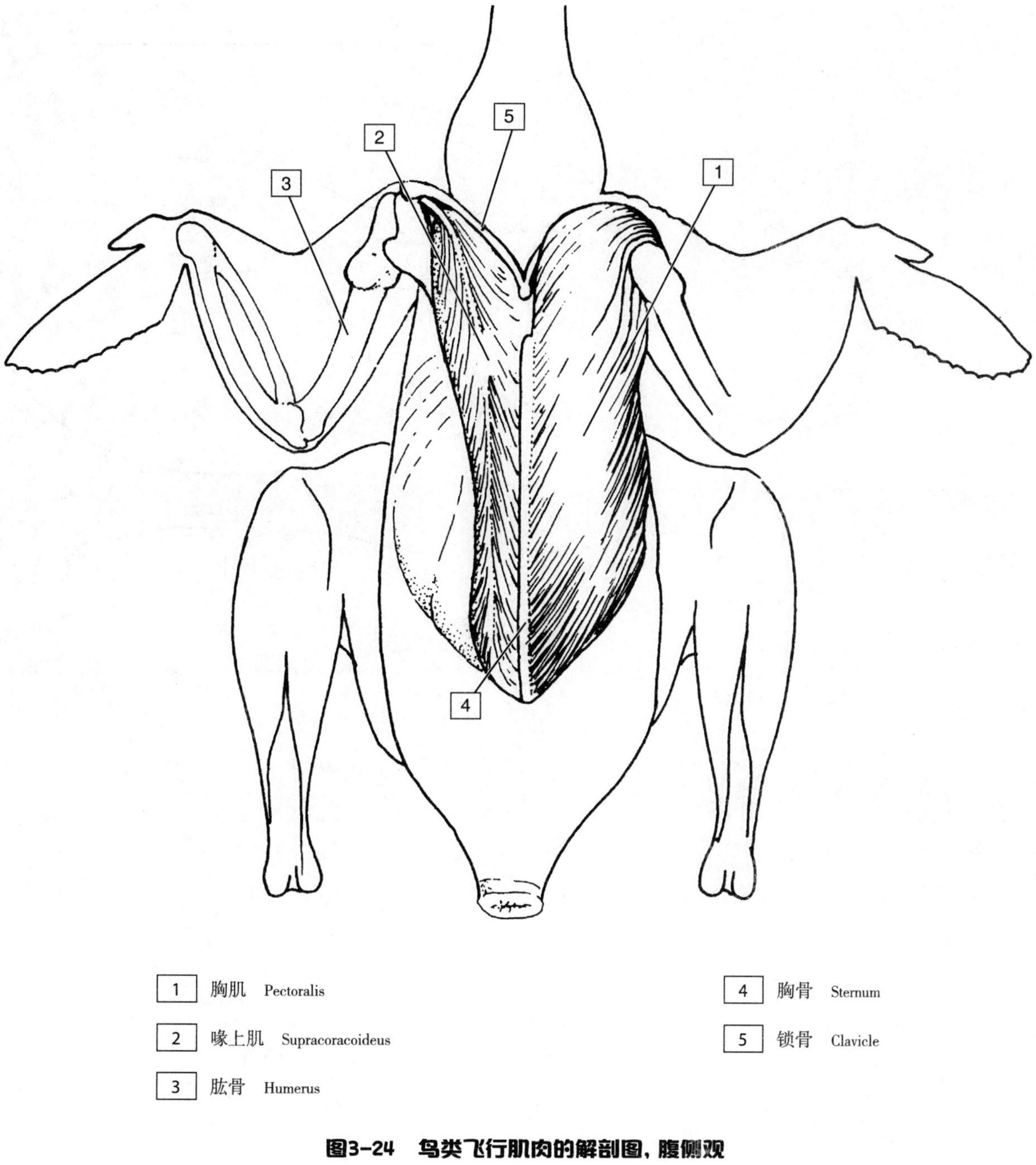

1 胸肌 Pectoralis

2 喙上肌 Supracoracoideus

3 肱骨 Humerus

4 胸骨 Sternum

5 锁骨 Clavicle

图3-24 鸟类飞行肌肉的解剖图，腹侧观

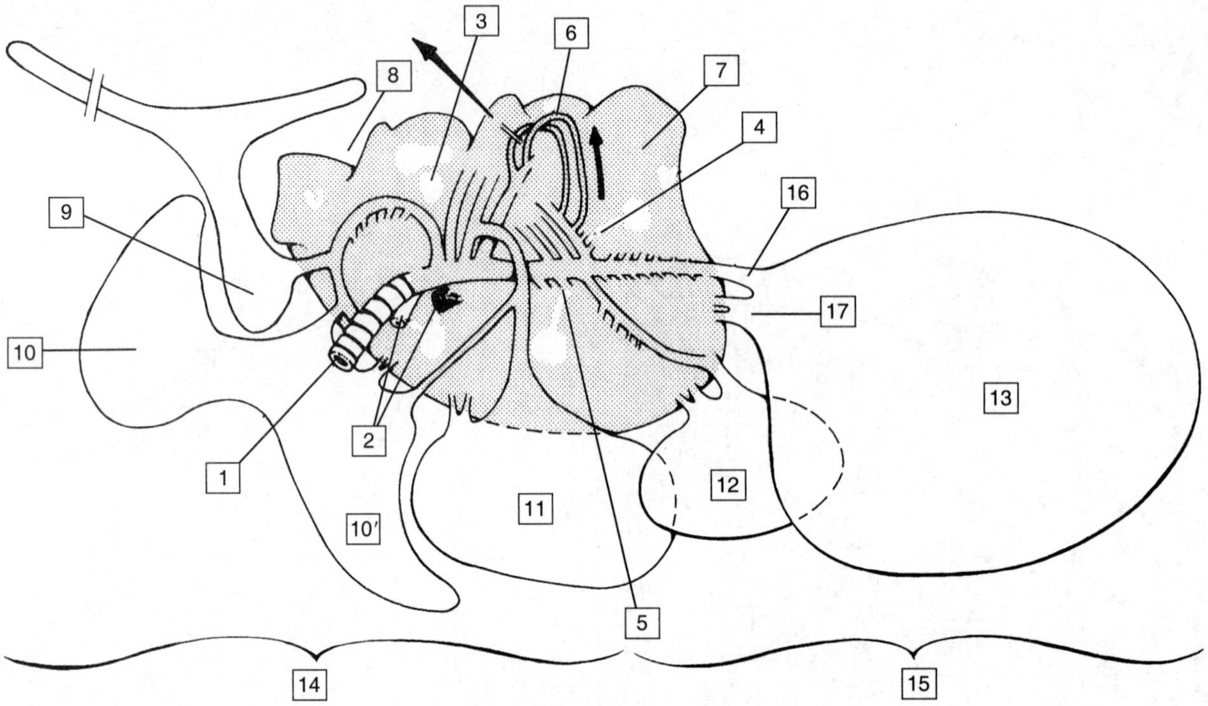

1 初级支气管 Primary bronchus

2 肺门处血管 Pulmonary vessels at hilus

3 腹中线的支气管 Medioventral bronchi

4 背中部的支气管 Mediodorsal bronchi

5 腹外侧支气管 Lateroventral bronchi

6 侧支气管环 Loops of parabronchi

7 肺 Lung

8 肋压迹 Indentations caused by ribs

9 颈气囊 Cervical air sac

10 胸外和胸内锁骨间气囊 Extrathoracic and intrathoracic parts of clavicular air sac

10′ 胸外和胸内锁骨间气囊 Extrathoracic and intrathoracic parts of clavicular air sac

11 胸前气囊 Cranial thoracic air sac

12 胸后气囊 Caudal thoracic air sac

13 腹气囊 Abdominal air sac

14 前气囊 Cranial air sacs

15 后气囊 Caudal air sacs

16 直接连接（气囊和支气管） Direct (saccobronchial) connection

17 间接连接气囊和肺（通过支气管） Indirect (recurrent bronchial) connection of air sac to lung

图3-25 鸟类右肺（腹中线观）及相关气囊

肺内结构已简化

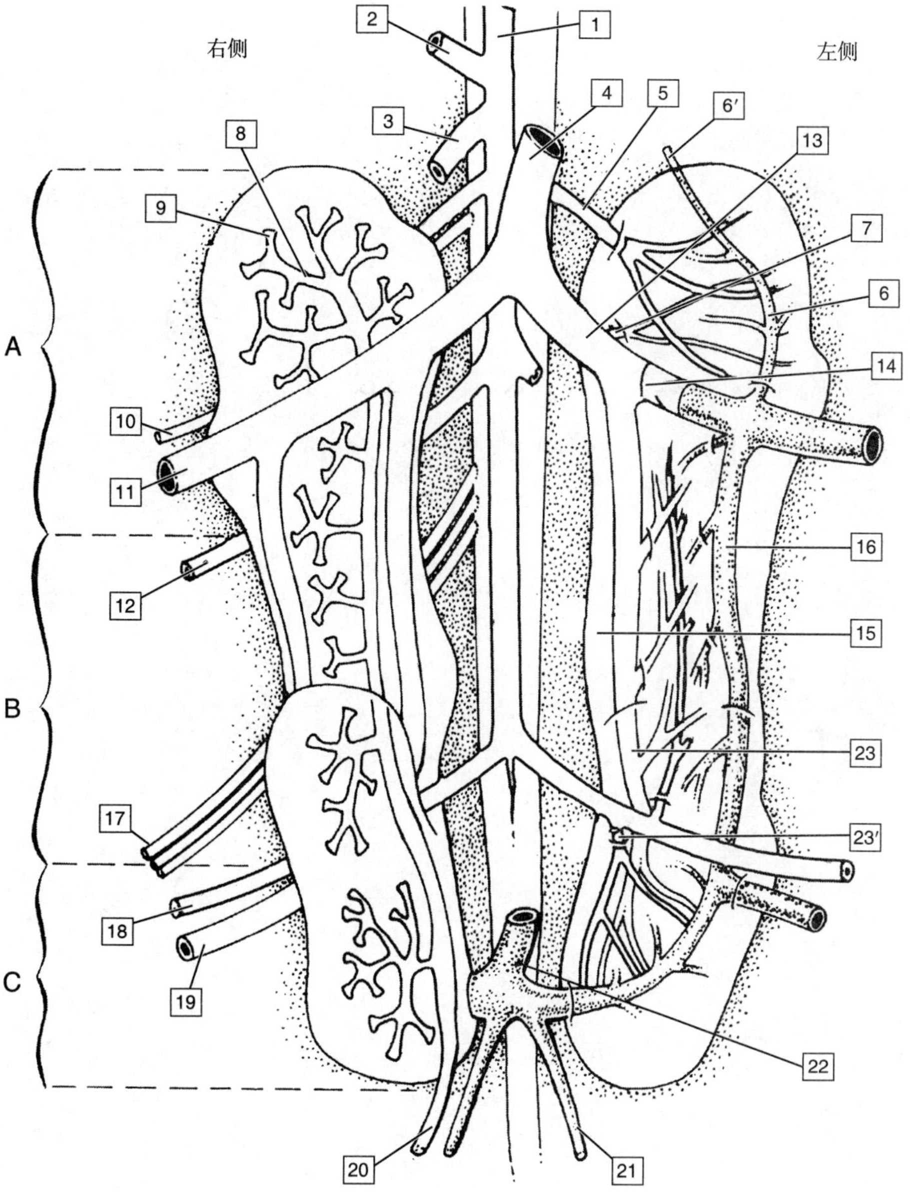

1	主动脉	Aorta
2	腹腔动脉	Celiac artery
3	肠系膜前动脉	Cranial mesenteric artery
4	后腔静脉	Caudal vena cava
5	肾前动脉	Cranial renal artery
6	肾门前静脉	Cranial renal portal vein
6′	椎静脉窦吻合支	Anastomosis with vertebral venous sinus
7	肾前静脉	Cranial renal vein
8	输尿管初级分支	Primary branch of ureter
9	输尿管次级分支	Secondary branch of ureter
10	股神经	Femoral nerve
11	髂外静脉	External iliac vein
12	髂外动脉	External iliac artery
13	髂总静脉	Common iliac vein
14	肾门静脉瓣	Portal valve
15	肾后静脉	Caudal renal vein
16	肾门后静脉	Caudal renal portal vein
17	荐神经	Sciatic nerve
18	坐骨动脉	Ischial artery
19	坐骨静脉	Ischial vein
20	输尿管	Ureter
21	髂内静脉	Internal iliac vein
22	肠系膜后静脉	Caudal mesenteric vein
23	肾中动脉	Middle renal artery
23′	肾后动脉	Caudal renal artery

图3-26 鸟类肾脏及其邻近的血管和神经，腹侧观

肾脏的前部（A），中部（B），后部（C）

4

腹部

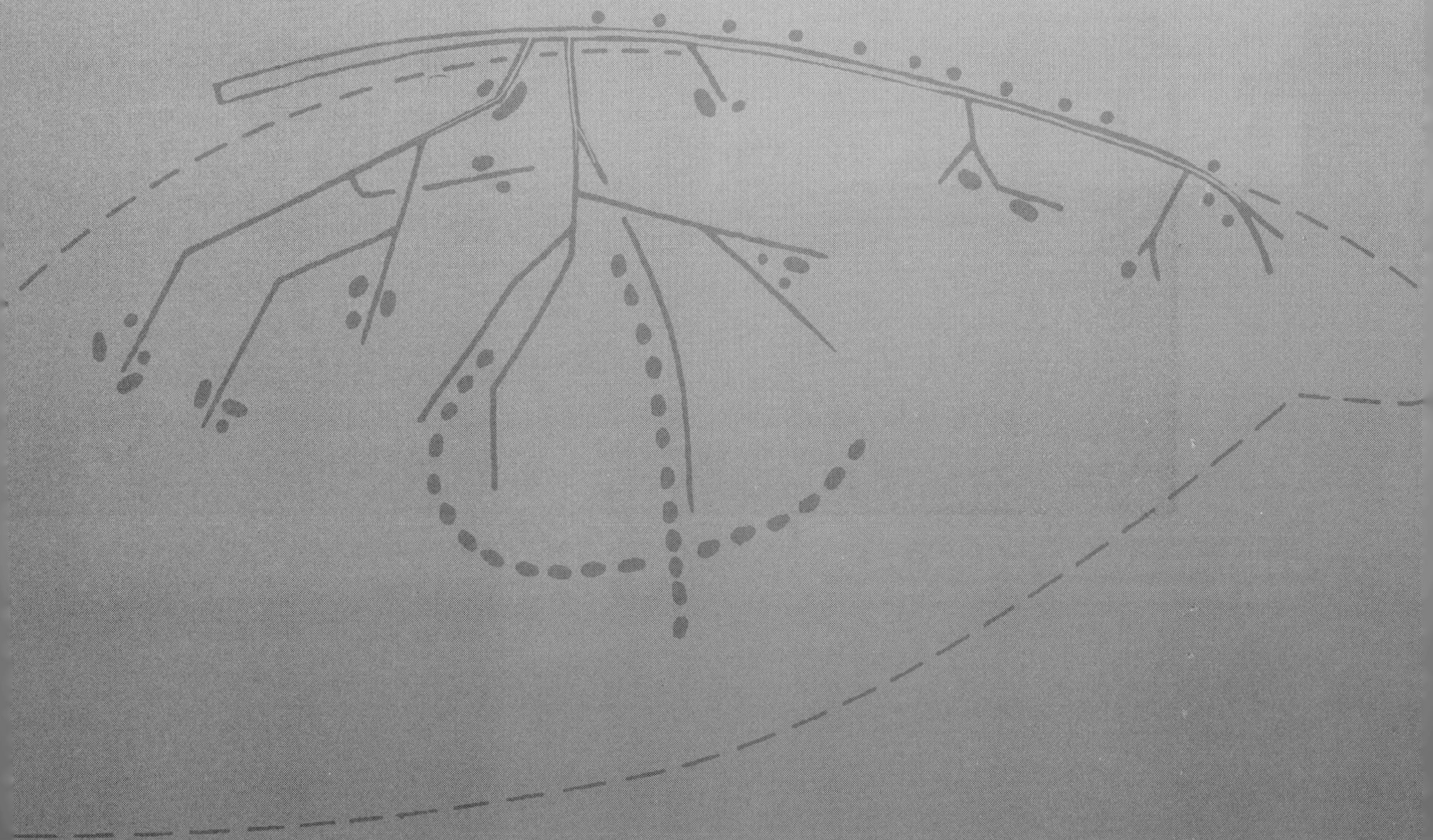

犬

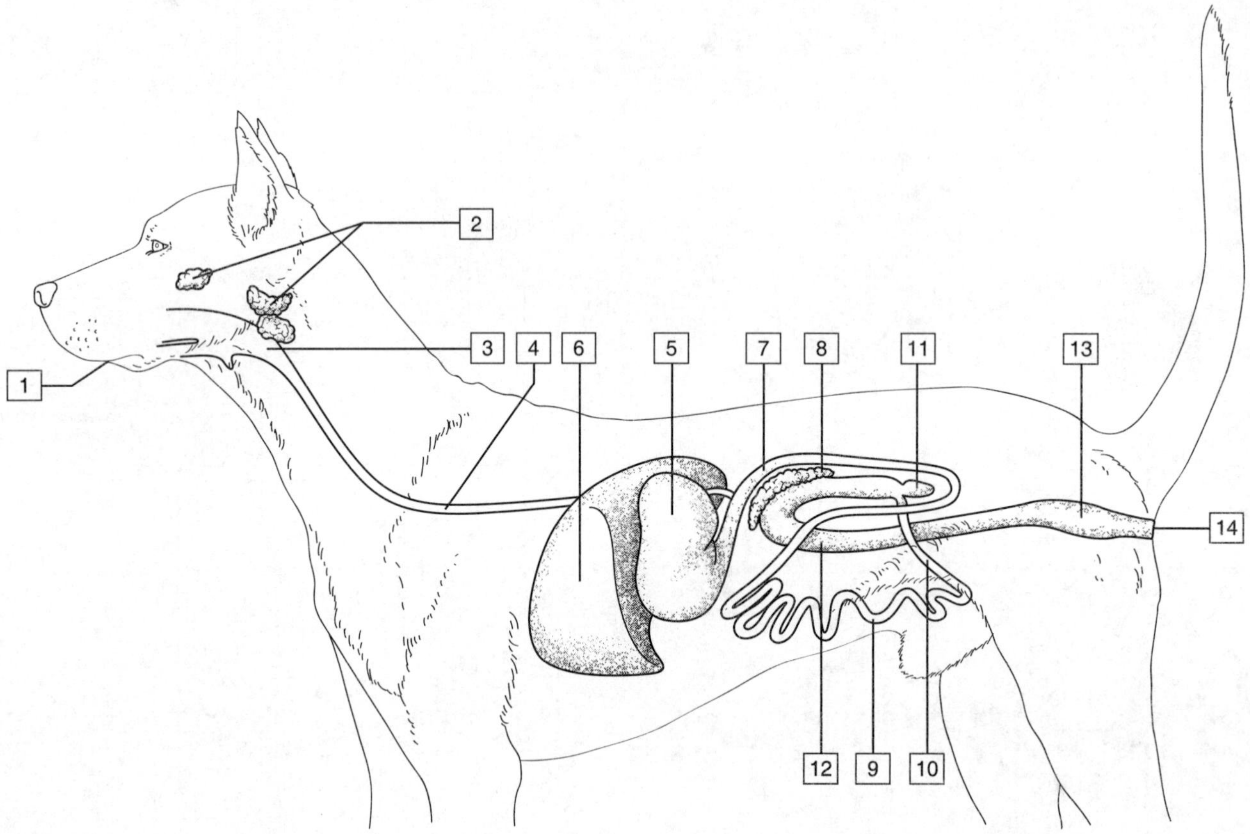

1	口腔 Mouth	5	胃 Stomach	9	空肠 Jejunum	13	直肠 Rectum
2	唾液腺 Salivary glands	6	肝脏 Liver	10	回肠 Ileum	14	肛门 Anus
3	咽 Pharynx	7	十二指肠 Duodenum	11	盲肠 Cecum		
4	食管 Esophagus	8	胰腺 Pancreas	12	结肠 Colon		

图4-1 犬消化器官示意图

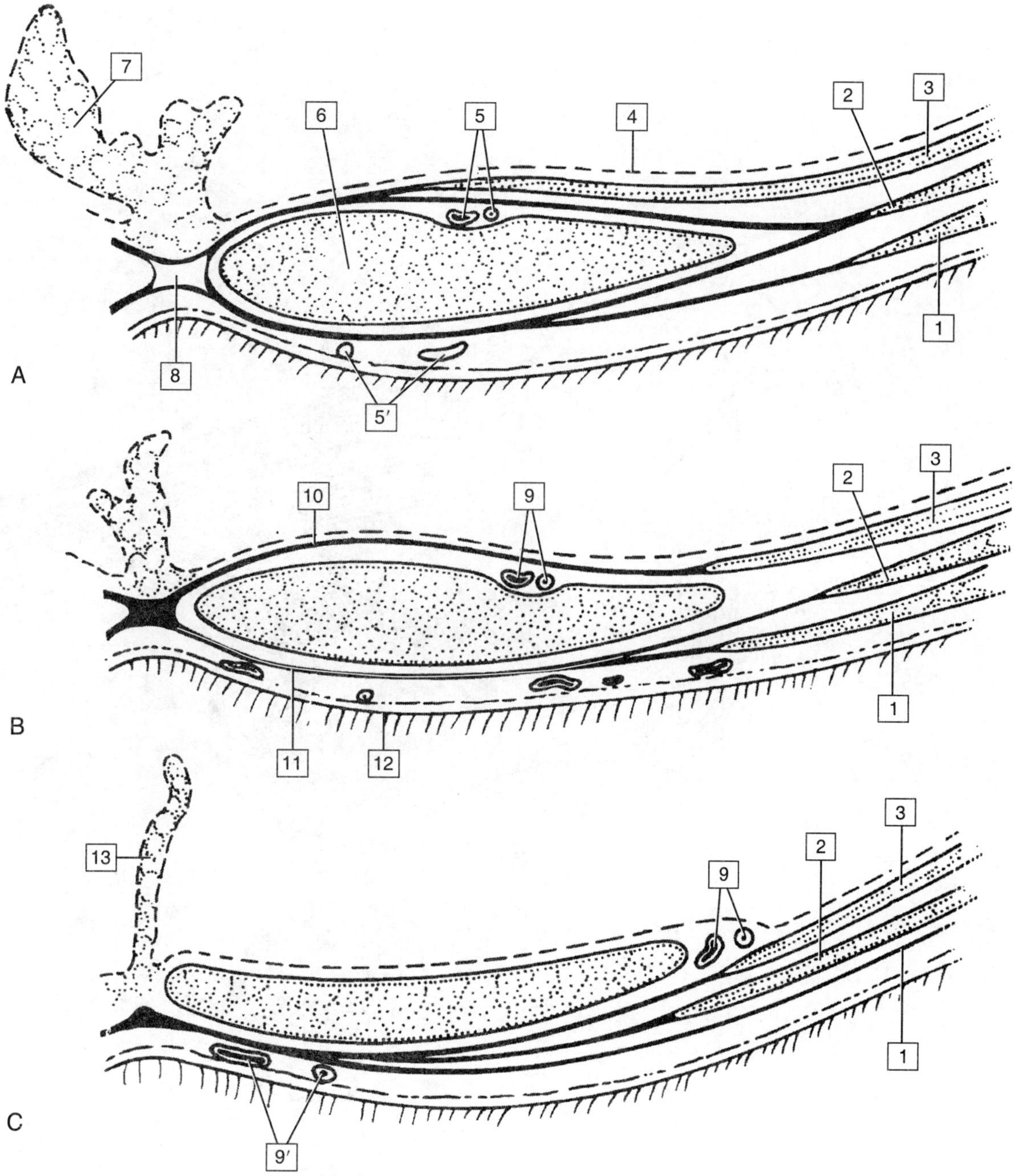

1 腹外斜肌 External abdominal oblique

2 腹内斜肌 Internal abdominal oblique

3 腹横肌 Transversus abdominis

4 腹膜 Peritoneum

5 腹壁前动、静脉 Cranial epigastric vessels

5′ 腹壁前浅动、静脉 Cranial superficial epigastric vessels

6 腹直肌 Rectus abdominis

7 充满脂肪的镰状韧带 Fat-filled falciform ligament

8 腹白线 Linea alba

9 腹壁后动、静脉 Caudal epigastric vessels

9′ 腹壁后浅动、静脉 Caudal superficial epigastric vessels

10 腹直肌鞘内板 Internal lamina of rectus sheath

11 腹直肌鞘外板 External lamina of rectus sheath

12 皮肤 Skin

13 膀胱正中韧带 Median ligament of the bladder

图4-2 犬腹直肌鞘横断面

A. 前部；B. 后部到脐部；C. 近耻骨部

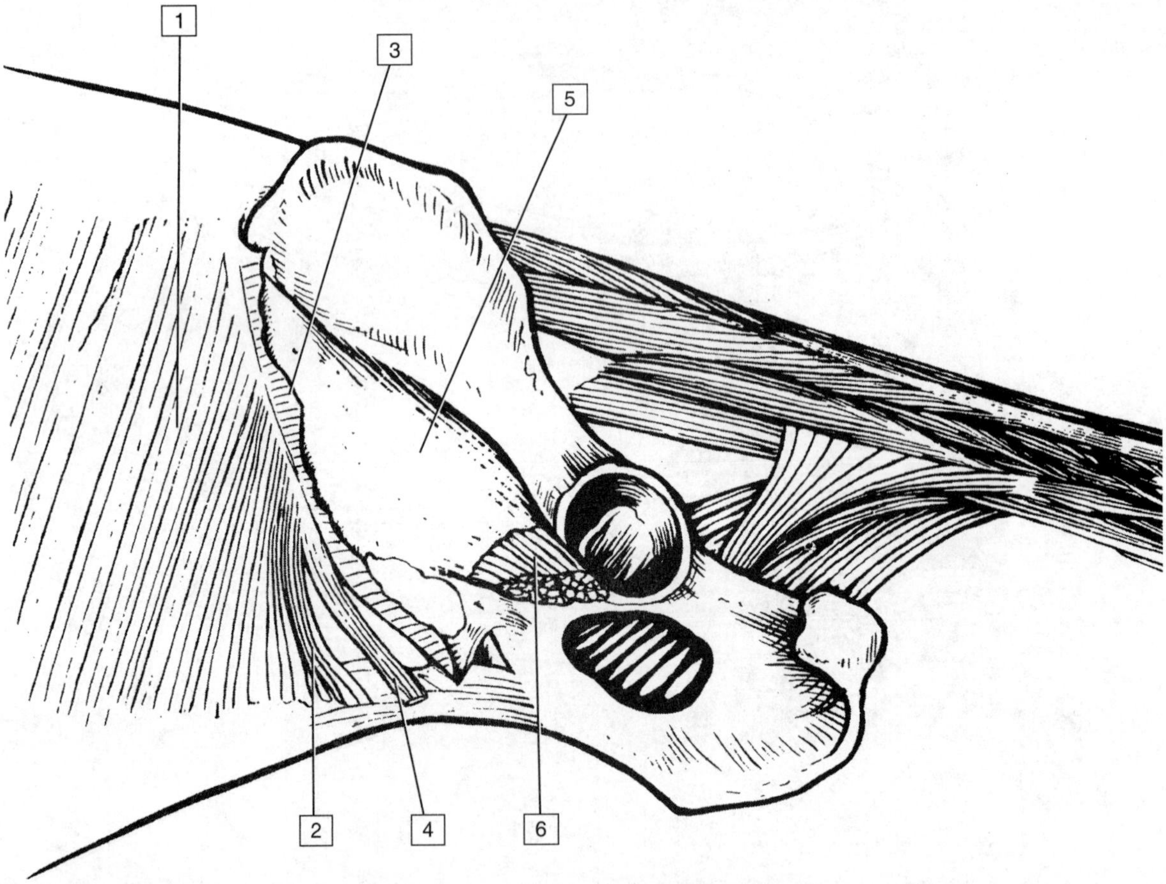

1 腹内斜肌 Internal abdominal oblique

2 游离的腹内斜肌后缘，形成腹股沟深环的边界
Free caudal edge of internal oblique, forming border of deep inguinal ring

3 腹外斜肌腱膜向后折转的断端
Stump of external oblique aponeurosis reflected caudally

4 从腹内斜肌分离的提睾肌
Cremaster muscle derived from internal oblique

5 覆盖在髂腰肌上的髂筋膜 Iliac fascia covering iliopsoas

6 髂腰肌 Iliopsoas

图4-3 犬腹股沟管及盆膈，左外侧观

腹外斜肌已切除

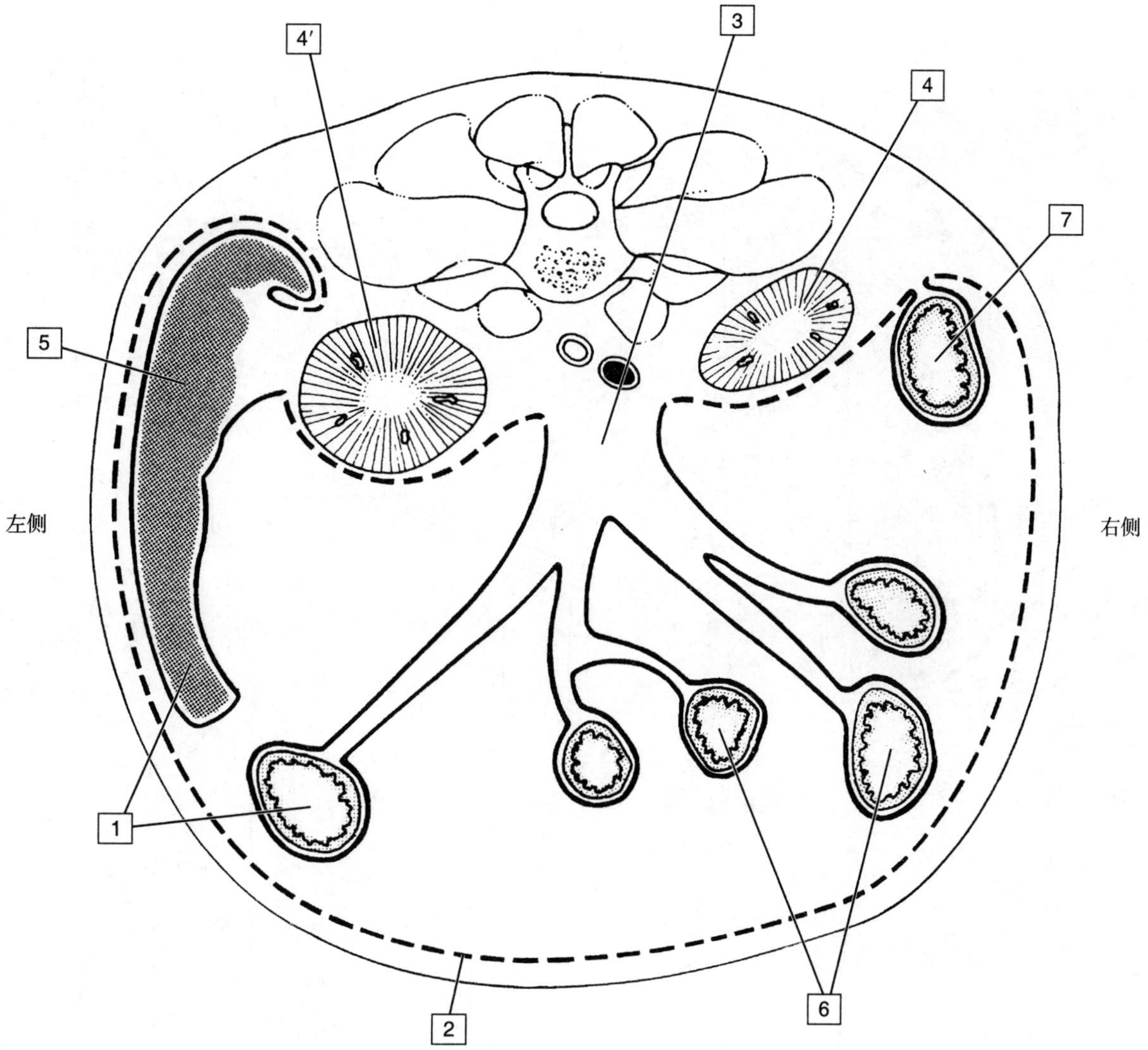

1 腹膜脏层（实线） Visceral peritoneum (continuous line)	4′ 左肾（腹膜外） Left kidney (retroperitoneal)
2 腹膜壁层（虚线） Parietal peritoneum (broken line)	5 脾 Spleen
3 肠系膜根部 Root of mesentery	6 空肠 Jejunum
4 右肾（腹膜外） Right kidney (retroperitoneal)	7 十二指肠降部 Descending duodenum

图4-4 犬腹部横断面示意图

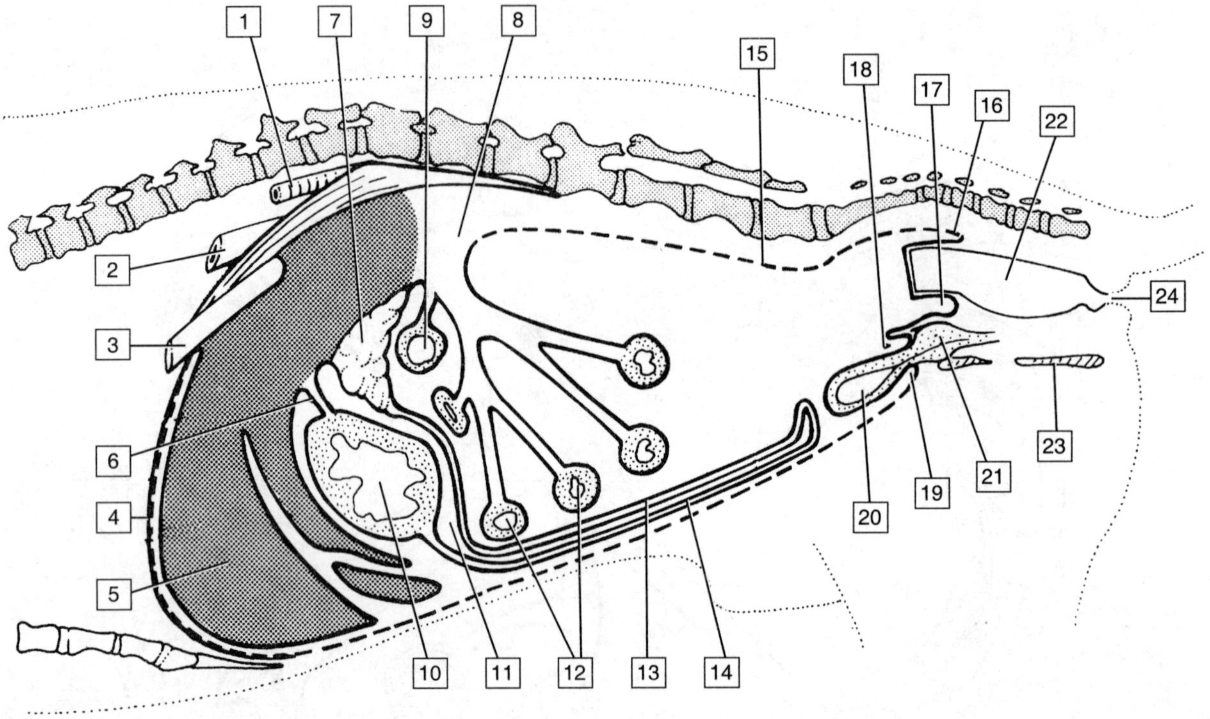

1	主动脉 Aorta	13	大网膜深层 Deep wall of greater omentum
2	食管 Esophagus	14	大网膜浅层 Superficial wall of greater omentum
3	后腔静脉 Caudal vena cava	15	腹膜壁层 Parietal peritoneum
4	膈 Diaphragm	16	直肠旁隐窝 Pararectal fossa
5	肝脏 Liver	17	直肠生殖腔 Rectogenital pouch
6	小网膜 Lesser omentum	18	膀胱生殖腔 Vesicogenital pouch
7	胰腺 Pancreas	19	耻骨膀胱腔 Pubovesical pouch
8	肠系膜根部 Root of mesentery	20	膀胱 Bladder
9	横结肠 Transverse colon	21	前列腺 Prostate
10	胃 Stomach	22	直肠 Rectum
11	网膜囊 Omental bursa	23	坐骨 Ischium
12	小肠 Small intestine	24	肛门 Anus

图4-5 犬腹腔靠近中央的切面，显示腹膜的分布

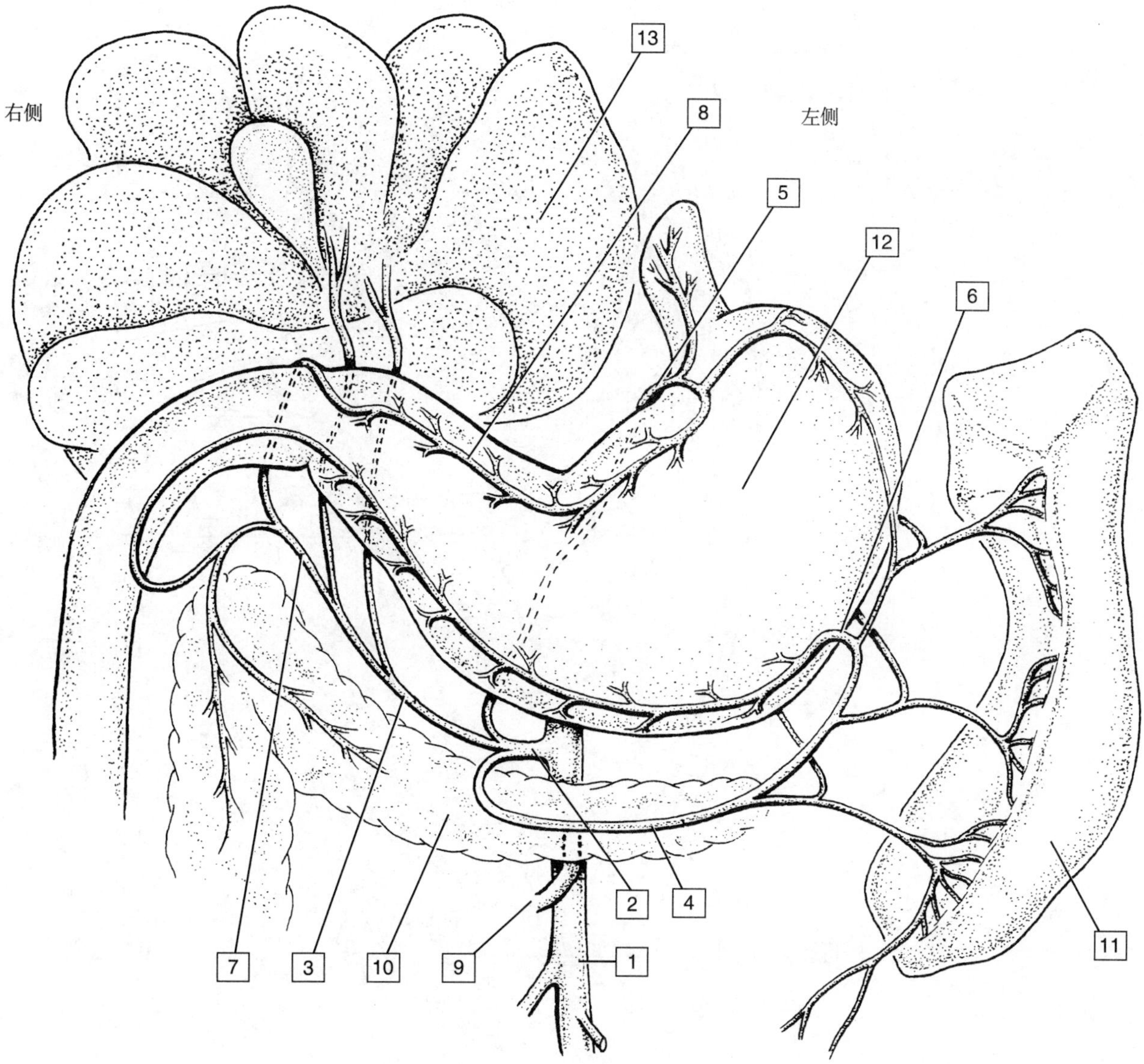

1	主动脉 Aorta	8	胃右动脉 Right gastric artery
2	腹腔动脉 Celiac artery	9	肠系膜前动脉 Cranial mesenteric artery
3	肝动脉 Hepatic artery	10	胰腺 Pancreas
4	脾动脉 Splenic artery	11	脾 Spleen
5	胃左动脉 Left gastric artery	12	胃 Stomach
6	胃网膜左动脉 Left gastroepiploic artery	13	肝脏 Liver
7	胃十二指肠动脉 Gastroduodenal artery		

图4-6 犬腹腔动脉分布，腹侧观

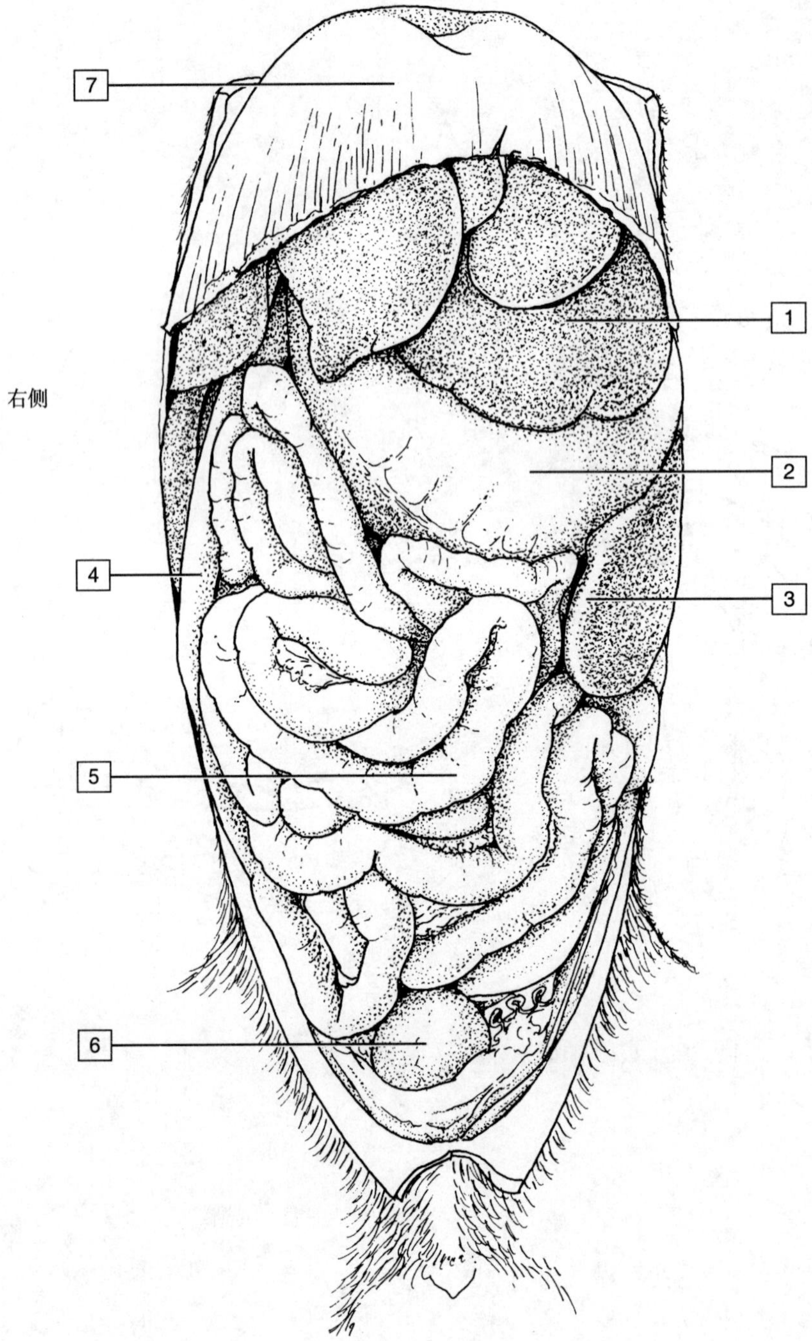

1 肝脏 Liver

2 胃 Stomach

3 脾 Spleen

4 十二指肠降部 Descending duodenum

5 空肠 Jejunum

6 膀胱 Bladder

7 膈 Diaphragm

图4-7 犬切除大网膜后的腹部器官，腹侧观

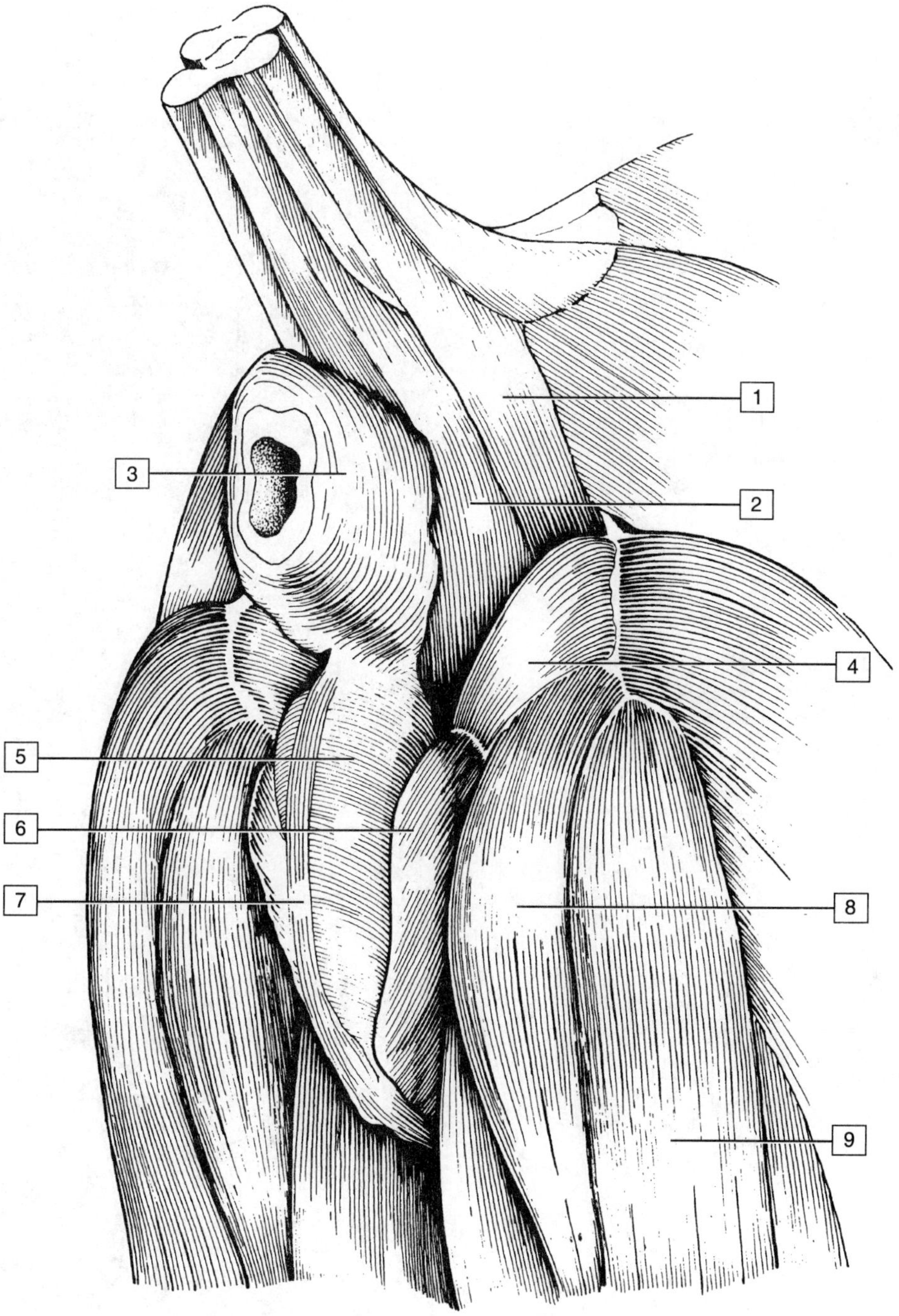

1	尾骨肌 Coccygeus	6	坐骨海绵体肌 Ischiocavernosus
2	肛提肌 Levator ani	7	阴茎牵缩肌 Retractor penis
3	肛门外括约肌 External anal sphincter	8	半膜肌 Semimembranosus
4	闭孔内肌 Internal obturator	9	半腱肌 Semitendinosus
5	球海绵体肌 Bulbospongiosus		

图4-8 犬会阴部的肌肉

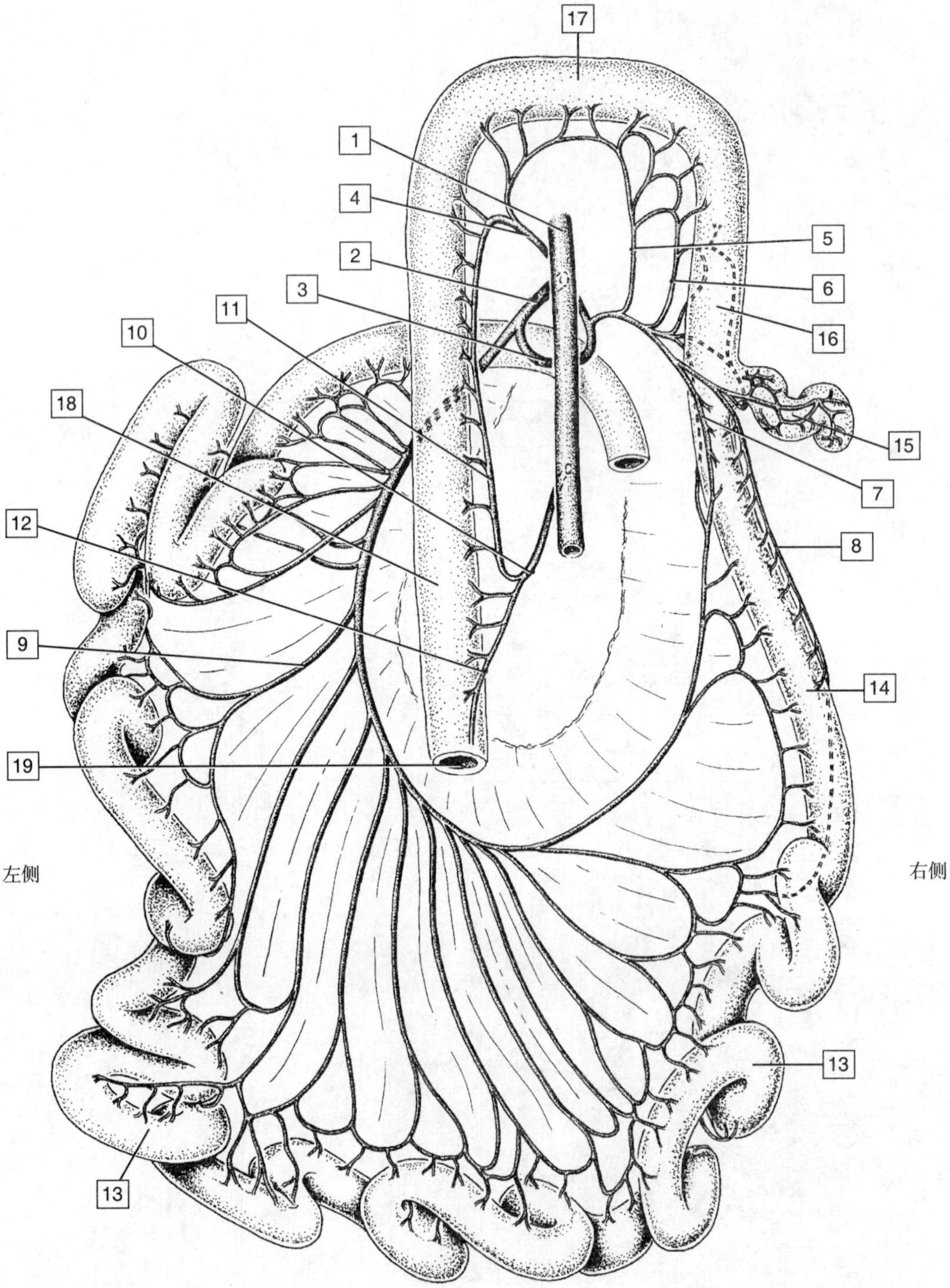

1	主动脉 Aorta
2	肠系膜前动脉 Cranial mesenteric artery
3	回结肠动脉 Ileocolic artery
4	结肠中动脉 Middle colic artery
5	右结肠动脉 Right colic artery
6	回结肠动脉结肠支 Colic branch of ileocolic artery
7	肠系膜回肠支 Mesenteric ileal branch
8	肠系膜小肠游离部回肠支 Antimesenteric ileal branch
9	空肠动脉 Jejunal aa.
10	肠系膜后动脉 Caudal mesenteric artery
11	左结肠动脉 Left colic artery
12	直肠前动脉 Cranial rectal artery
13	空肠 Jejunum
14	回肠 Ileum
15	盲肠 Cecum
16	升结肠 Ascending colon
17	横结肠 Transverse colon
18	降结肠 Descending colon
19	直肠 Rectum

图4-9 大肠系膜前动脉和肠系膜后动脉的分布，背侧观

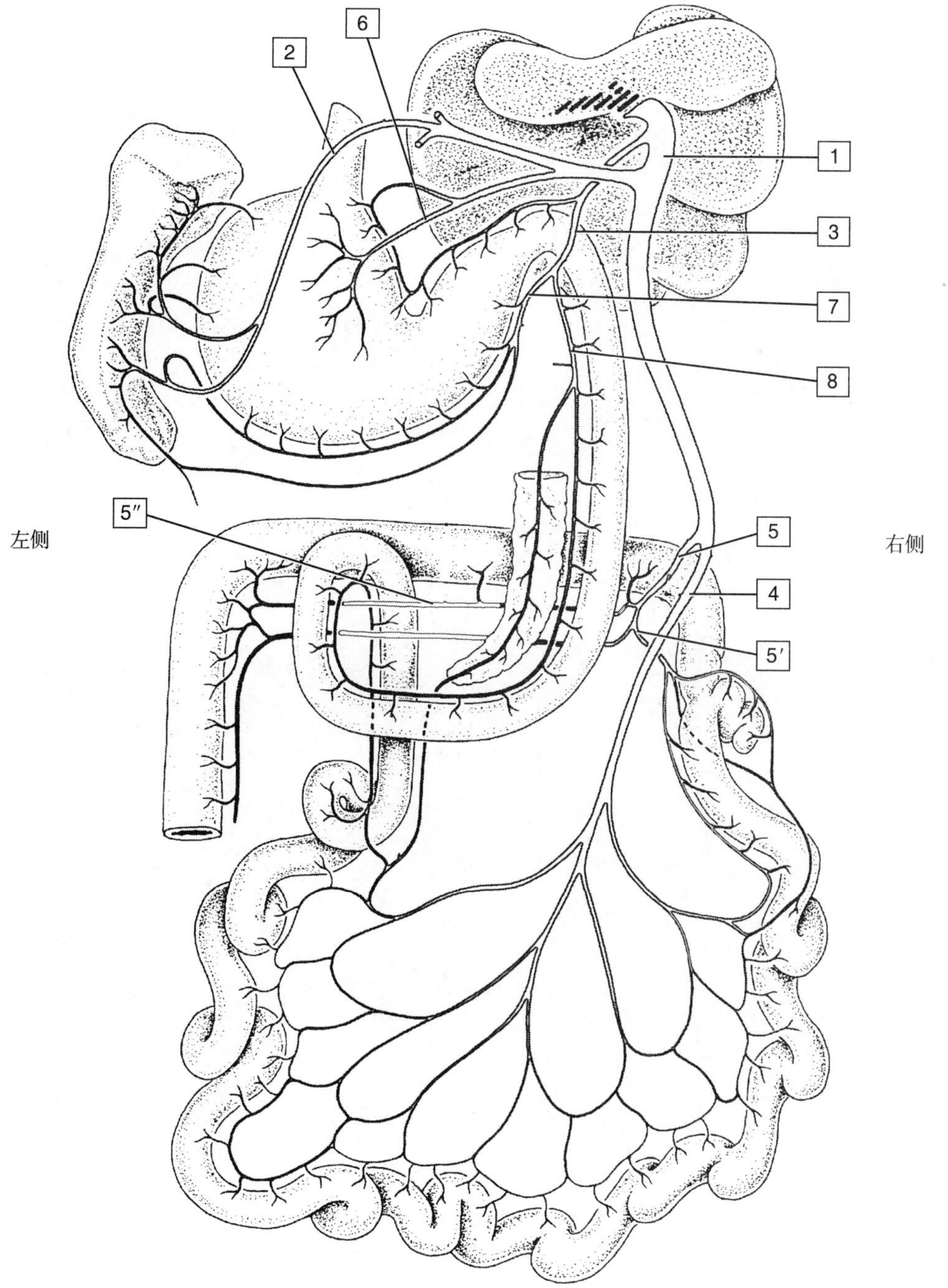

1 门静脉 Portal vein	5′ 回结肠静脉 Ileocolic vein
2 脾静脉 Splenic vein	5″ 结肠中静脉 Middle colic vein
3 胃十二指肠静脉 Gastroduodenal vein	6 胃左静脉 Left gastric vein
4 肠系膜前静脉 Cranial mesenteric vein	7 胃网膜右静脉 Right gastroepiploic vein
5 肠系膜后静脉 Caudal mesenteric vein	8 胰十二指肠前静脉 Cranial pancreaticoduodenal vein

图4-10 犬门静脉构造的部分示意图，背侧观

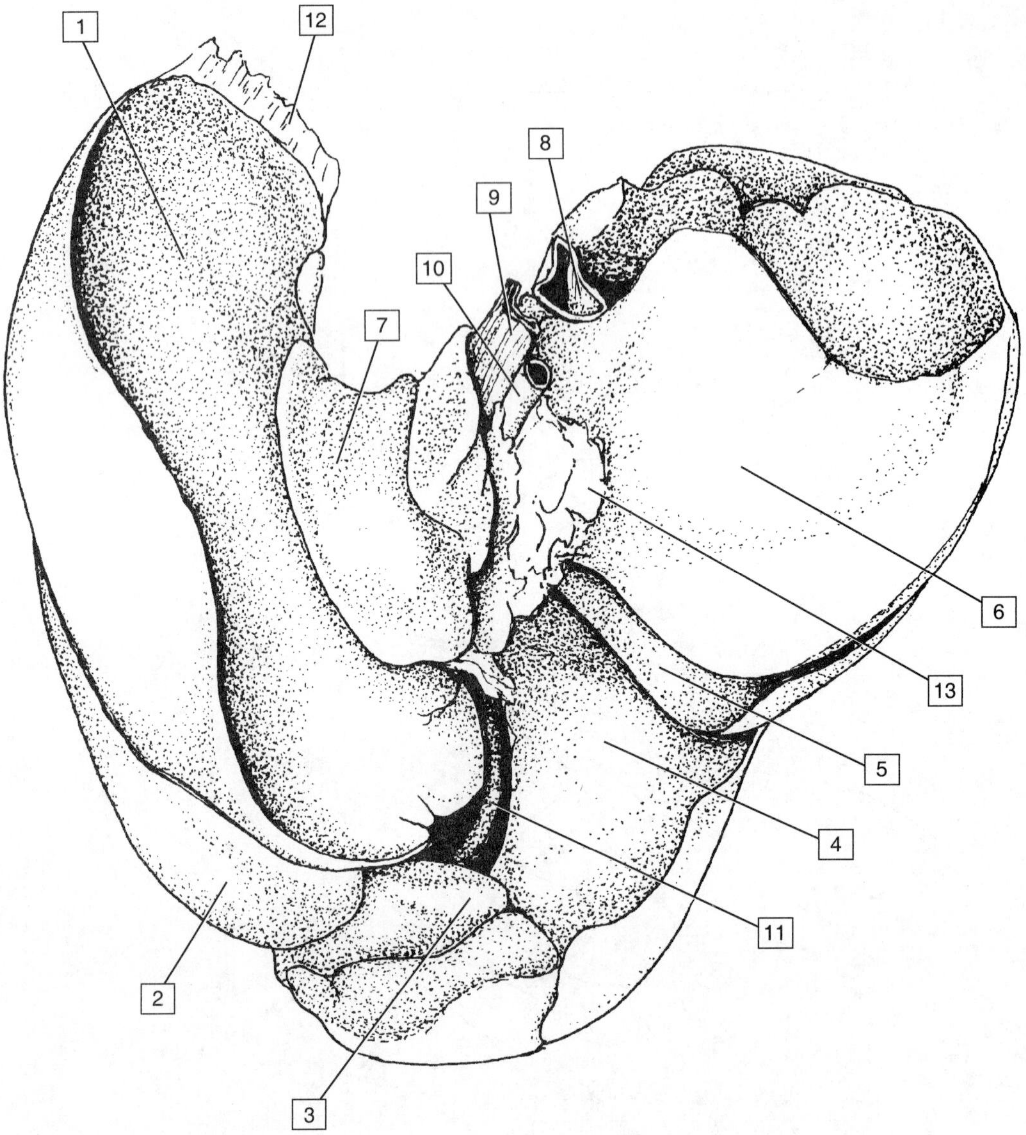

1	左外叶 Left lateral lobe	8	腔静脉后端 Caudal vena cava
2	左内叶 Left medial lobe	9	门静脉 Portal vein
3	方叶 Quadrate lobe	10	肝动脉 Hepatic artery
4	右内叶 Right medial lobe	11	胆囊 Gallbladder
5	右外叶 Right lateral lobe	12	左三角韧带 Left triangular ligament
6	（尾状叶的）尾状突 Caudate process (of caudate lobe)	13	小网膜 Lesser omentum
7	（尾状叶的）乳突 Papillary process (of caudate lobe)		

图4-11 犬肝脏的脏面

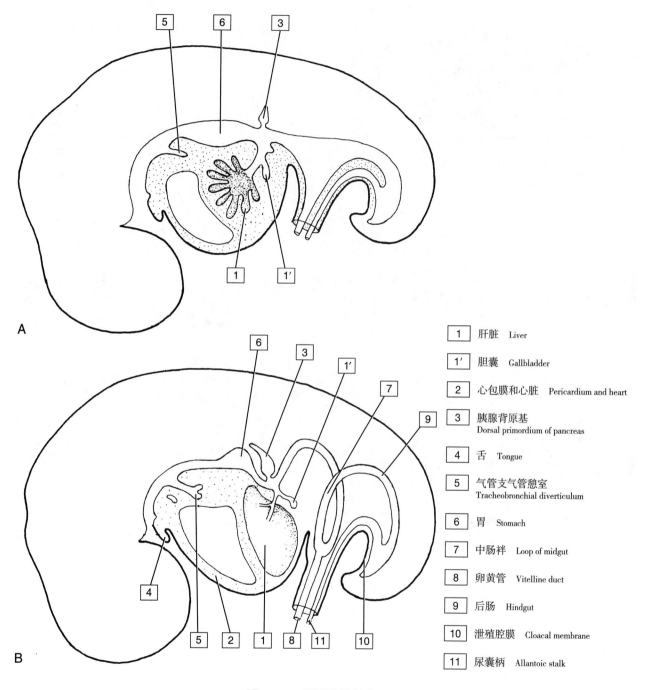

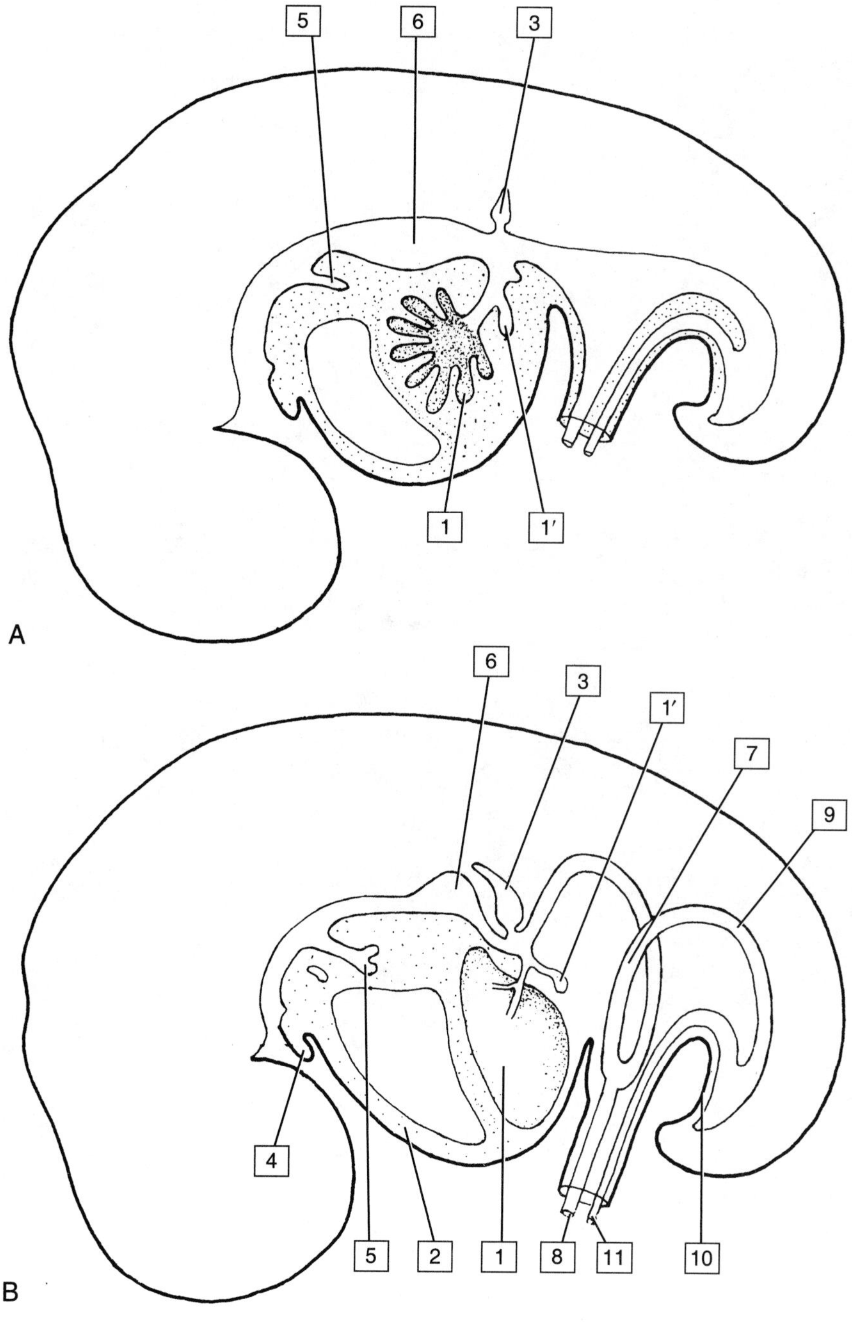

1 肝脏 Liver
1′ 胆囊 Gallbladder
2 心包膜和心脏 Pericardium and heart
3 胰腺背原基 Dorsal primordium of pancreas
4 舌 Tongue
5 气管支气管憩室 Tracheobronchial diverticulum
6 胃 Stomach
7 中肠袢 Loop of midgut
8 卵黄管 Vitelline duct
9 后肠 Hindgut
10 泄殖腔膜 Cloacal membrane
11 尿囊柄 Allantoic stalk

图4-12 犬肝脏的发育

A. 在早期发育阶段，内胚层憩室的前支侵入横膈，尾支形成胆囊和胆囊管；

B. 在晚期发育阶段，发育中的肝脏向腹腔扩展

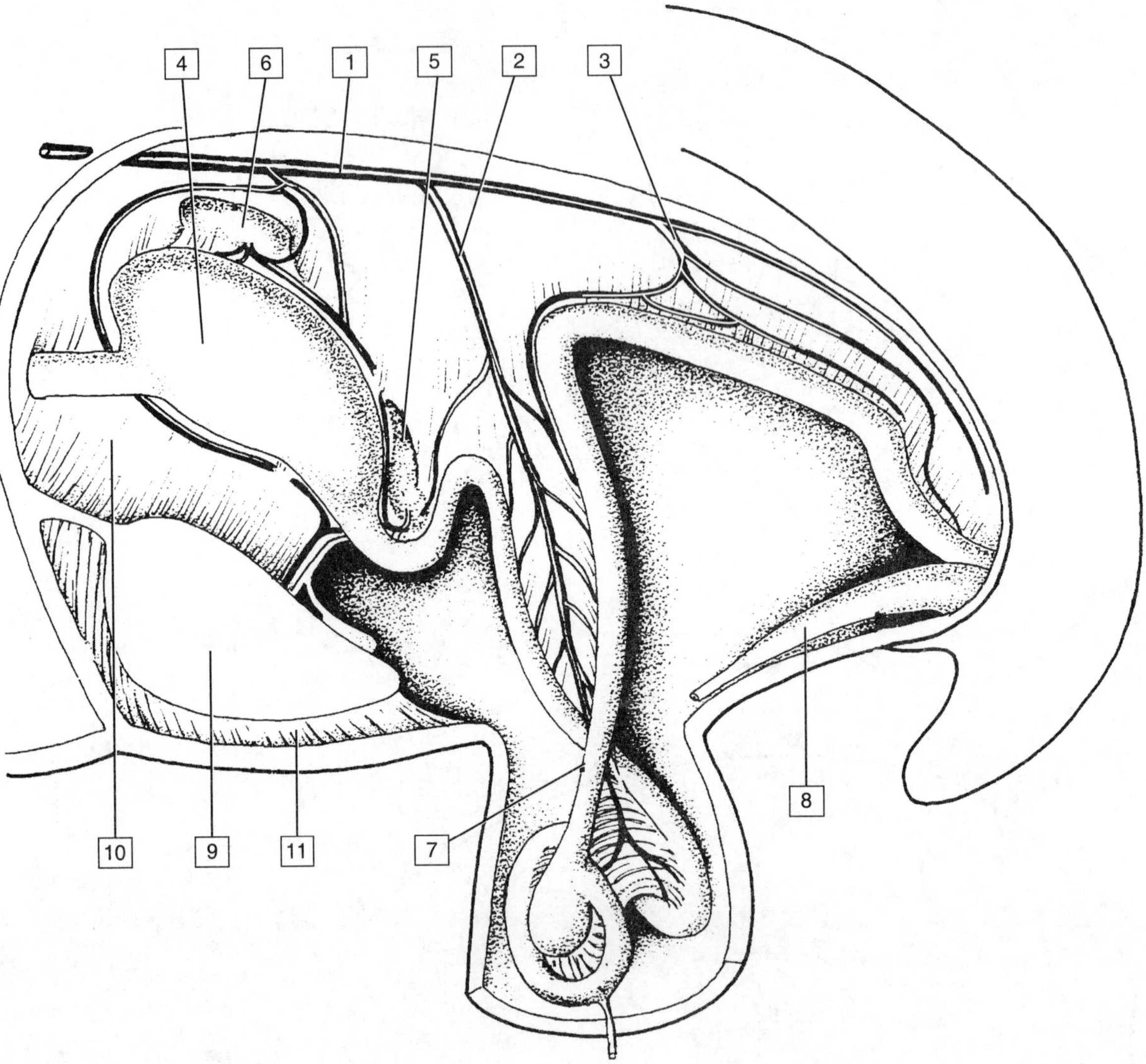

1	腹腔动脉 Celiac artery		7	中肠袢 Loop of midgut
2	肠系膜前动脉 Cranial mesenteric artery		8	泌尿生殖窦的膀胱扩张 Bladder expansion of the urogenital sinus
3	肠系膜后动脉 Caudal mesenteric artery		9	肝脏 Liver
4	胃 Stomach		10	小网膜 Lesser omentum
5	胰腺 Pancreas		11	镰状韧带 Falciform ligament
6	脾 Spleen			

图4-13 犬肠管在形成肠袢时的发育
中肠袢突入胚胎外的体腔内

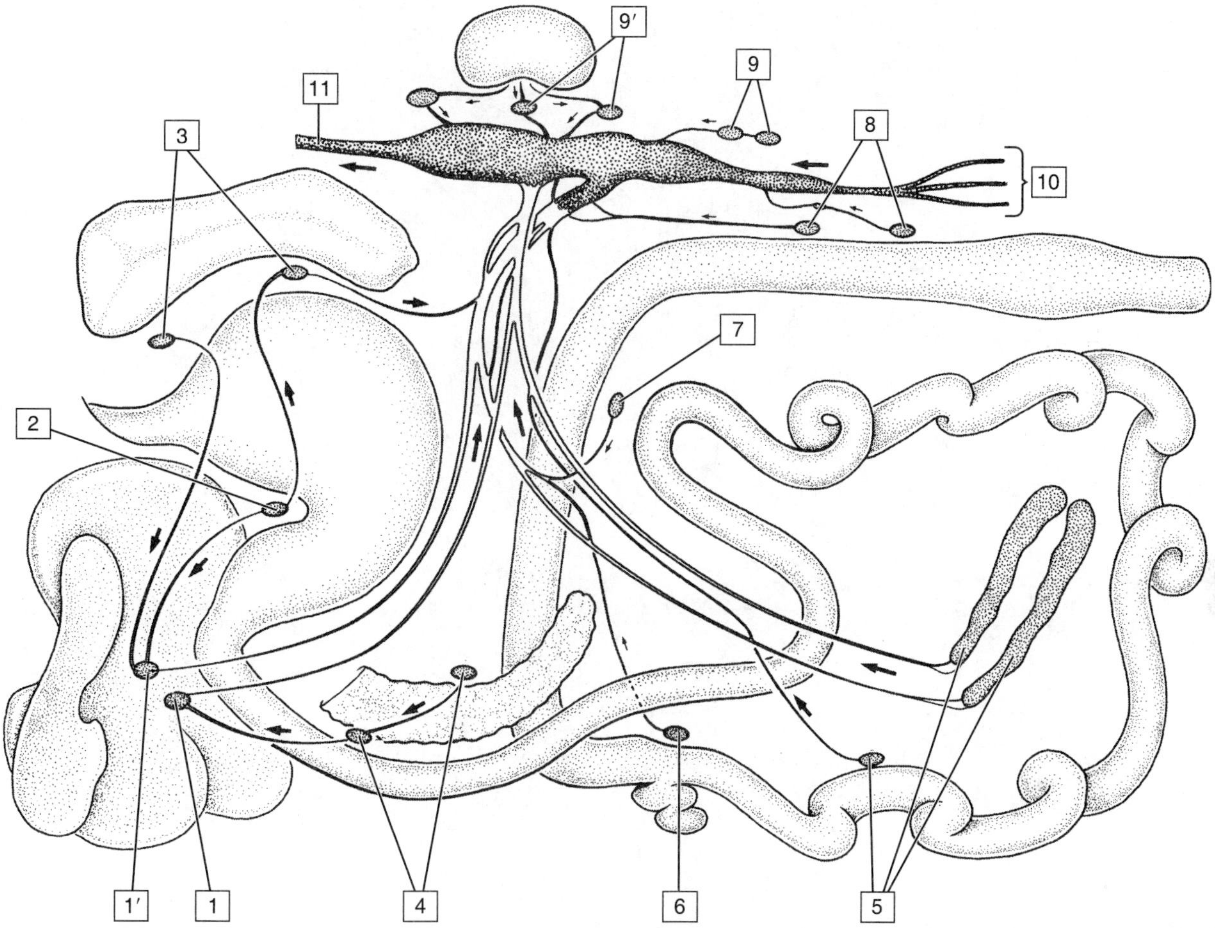

1 肝右淋巴结 Right hepatic node	7 结肠中淋巴结 Middle colic node
1′ 肝左淋巴结 Left hepatic node	8 肠系膜后淋巴结 Caudal mesenteric nodes
2 胃淋巴结 Gastric node	9 腰主动脉淋巴结 Lumbar aortic nodes
3 脾淋巴结 Splenic nodes	9′ 肾淋巴结 Renal nodes
4 胰十二指肠淋巴结 Pancreaticoduodenal nodes	10 荐髂输出淋巴管 Efferents from the iliosacral region
5 空肠淋巴结 Jejunal nodes	11 乳糜池延续为胸导管 Continuation of cisterna chyli as thoracic duct
6 结肠右淋巴结 Right colic node	

图4-14 犬腹腔和盆腔器官的淋巴引流

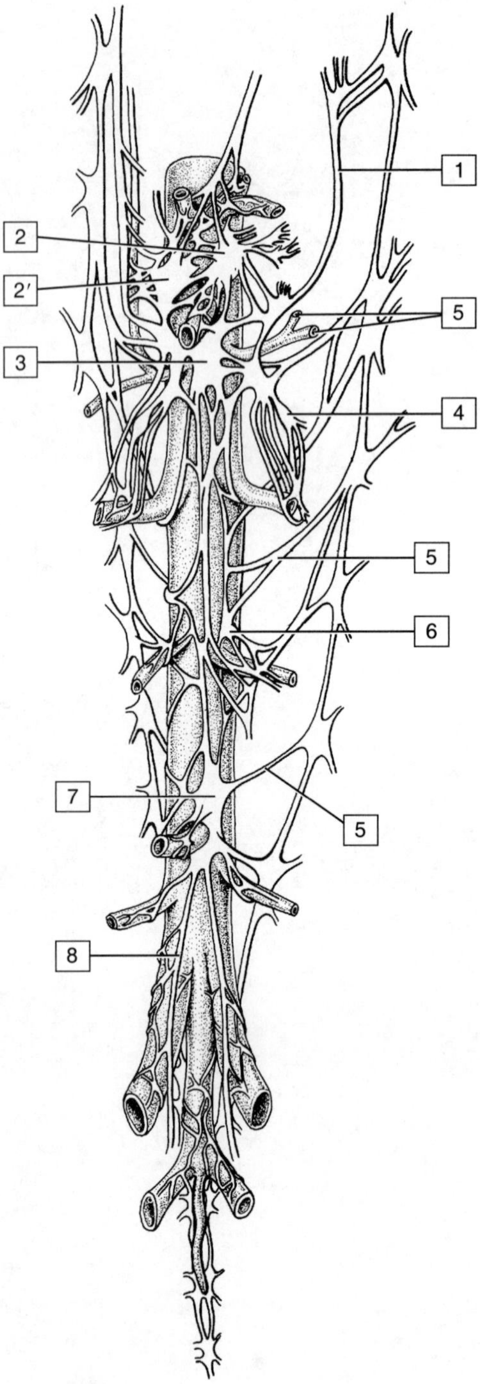

1	内脏大神经 Greater splanchnic nerve	5	腰内脏神经 Lumbar splanchnic nerve
2	左腹腔神经节 Left celiac ganglion	6	性腺神经节 Gonadal ganglion
2′	右腹腔神经节 Right celiac ganglion	7	肠系膜后神经节 Caudal mesenteric ganglion
3	肠系膜前神经节 Cranial mesenteric ganglion	8	右腹下神经 Right hypogastric nerve
4	肾神经节 Renal ganglion		

图4-15 犬腹腔神经节和神经丛，腹侧观

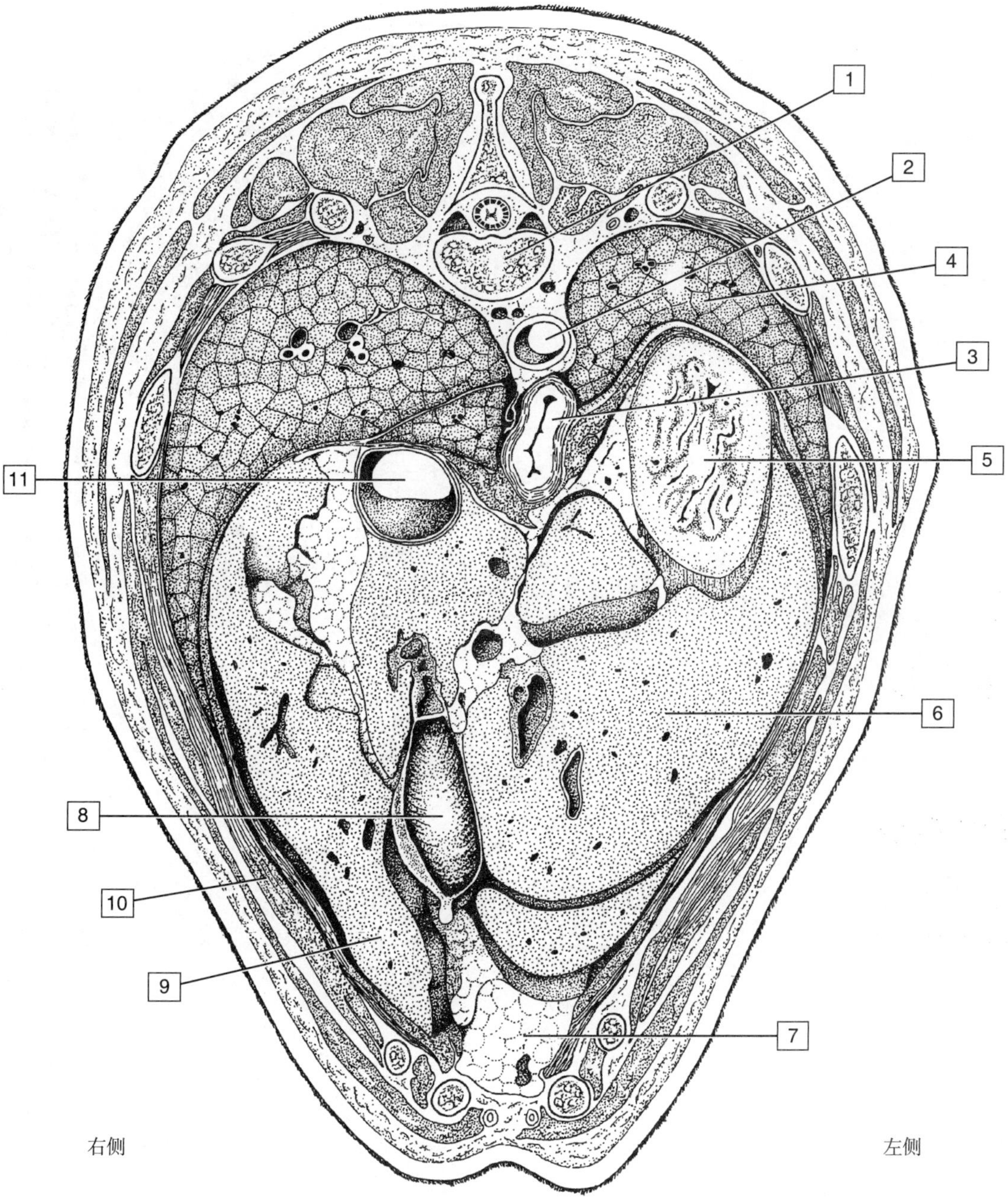

1	第11胸椎 Eleventh thoracic vertebra	5	胃底 Fundus of stomach	9	肝右内叶 Right medial lobe of liver
2	主动脉 Aorta	6	肝左外叶 Left lateral lobe of liver	10	膈 Diaphragm
3	食管 Esophagus	7	充满脂肪的镰状韧带 Fat-filled falciform ligament	11	后腔静脉 Caudal vena cava
4	左肺 Left lung	8	胆囊 Gallbladder		

图4-16 犬躯干的第11胸椎水平横断面

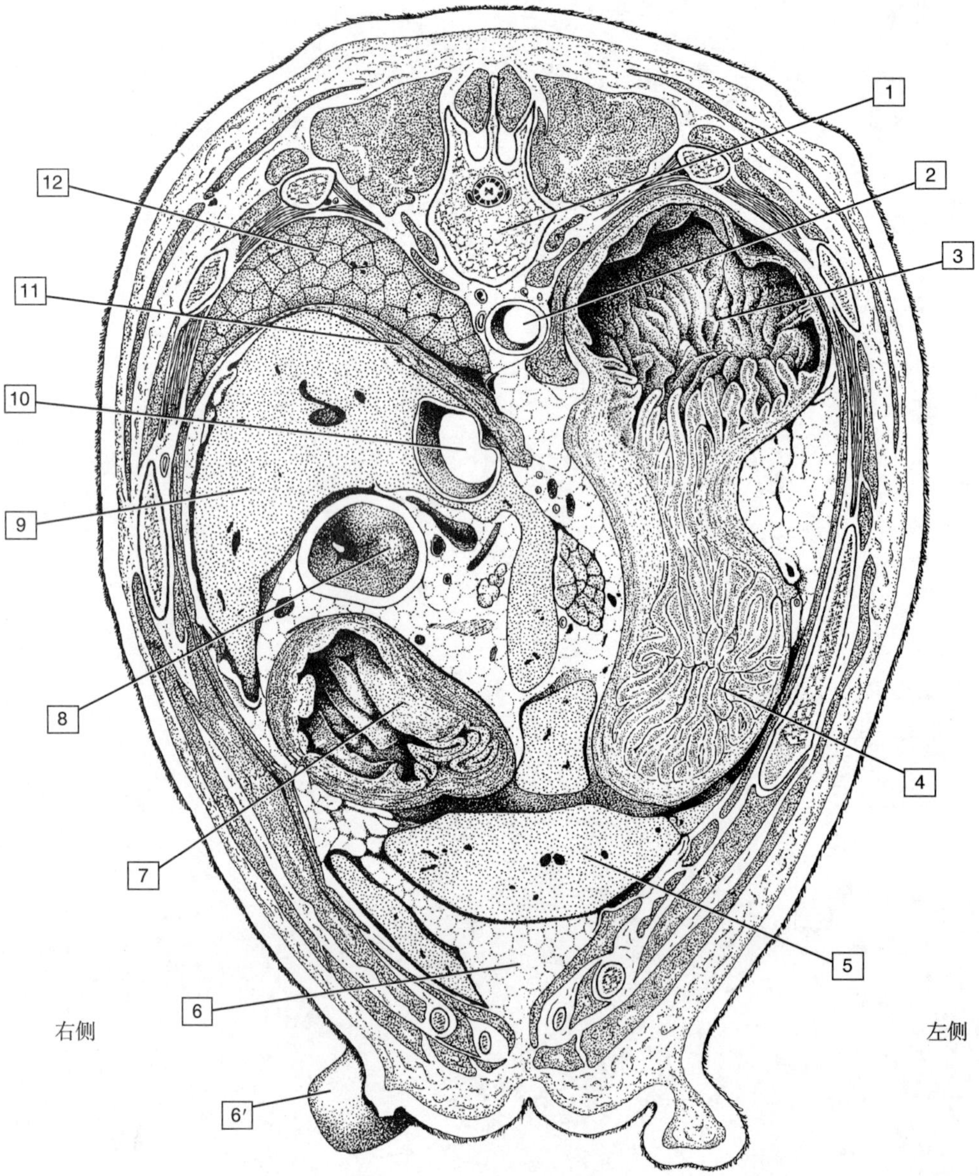

1	第12胸椎 Twelfth thoracic vertebra	6	充满脂肪的镰状韧带 Fat-filled falciform ligament	10	后腔静脉 Caudal vena cava
2	主动脉 Aorta	6′	乳头 Teat	11	膈 Diaphragm
3	胃底 Fundus of stomach	7	胃幽门部 Pyloric part of stomach	12	右肺 Right lung
4	胃体 Body of stomach	8	十二指肠降部 Descending duodenum		
5	肝脏 Liver	9	肝脏的尾状突 Caudate process of liver		

图4-17 犬躯干的第12胸椎水平横断面

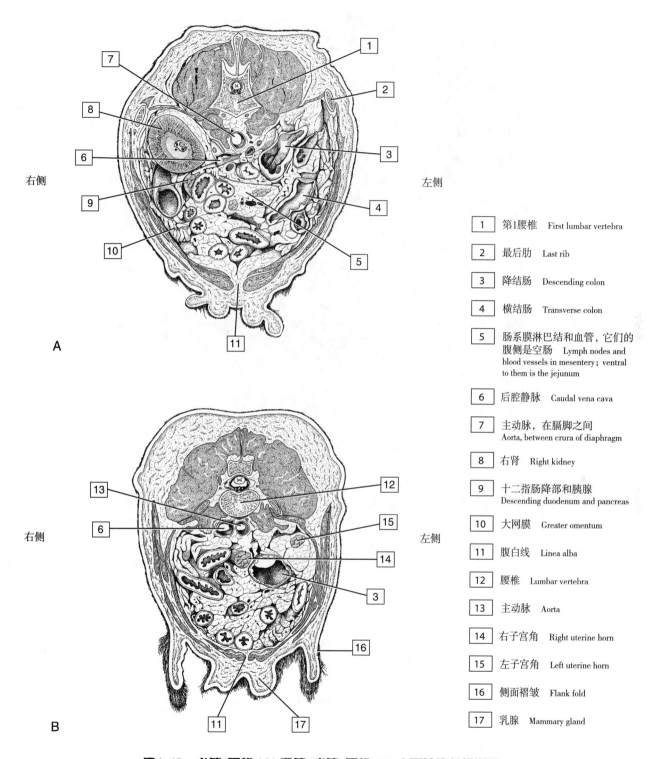

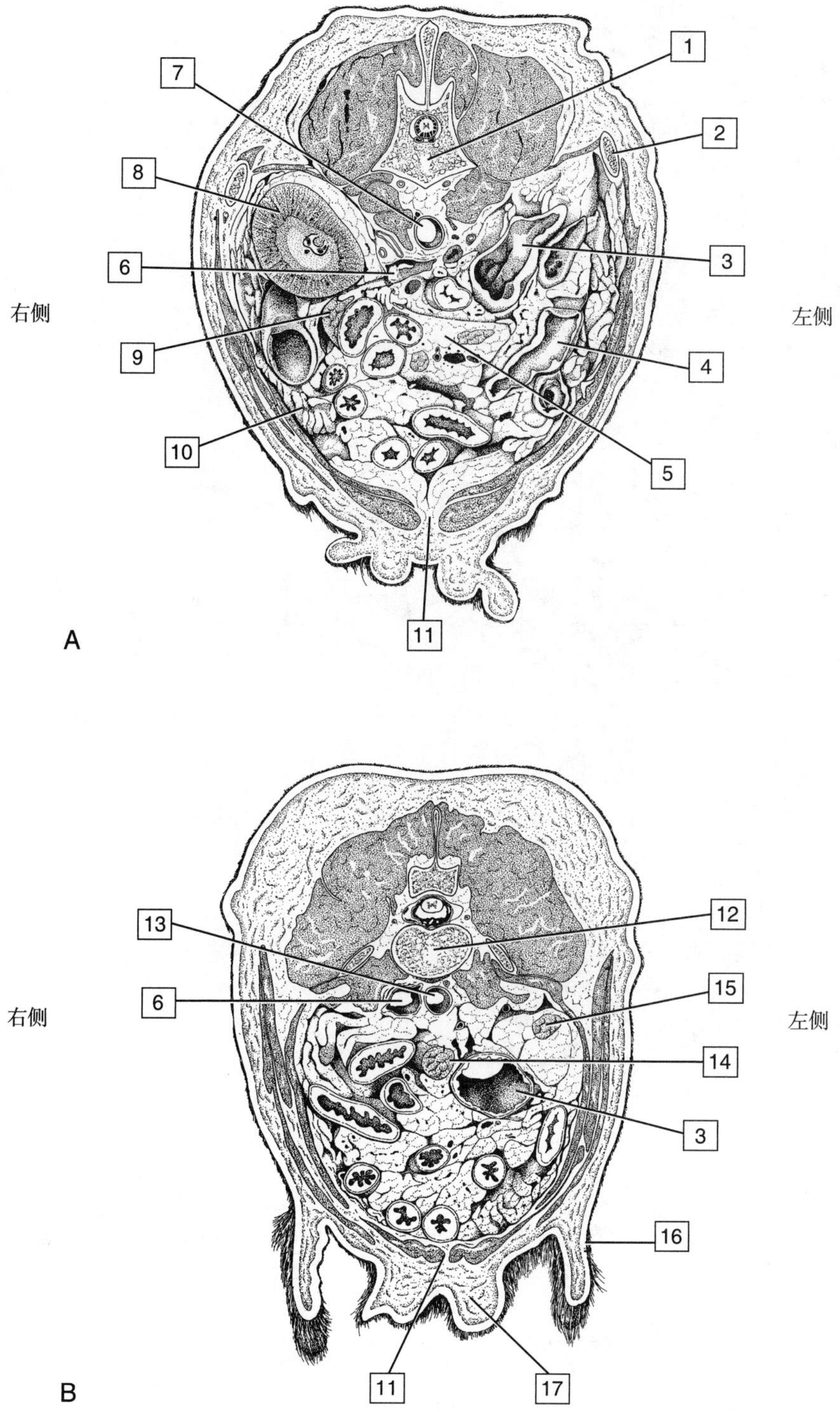

1 第1腰椎 First lumbar vertebra

2 最后肋 Last rib

3 降结肠 Descending colon

4 横结肠 Transverse colon

5 肠系膜淋巴结和血管，它们的腹侧是空肠 Lymph nodes and blood vessels in mesentery; ventral to them is the jejunum

6 后腔静脉 Caudal vena cava

7 主动脉，在膈脚之间 Aorta, between crura of diaphragm

8 右肾 Right kidney

9 十二指肠降部和胰腺 Descending duodenum and pancreas

10 大网膜 Greater omentum

11 腹白线 Linea alba

12 腰椎 Lumbar vertebra

13 主动脉 Aorta

14 右子宫角 Right uterine horn

15 左子宫角 Left uterine horn

16 侧面褶皱 Flank fold

17 乳腺 Mammary gland

图4-18 大第1腰椎（A）和第4或第5腰椎（B）水平的腹部横断面

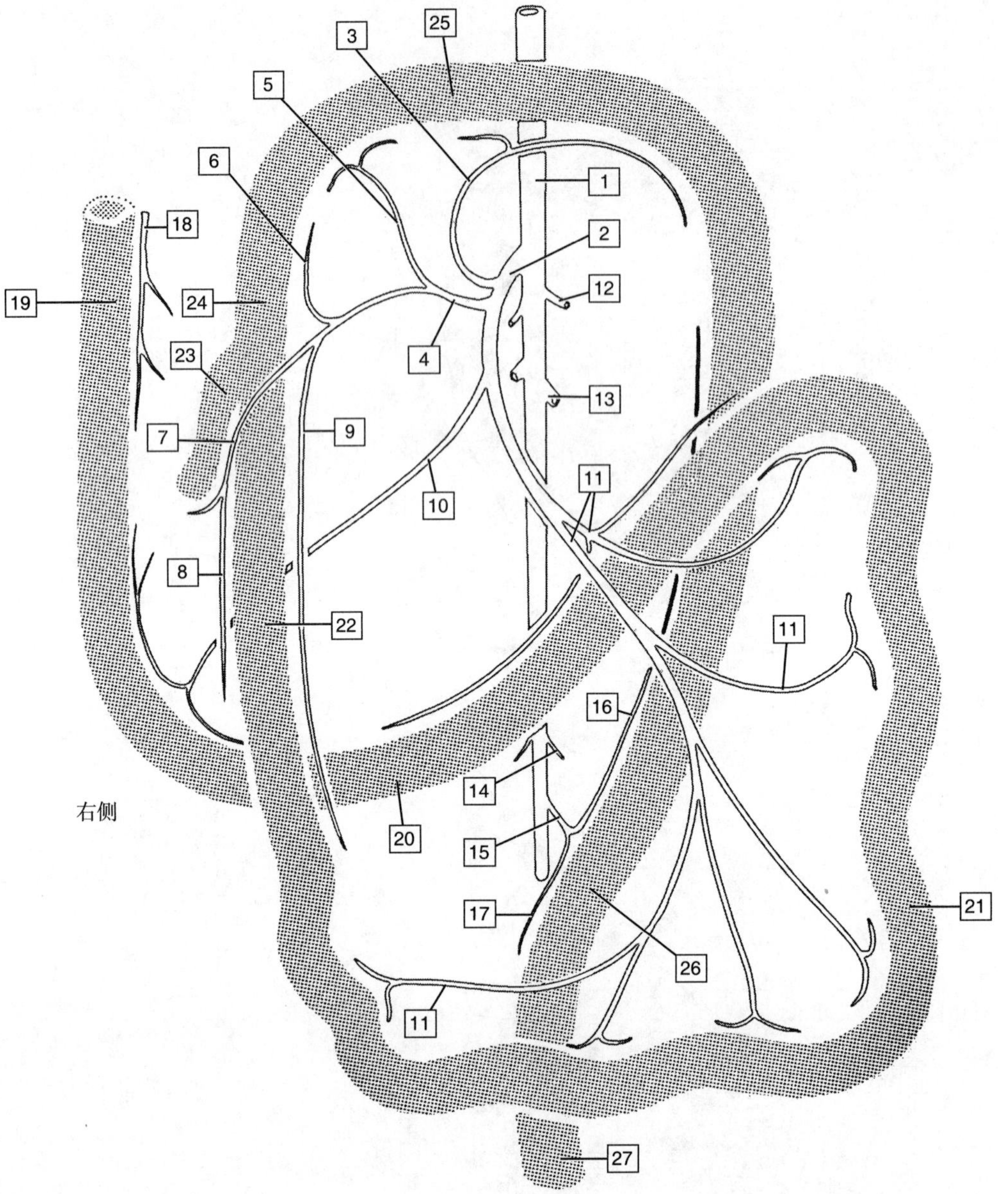

1	腹主动脉 Abdominal aorta
2	肠系膜前动脉 Cranial mesenteric artery
3	结肠中动脉 Middle colic artery
4	回结肠动脉 Ileocolic artery
5	结肠右动脉 Right colic artery
6	回结肠动脉结肠支 Colic branch of ileocolic artery
7	盲肠动脉 Cecal artery
8	肠系膜小肠游离的回肠支 Antimesenteric ileal branch
9	肠系膜回肠支 Mesenteric ileal branch
10	胰十二指肠后动脉 Caudal pancreaticoduodenal artery
11	空肠微动脉 Jejunal artery
12	膈腹腔微动脉 Phrenicoabdominal artery
13	肾动脉 Renal artery
14	睾丸（卵巢）微动脉 Testicular (ovarian) artery
15	肠系膜后动脉 Caudal mesenteric artery
16	结肠左动脉 Left colic artery
17	直肠前动脉 Cranial rectal artery
18	胰十二指肠上动脉 Cranial pancreaticoduodenal artery
19	十二指肠降部 Descending duodenum
20	十二指肠升部 Ascending duodenum
21	空肠 Jejunum
22	回肠 Ileum
23	盲肠 Cecum
24	升结肠 Ascending colon
25	横结肠 Transverse colon
26	降结肠 Descending colon
27	直肠 Rectum

图4-19　大肠道的血液供应，腹侧观

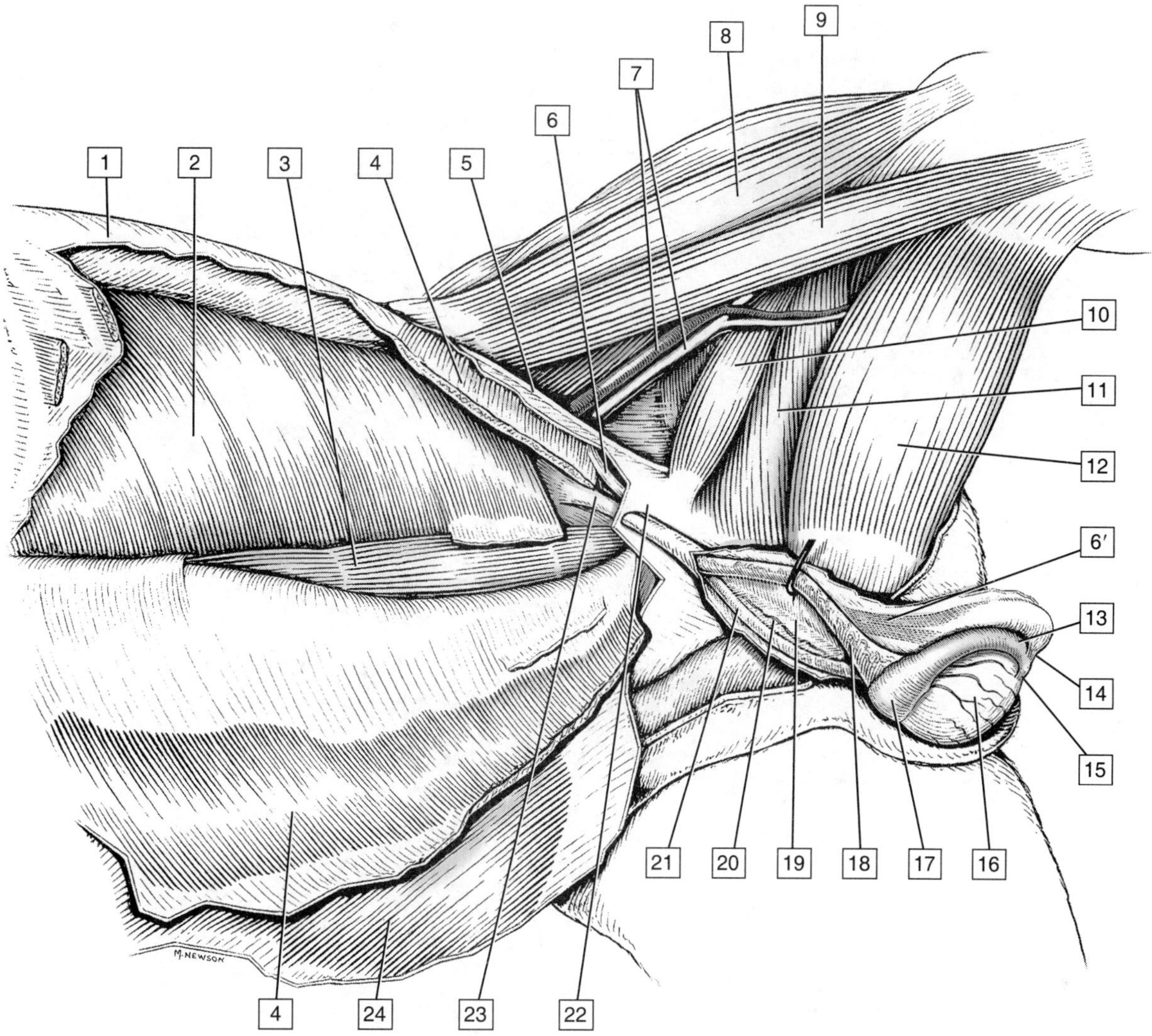

1 胸腰筋膜 Thoracolumbar fascia

2 腹横肌 Transversus abdominis

3 腹直肌 Rectus abdominis

4 腹内斜肌（横断并掀开）Internal abdominal oblique (transected and reflected)

5 腹股沟韧带（腹外斜肌的腱膜后缘）Inguinal ligament (caudal border of aponeurosis of external abdominal oblique muscle)

6 提睾肌的起点 Cremaster muscle at its origin

6′ 鞘膜壁层外表面的提睾肌 Cremaster muscle on external surface of parietal layer of vaginal tunic

7 股动脉和股静脉 Femoral artery and vein

8 缝匠肌前部 Cranial part of sartorius

9 缝匠肌后部 Caudal part of sartorius

10 耻骨肌 Pectineus

11 内收肌 Adductor

12 股薄肌 Gracilis

13 附睾尾 Tail of epididymis

14 附睾尾部韧带 Ligament of tail of epididymis

15 睾丸固有韧带 Proper ligament of testis

16 位于鞘膜脏层内的睾丸 Testis in visceral vaginal tunic

17 附睾头 Head of epididymis

18 鞘膜脏层内的睾丸动脉和睾丸静脉 Testicular artery and vein in visceral vaginal tunic (mesorchium)

19 睾丸系膜 Mesorchium

20 输精管 Mesoductus deferens

21 鞘膜脏层内的输精管 Ductus deferens in visceral vaginal tunic

22 腹股沟浅环，外侧脚 Superficial inguinal ring, lateral crus

23 腹股沟管内的鞘膜壁层 Parietal vaginal tunic in the inguinal canal

24 腹外斜肌（掀开）External abdominal oblique (reflected)

图4-20 犬（雄性）腹部肌肉和腹股沟深层解剖图，左侧观

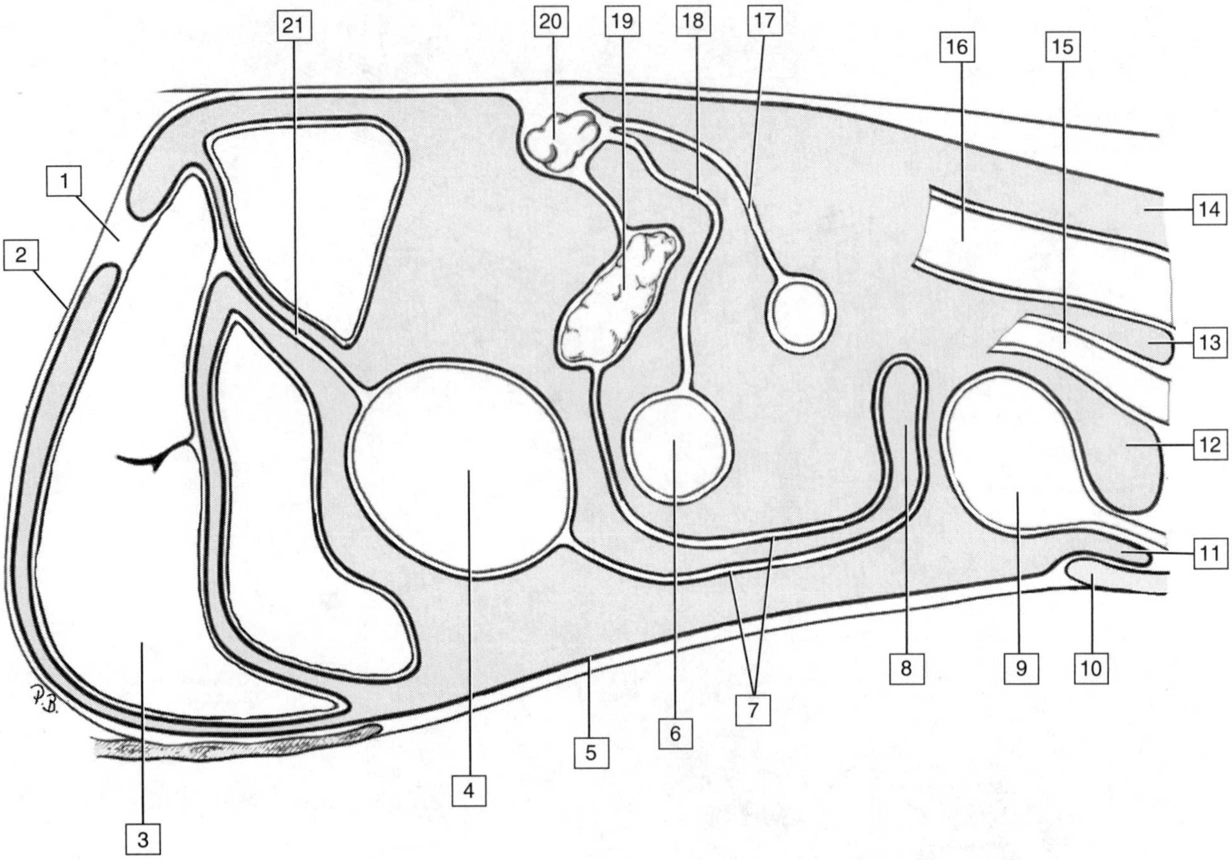

1 冠状韧带 Coronary ligament	12 膀胱生殖腔 Vesicogenital pouch
2 膈 Diaphragm	13 直肠生殖腔 Rectogenital pouch
3 肝脏 Liver	14 直肠旁隐窝 Pararectal fossa
4 胃 Stomach	15 子宫 Uterus
5 壁腹膜 Parietal peritoneum	16 降结肠 Descending colon
6 横结肠 Transverse colon	17 肠系膜 Mesentery
7 大网膜深层和浅层 Greater omentum, deep and superficial leaves	18 横结肠系膜 Transverse mesocolon
8 网膜囊 Omental bursa	19 胰腺左叶 Left lobe, pancreas
9 膀胱 Bladder	20 淋巴结 Lymph nodes
10 骨盆联合 Symphysis	21 小网膜 Lesser omentum
11 耻骨膀胱陷窝 Pubovesical pouch	

图4-21 犬腹膜的折转，矢状面

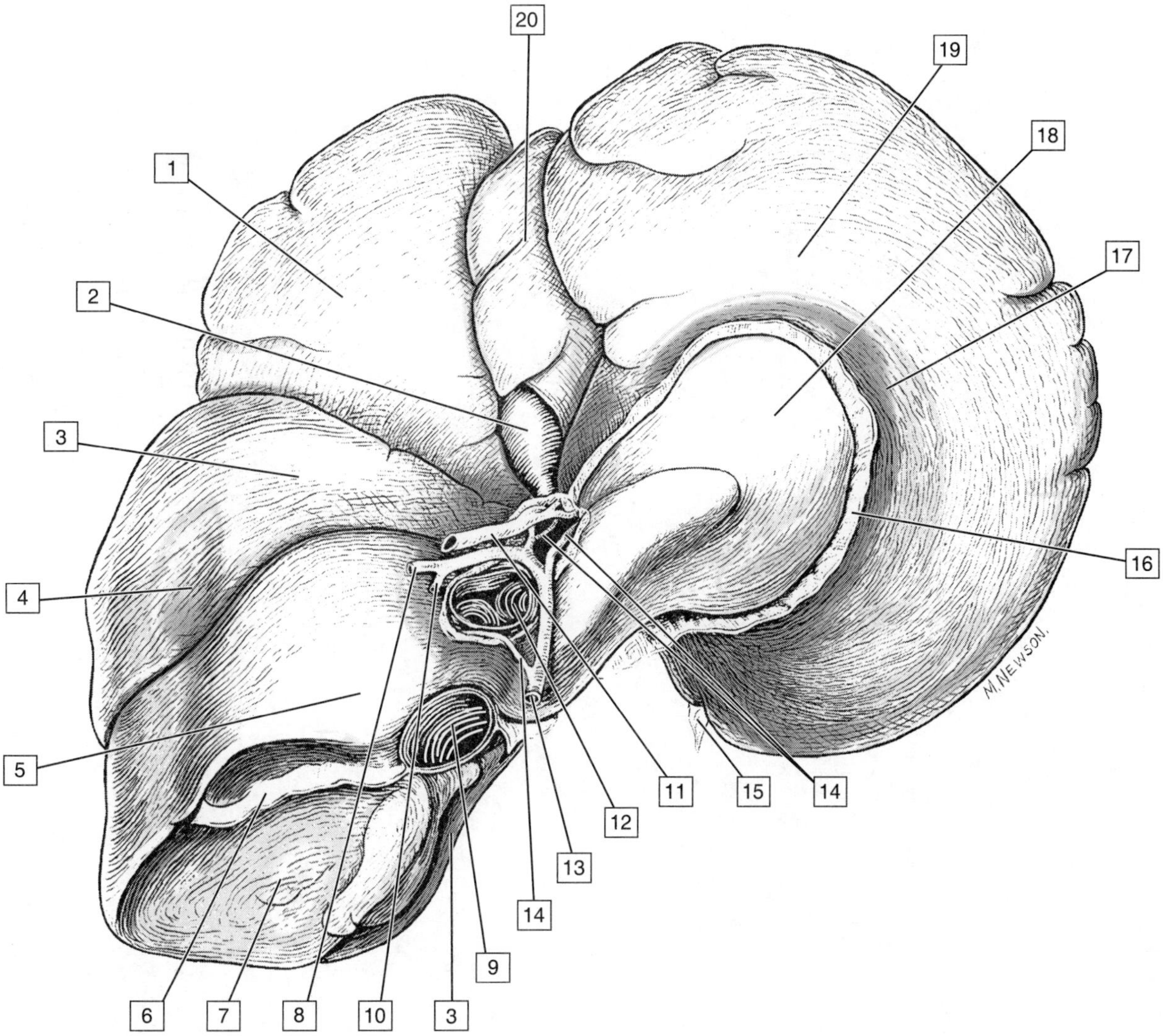

1	右内叶 Right medial lobe	8	胃十二指肠动脉 Gastroduodenal artery	15	左三角韧带 Left triangular ligament
2	胆囊 Gallbladder	9	后腔静脉 Postcava	16	小网膜 Lesser omentum
3	右外叶 Right lateral lobe	10	胃右动脉 Right gastric artery	17	胃压迹 Gastric impression
4	十二指肠压迹 Duodenal impression	11	胆管 Bile duct	18	尾叶乳突 Papillary process of caudate lobe
5	尾叶的尾状突 Caudate process of caudate lobe	12	门静脉 Portal vein	19	左外叶 Left lateral lobe
6	肝肾韧带 Hepatorenal ligament	13	肝动脉 Hepatic artery	20	尾叶 Caudate lobe
7	肾窝 Renal fossa	14	肝动脉分支 Hepatic branches		

图4-22 犬肝脏，脏面

犬背卧位，从后往前看

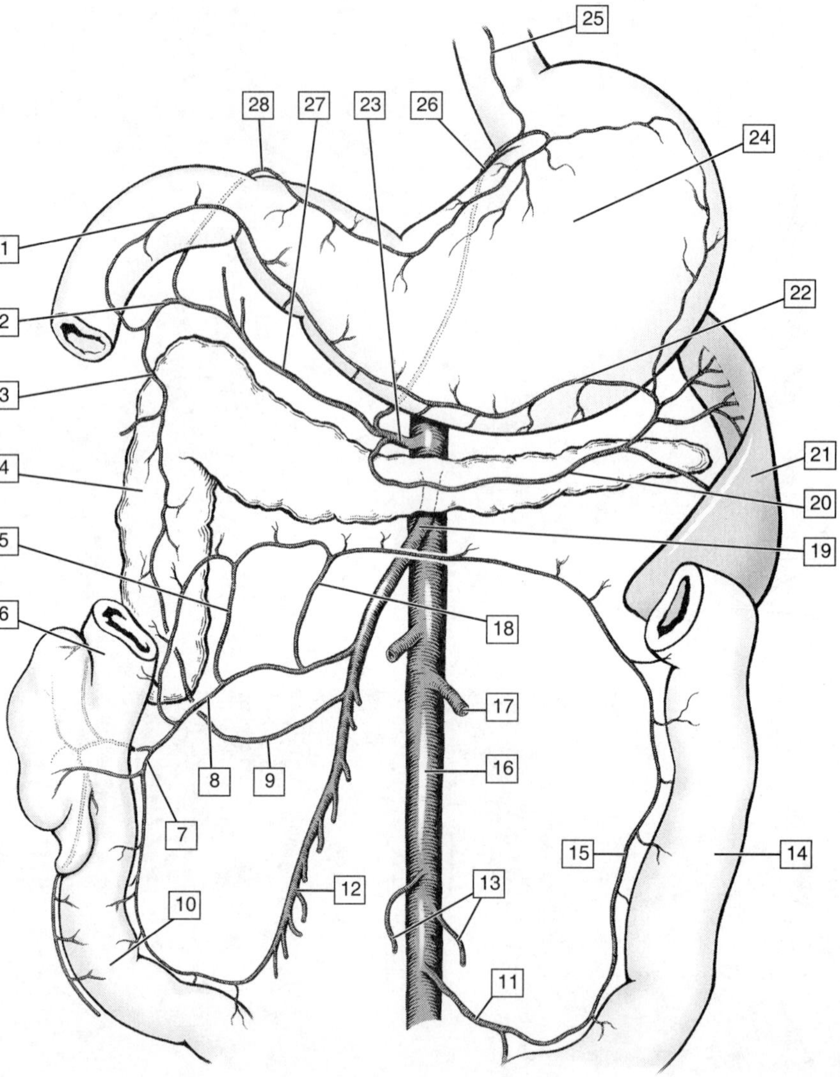

- 1 胃网膜右动脉 Right gastroepiploic artery
- 2 胃十二指肠动脉 Gastroduodenal artery
- 3 胰十二指肠前动脉 Cranial pancreaticoduodenal artery
- 4 胰腺 Pancreas
- 5 结肠右动脉 Right colic artery
- 6 升结肠 Ascending colon
- 7 盲肠动脉 Cecal artery
- 8 回结肠动脉 Ileocolic artery
- 9 胰十二指肠后动脉 Caudal pancreaticoduodenal artery
- 10 回肠 Ileum
- 11 肠系膜后动脉 Caudal mesenteric artery
- 12 空肠微动脉 Jejunal aa.
- 13 睾丸（卵巢）微动脉 Testucular（ovarian）aa.
- 14 降结肠 Descending colon
- 15 结肠左动脉 Left colic artery
- 16 主动脉 Aorta artery
- 17 肾左动脉 Left renal artery
- 18 中结肠动脉 Middle colic artery
- 19 肠系膜前动脉 Cranial mesenteric artery
- 20 脾动脉 Splenic artery
- 21 脾 Spleen
- 22 胃网膜左动脉 Left gastroepiploic artery
- 23 腹腔动脉 Celiac artery
- 24 胃 Stomach
- 25 食管动脉 Esophageal artery
- 26 胃左动脉 Left gastric artery
- 27 肝动脉 Hepatic artery
- 28 胃右动脉 Right gastric artery

图4-23 伴有主要吻合支的大腹腔动脉和肠系膜前动脉分支

这是背侧观还是腹侧观？

马

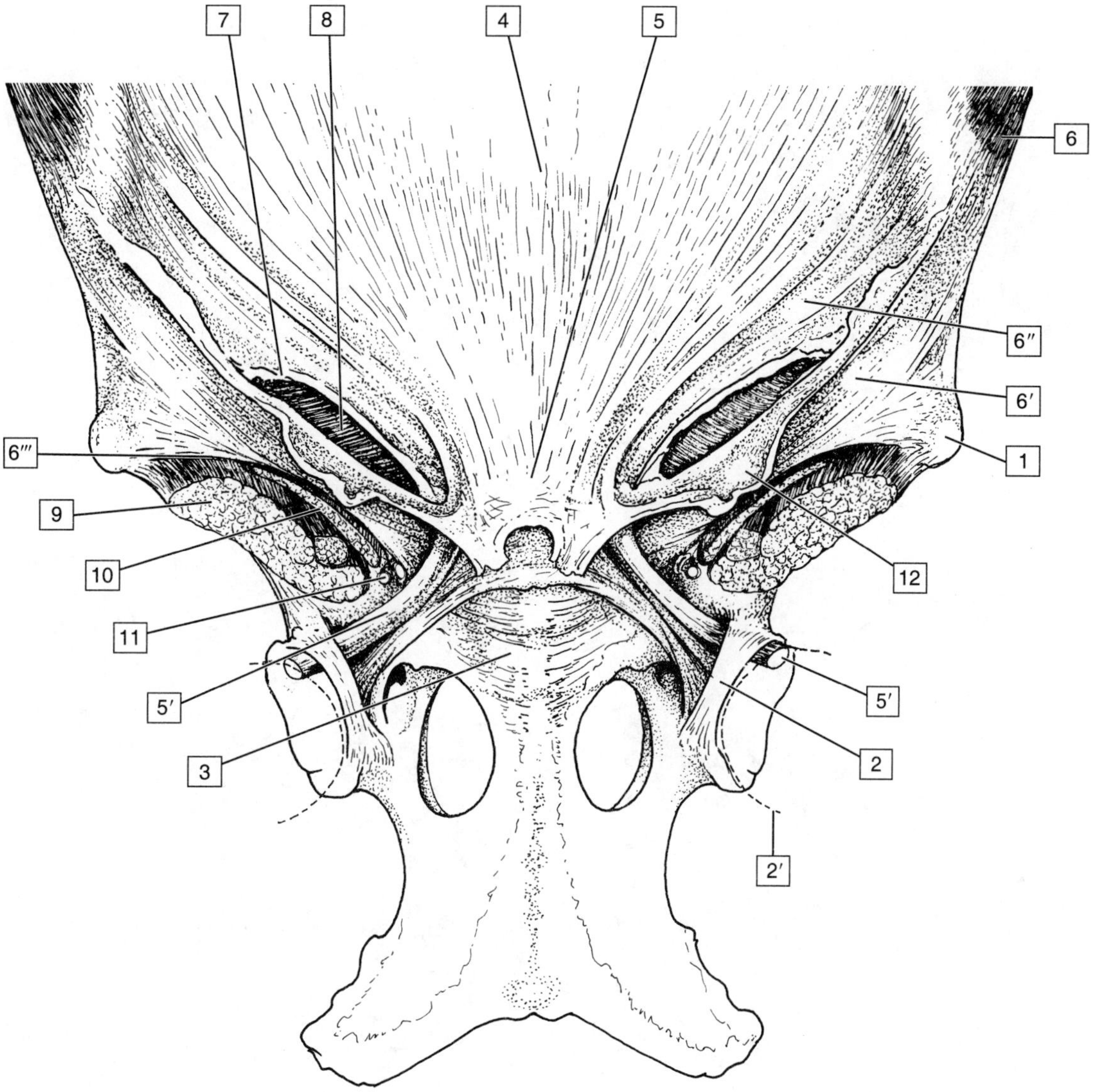

1	髋结节 Coxal tuber	6″	腹外斜肌腱膜的盆肌腱 Pelvic tendon of external oblique aponeuros
2	髋臼横韧带 Transverse acetabular ligament	6‴	腹外斜肌腱膜的盆肌腱附着于缝匠肌和髂腰肌上（腹股沟韧 Attachment of pelvic tendon of external oblique aponeurosis on sartorius and iliopsoas（"inguinal ligament"）
2′	股骨头 Femoral head	7	腹股沟管浅环 Superficial inguinal ring
3	耻骨 Pubis	8	腹内斜肌 Internal abdominal oblique
4	腹白线上的腹黄膜 Tunica flava over linea alba	9	髂腰肌 Iliopsoas
5	耻前腱 Prepubic tendon	10	缝匠肌 Sartorius
5′	副韧带 Accessory ligament	11	含股血管的血管腔 Vascular lacuna containing femoral vessels
6	腹外斜肌 External abdominal oblique		

图4-24 附着于骨盆和耻前腱的马腹部肌肉

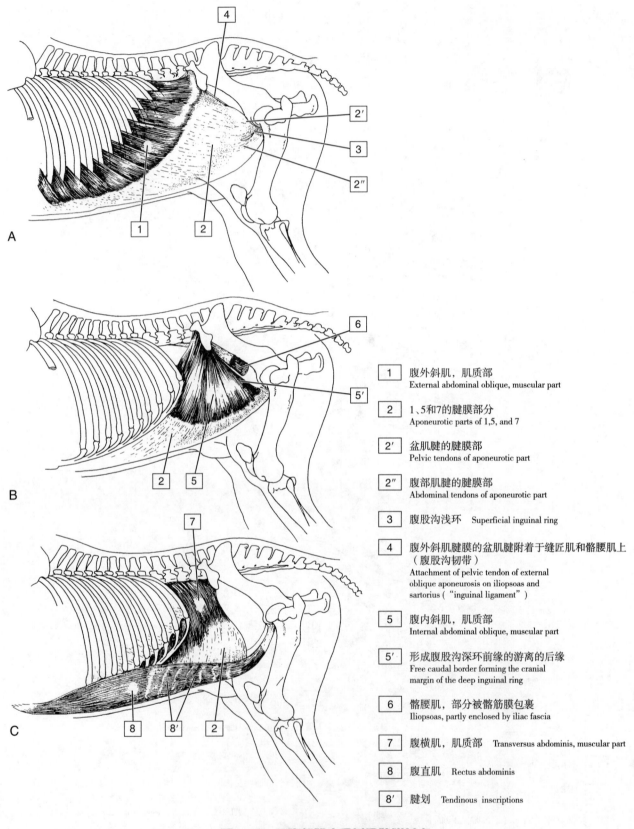

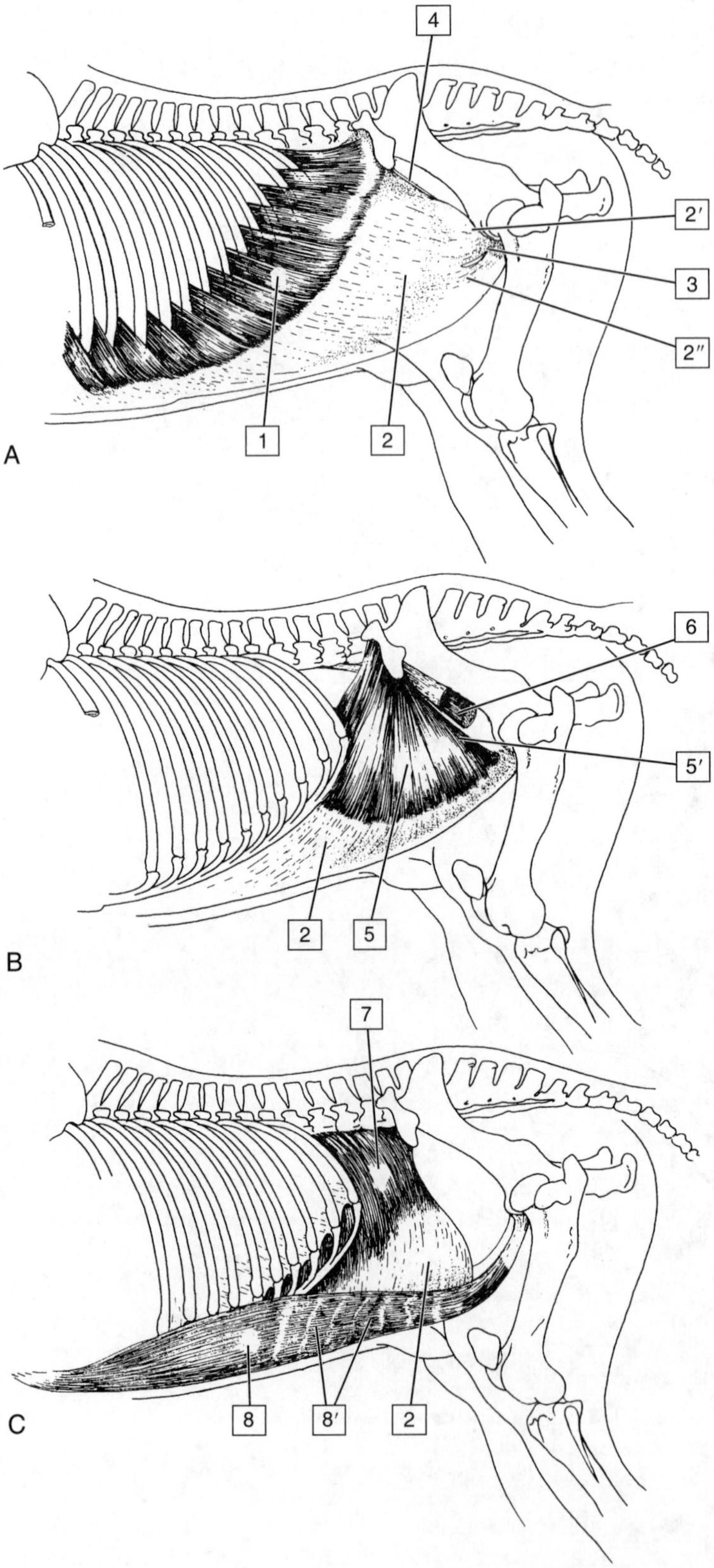

1	腹外斜肌，肌质部 External abdominal oblique, muscular part
2	1、5和7的腱膜部分 Aponeurotic parts of 1,5, and 7
2′	盆肌腱的腱膜部 Pelvic tendons of aponeurotic part
2″	腹部肌腱的腱膜部 Abdominal tendons of aponeurotic part
3	腹股沟浅环 Superficial inguinal ring
4	腹外斜肌腱膜的盆肌腱附着于缝匠肌和髂腰肌上（腹股沟韧带） Attachment of pelvic tendon of external oblique aponeurosis on iliopsoas and sartorius (“inguinal ligament”)
5	腹内斜肌，肌质部 Internal abdominal oblique, muscular part
5′	形成腹股沟深环前缘的游离的后缘 Free caudal border forming the cranial margin of the deep inguinal ring
6	髂腰肌，部分被髂筋膜包裹 Iliopsoas, partly enclosed by iliac fascia
7	腹横肌，肌质部 Transversus abdominis, muscular part
8	腹直肌 Rectus abdominis
8′	腱划 Tendinous inscriptions

图4-25 马腹部肌肉及其骨骼附着点

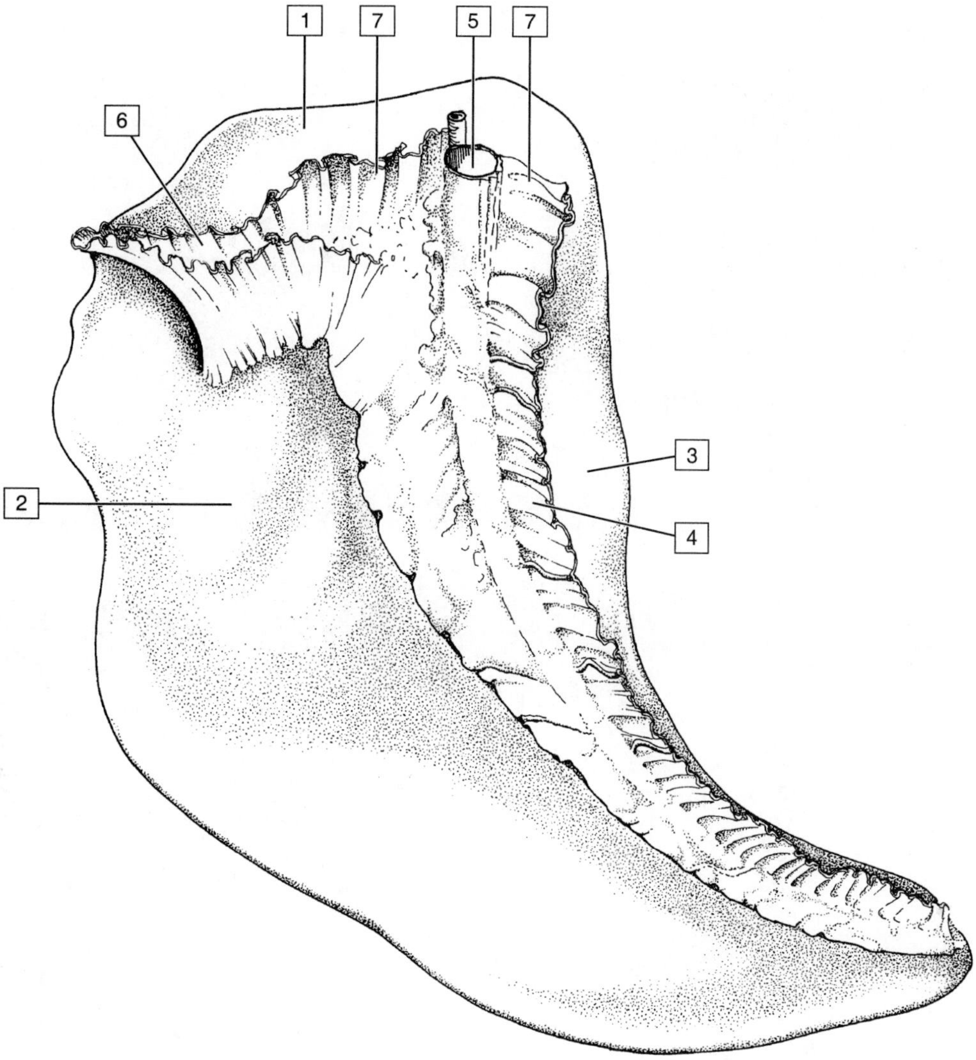

1	肾面	Renal surface
2	肠面	Intestinal surface
3	胃面	Gastric surface
4	大网膜（胃脾韧带）	Greater omentum (gastrosplenic ligament)
5	脾动脉和脾静脉	Splenic artery and vein
6	脾肾韧带	Renosplenic ligament
7	脾膈韧带	Phrenicosplenic ligament

图4-26 马脾，脏面

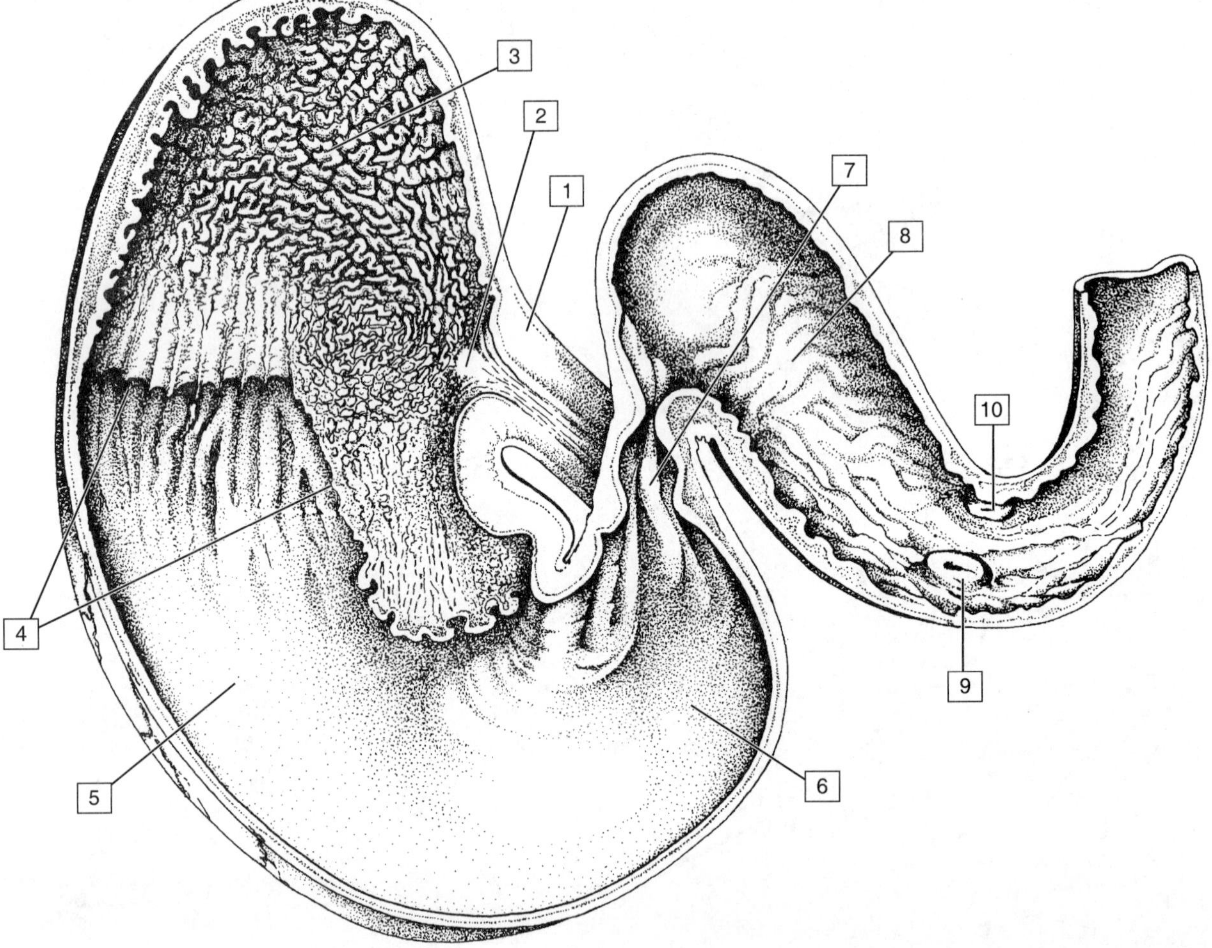

1	食管 Esophagus	6	幽门部 Pyloric part
2	贲门口 Cardiac opening	7	幽门 Pylorus
3	胃底（盲囊） Fundus (blind sac)	8	十二指肠前部 Cranial part of duodenum
4	褶缘 Margo plicatus	9	肝胰壶腹共同开口于十二指肠大乳头 Major duodenal papilla within hepatopancreatic ampulla
5	胃体 Body	10	十二指肠小乳头 Minor duodenal papilla

图4-27 马胃和十二指肠的内部结构

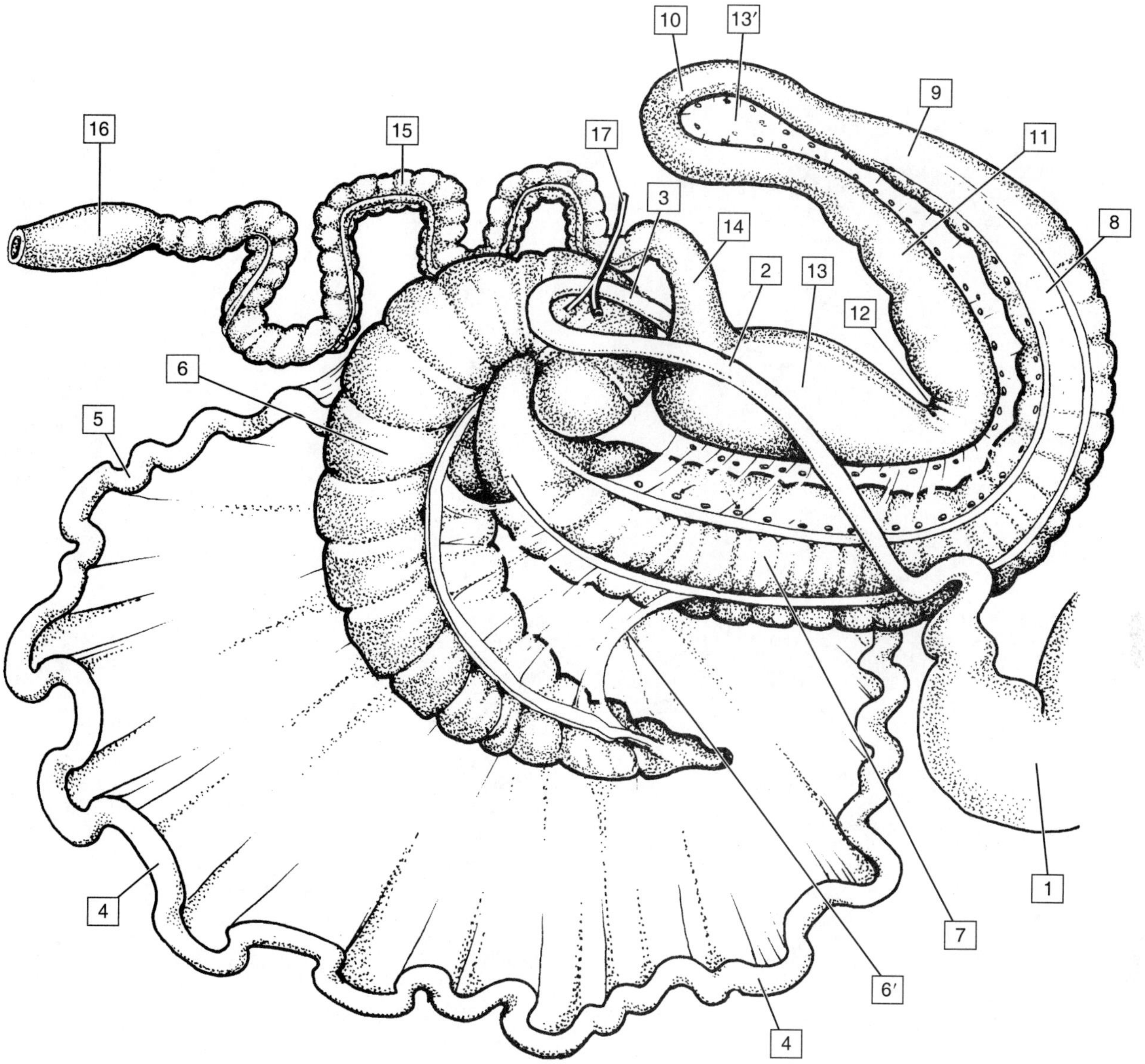

1	胃	Stomach	7	右下大结肠	Right ventral colon	13′	升结肠系膜	Ascending mesocolon
2	十二指肠降部	Descending duodenum	8	胸骨曲	Ventral diaphragmatic flexure	14	横结肠	Transverse colon
3	十二指肠升部	Ascending duodenum	9	左下大结肠	Left ventral colon	15	降（小）结肠	Descending (small) colon
4	空肠	Jejunum	10	骨盆曲	Pelvic flexure	16	直肠	Rectum
5	回肠	Ileum	11	左上大结肠	Left dorsal colon	17	肠系膜前动脉	Cranial mesenteric artery
6	盲肠	Cecum	12	膈曲	Dorsal diaphragmatic flexure			
6′	盲结肠韧带	Cecocolic fold	13	右上大结肠	Right dorsal colon			

图4-28 从右侧观察的马肠管

十二指肠的后曲和肠系膜前动脉被移到动物右侧的盲肠底部

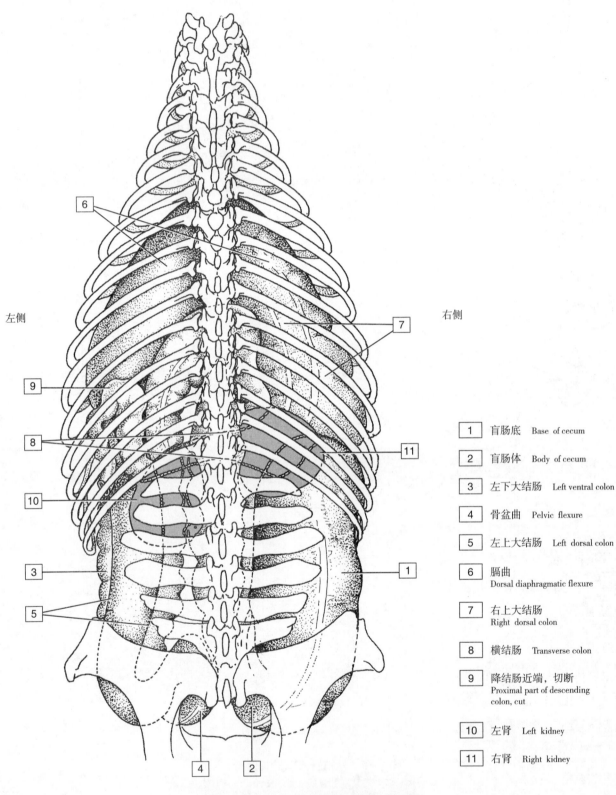

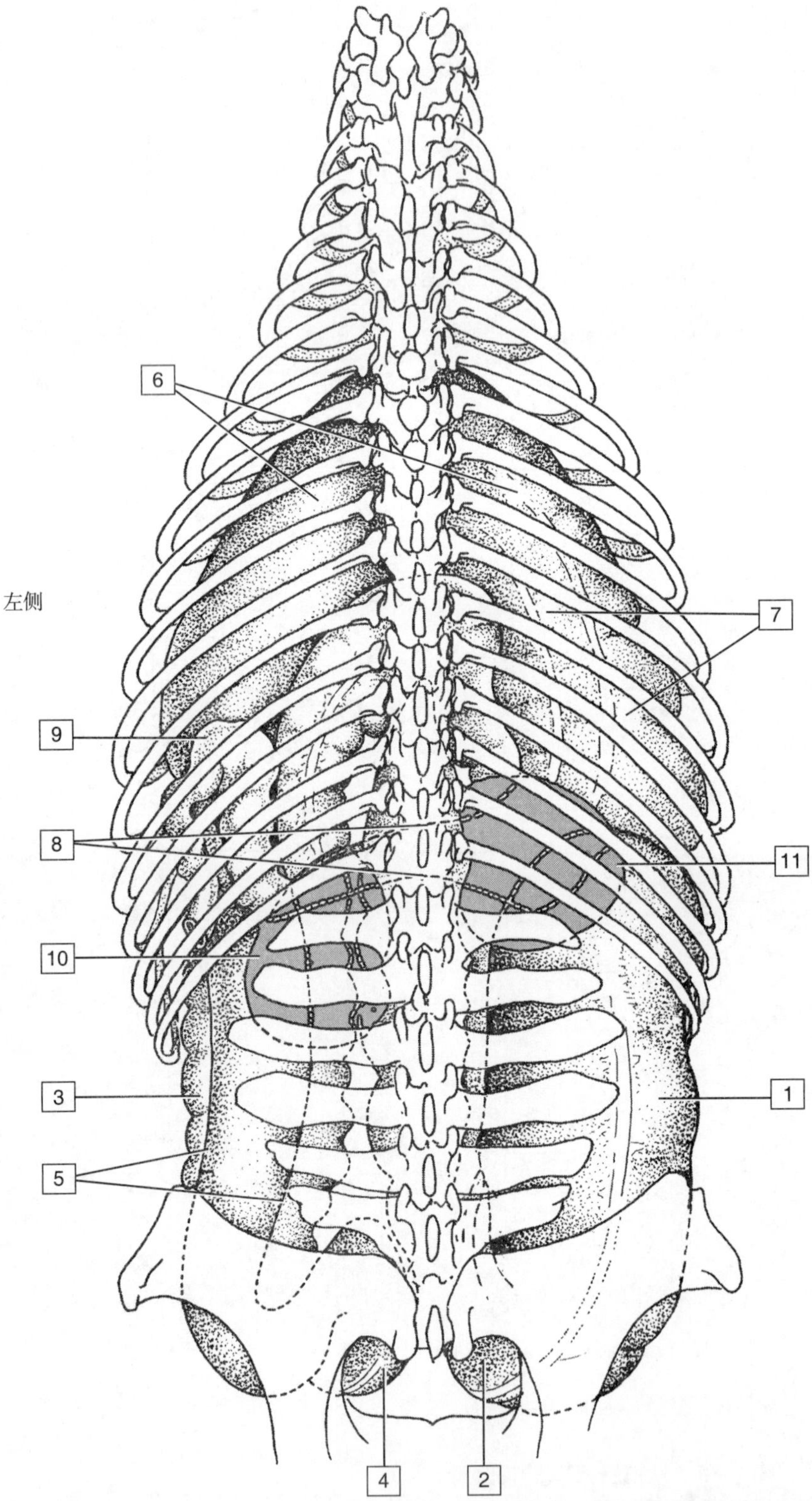

1 盲肠底 Base of cecum
2 盲肠体 Body of cecum
3 左下大结肠 Left ventral colon
4 骨盆曲 Pelvic flexure
5 左上大结肠 Left dorsal colon
6 膈曲 Dorsal diaphragmatic flexure
7 右上大结肠 Right dorsal colon
8 横结肠 Transverse colon
9 降结肠近端，切断 Proximal part of descending colon, cut
10 左肾 Left kidney
11 右肾 Right kidney

图4-29 马大肠和肾脏的位置，背侧观

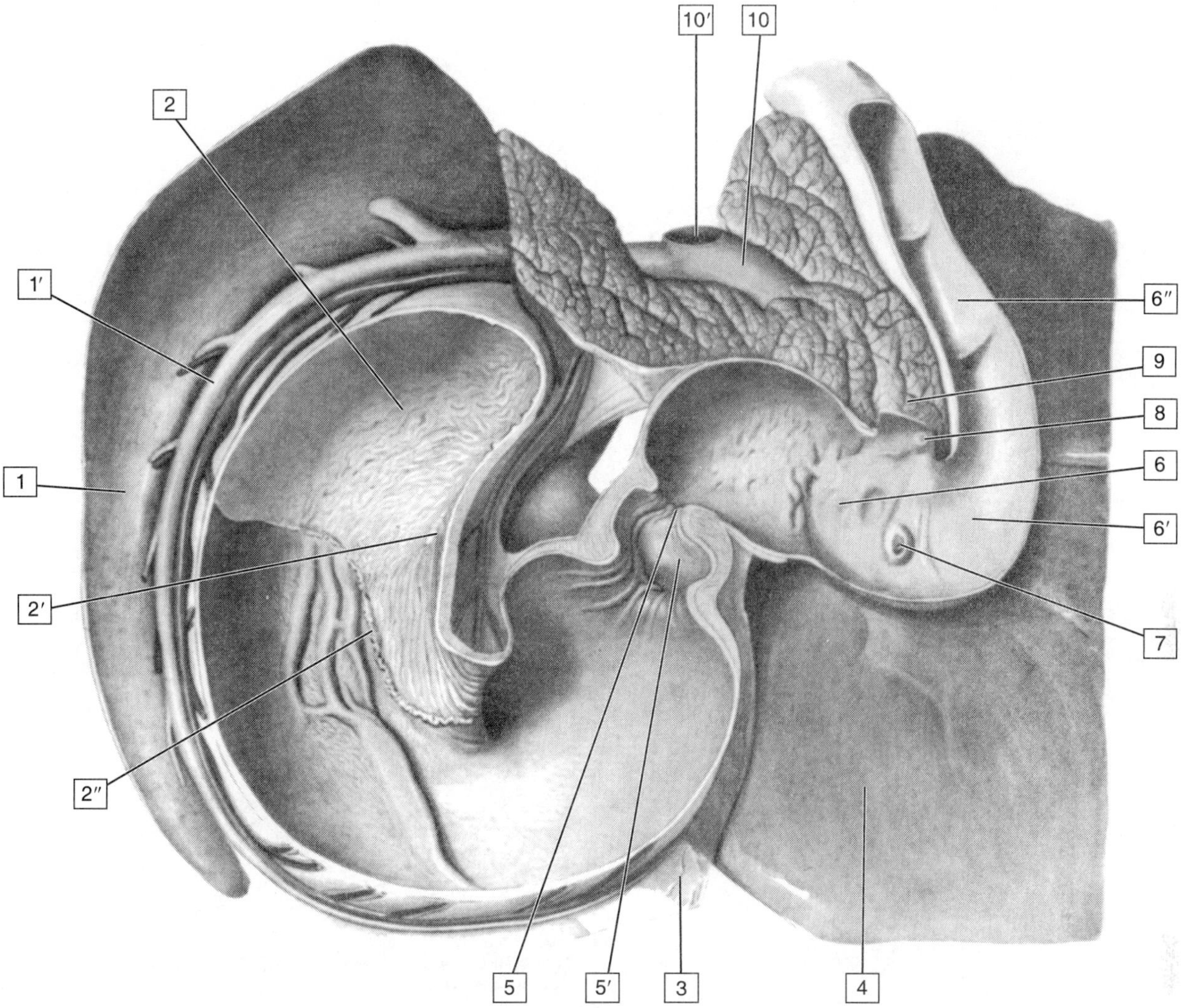

1	脾肠面 Intestinal surface of spleen	6	S状十二指肠前部 S-shaped cranial part of duodenum
1′	脾动脉和脾静脉 Splenic artery and vein	6′	十二指肠前曲 Cranial flexure of duodenum
2	胃底（盲囊） Fundus (blind sac) of stomach	6″	十二指肠降部 Descending duodenum
2′	贲门 Cardia	7	十二指肠大乳头 Major duodenal papilla
2″	褶缘 Margo plicatus	8	十二指肠小乳头 Minor duodenal papilla
3	大网膜 Greater omentum	9	胰体 Body of pancreas
4	肝脏 Liver	10	门静脉 Portal vein
5	幽门孔 Pyloric orifice	10′	肠系膜前静脉断端 Stump of cranial mesenteric vein
5′	幽门窦 Pyloric antrum		

图4-30 马脾、胃、胰腺和肝脏的局部解剖图，腹后侧观

牛

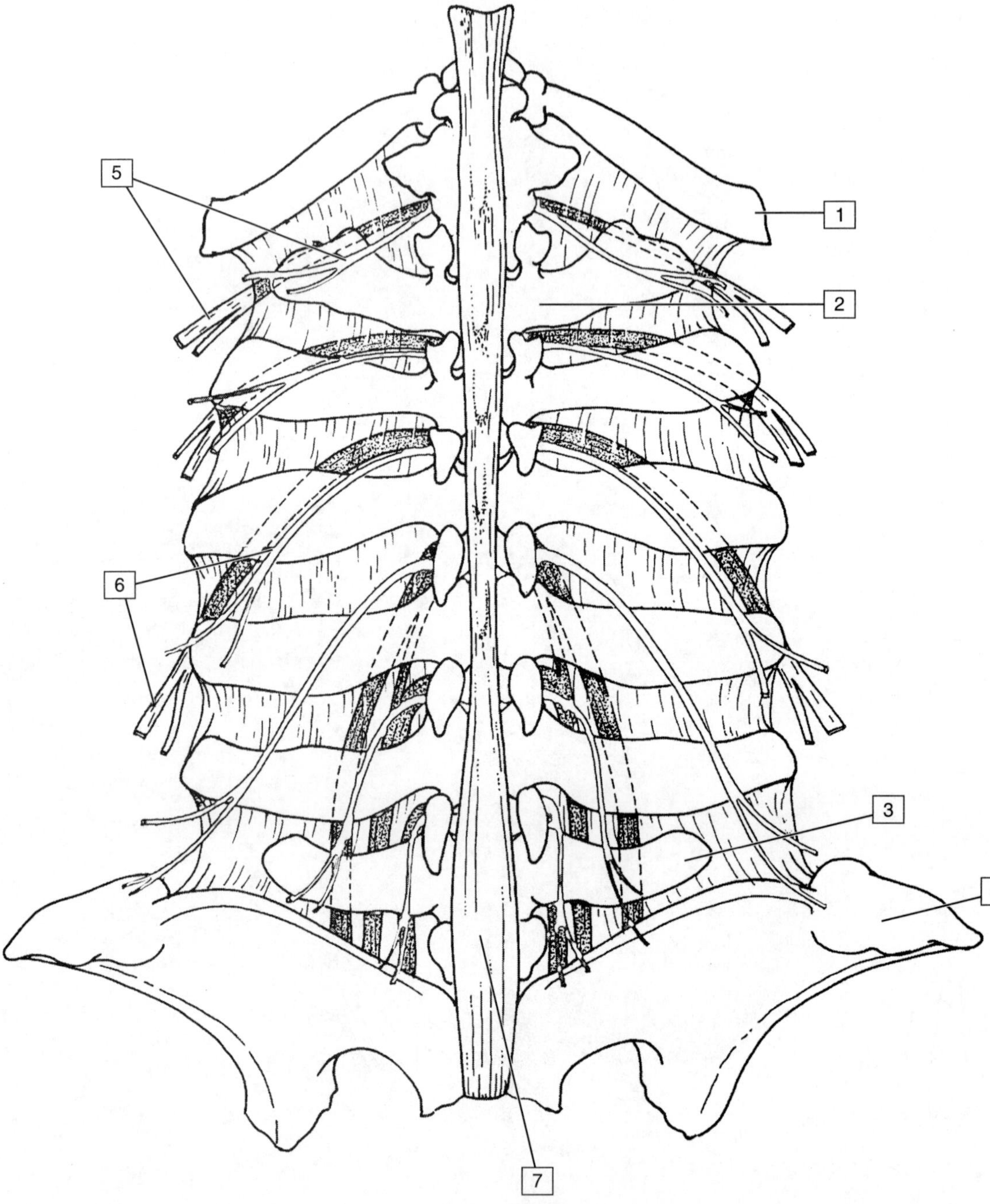

1	最后肋 Last rib
2	第1腰椎 First lumbar vertebra
3	第6腰椎 Sixth lumbar vertebra
4	髋结节 Coxal tuber
5	第13胸神经背侧支和腹侧支（腹侧支部分被虚线标出）Dorsal and ventral branches of 13th thoracic nerve (the ventral branch is partly stippled)
6	第2腰椎神经的背侧支和腹侧支 Dorsal and ventral branches of second lumbar nerve
7	棘上韧带 Supraspinous ligament

图4-31 牛腰椎神经与腰椎横突的关系

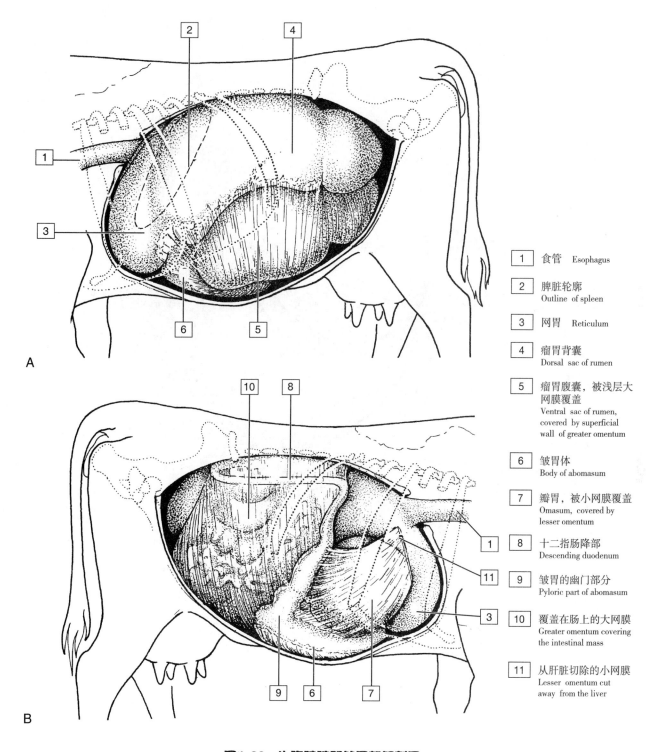

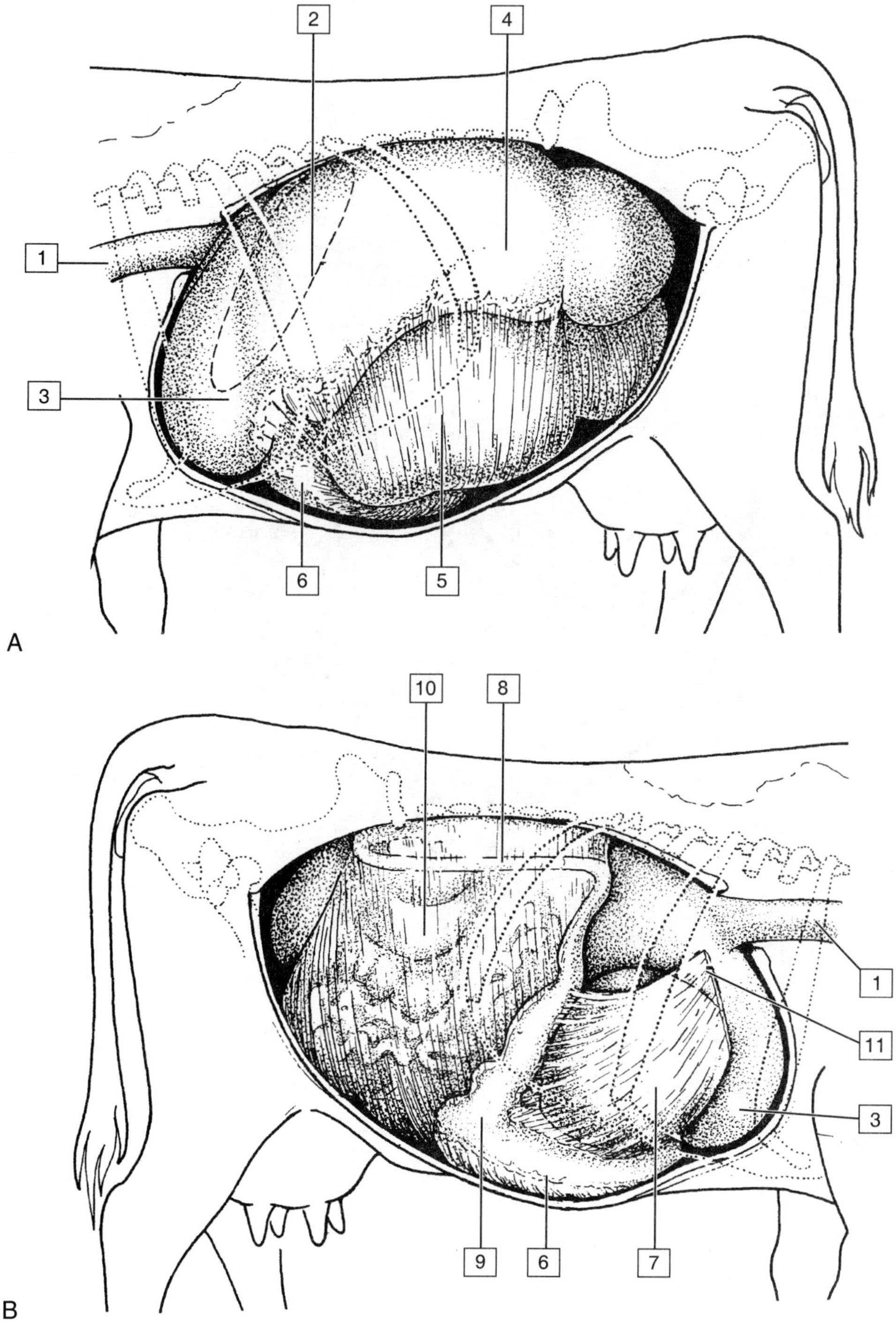

1 食管 Esophagus

2 脾脏轮廓 Outline of spleen

3 网胃 Reticulum

4 瘤胃背囊 Dorsal sac of rumen

5 瘤胃腹囊，被浅层大网膜覆盖 Ventral sac of rumen, covered by superficial wall of greater omentum

6 皱胃体 Body of abomasum

7 瓣胃，被小网膜覆盖 Omasum, covered by lesser omentum

8 十二指肠降部 Descending duodenum

9 皱胃的幽门部分 Pyloric part of abomasum

10 覆盖在肠上的大网膜 Greater omentum covering the intestinal mass

11 从肝脏切除的小网膜 Lesser omentum cut away from the liver

图4-32 牛腹腔脏器的局部解剖图

A. 腹腔脏器与左腹壁的关系；B. 腹腔脏器与右腹壁的关系。肝脏已被切除

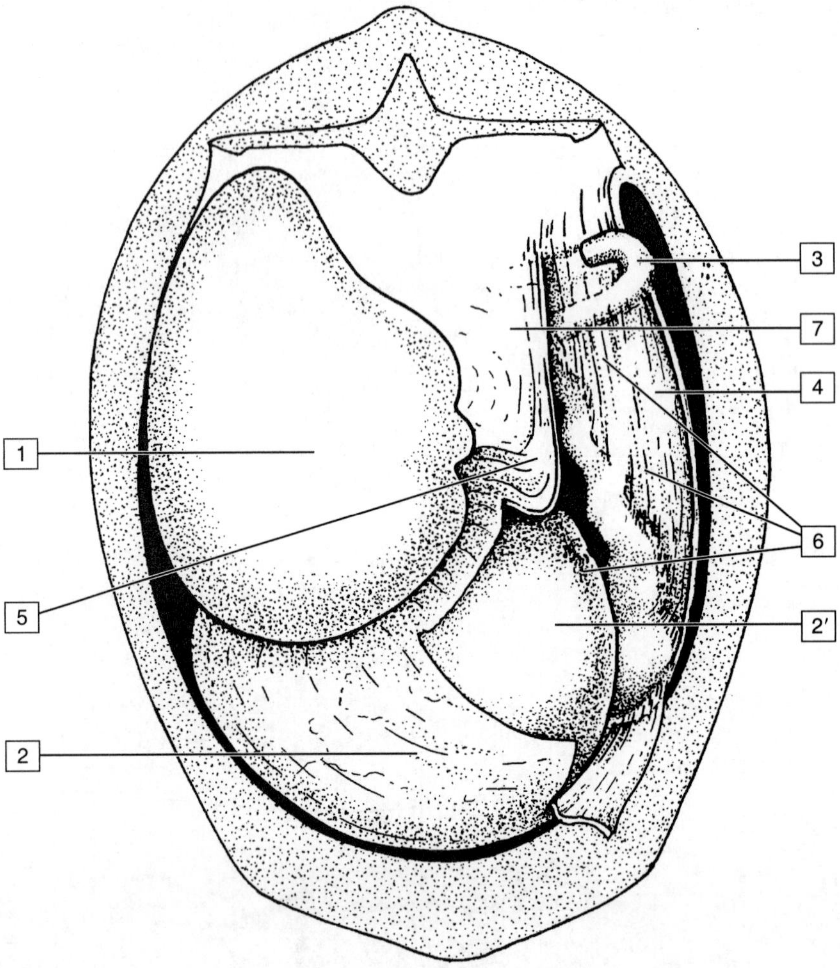

1	瘤胃背囊 Dorsal sac of rumen
2	瘤胃腹囊，被浅层大网膜覆盖 Ventral sac of rumen, covered by superficial wall of greater omentum
2′	瘤胃腹囊进入大网膜囊 Ventral sac of rumen projecting into omental bursa
3	十二指肠后曲 Caudal flexure of duodenum
4	浅层大网膜 Superficial wall of greater omentum
5	深层大网膜 Deep wall of greater omentum
6	网膜囊 Omental bursa
7	网膜上隐窝 Supraomental recess

图4-33 牛网膜囊，后面观

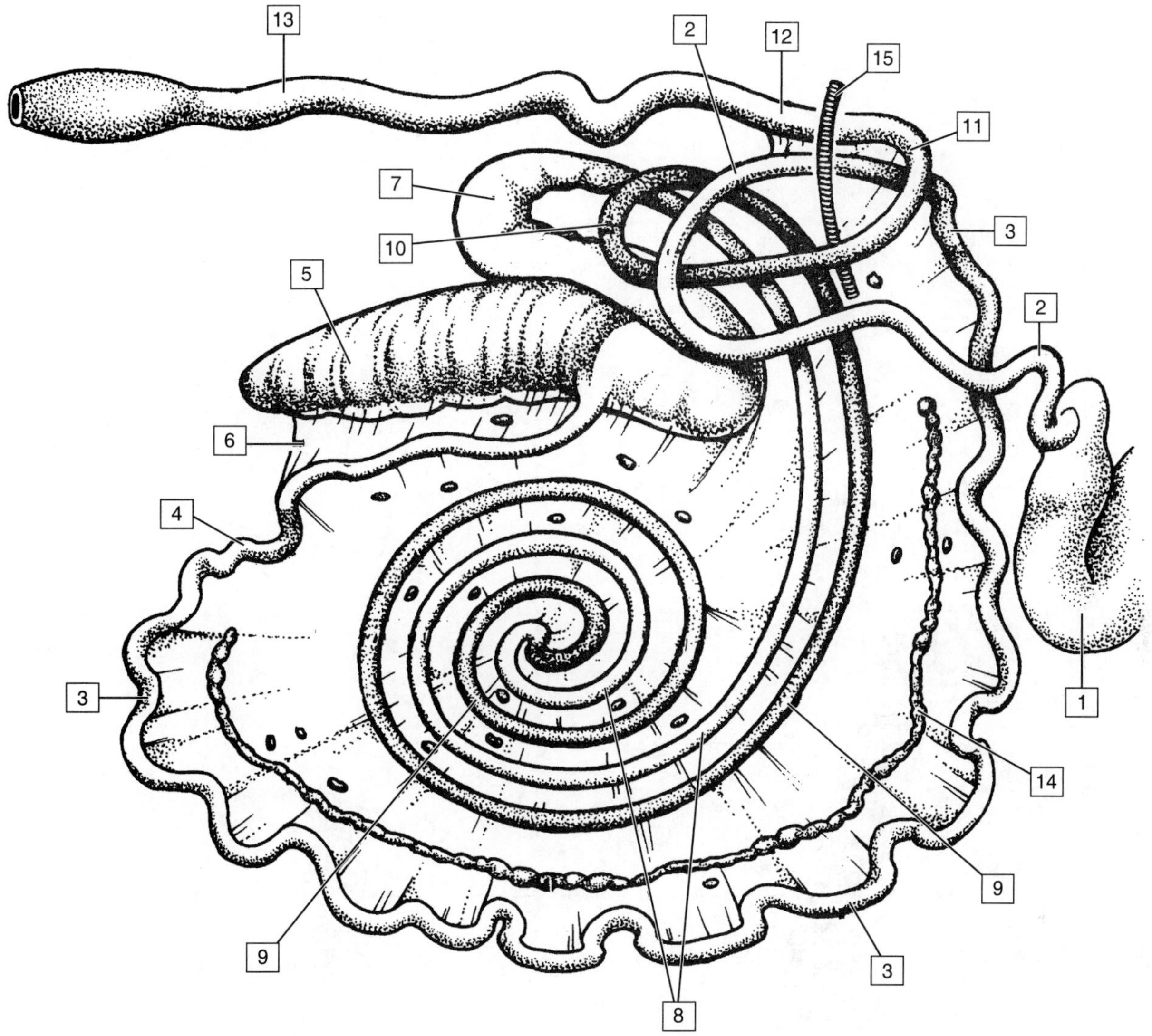

1 皱胃的幽门部 Pyloric part of abomasum	8 结肠旋袢向心回 Centripetal turns of spiral colon
2 十二指肠 Duodenum	9 结肠旋袢离心回 Centrifugal turns of spiral colon
3 空肠 Jejunum	10 升结肠终袢 Distal loop of ascending colon
4 回肠 Ileum	11 横结肠 Transverse colon
5 盲肠 Cecum	12 降结肠 Descending colon
6 回盲韧带 Ileocecal fold	13 直肠 Rectum
7 ~ 10 升结肠	14 空肠淋巴结 Jejunal lymph nodes
7 升结肠初袢 Proximal loop of ascending colon	15 肠系膜前动脉 Cranial mesenteric artery

图4-34 牛肠道，右侧观

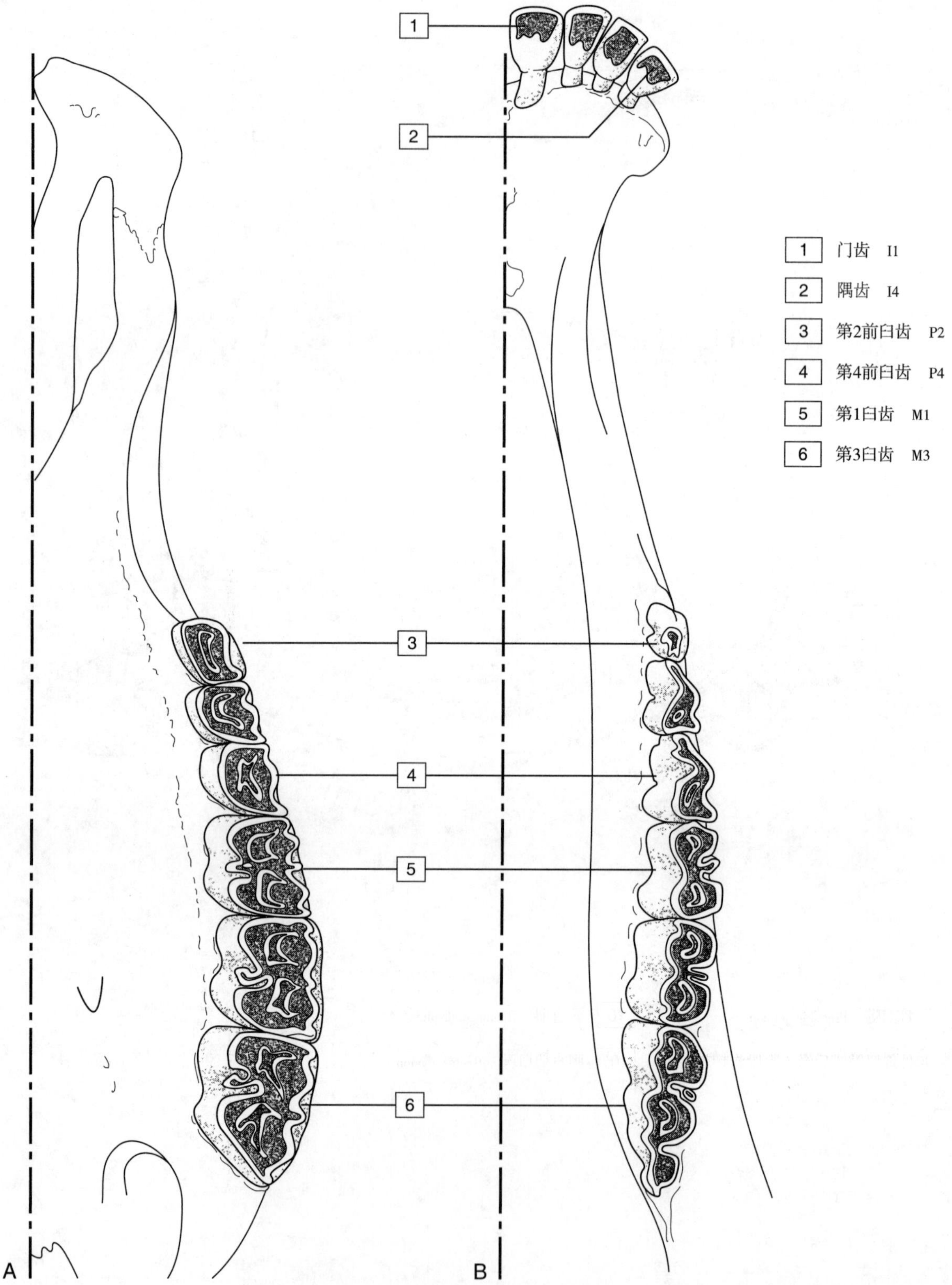

图4-35 牛的恒齿，上颌（A）和下颌（B）

猪

1	骨盆联合 Pelvic symphysis
2	耻骨前腱 Prepubic tendon
3	腹外斜肌腱膜后缘（腹股沟韧带）Caudal border of external oblique aponeurosis ("inguinal ligament")
4	髂外动脉 External iliac artery
5	股动脉 Femoral artery
6	股深动脉 Deep femoral artery
7	腹直肌腱外侧缘 Lateral border of rectus tendon
8	阴部外动脉 External pudendal artery
9	后上腹动脉 Caudal epigastric artery
10	腹直肌 Rectus abdominis
10'	腹直肌肌腱 Rectus tendon
11	腹内斜肌肌质部 Muscular part of internal abdominal oblique
11'	腹内斜肌腱膜部 Aponeurotic part of internal abdominal oblique
12	腹内斜肌后部游离缘 Caudal free border of internal abdominal oblique
13	提睾肌 Cremaster muscle
14	鞘膜和精索 Tunica vaginalis and spermatic cord
15	腹外斜肌肌质部 Muscular part of external abdominal oblique
15'	腹外斜肌腱膜部 Aponeurotic part of external abdominal oblique
16	腹股沟浅环（腹环）Superficial inguinal ring
17	腹股沟深环（皮下环）（箭头）Deep inguinal ring (arrows)
18	腹白线 Linea alba

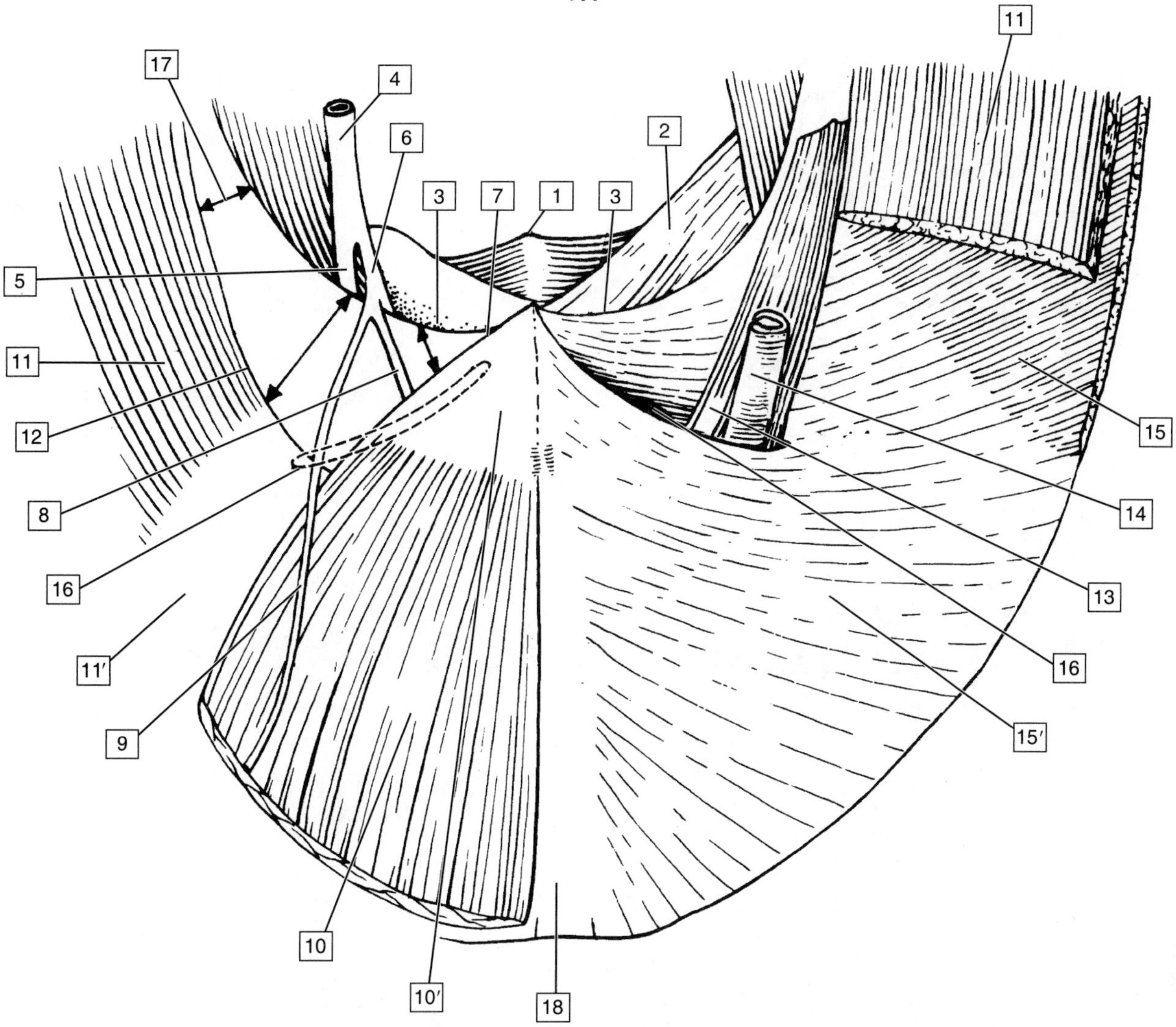

图4-36 公猪腹股沟管，前面观

后腹壁内部（深层）可见

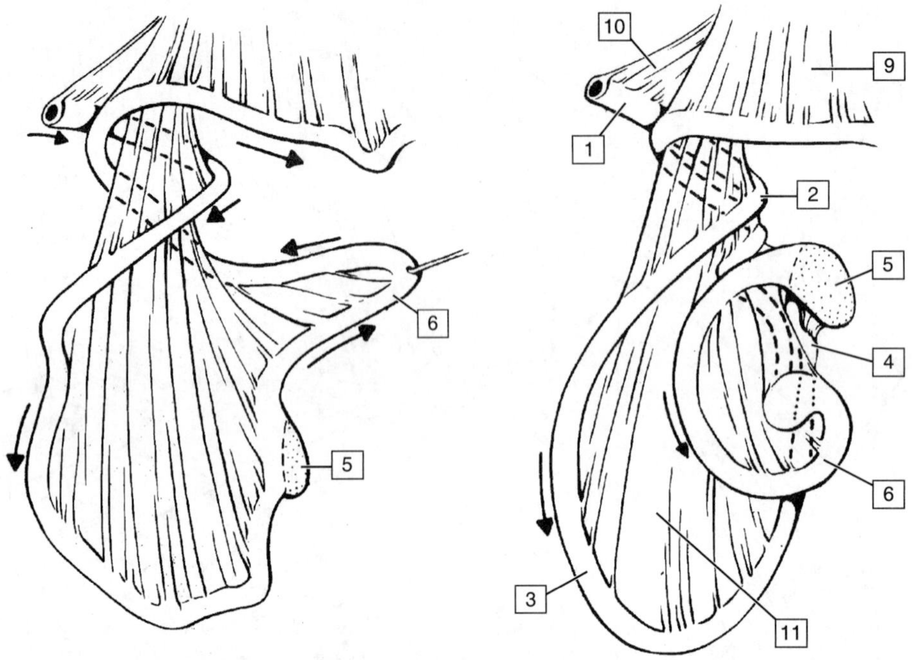

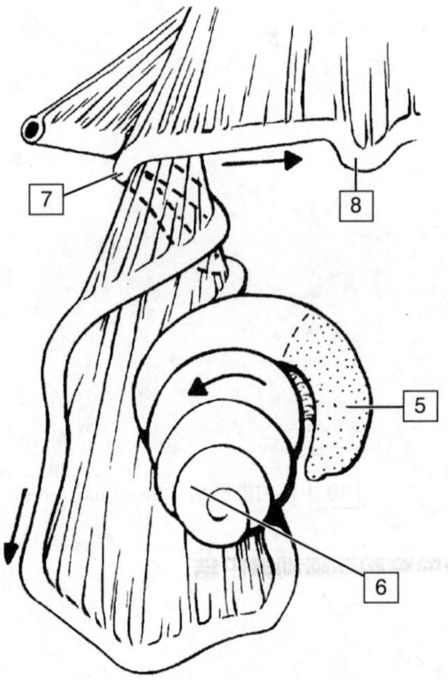

1 十二指肠降部 Descending duodenum

2 十二指肠后曲 Caudal flexure of duodenum

3 空肠 Jejunum

4 回肠 Ileum

5 盲肠 Cecum

6 升结肠 Ascending colon

7 横结肠 Transverse colon

8 降结肠 Descending colon

9 降结肠系膜 Descending mesocolon

10 十二指肠系膜 Mesoduodenum

11 肠系膜 Mesentery

图4-37 猪升结肠的发育，左外侧观

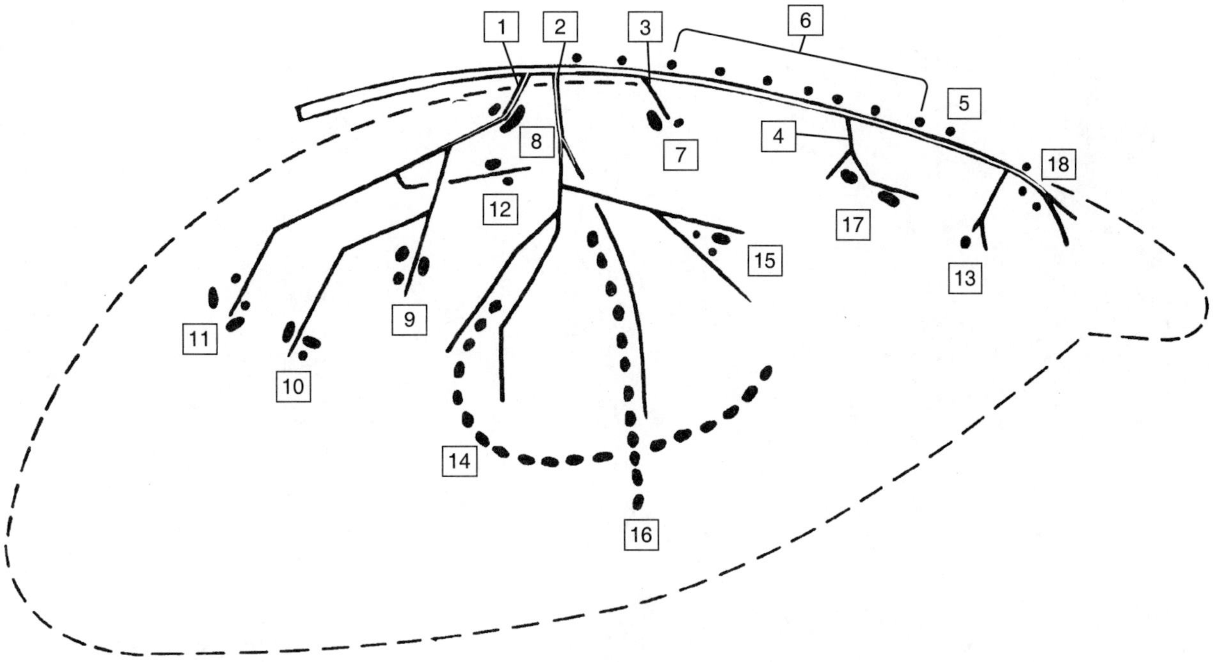

1 腹腔动脉 Celiac artery

2 肠系膜前动脉 Cranial mesenteric artery

3 肾动脉 Renal artery

4 肠系膜后动脉 Caudal mesenteric artery

5 旋髂深动脉 Deep circumflex iliac artery

6 腰主动脉淋巴结 Lumbar aortic nodes

7 肾淋巴结 Renal nodes

8 腹腔淋巴结 Celiac nodes

9 脾淋巴结 Splenic nodes

10 胃淋巴结 Gastric nodes

11 肝淋巴结 Hepatic nodes

12 胰十二指肠淋巴结 Pancreaticoduodenal nodes

13 髂外淋巴结 Lateral iliac nodes

14 空肠淋巴结 Jejunal nodes

15 回结肠淋巴结 Ileocolic nodes

16 腹腔淋巴结 Colic nodes

17 肠系膜后淋巴结 Caudal mesenteric nodes

18 髂内淋巴结 Medial iliac nodes

图4-38 猪主要的腹腔动脉和淋巴结

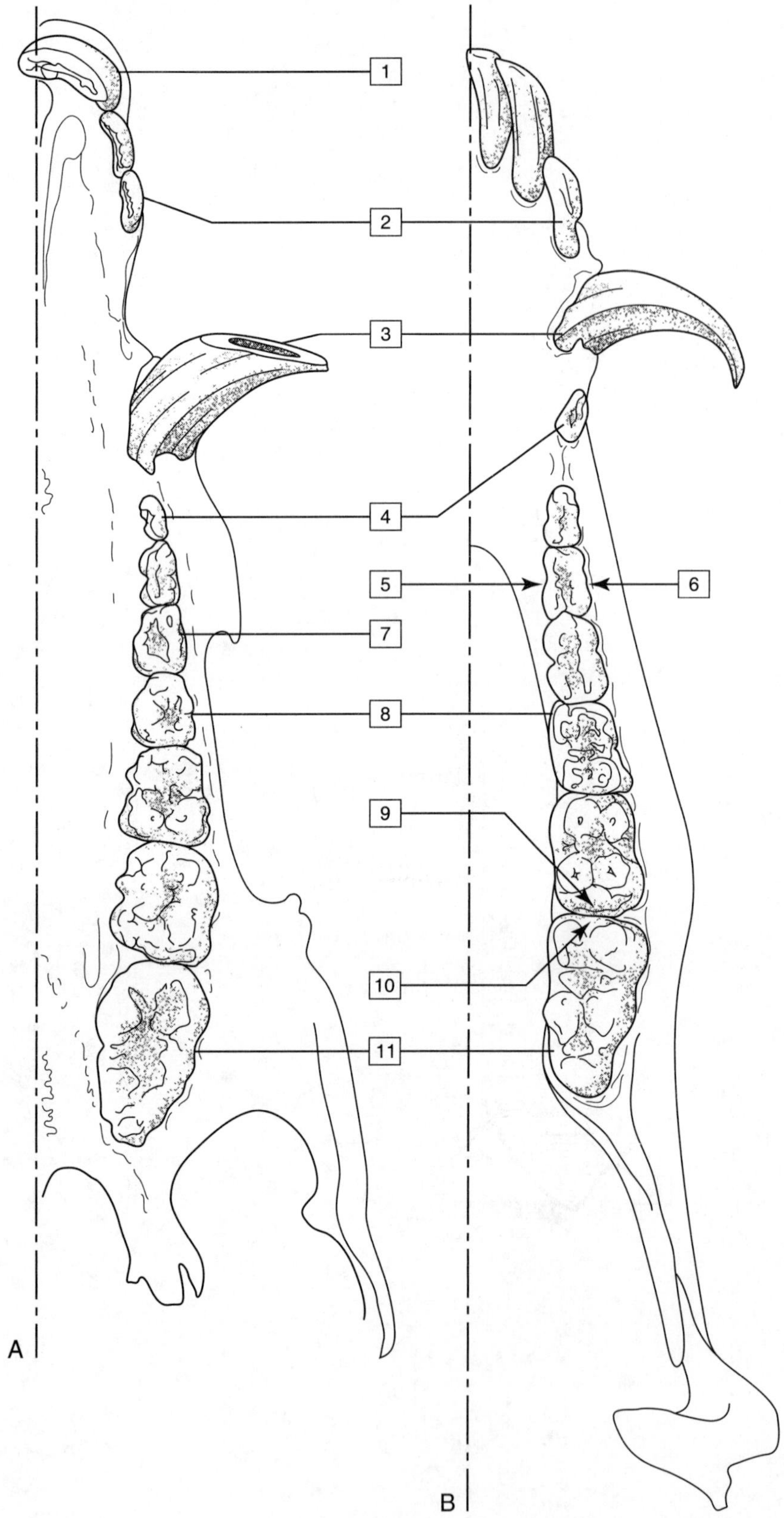

1	第1切齿 I1
2	第3切齿 I3
3	犬齿 C
4	第1前臼齿 P1
5	舌面 Lingual surface
6	前庭面 Vestibular surface
7	第4前臼齿 P4
8	第1臼齿 M1
9	远侧面 Distal surface
10	近中面 Mesial surface
11	第3臼齿 M3

图4-39 猪的恒齿，上颌（A）和下颌（B）

鸟类

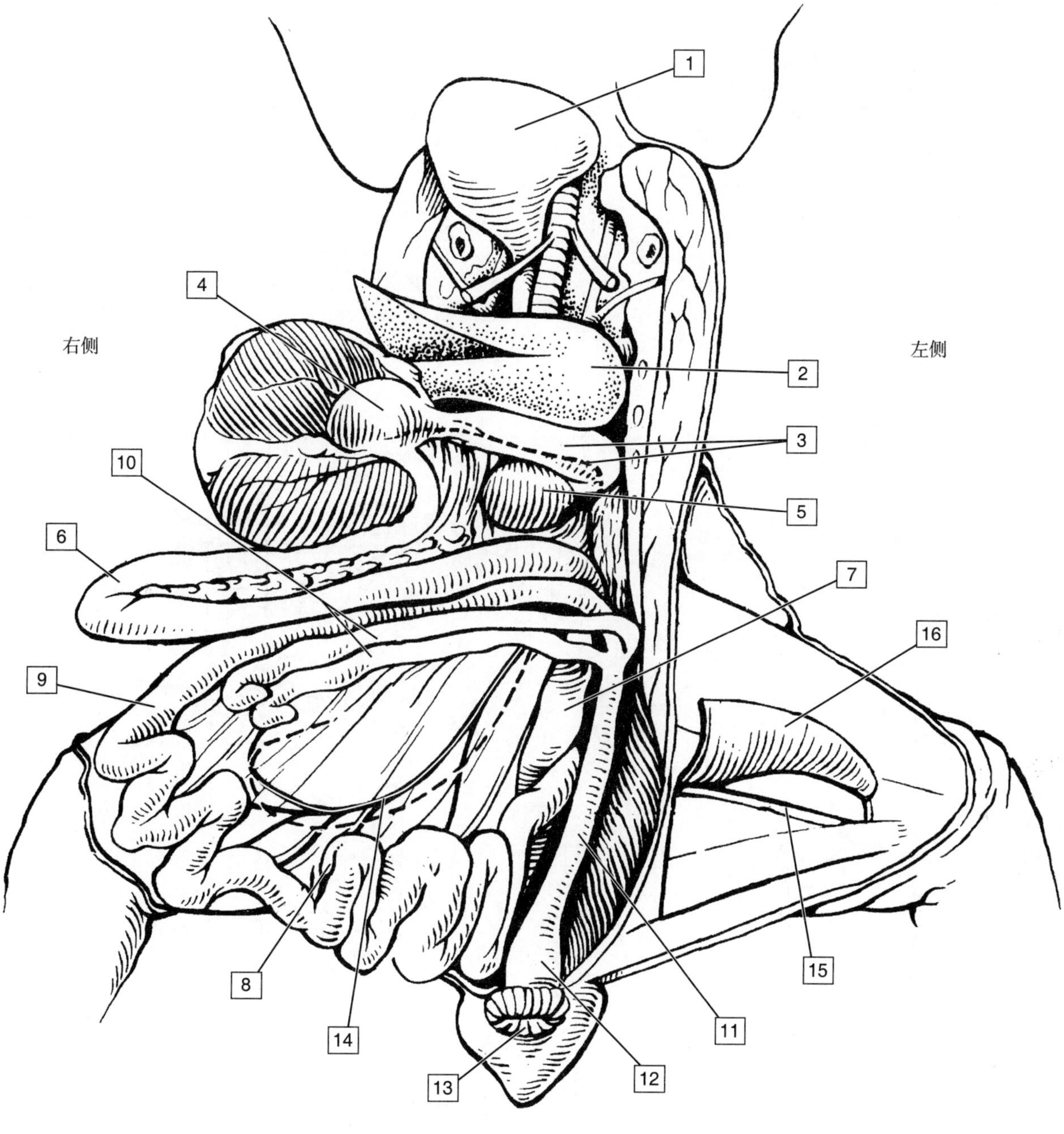

1	嗉囊 Crop
2	肝左叶 Left lobe of liver
3	背侧有迷走神经的腺胃 Proventriculus with vagus on dorsal surface
4	前盲囊位于掀开的肌胃右侧 Cranial blind sac on right side of reflected gizzard
5	脾 Spleen
6	包围胰腺的十二指肠袢 Duodenal loop enclosing pancreas
7	空肠 Jejunum
8	卵黄囊憩室 Vitelline diverticulum
9	回肠 Ileum
10	盲肠 Ceca
11	结肠 Colon
12	泄殖腔 Cloaca
13	肛门 Vent
14	肠系膜中的肠系膜前血管和肠神经 Cranial mesenteric vessels and intestinal nerve in mesentery
15	荐神经和坐骨动脉 Sciatic nerve and ischial artery
16	股薄肌和内收肌 Gracilis and adductor

图4-40 肝脏、胃和小肠向前掀开可见禽类的胃肠道，腹侧观

不同动物的解剖结构比较

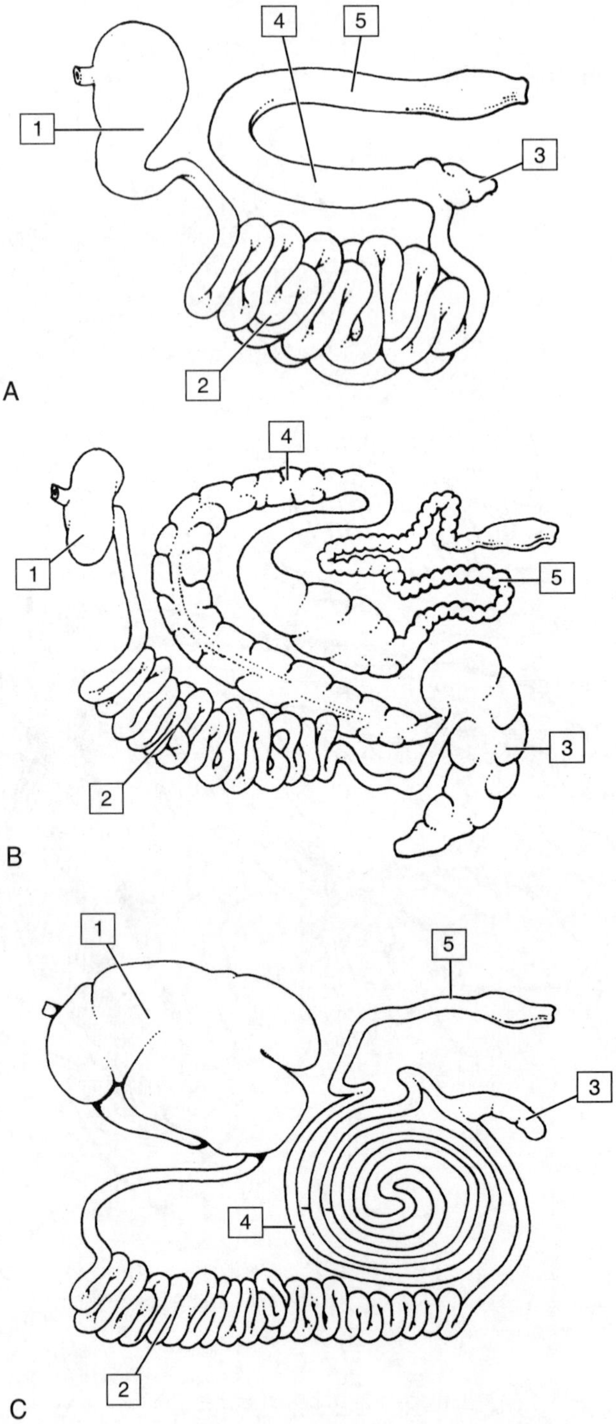

1 胃 Stomach

2 小肠 Small intestine

3 盲肠 Cecum

4 升结肠 Ascending colon

5 降结肠 Descending colon

图4-41 置于同一平面的犬（A）、马（B）和牛（C）的胃肠胃

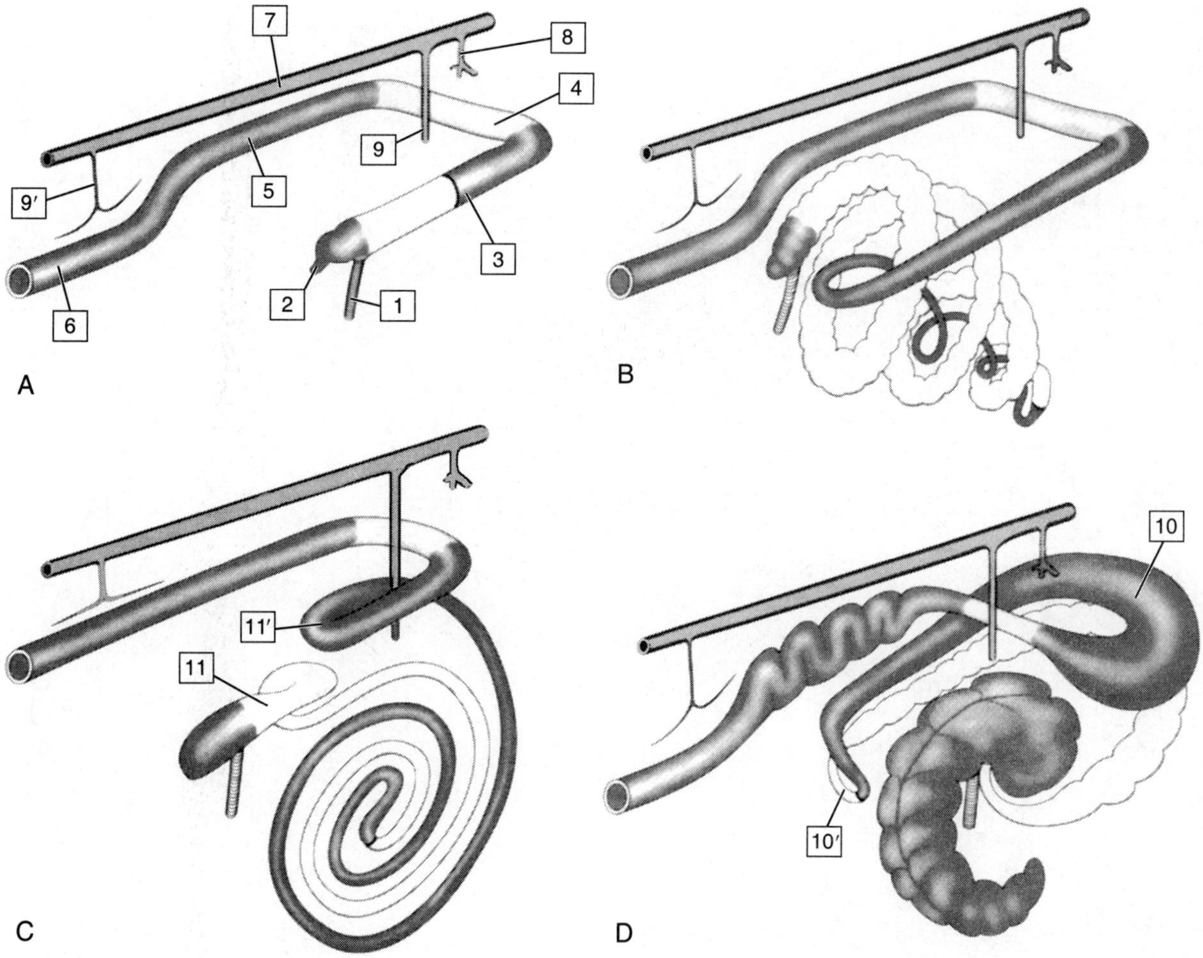

1	回肠	Ileum
2	盲肠	Cecum
3	升结肠	Ascending colon
4	横结肠	Transverse colon
5	降结肠	Descending colon
6	直肠和肛门	Rectum and anus
7	主动脉	Aorta
8	腹腔动脉	Celiac artery
9	肠系膜前动脉	Cranial mesenteric artery
9′	肠系膜后动脉	Caudal mesenteric artery
10	升结肠的背侧膈曲	Dorsal diaphragmatic flexures of ascending colon
10′	升结肠的盆曲	Pelvic flexures of ascending colon
11	升结肠近袢	Proximal loop of ascending colon
11′	升结肠远袢	Distal loop of ascending colon

图4-42 犬和猫(A)、猪(B)、牛(C)、马(D)的大肠

头位于右上方

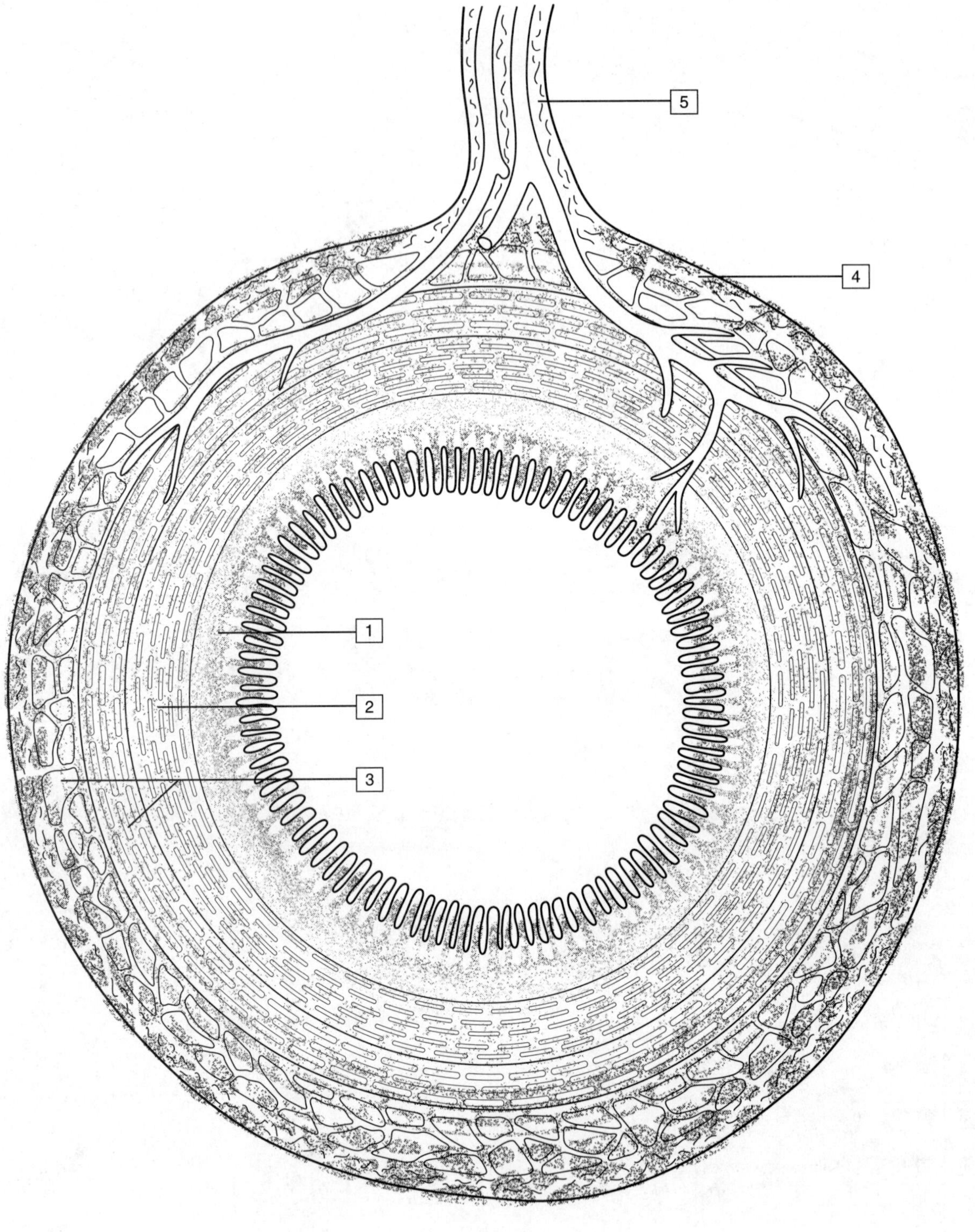

1 黏膜 Mucosa	3 肌肉层 Muscle layers	5 肠系膜 Mesentery
2 黏膜下层 Submucosa	4 浆膜 Serosa	

图4-43　肠的横断面（一般解剖）

5

骨盆和生殖器官

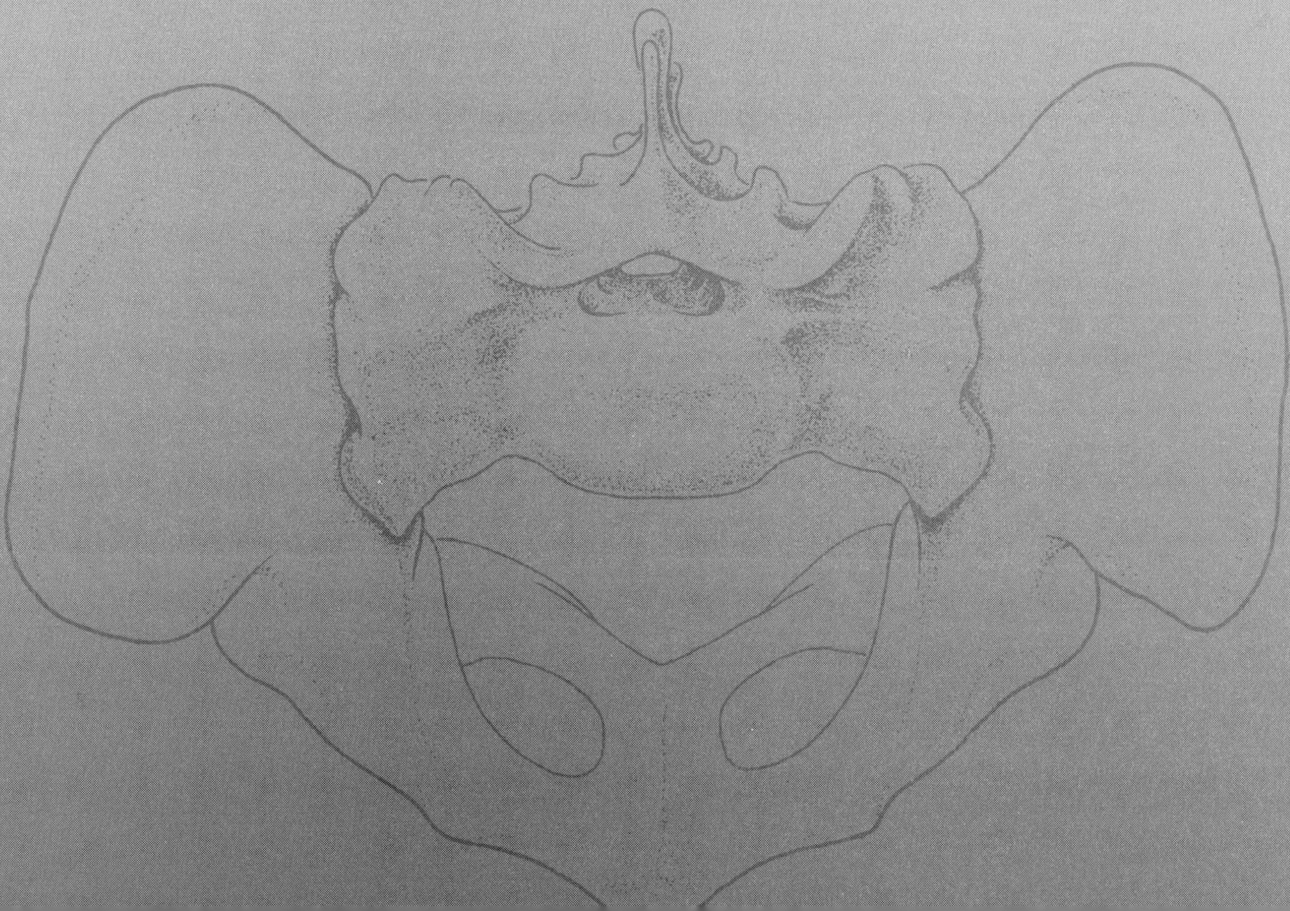

犬

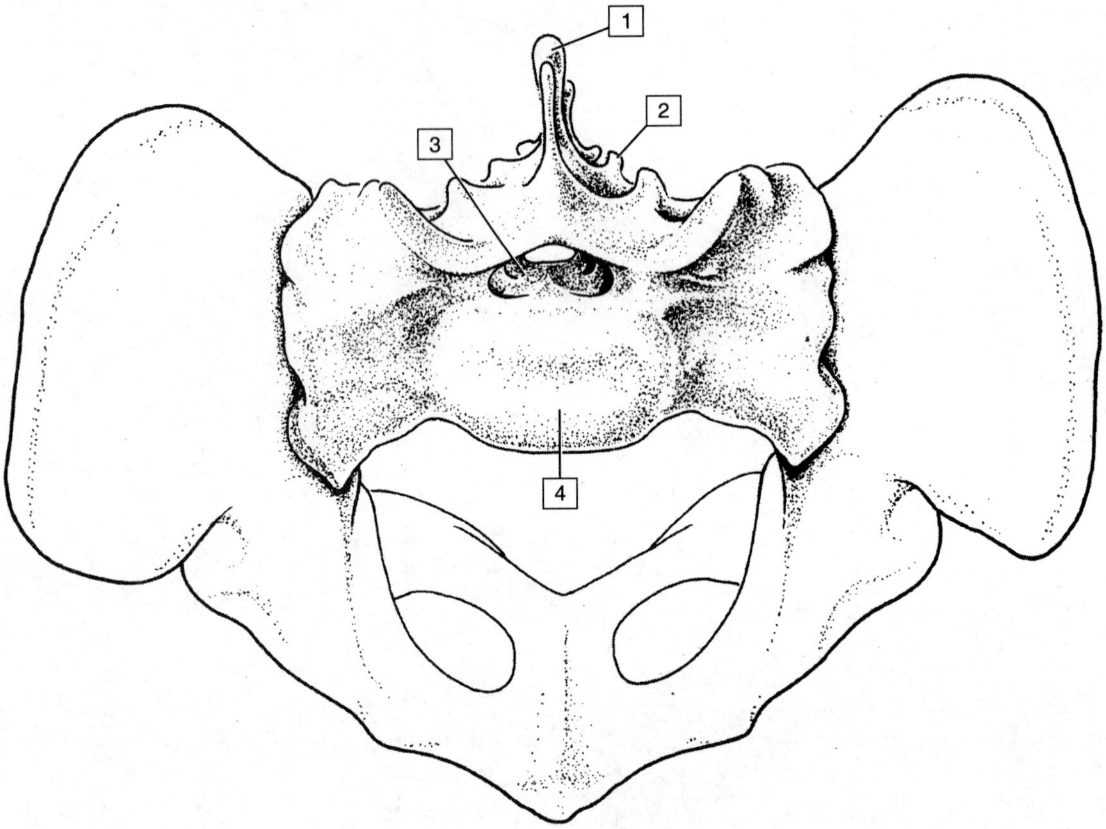

1 棘突 Spinous process	3 椎管 Vertebral canal
2 基本关节突 Rudimentary articular process	4 椎体 Body of vertebra

图5-1 犬的荐骨，前面观

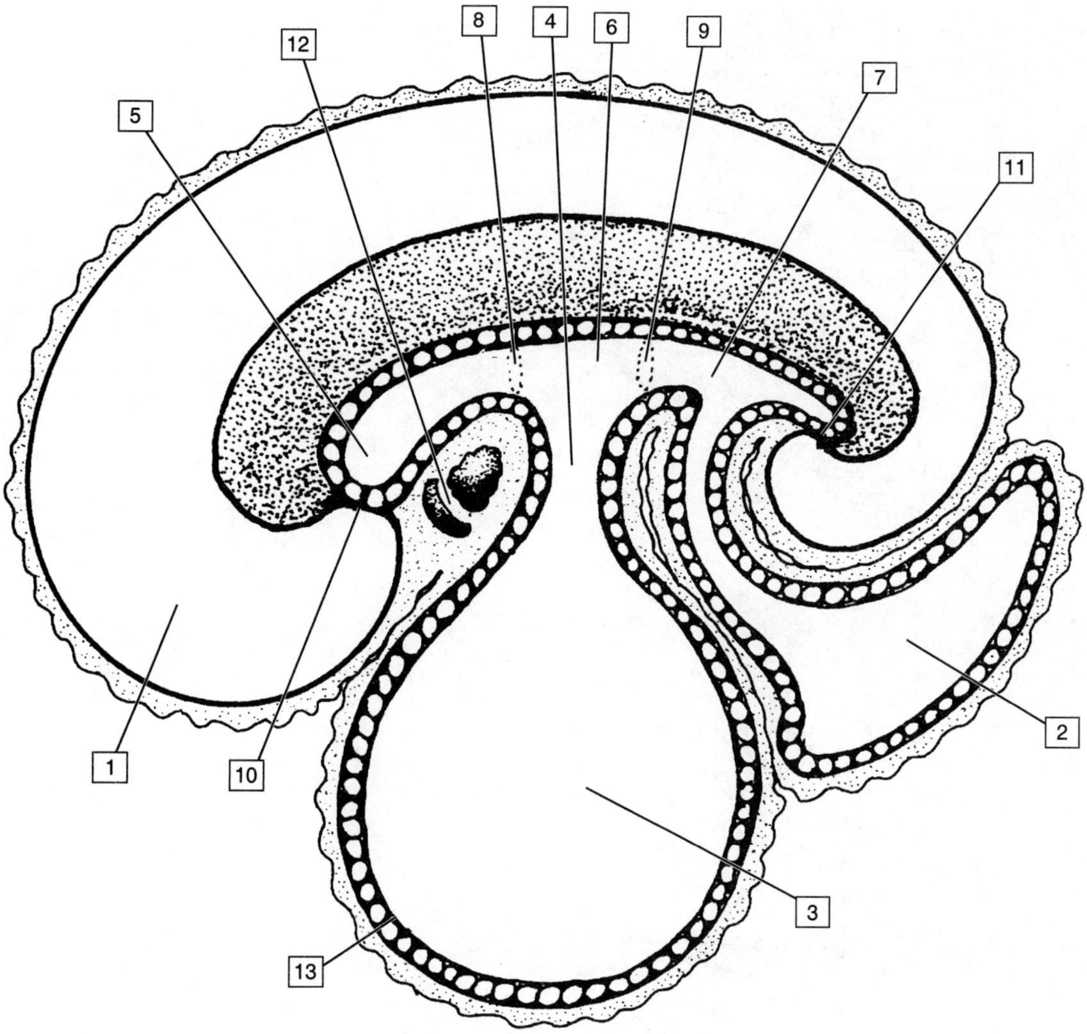

1 羊膜腔 Amniotic cavity

2 尿囊腔 Allantoic cavity

3 卵黄囊 Yolk sac

4 卵黄囊柄 Stalk of yolk sac

5 前肠 Foregut

6 中肠 Midgut

7 后肠 Hindgut

8 前肠口 Cranial intestinal portal

9 后肠口 Caudal intestinal portal

10 椭圆盘 Oral plate

11 泄殖腔板 Cloacal plate

12 心脏和心包腔 Heart and pericardial cavity

13 内胚层 Endoderm

图5-2 犬胚胎早期的矢状面

部分卵黄囊在折叠过程中被带入体内

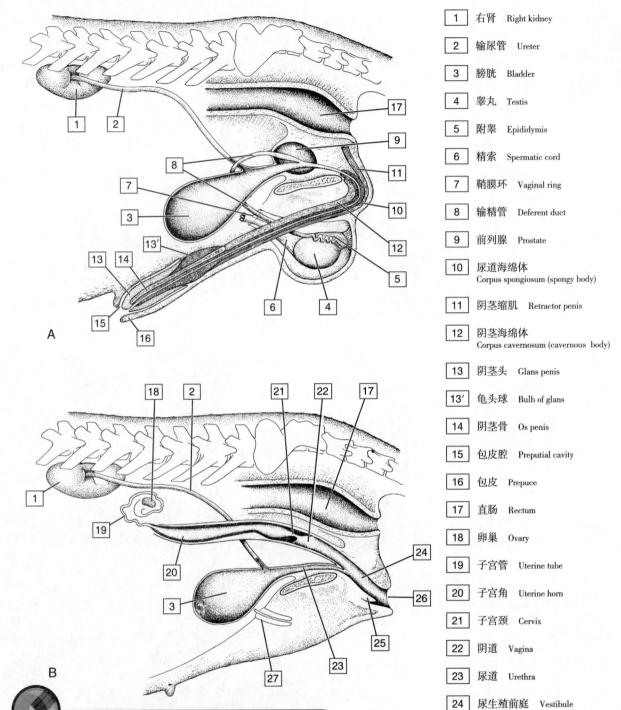

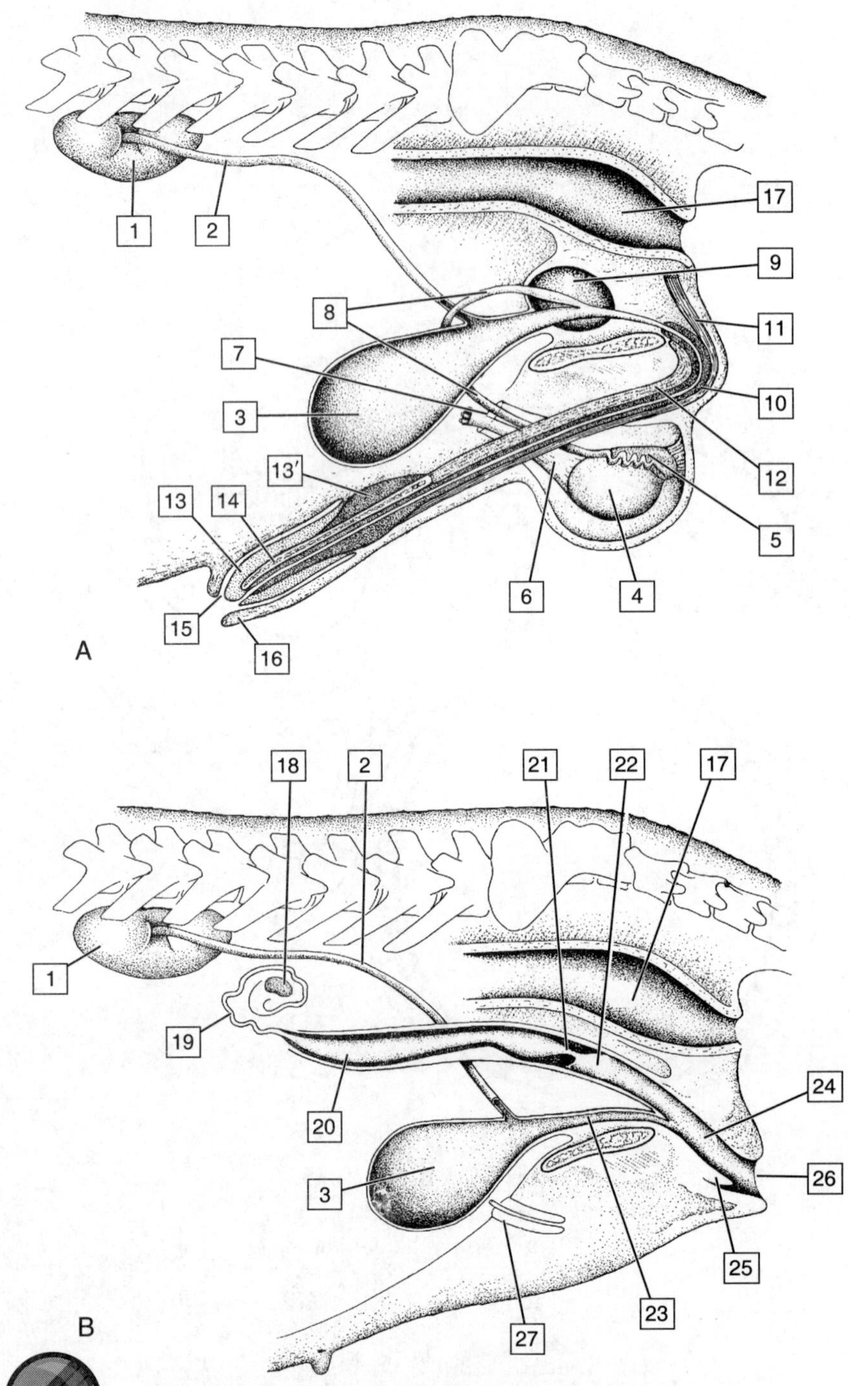

请说出为雌性犬子宫提供养分的动脉名称。

- 1 右肾 Right kidney
- 2 输尿管 Ureter
- 3 膀胱 Bladder
- 4 睾丸 Testis
- 5 附睾 Epididymis
- 6 精索 Spermatic cord
- 7 鞘膜环 Vaginal ring
- 8 输精管 Deferent duct
- 9 前列腺 Prostate
- 10 尿道海绵体 Corpus spongiosum (spongy body)
- 11 阴茎缩肌 Retractor penis
- 12 阴茎海绵体 Corpus cavernosum (cavernous body)
- 13 阴茎头 Glans penis
- 13′ 龟头球 Bulb of glans
- 14 阴茎骨 Os penis
- 15 包皮腔 Preputial cavity
- 16 包皮 Prepuce
- 17 直肠 Rectum
- 18 卵巢 Ovary
- 19 子宫管 Uterine tube
- 20 子宫角 Uterine horn
- 21 子宫颈 Cervix
- 22 阴道 Vagina
- 23 尿道 Urethra
- 24 尿生殖前庭 Vestibule
- 25 阴蒂 Clitoris
- 26 外阴 Vulva
- 27 鞘突 Vaginal process

图5-3 犬泌尿系统及生殖器官

A. 雄性；B. 雌性

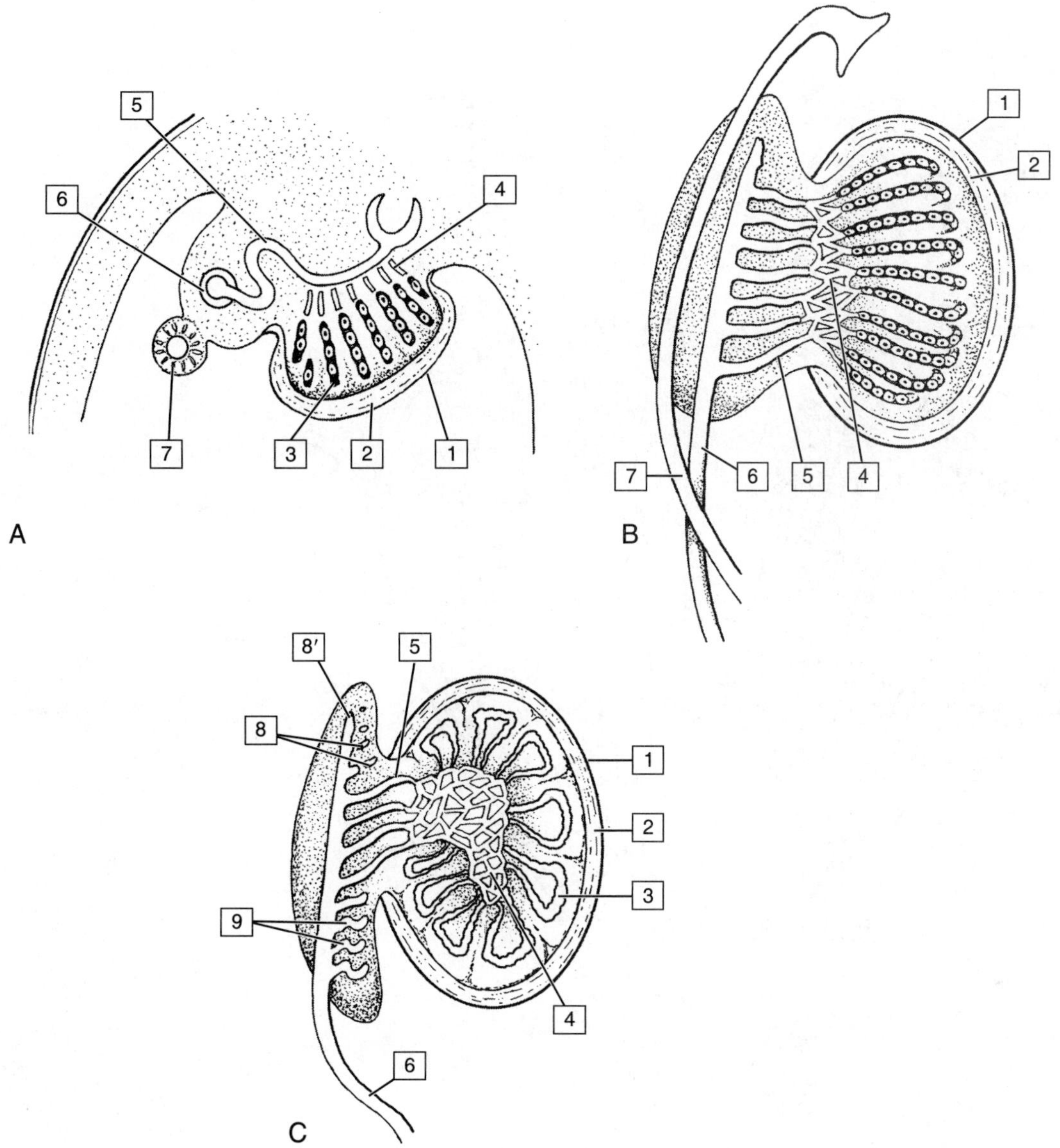

1 体腔 Celom epithelium

2 白膜 Tunica albuginea

3 上皮索，曲细精管 Epithelial cords, seminiferous tubules

4 睾丸网 Rete testis

5 中肾小管，输出管 Mesonephric tubules, efferent ductules

6 中肾管（后输出管） Mesonephric (later deferent) duct

7 副中肾管 Paramesonephric duct

8 中肾小管的残迹（畸变小管）
Cranial remnant of mesonephric tubules (aberrant ductules)

8′ 中肾管残迹（近附睾） Remnant of 6 (appendix of epididymis)

9 后部残迹（旁睾） Caudal remnant (paradidymis)

图5-4 犬睾丸发育的三个阶段

A. 上皮索通过形成白膜与上皮分离；B. 上皮索、睾丸网和中肾小管相互连接；

C. 上皮索形成精小管，中肾逐渐转变为附睾的一部分

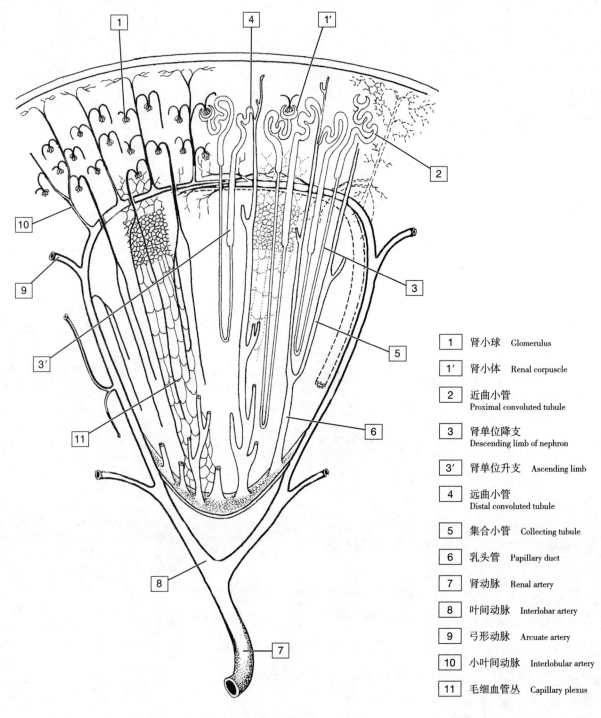

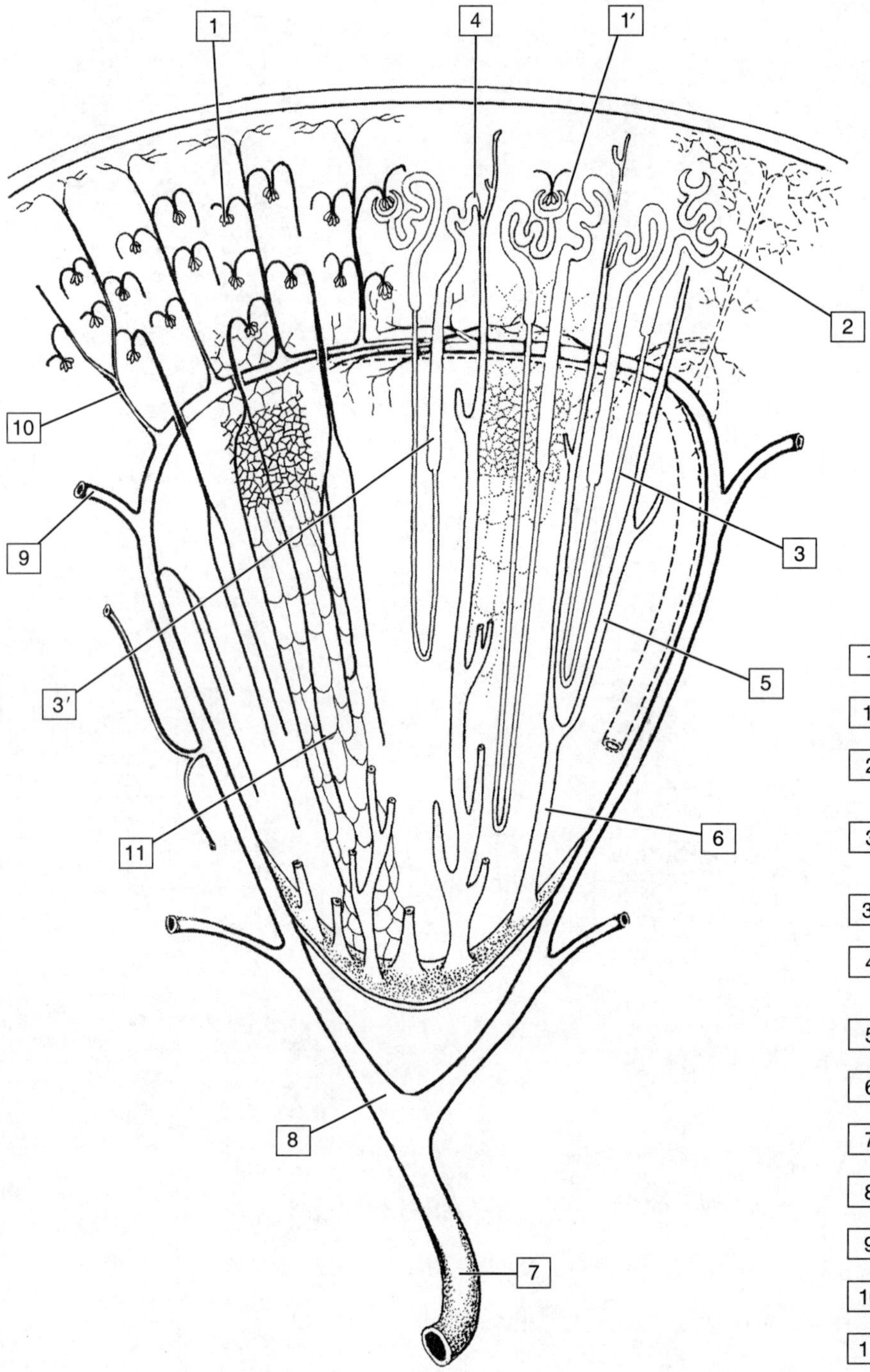

1	肾小球	Glomerulus
1′	肾小体	Renal corpuscle
2	近曲小管	Proximal convoluted tubule
3	肾单位降支	Descending limb of nephron
3′	肾单位升支	Ascending limb
4	远曲小管	Distal convoluted tubule
5	集合小管	Collecting tubule
6	乳头管	Papillary duct
7	肾动脉	Renal artery
8	叶间动脉	Interlobar artery
9	弓形动脉	Arcuate artery
10	小叶间动脉	Interlobular artery
11	毛细血管丛	Capillary plexus

图5-5　犬肾叶

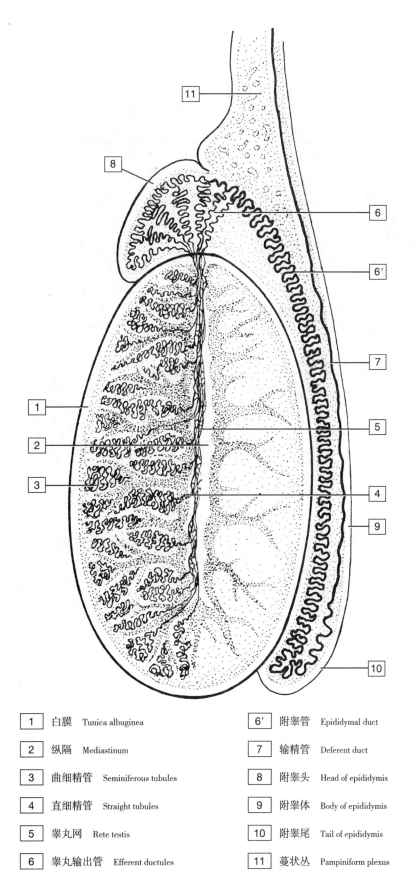

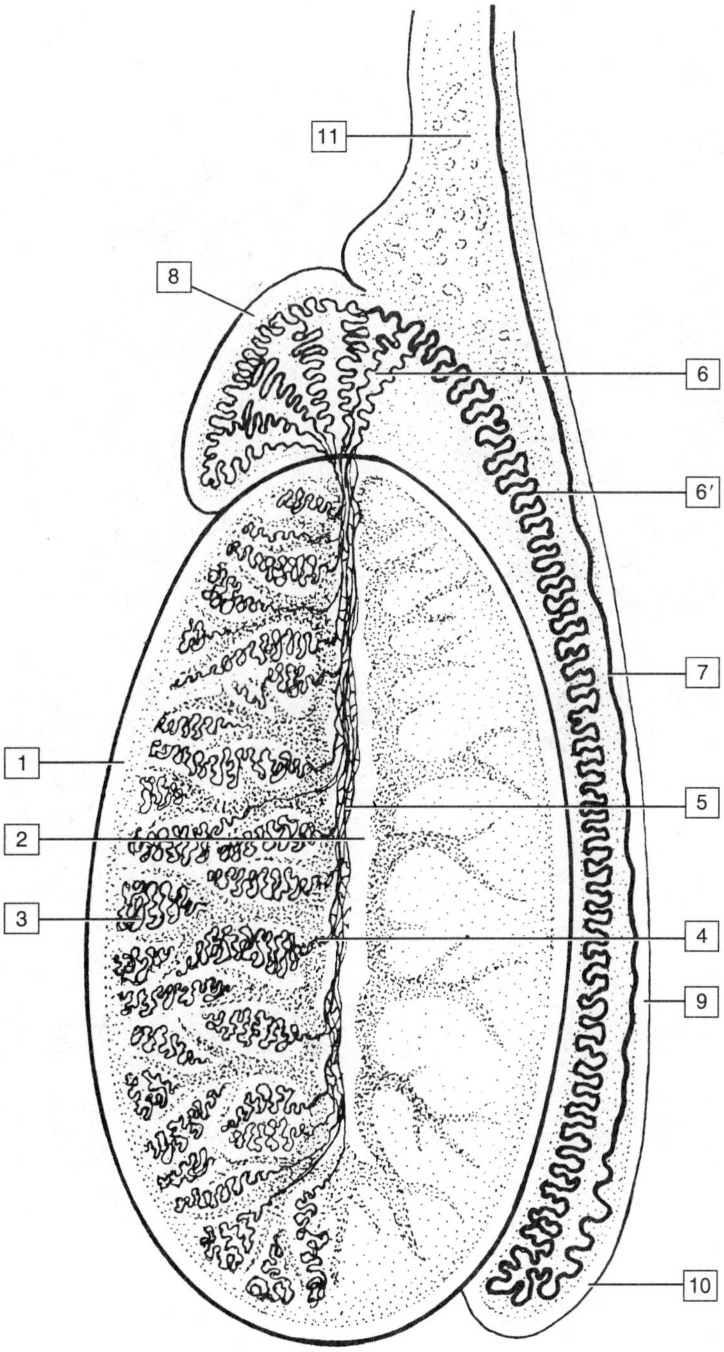

1	白膜	Tunica albuginea	6′	附睾管	Epididymal duct
2	纵隔	Mediastinum	7	输精管	Deferent duct
3	曲细精管	Seminiferous tubules	8	附睾头	Head of epididymis
4	直细精管	Straight tubules	9	附睾体	Body of epididymis
5	睾丸网	Rete testis	10	附睾尾	Tail of epididymis
6	睾丸输出管	Efferent ductules	11	蔓状丛	Pampiniform plexus

图5-6 犬睾丸和附睾纵切面

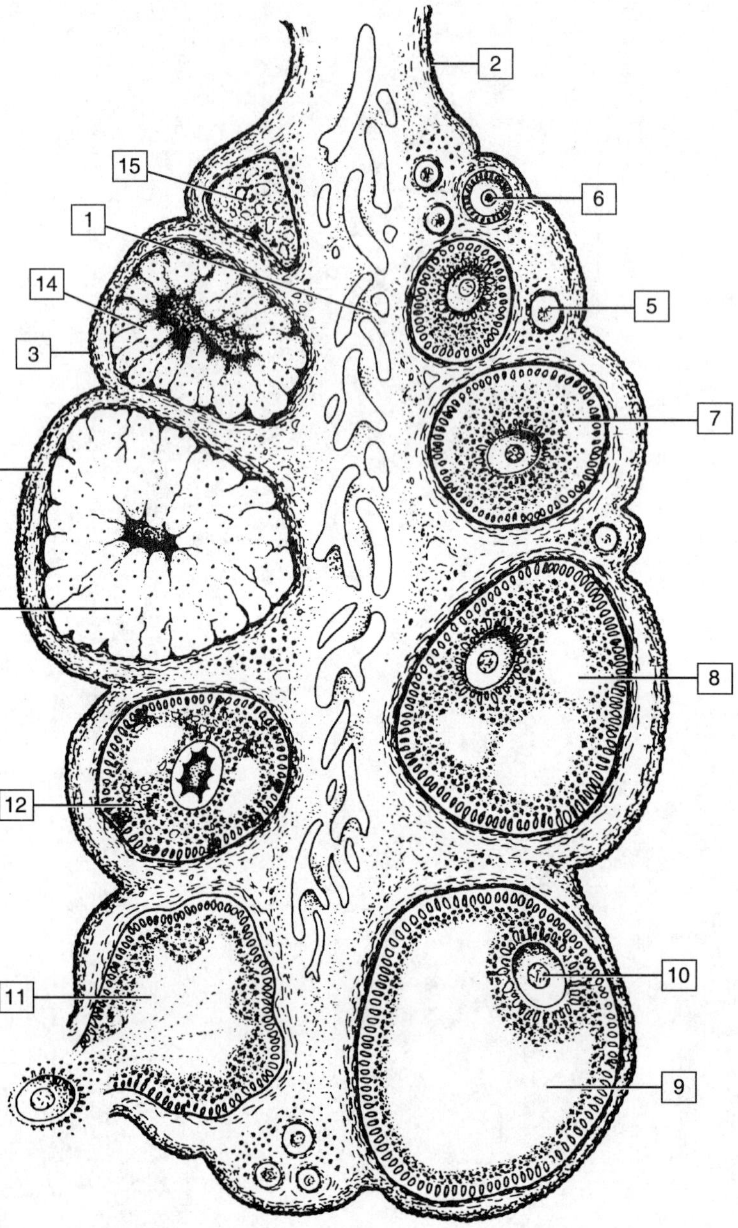

1	髓质	Medulla	5	原始卵泡	Primordial follicle	9	成熟卵泡	Mature follicle	13	黄体	Corpus luteum
2	卵巢系膜	Mesovarium	6	初级卵泡	Primary follicle	10	卵母细胞	Oocyte	14	闭锁黄体	Atretic corpus luteum
3	上皮表面	Surface epithelium	7	次级卵泡	Secondary follicle	11	破裂卵泡	Ruptured follicle	15	白体	Corpus albicans
4	白膜（发育不良）	Tunica albuginea (poorly developed)	8	早期三级卵泡	Early tertiary follicle	12	闭锁卵泡	Atretic follicle			

图5-7 犬卵巢活动的不同功能阶段

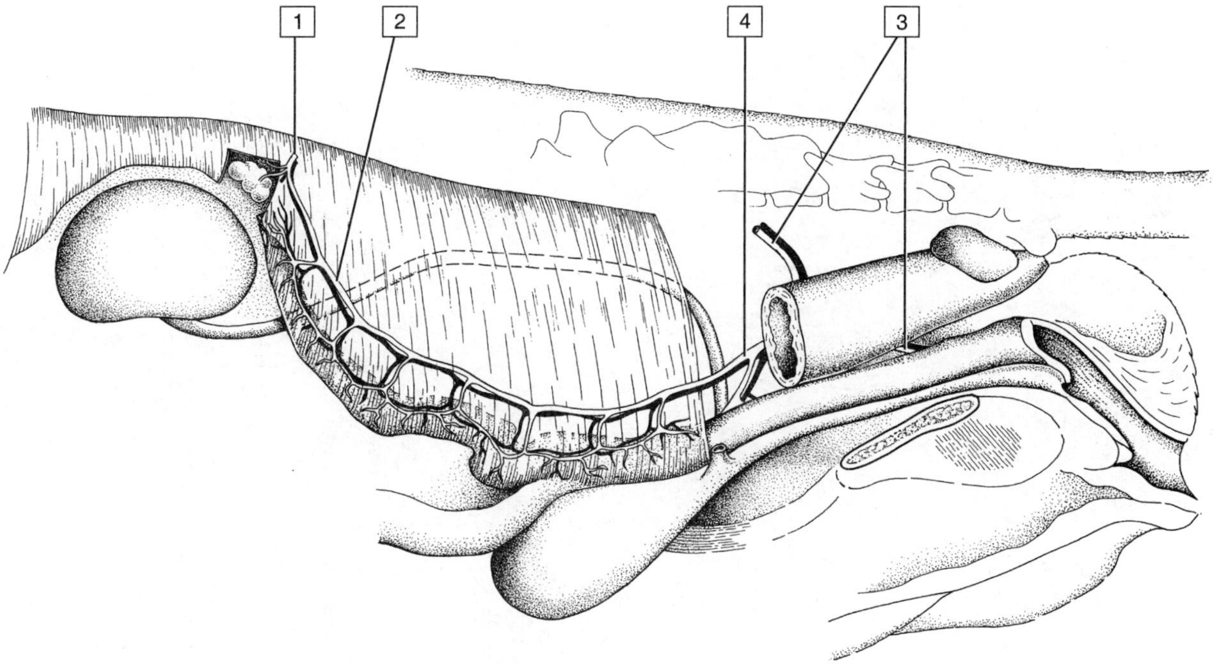

1	卵巢动脉 Ovarian artery	3	阴道动脉 Vaginal artery
2	卵巢动脉子宫支 Uterine branch of the ovarian artery	4	子宫动脉 Uterine artery

图5-8 雌性犬生殖道的血液供应

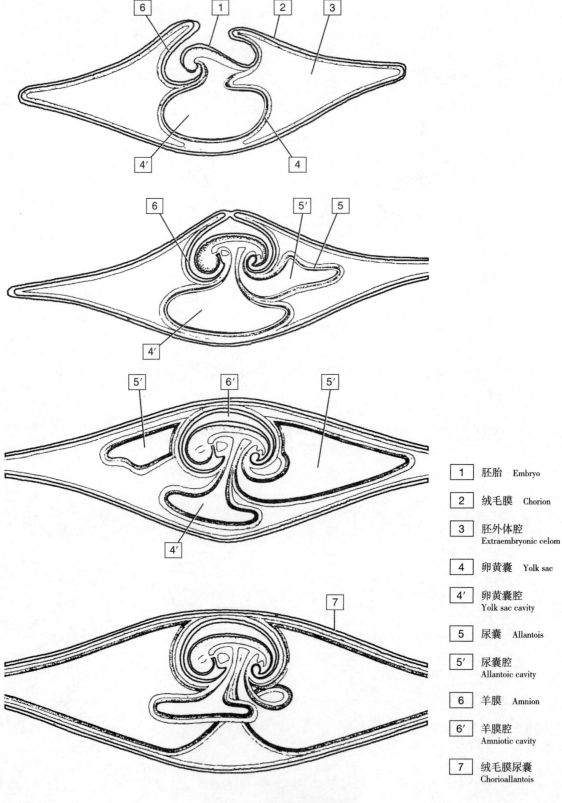

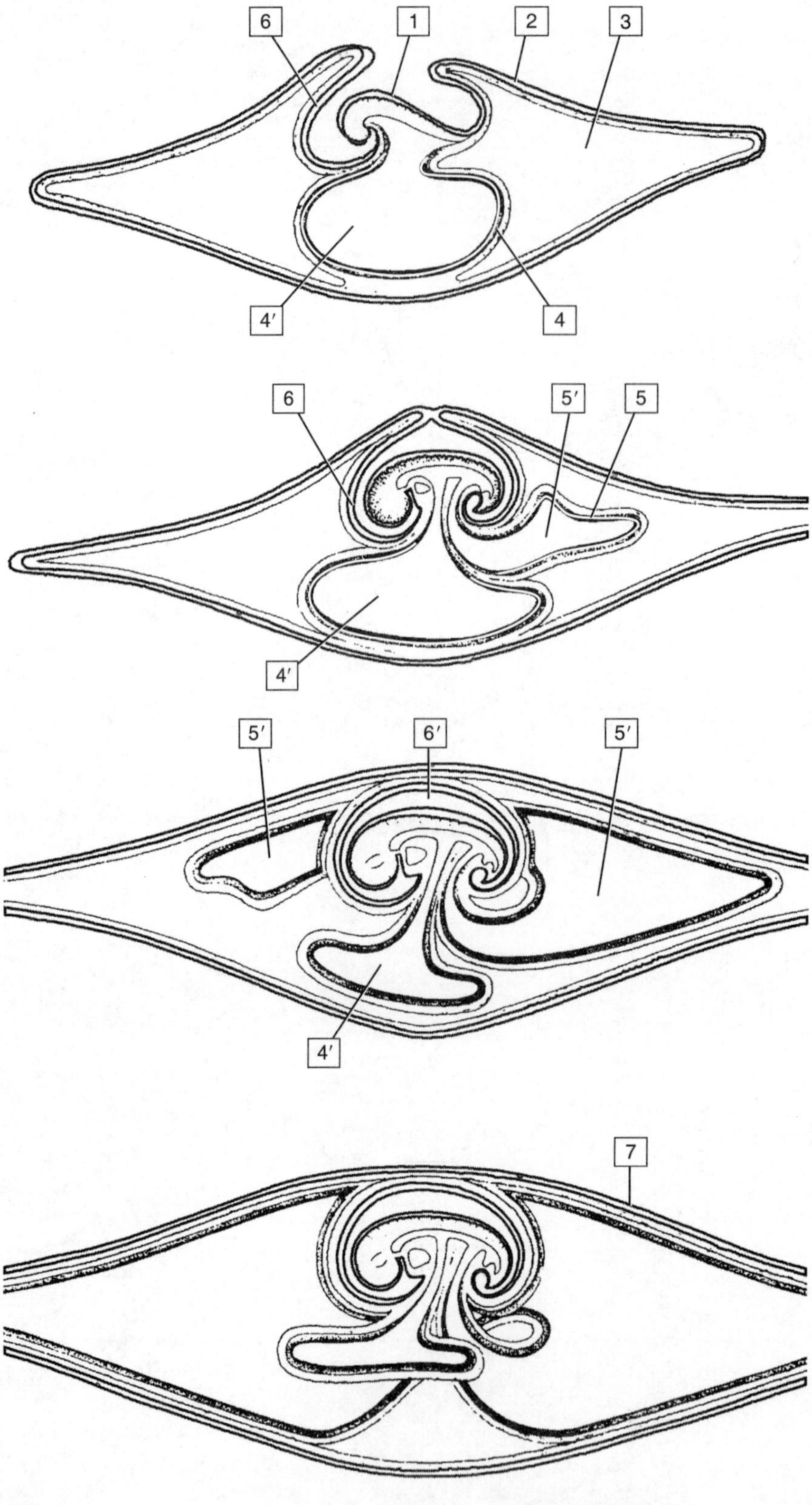

1	胚胎	Embryo
2	绒毛膜	Chorion
3	胚外体腔	Extraembryonic celom
4	卵黄囊	Yolk sac
4′	卵黄囊腔	Yolk sac cavity
5	尿囊	Allantois
5′	尿囊腔	Allantoic cavity
6	羊膜	Amnion
6′	羊膜腔	Amniotic cavity
7	绒毛膜尿囊	Chorioallantois

图5-9　犬胚胎外膜的形成

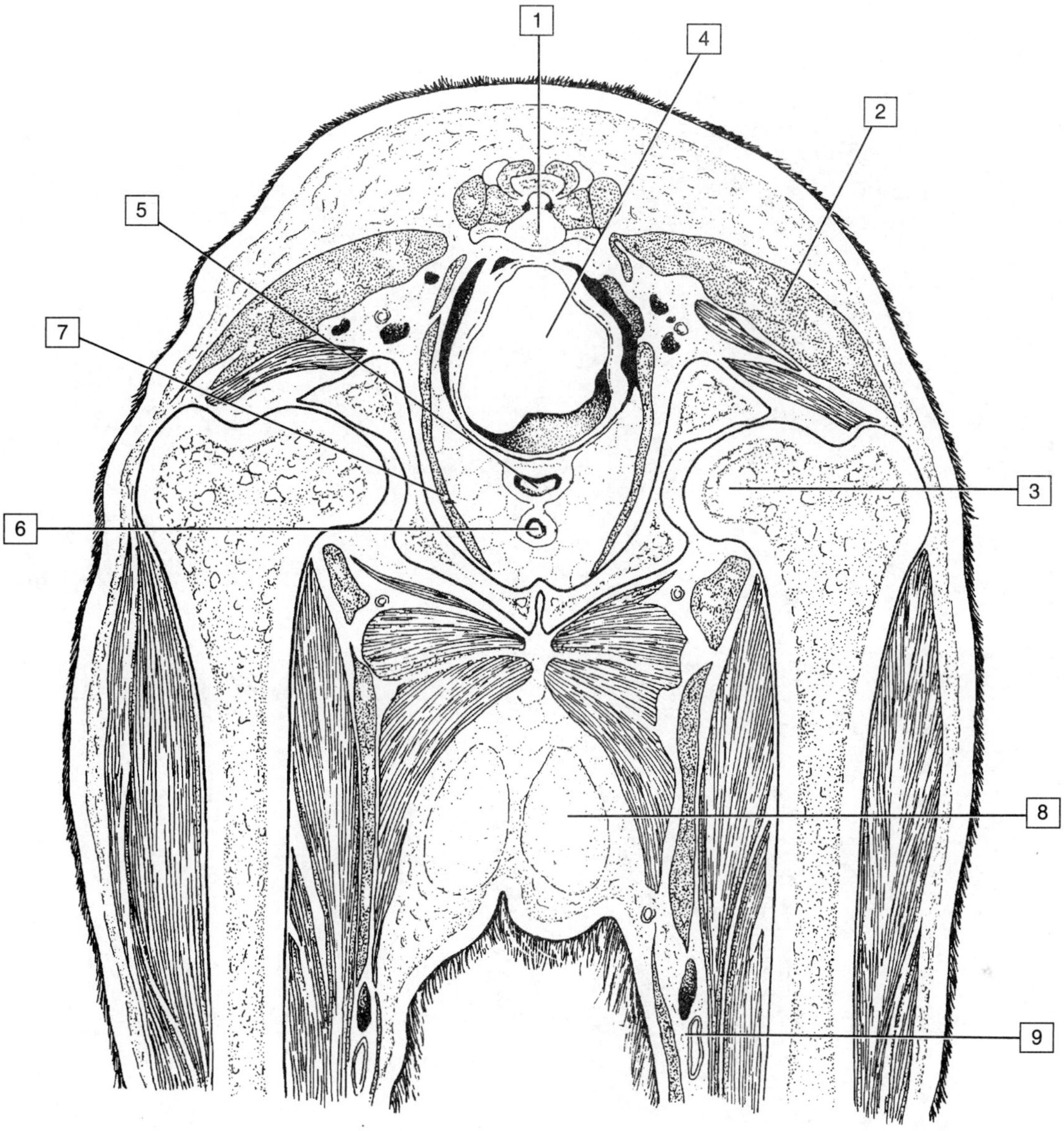

1	尾椎 Caudal vertebra	6	尿道 Urethra
2	臀浅肌 Superficial gluteal muscle	7	肛提肌 Levator ani
3	髋臼股骨头 Head of femur in acetabulum	8	腹股沟乳腺 Inguinal mammary gland
4	直肠被短的直肠系膜悬吊 Rectum suspended by a short mesorectum	9	股动脉和股静脉 Femoral artery and vein
5	阴道 Vagina		

图5-10 髋关节水平的犬骨盆横断面

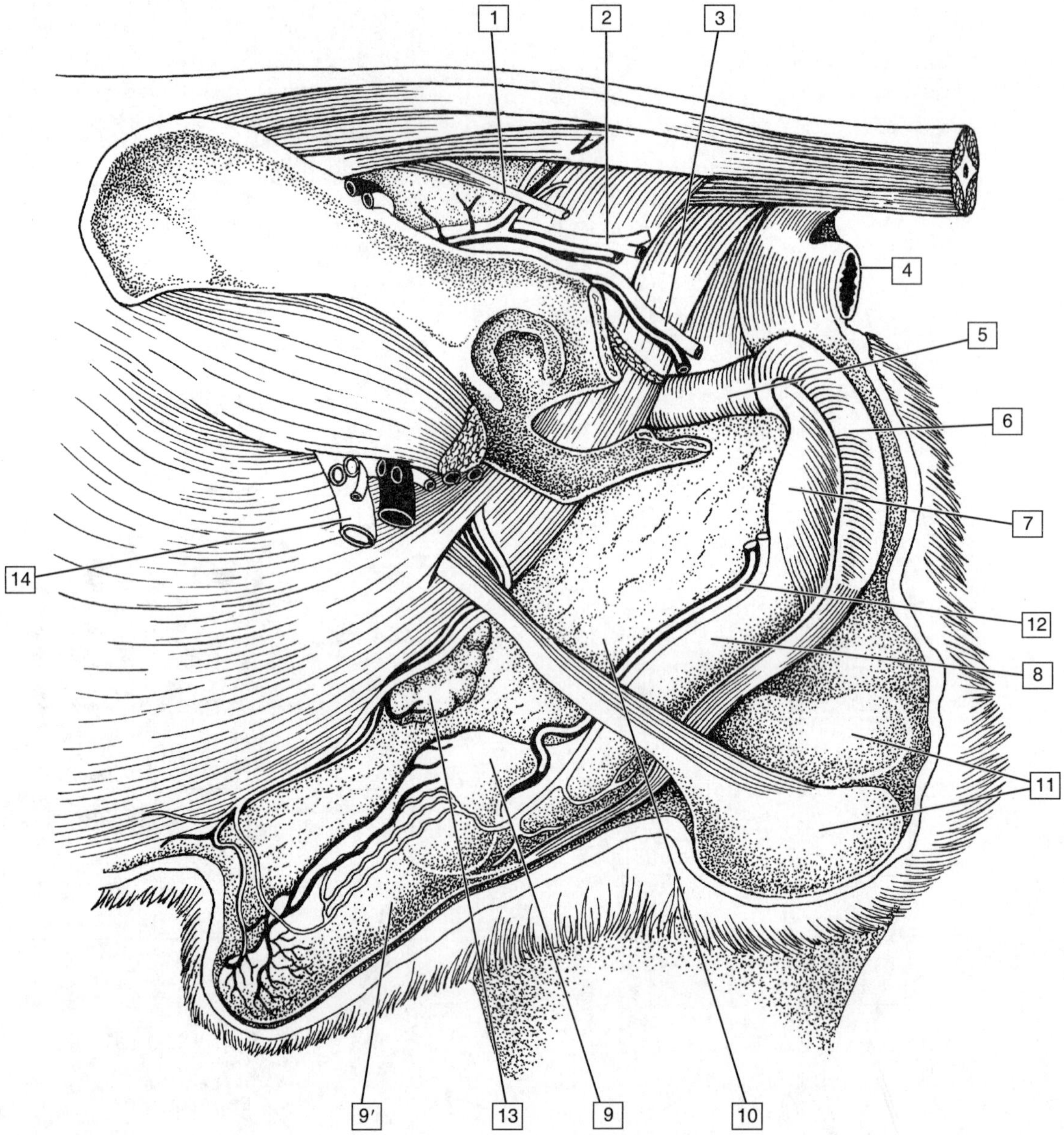

1 骶结节韧带 Sacrotuberous ligament	9 阴茎头球 Bulbus glandis
2 臀后血管 Caudal gluteal vessels	9′ 阴茎头长部 Pars longa glandis
3 阴部内血管 Internal pudendal vessels	10 精索 Spermatic cord
4 肛门 Anus	11 阴囊内的睾丸 Testes in scrotum
5 盆腔尿道 Pelvic urethra	12 阴茎背侧动静脉 Dorsal artery and vein of the penis
6 阴茎海绵球体 Bulb of penis enclosed by Bulbospongiosus	13 腹股沟浅淋巴结和后腹壁浅层血管 Superficial inguinal lymph nodes and caudal superficial epigastric vessels
7 左小腿上坐骨海绵体肌 Ischiocavernosus over left crus	14 股部血管 Femoral vessels
8 阴茎体 Body of penis	

图5-11 雄性犬外生殖器官的深层解剖图

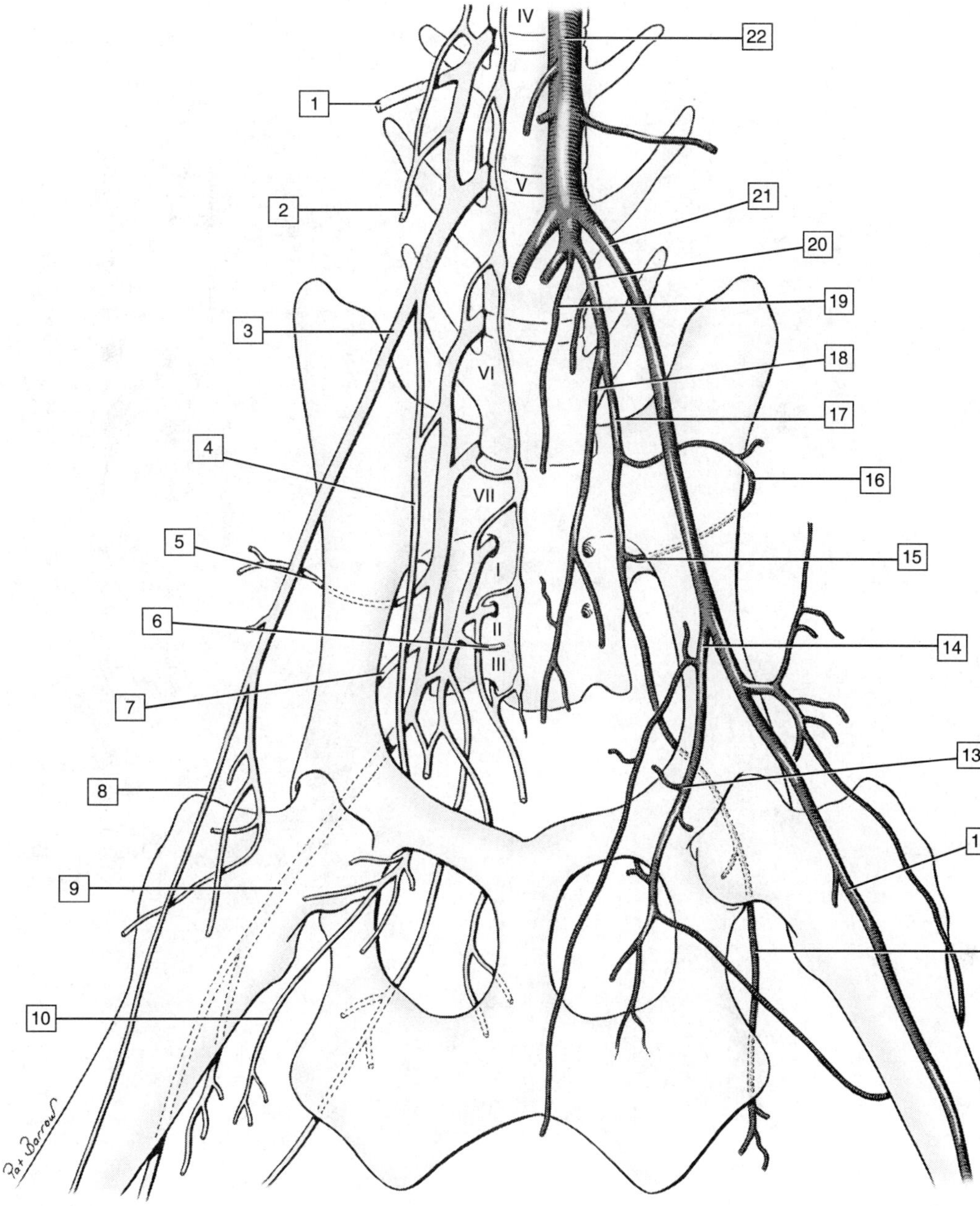

1 股外侧皮神经 Lateral cutaneous femoral nerve

2 生殖股神经 Genitofemoral nerve

3 股神经 Femoral nerve

4 闭孔神经 Obturator nerve

5 臀前神经 Cranial gluteal nerve

6 盆神经 Pelvic nerve

7 臀后神经 Caudal gluteal nerve

8 隐神经 Saphenous nerve

9 坐骨神经 Sciatic nerve

10 闭孔神经 Obturator nerve

11 臀后神经 Caudal gluteal nerve

12 股动脉 Femoral artery

13 旋股内侧动脉 Medial circumflex femoral artery

14 股深动脉 Deep femoral artery

15 臀前动脉 Cranial gluteal artery

16 髂腰动脉 Iliolumbar artery

17 臀后动脉 Caudal gluteal artery

18 阴部内动脉 Internal pudendal artery

19 荐中动脉 Median sacral artery

20 髂内动脉 Internal iliac artery

21 髂外动脉 External iliac artery

22 主动脉 Aorta

图5-12 犬右侧腰荐神经和左侧动脉，腹侧观

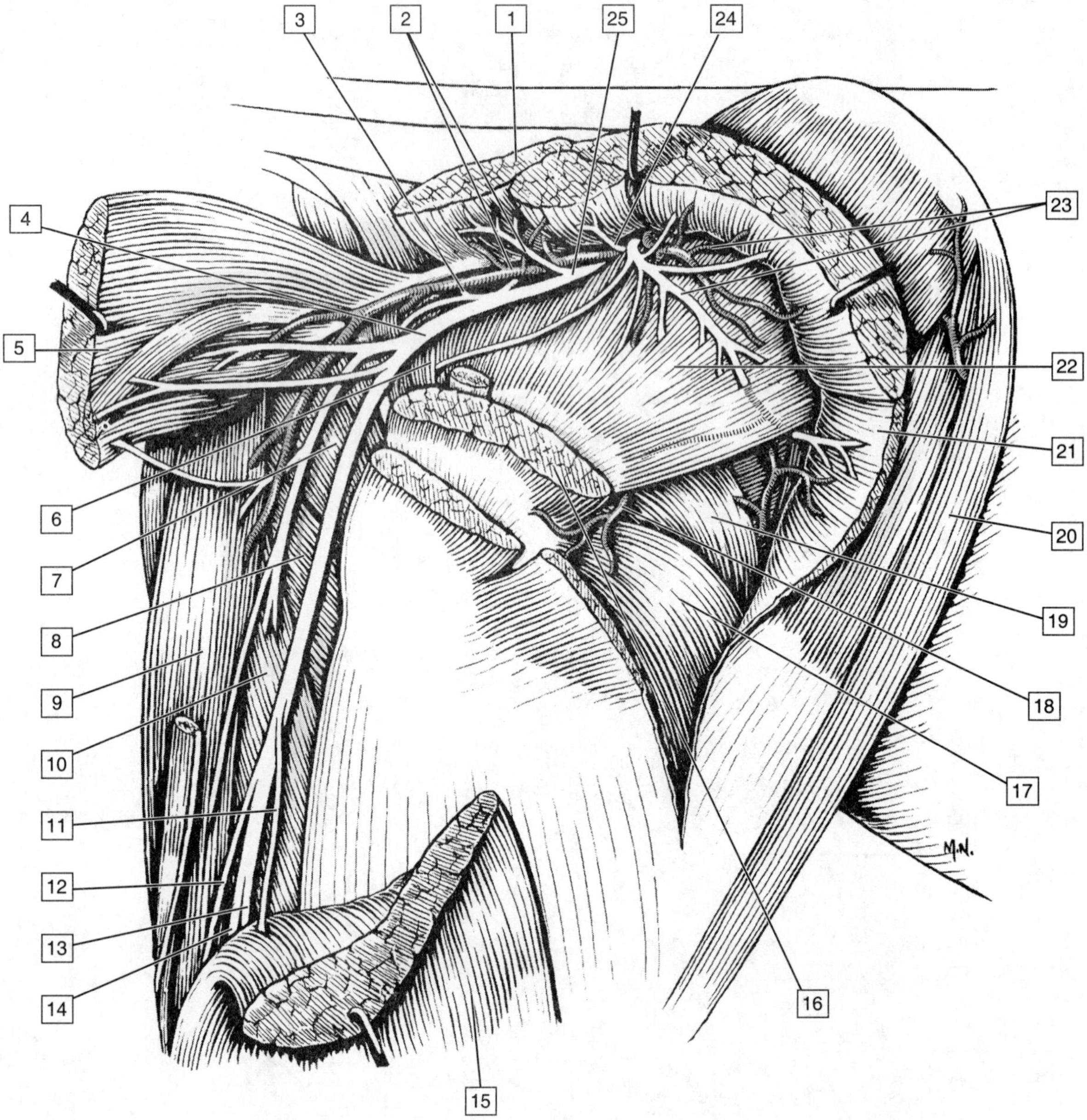

1 浅臀肌 Superficial gluteal muscle

2 臀后动脉和神经 Caudal gluteal artery and nerve

3 神经到内闭孔，神经束 Nerve to internal obturator, gemelli, and guadratus femoris

4 坐骨神经 Sciatic nerve

5 股二头肌 Biceps femoris

6 孖肌 Gemelli

7 股方肌 Quadratus femoris

8 内收肌 Adductor

9 半腱肌 Semitendinosus

10 半膜肌 Semimembranosus

11 腓肠外侧皮神经 Lateral cutaneous sural nerve

12 小腿后侧皮神经 Caudal cutaneous sural nerve

13 腓总神经 Common fibular nerve

14 胫神经 Tibial nerve

15 股二头肌 Biceps femoris

16 臀中肌 Middle gluteal

17 股外侧肌 Vastus lateralis

18 旋股外侧动脉 Lateral circumflex femoral artery

19 股直肌 Rectus femoris

20 缝匠肌 Sartorius

21 阔筋膜张肌 Tensor fasciae latae

22 臀深肌 Deep gluteal

23 臀前动脉和神经 Cranial gluteal artery and nerve

24 支配到梨状肌的神经 Nerve to piriformis

25 腰荐神经 Lumbosacral trunk

图5-13 犬右侧臀部神经、动脉和肌肉，外侧观

猫

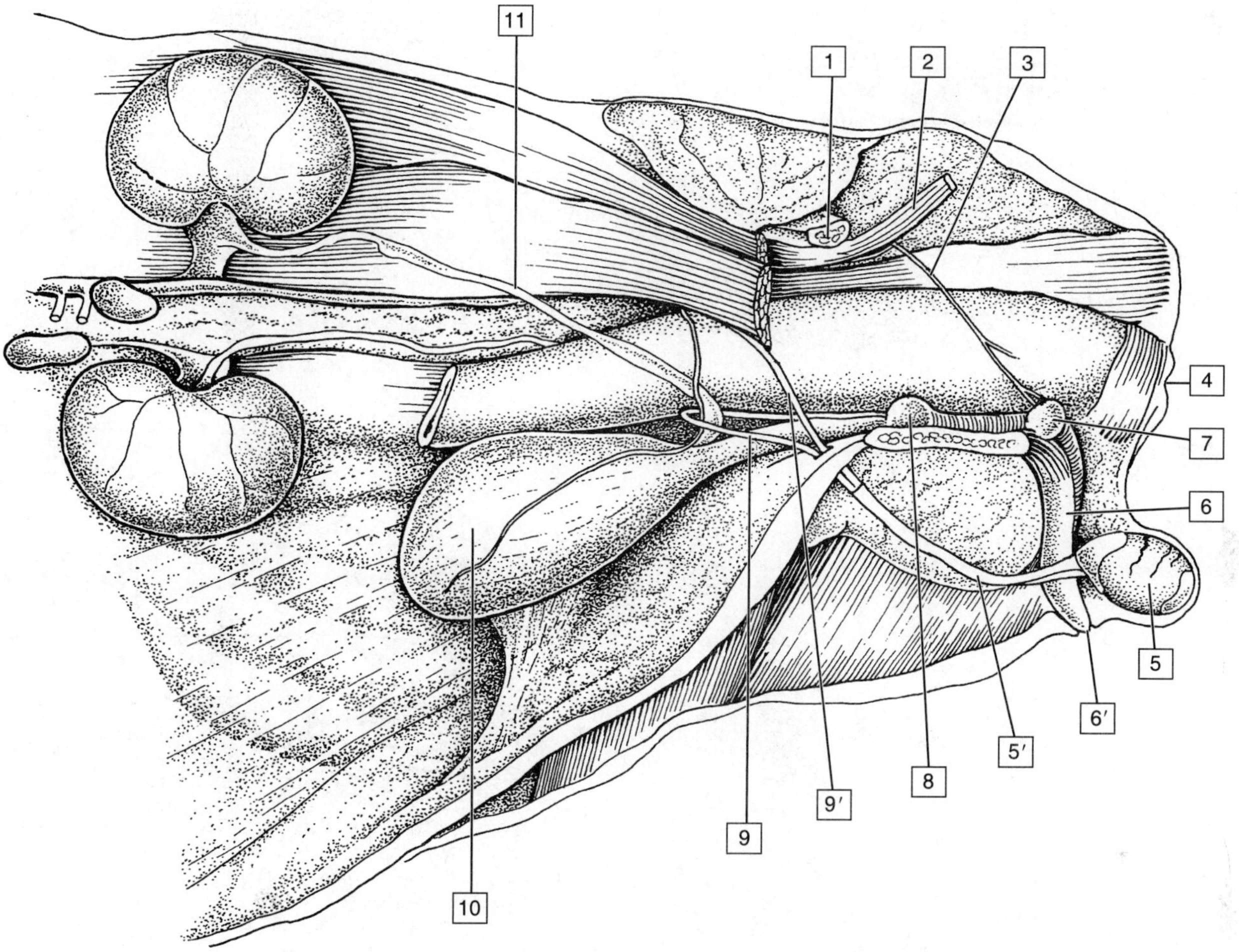

1	髂骨干 Shaft of ilium	6′	包皮 Prepuce
2	坐骨神经 Sciatic nerve	7	尿道球腺 Bulbourethral gland
3	阴部神经 Pudendal nerve	8	前列腺 Prostate
4	肛门 Anus	9	输精管 Deferent duct
5	左侧阴囊中的睾丸 Left testis in scrotum	9′	睾丸血管 Testicular vessels
5′	精索 Spermatic cord	10	膀胱 Bladder
6	阴茎 Penis	11	左侧输尿管 Left ureter

图5-14 雄性猫原位生殖器官，左外侧观

马

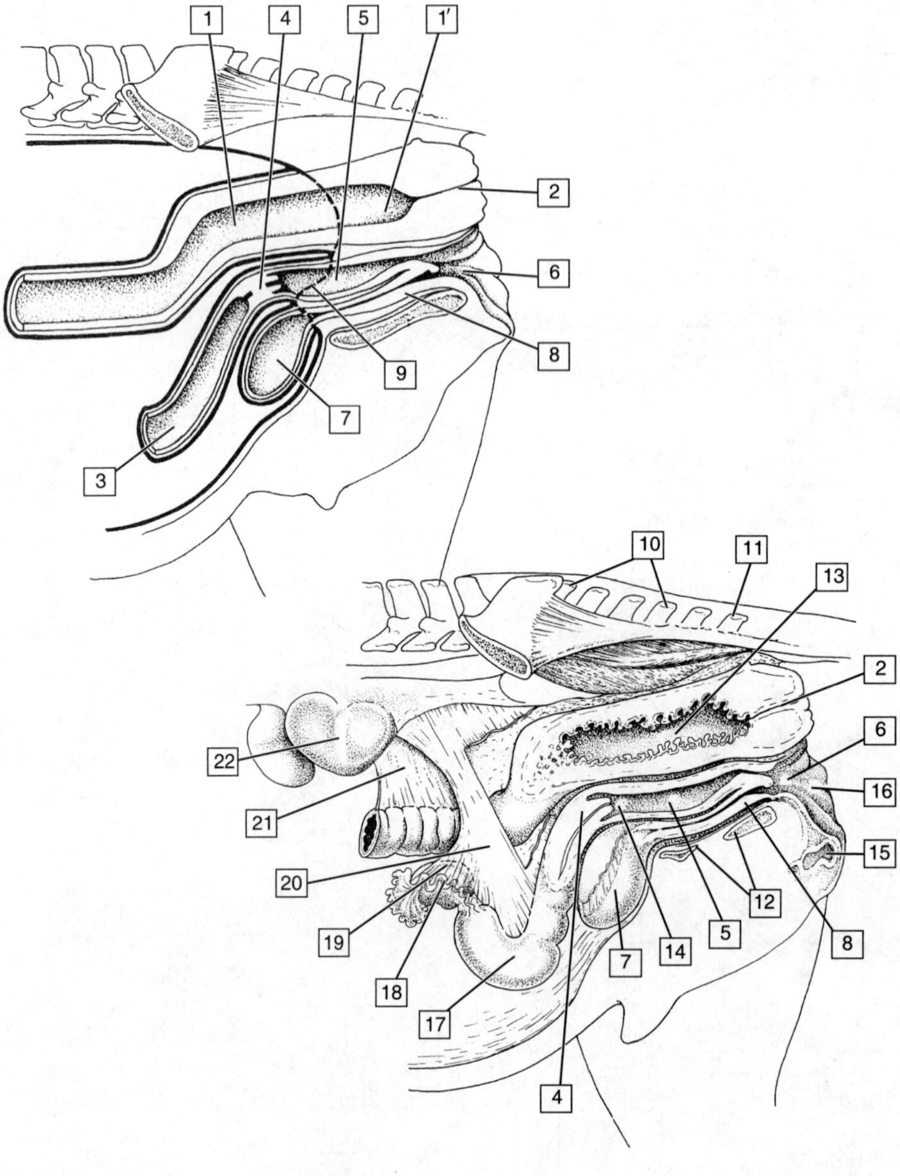

1 直肠的腹膜部分 Peritoneal part of the rectum	6 尿生殖前庭 Vestibule	12 盆底 Floor of pelvis	18 输卵管 Uterine tube
1′ 直肠的腹膜后部分 Retroperitoneal parts of the rectum	7 膀胱 Bladder	13 直肠 Rectum	19 卵巢 Ovary
2 肛道 Proctodeum	8 尿道 Urethra	14 子宫颈阴道部 Vaginal part of cervix	20 阔韧带（大部分已被切除） Broad ligament (largely cut away)
3 子宫 Uterus	9 腹膜后部分 Caudal extent of peritoneum	15 阴蒂 Clitoris	21 降结肠系膜 Descending mesocolon
4 子宫颈 Cervix	10 荐骨 Sacrum	16 外阴 Vulva	22 左肾 Left kidney
5 阴道 Vagina	11 第2尾椎 Cd2	17 左侧子宫角 Left uterine horn	

图5-15　母马后腹部和盆腔器官原位图，正中面观

内脏器官在骨盆的旁正中面被切开

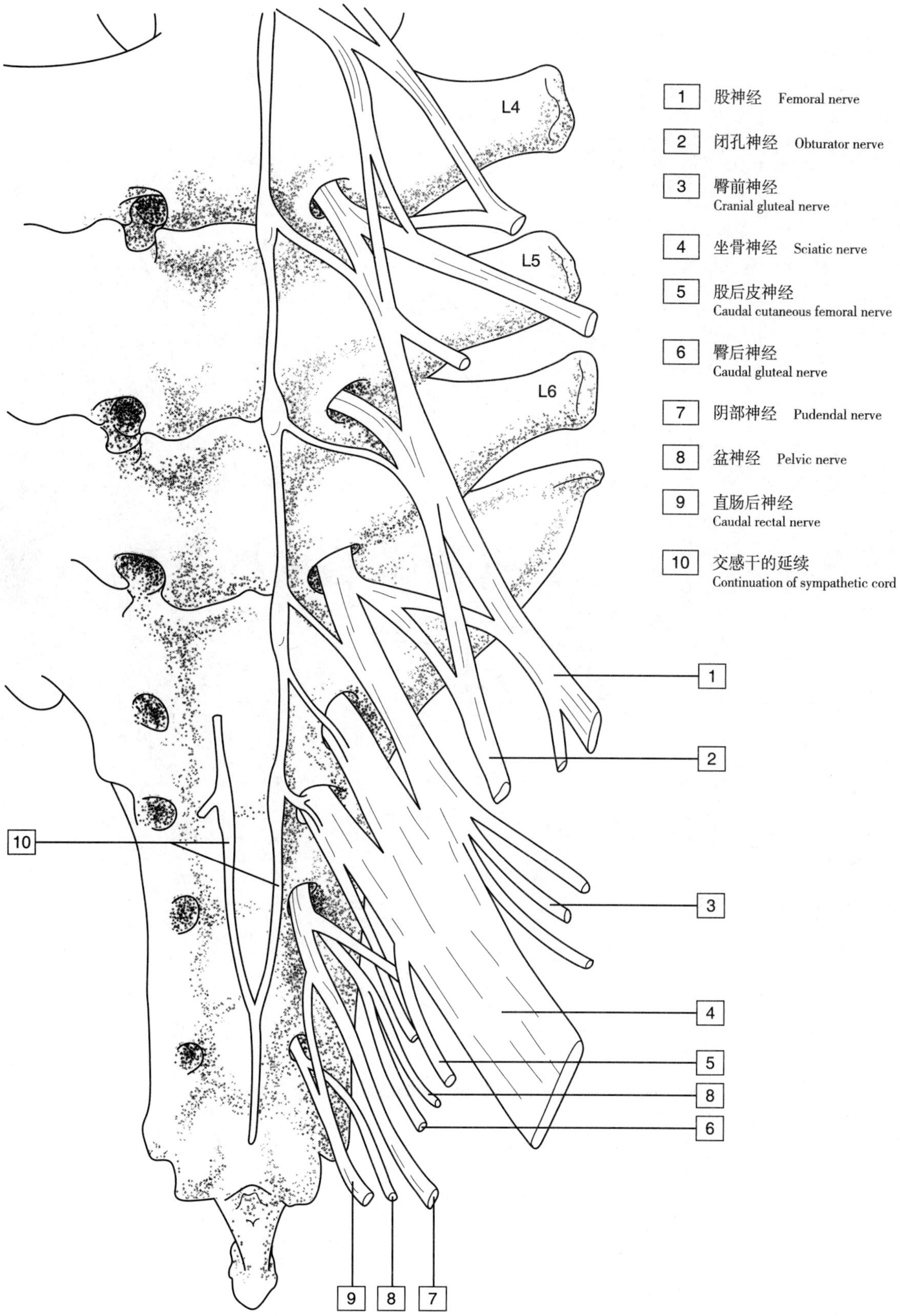

图5-16 由马荐椎和腰椎后部发出的腹侧支形成腰间神经丛，腹侧观

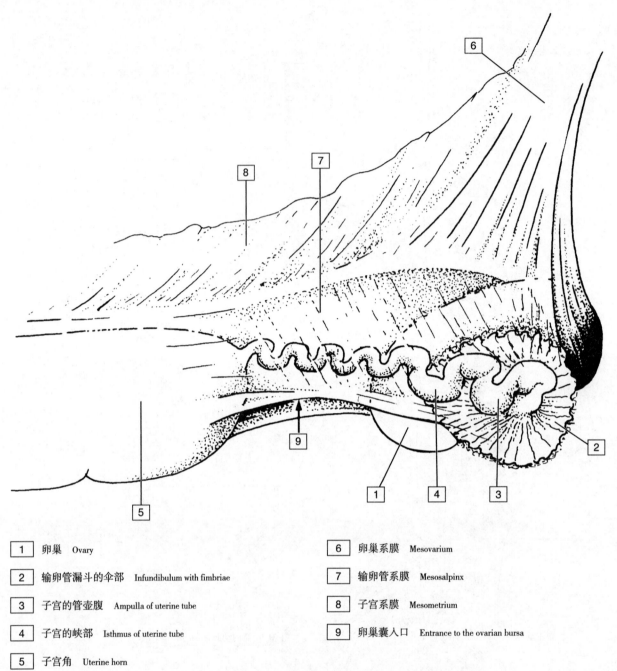

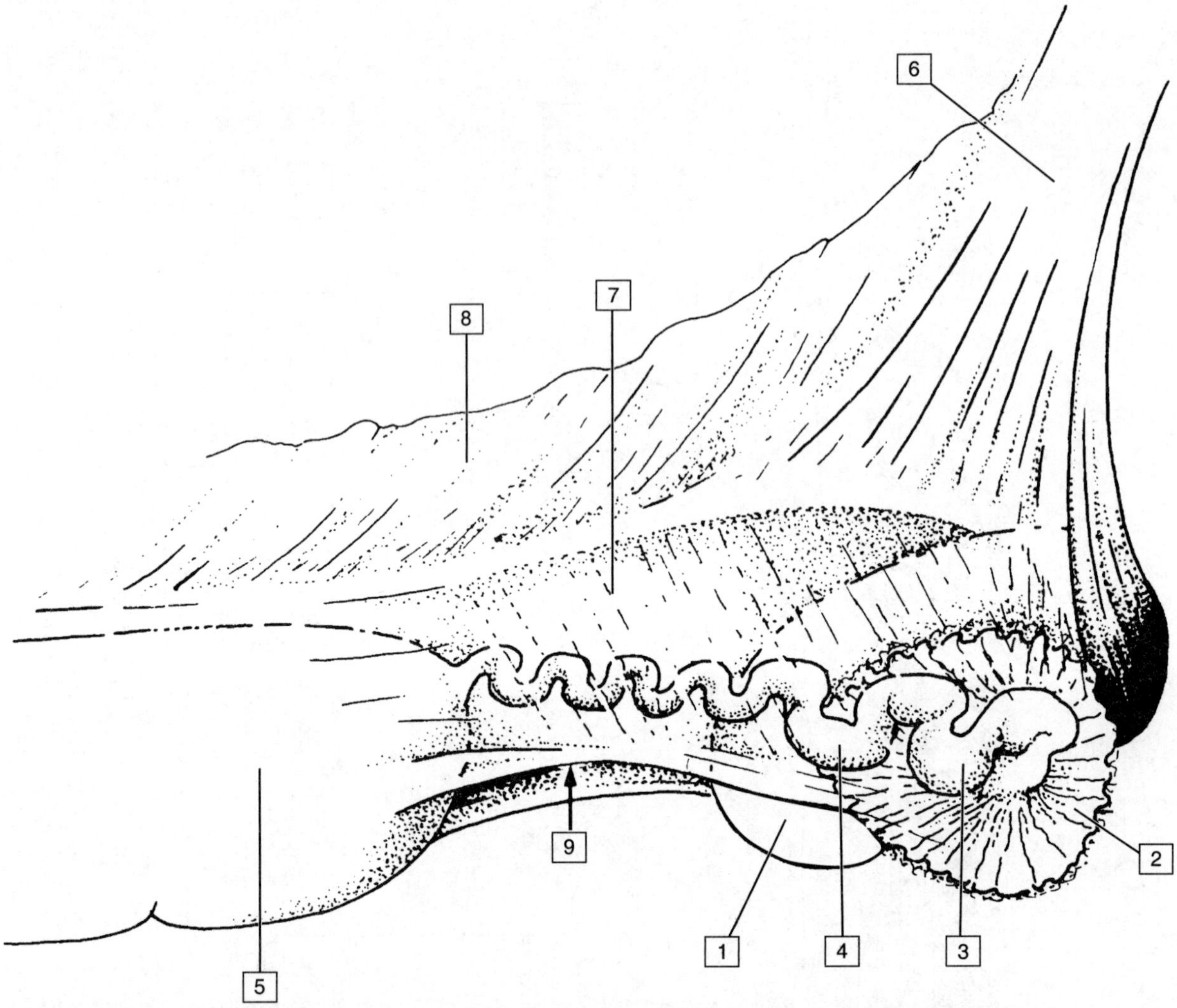

1	卵巢 Ovary	6	卵巢系膜 Mesovarium
2	输卵管漏斗的伞部 Infundibulum with fimbriae	7	输卵管系膜 Mesosalpinx
3	子宫的管壶腹 Ampulla of uterine tube	8	子宫系膜 Mesometrium
4	子宫的峡部 Isthmus of uterine tube	9	卵巢囊入口 Entrance to the ovarian bursa
5	子宫角 Uterine horn		

图5-17 马的右侧卵巢、输卵管和子宫角，外侧观

牛

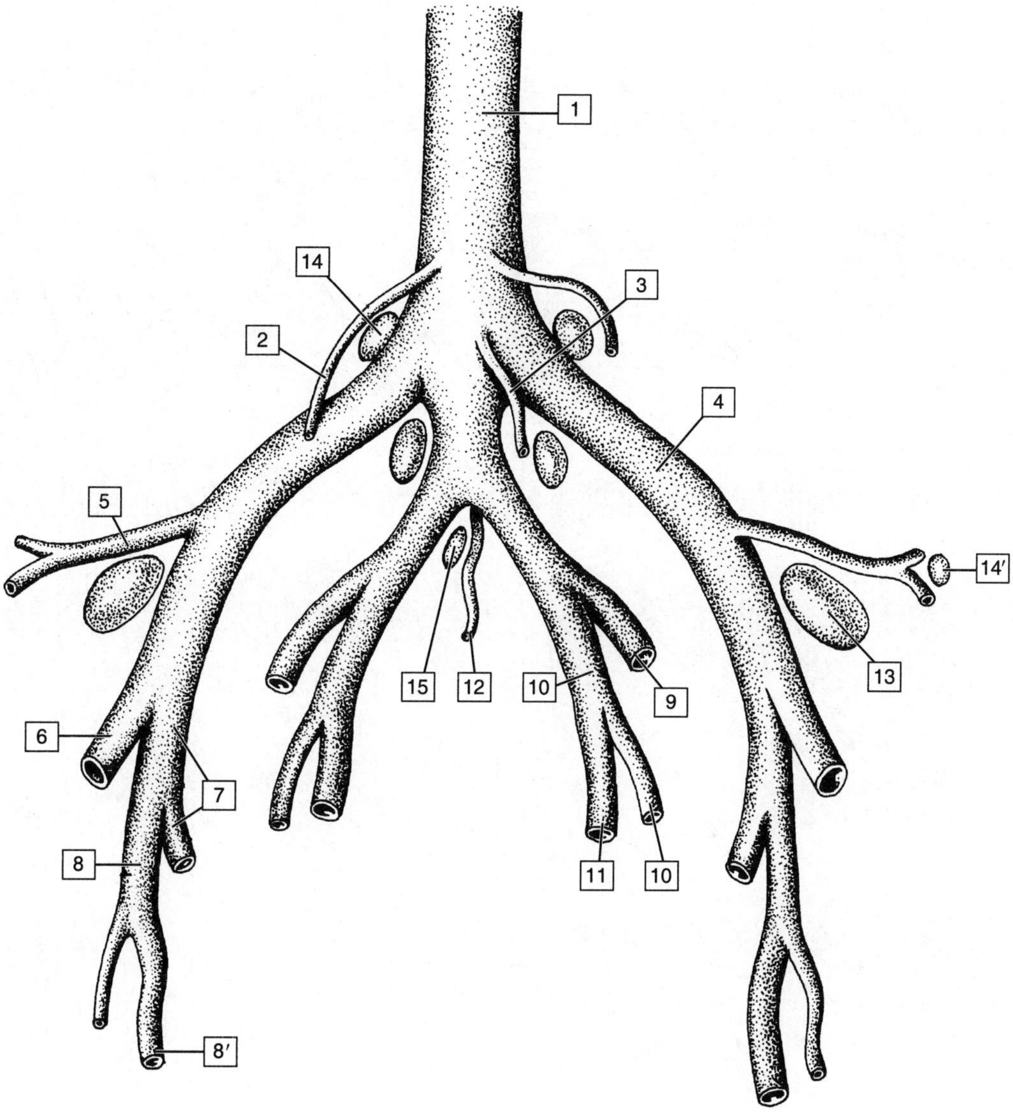

1 腹主动脉 Abdominal aorta	9 髂内动脉 Internal iliac artery
2 卵巢动脉 Ovarian artery	10 脐动脉 Umbilical artery
3 肠系膜后动脉 Caudal mesenteric artery	11 子宫动脉 Uterine artery
4 髂外动脉 External iliac artery	12 荐中动脉 Median sacral artery
5 旋髂深动脉 Deep circumflex iliac artery	13 腹股沟深淋巴结（髂股淋巴结） Deep inguinal (iliofemoral) lymph node
6 股动脉 Femoral artery	14 髂内淋巴结 Medial iliac lymph node
7 股深动脉 Deep femoral artery	14′ 髂外淋巴结 Lateral iliac lymph node
8 阴部腹壁动脉干 Pudendoepigastric trunk	15 荐淋巴结 Sacral lymph node
8′ 阴部外动脉 External pudendal artery	

图5-18 牛腹主动脉后部分支和局部淋巴结

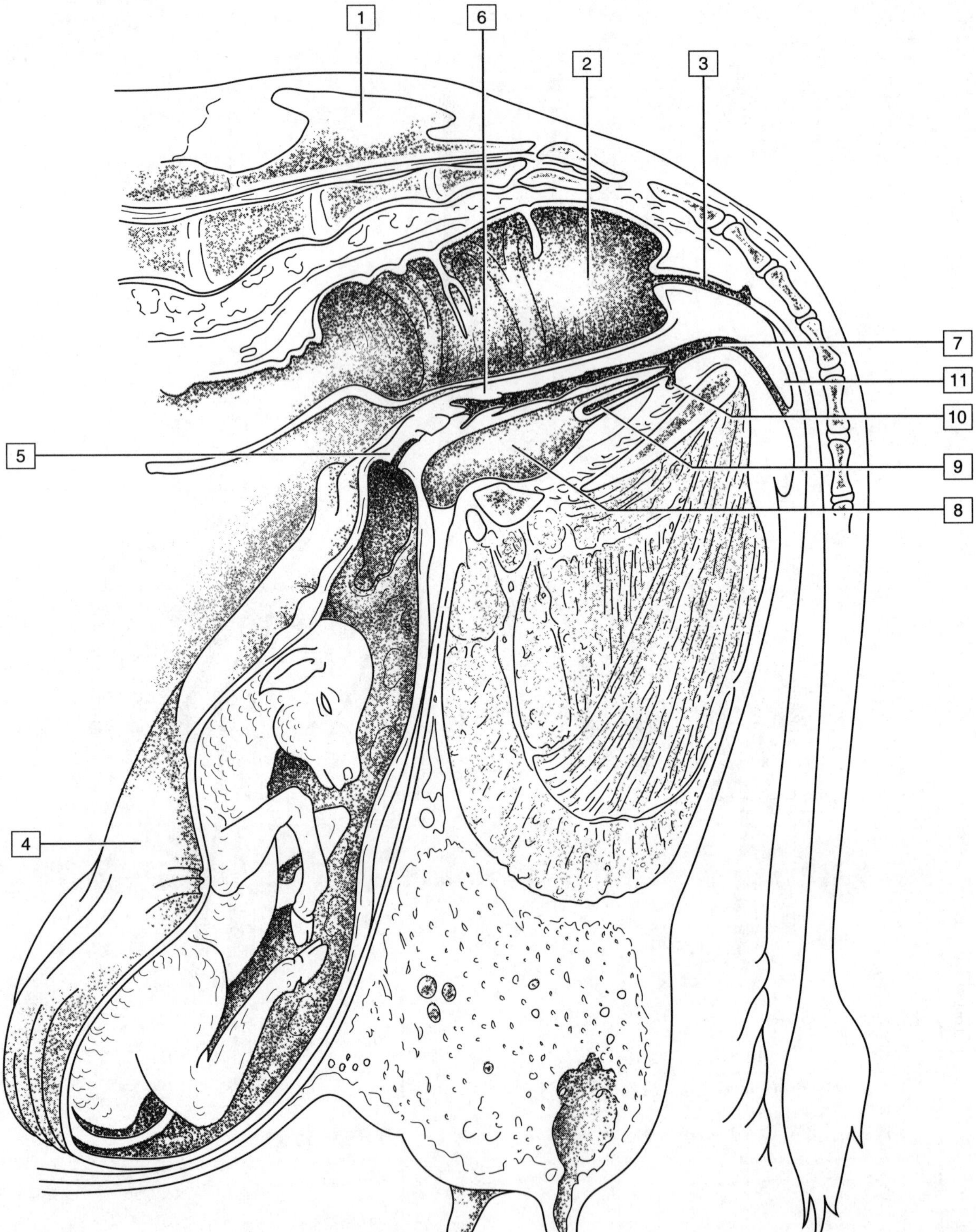

1	荐骨 Sacrum	4	子宫 Uterus	7	尿生殖前庭 Vestibule	10	尿道下憩室 Suburethral diverticulum
2	直肠 Rectum	5	子宫颈 Cervix	8	膀胱 Bladder	11	阴门 Vulva
3	肛道 Proctodeum	6	阴道 Vagina	9	尿道 Urethra		

图5-19 妊娠母牛后腹部和骨盆的旁正中面

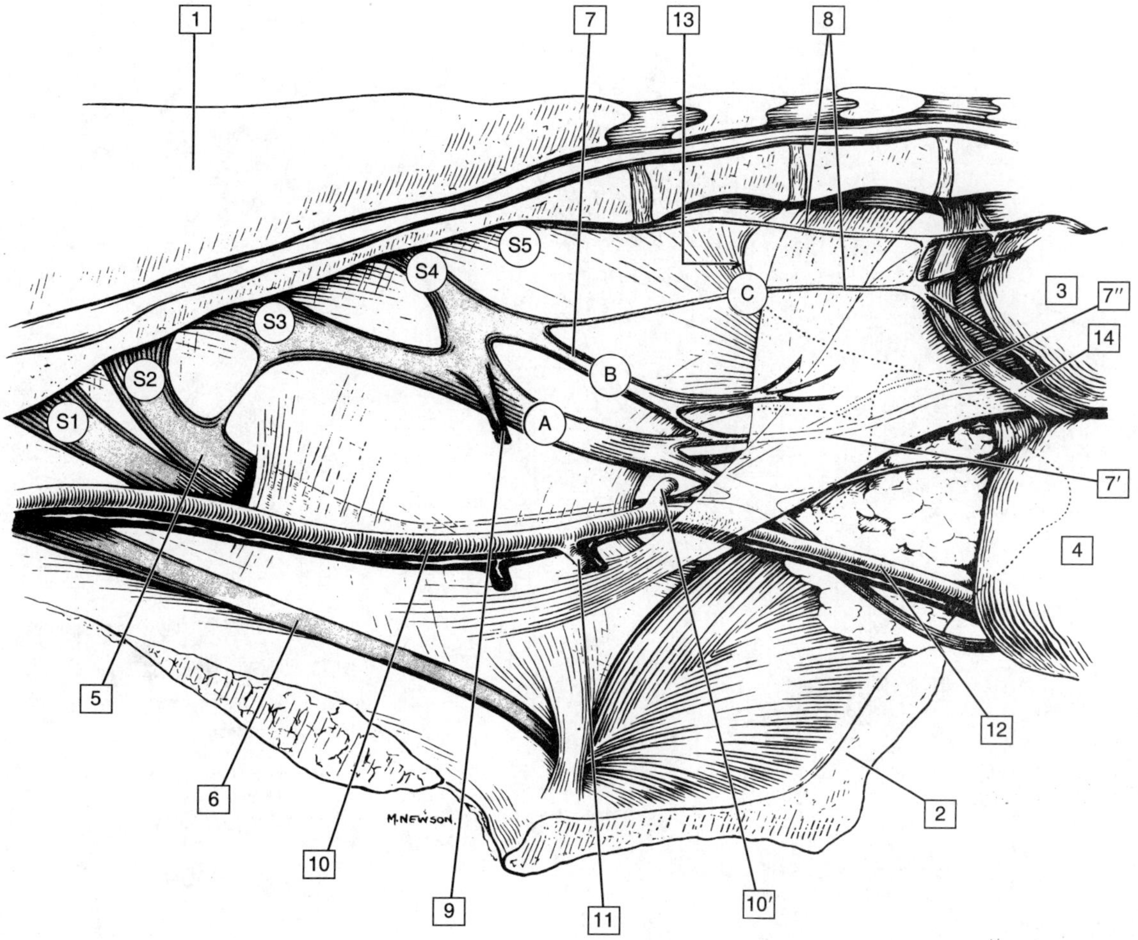

1 荐骨 Sacrum	8 直肠后神经 Caudal rectal nerve
2 骨盆联合 Pelvic symphysis	9 腔神经 Pelvic nerve
3 直肠（掀开） Rectum (reflected)	10 髂内动脉 Internal iliac artery
4 阴道（掀开） Vagina (reflected)	10′ 臀后动脉 Caudal gluteal artery
5 坐骨神经 Sciatic nerve	11 阴部动脉 Vaginal artery
6 闭孔神经 Obturator nerve	12 阴部内动脉 Internal pudendal artery
7 阴部神经 Pudendal nerve	13 荐髂韧带后缘 Caudal border of sacrosciatic ligament
7′ 阴部神经远侧皮支 Distal cutaneous branch of pudendal nerve	14 阴蒂缩肌 Retractor clitoridis
7′′ 阴部神经近侧皮支 Proximal cutaneous branch of pudendal nerve	

图5-20 牛盆腔血管和神经的内侧面

在A和B注射局麻药以麻醉阴部神经，在C注射局麻药以麻醉直肠后神经

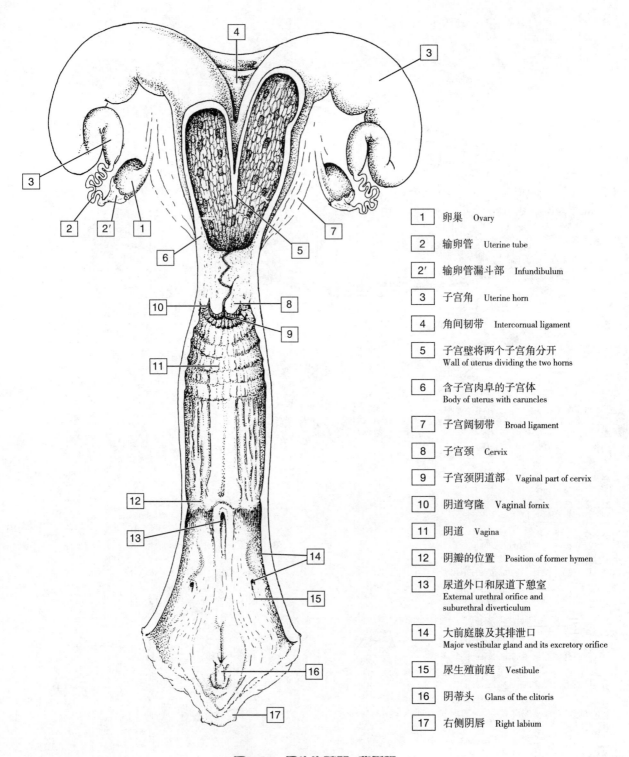

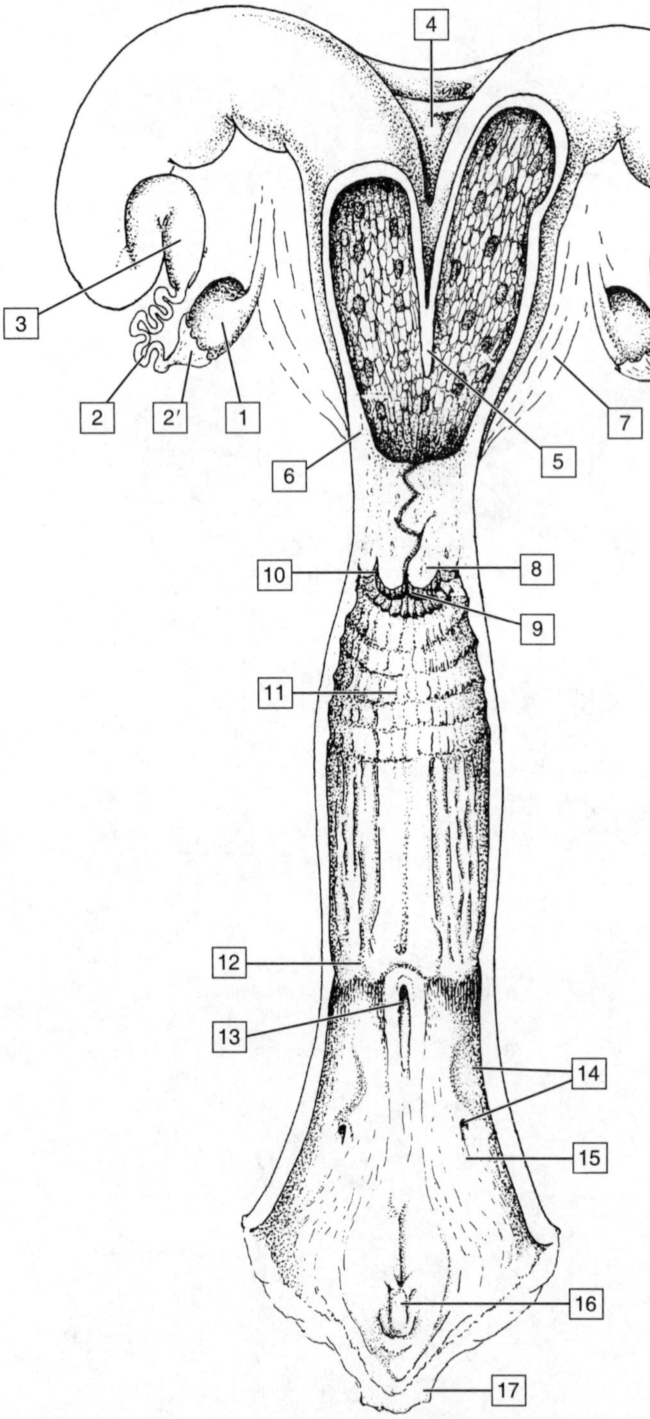

1	卵巢	Ovary
2	输卵管	Uterine tube
2′	输卵管漏斗部	Infundibulum
3	子宫角	Uterine horn
4	角间韧带	Intercornual ligament
5	子宫壁将两个子宫角分开	Wall of uterus dividing the two horns
6	含子宫肉阜的子宫体	Body of uterus with caruncles
7	子宫阔韧带	Broad ligament
8	子宫颈	Cervix
9	子宫颈阴道部	Vaginal part of cervix
10	阴道穹隆	Vaginal fornix
11	阴道	Vagina
12	阴瓣的位置	Position of former hymen
13	尿道外口和尿道下憩室	External urethral orifice and suburethral diverticulum
14	大前庭腺及其排泄口	Major vestibular gland and its excretory orifice
15	尿生殖前庭	Vestibule
16	阴蒂头	Glans of the clitoris
17	右侧阴唇	Right labium

图5-21　母牛生殖器，背侧观

子宫、子宫颈、阴道和尿生殖前庭已被切开

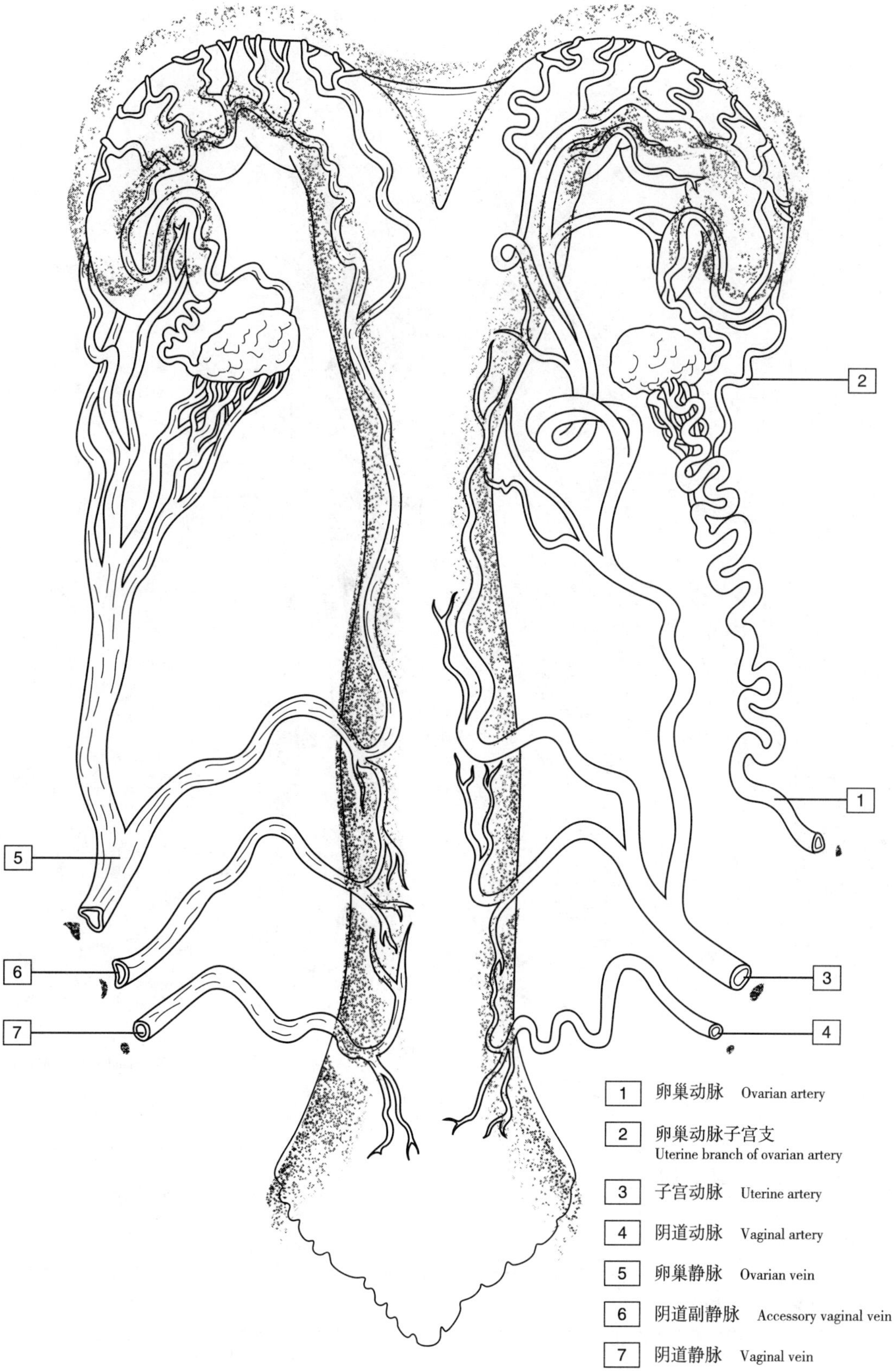
1
2
3
4
5
6
7
1 卵巢动脉 Ovarian artery
2 卵巢动脉子宫支 Uterine branch of ovarian artery
3 子宫动脉 Uterine artery
4 阴道动脉 Vaginal artery
5 卵巢静脉 Ovarian vein
6 阴道副静脉 Accessory vaginal vein
7 阴道静脉 Vaginal vein

图5-22 牛生殖道血供示意图，腹侧观

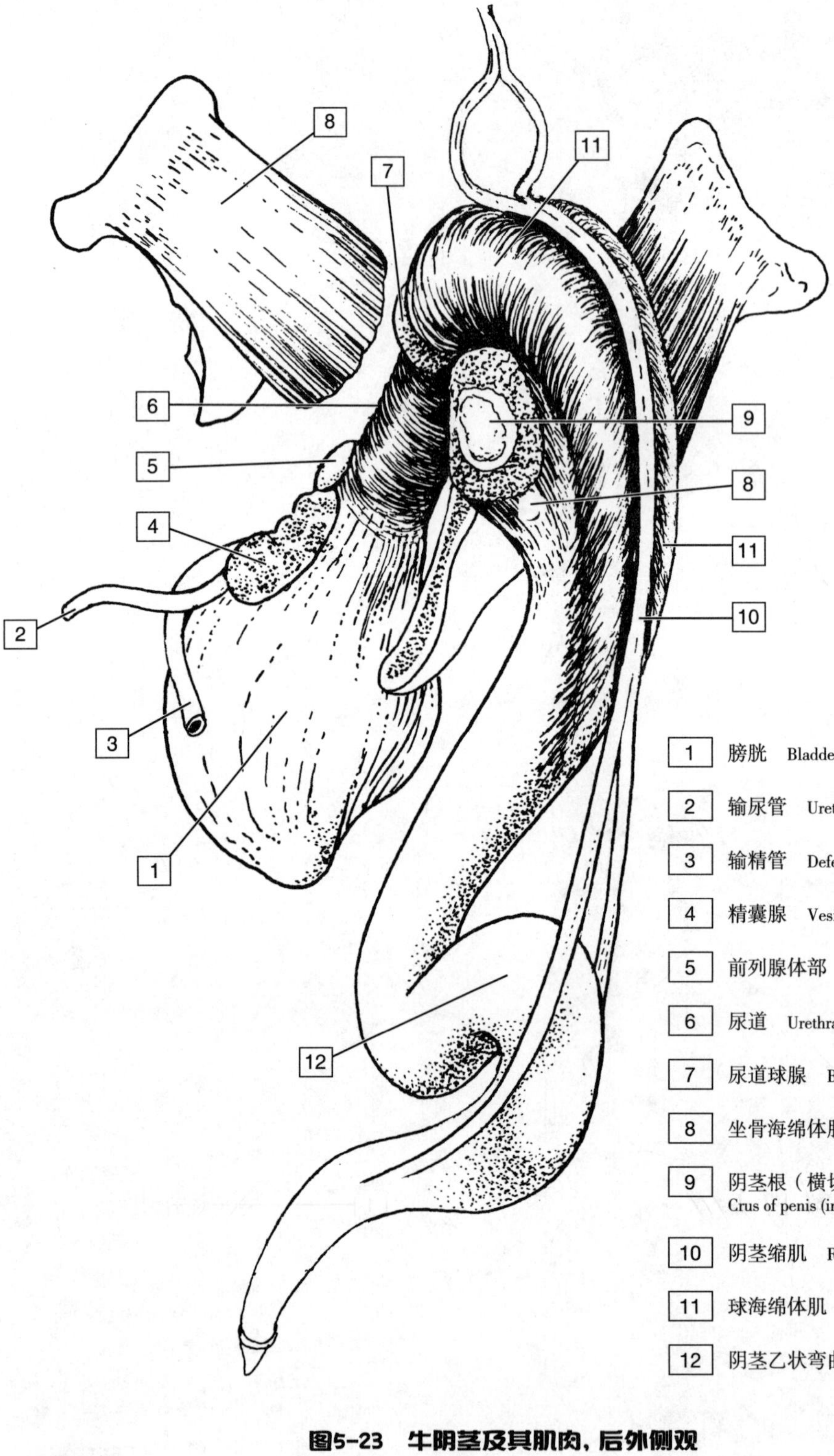

1	膀胱 Bladder
2	输尿管 Ureter
3	输精管 Deferent duct
4	精囊腺 Vesicular gland
5	前列腺体部 Body of prostate
6	尿道 Urethralis
7	尿道球腺 Bulbourethral gland
8	坐骨海绵体肌 Ischiocavernosus
9	阴茎根（横切面）Crus of penis (in transverse section)
10	阴茎缩肌 Retractor penis
11	球海绵体肌 Bulbospongiosus
12	阴茎乙状弯曲 Sigmoid flexure

图5-23 牛阴茎及其肌肉，后外侧观

阻断哪些神经可以使阴茎缩肌松弛？对阴茎有何影响？

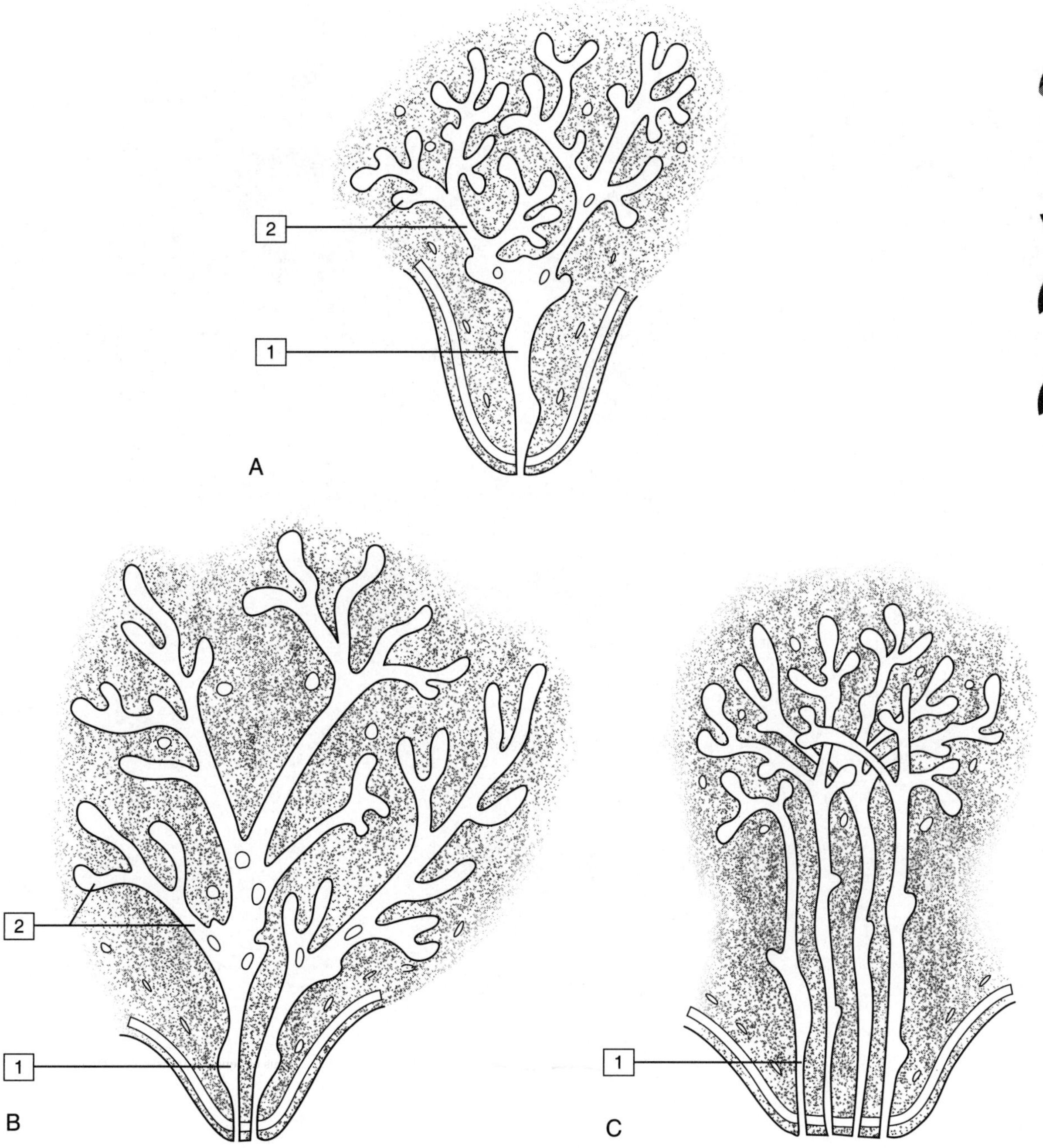

1 初生芽 Primary sprout 2 第二次芽、第三次芽 Secondary and tertiary sprouts

图5-24 从胎儿乳头顶端开始的发育中的导管系统

A. 母牛和母羊；B. 母马和母猪；C. 雌犬和雌猫（只显示四个初生芽）

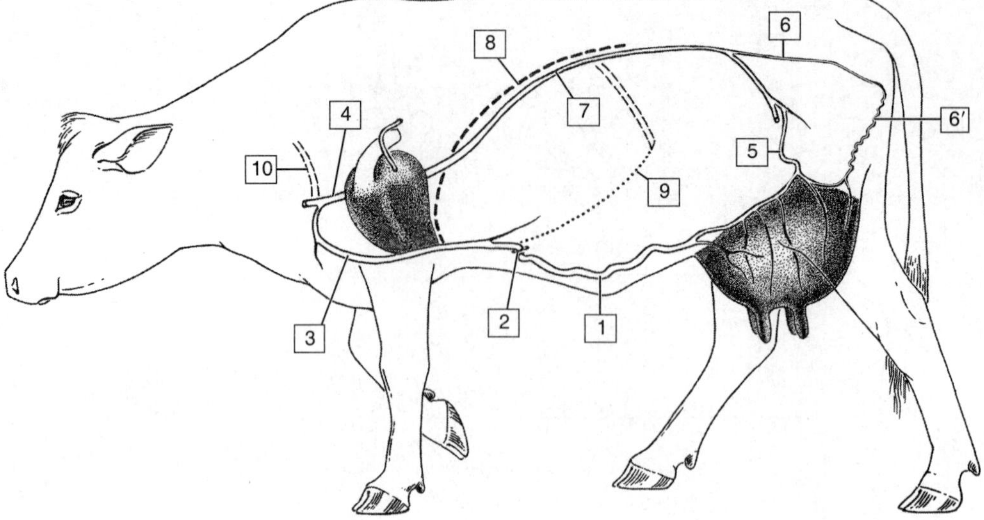

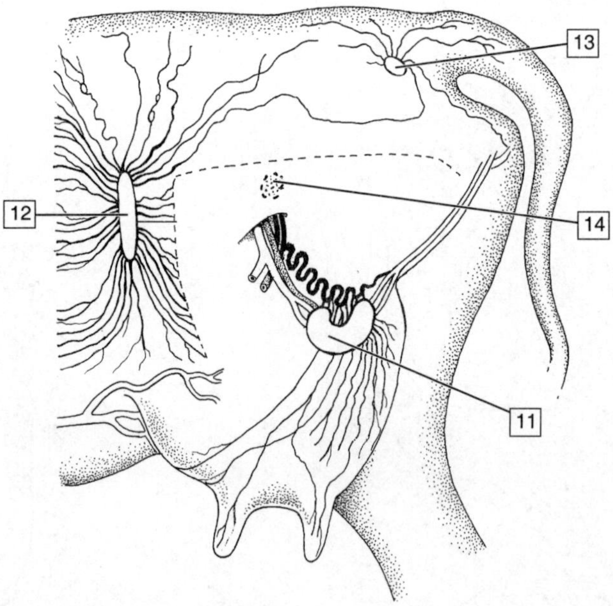

1 腹部皮下静脉（收集乳房静脉血）
Subcutaneous abdominal (milk) vein

2 乳井 Milk "well"

3 胸廓内静脉 Internal thoracic vein

4 前腔静脉 Cranial vena cava

5 阴部外静脉 External pudendal vein

6 阴部内静脉 Internal pudendal vein

6′ 阴唇腹侧静脉（连接会阴腹侧静脉和乳房后静脉）
Ventral labial vein (connecting ventral perineal vein with caudal mammary veins)

7 后腔静脉 Caudal vena cava

8 膈 Diaphragm

9 肋弓 Costal arch

10 第1肋 First rib

11 乳上淋巴结（腹股沟浅淋巴结）
Mammary (superficial inguinal) lymph node

12 髂下淋巴结 Subiliac lymph node

13 坐骨淋巴结 Ischial lymph node

14 腹股沟深淋巴结的位置（髂股淋巴结）
Position of deep inguinal (iliofemoral) node

图5-25 牛的静脉和乳房淋巴引流

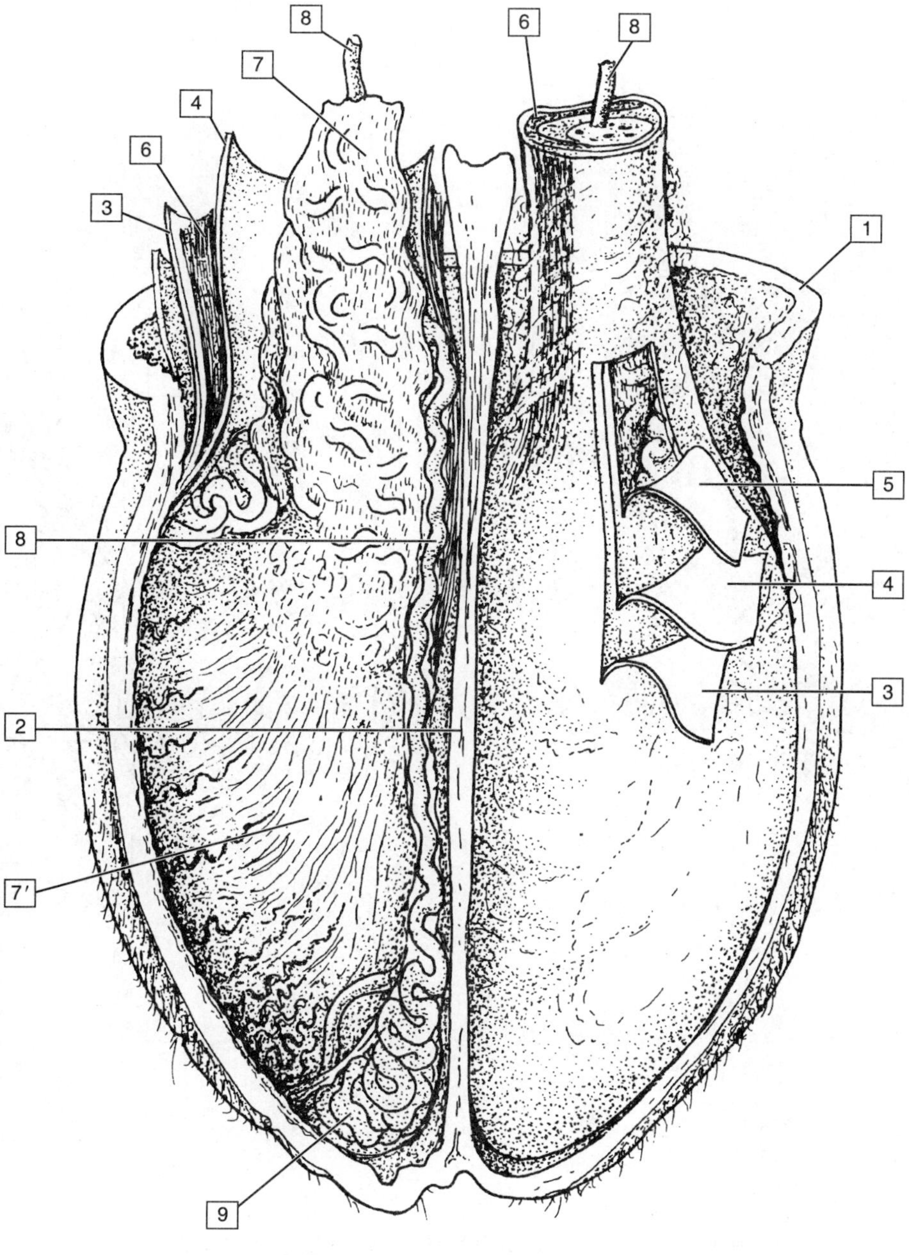

1	阴囊皮肤和肉膜	Scrotal skin and dartos
2	阴囊中隔	Scrotal septum
3	精索外筋膜	External spermatic fascia
4	鞘膜壁层	Parietal layer of vaginal tunic
5	鞘膜脏层（从睾丸表面切开）	Visceral layer (dissected from surface of testis)
6	提睾肌	Cremaster muscle
7	覆盖在精索外的鞘膜脏层	Visceral layer of vaginal tunic covering structures in spermatic cord
7′	睾丸鞘膜脏层	Visceral layer on testis
8	输精管	Deferent duct
9	附睾尾	Tail of epididymis

图5-26 剖开的公牛阴囊，前面观

睾丸已被部分切开

猪

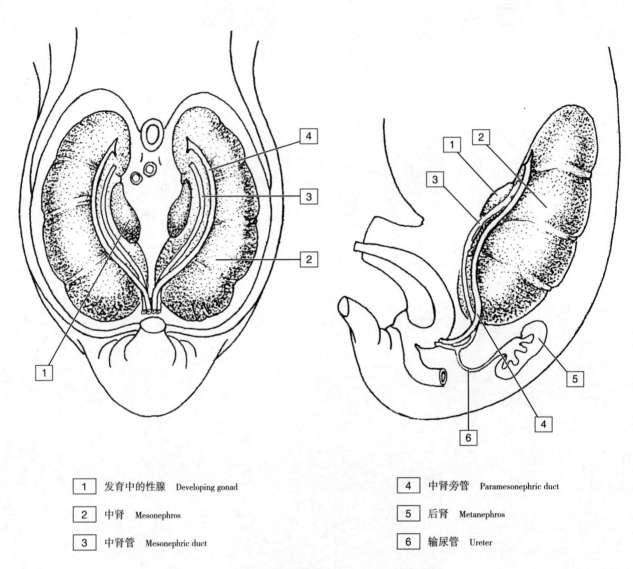

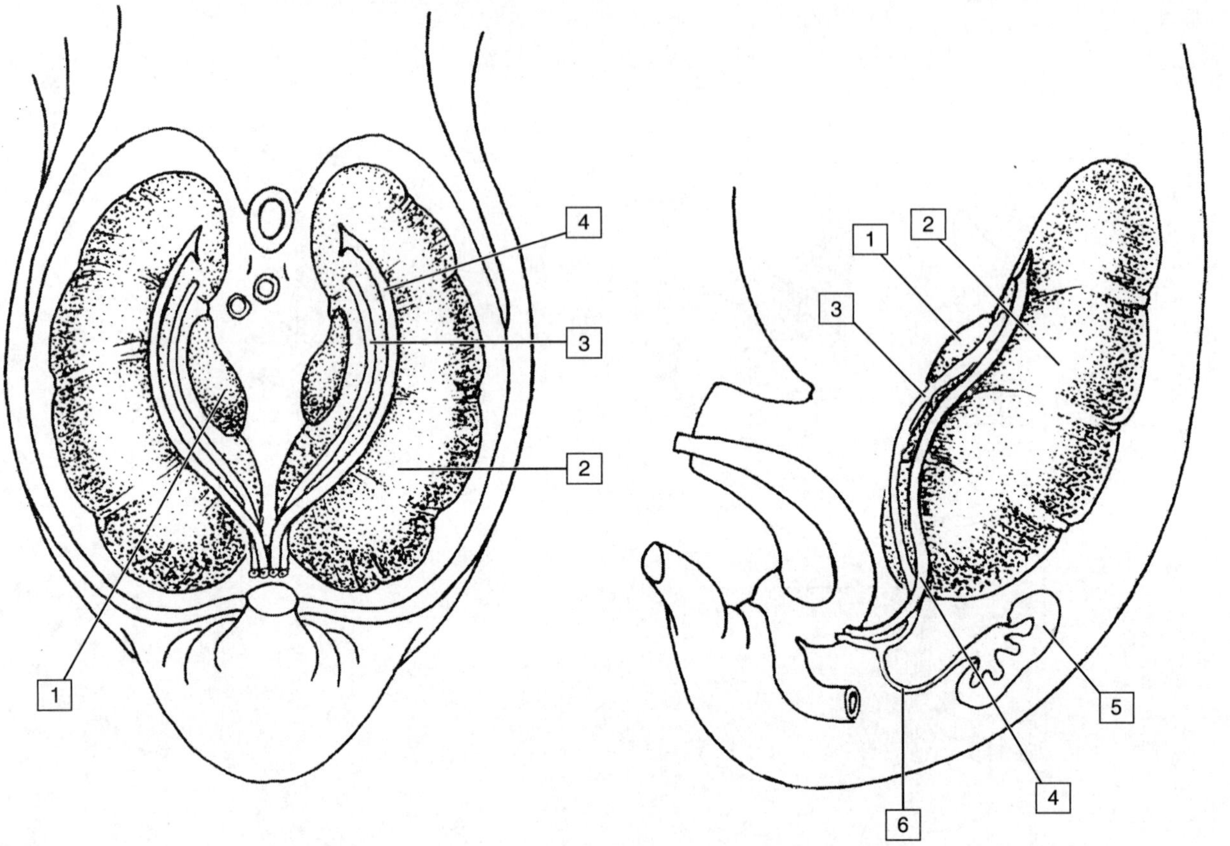

1	发育中的性腺 Developing gonad	4	中肾旁管 Paramesonephric duct
2	中肾 Mesonephros	5	后肾 Metanephros
3	中肾管 Mesonephric duct	6	输尿管 Ureter

图5-27　2.5厘米的猪胚胎腹顶部，腹侧观和外侧观

前肾管引流至中肾，现在称为中肾管

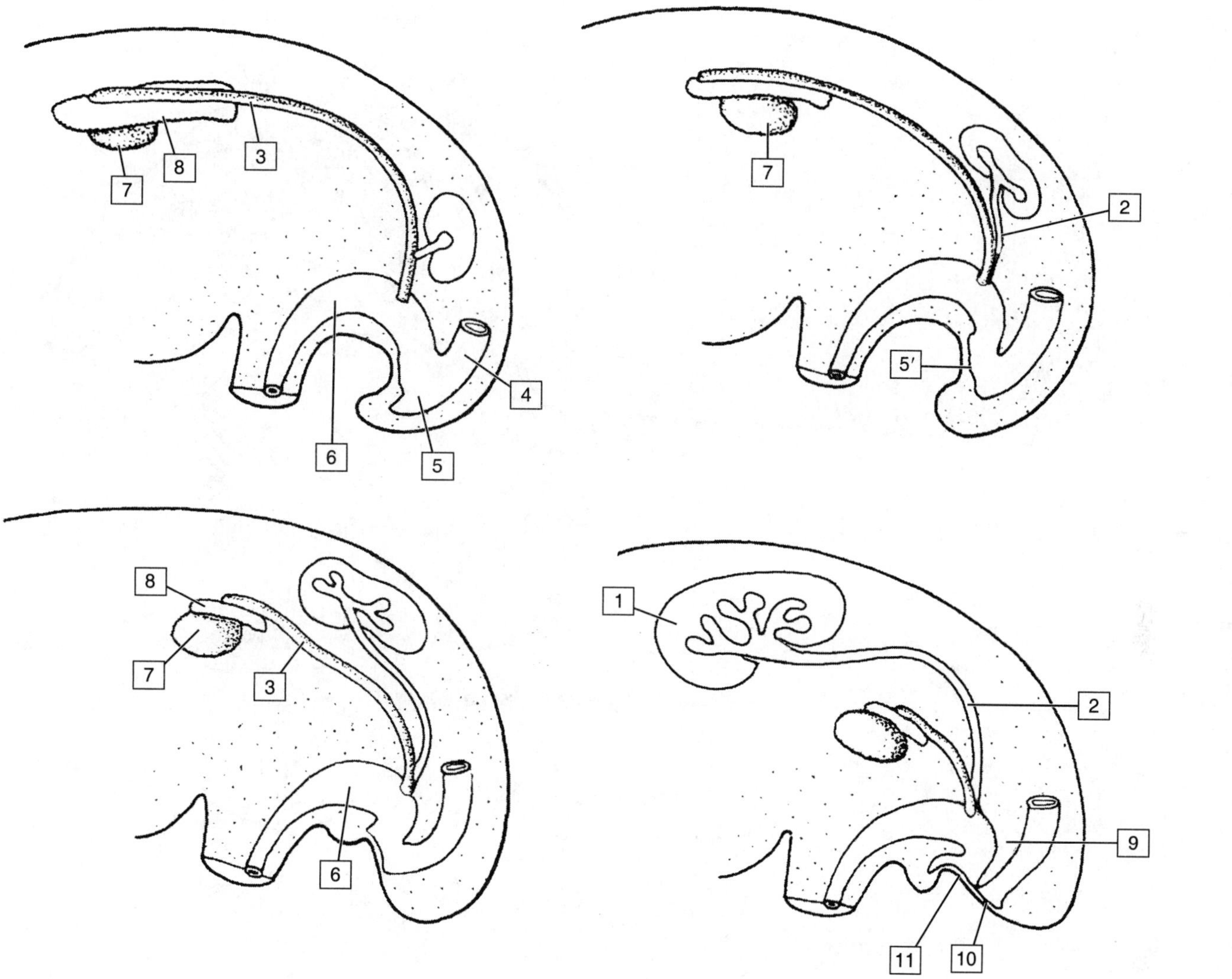

1	后肾 Metanephros	6	泌尿生殖窦 Urogenital sinus
2	输尿管芽（输尿管） Ureteric bud (future ureter)	7	性腺 Gonad
3	中肾管（输出管） Mesonephric (deferent) duct	8	中肾残迹（附睾） Remnant of mesonephros (future epididymis)
4	直肠 Rectum	9	尿道直肠隔 Urorectal septum
5	泄殖腔 Cloaca	10	肛膜 Anal membrane
5′	泄殖腔膜 Cloacal membrane	11	泌尿生殖膜 Urogenital membrane

图5-28 来自两个猪原基（后肾索和输尿管芽）的后肾发育

注意中肾逐渐退化

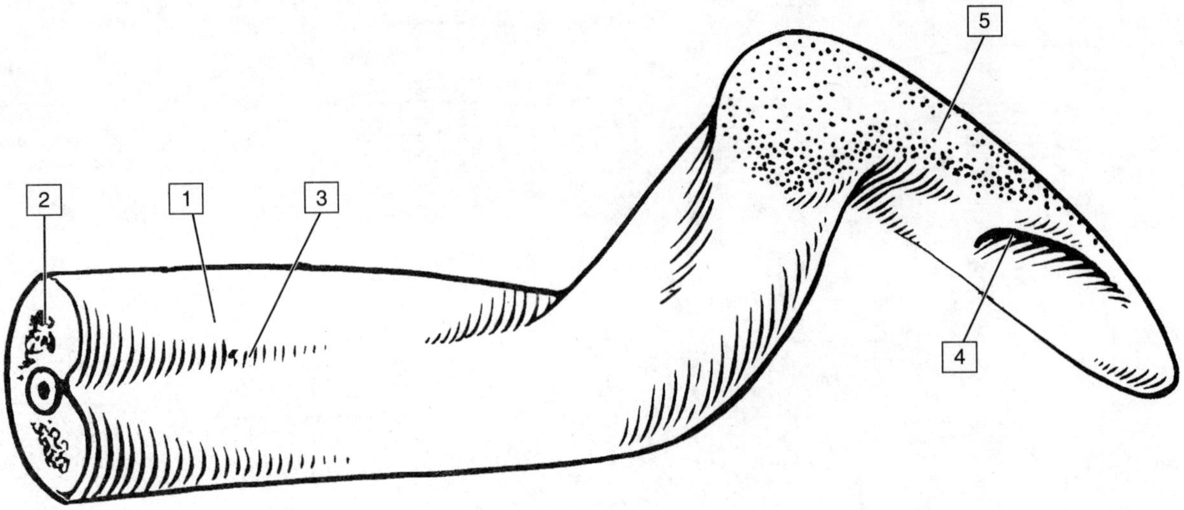

1	白膜	Tunica albuginea
2	阴茎海绵体	Corpus cavernosum
3	尿道沟	Urethral groove
4	尿道外口	External urethral orifice
5	细阴茎头	Thin glans penis

图5-29　猪阴茎游离端（横断面）

6

前　肢

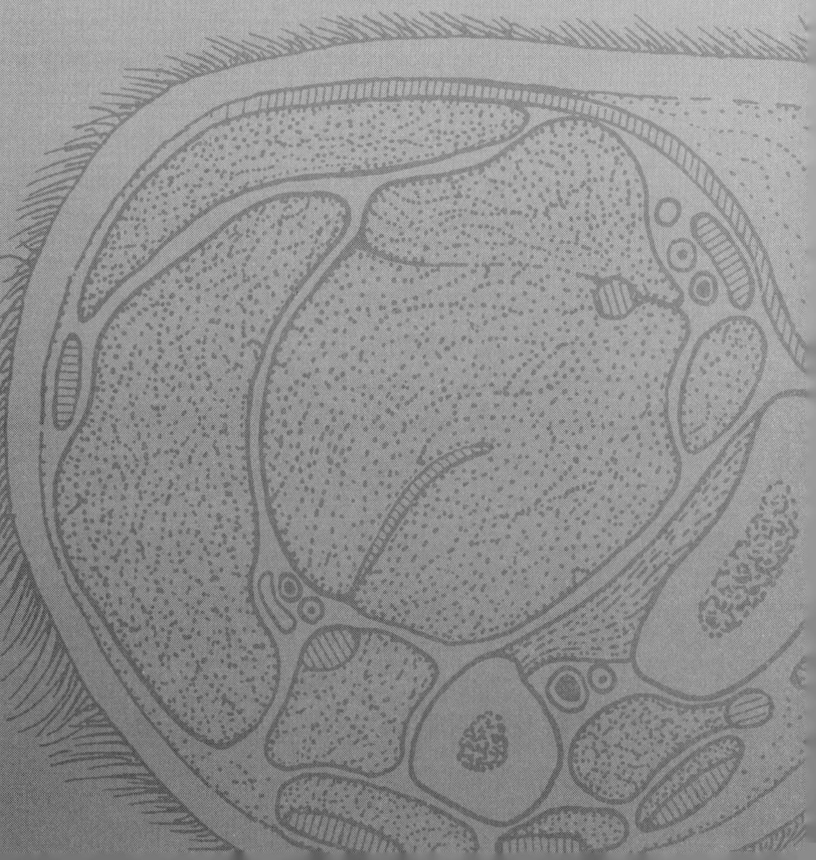

犬

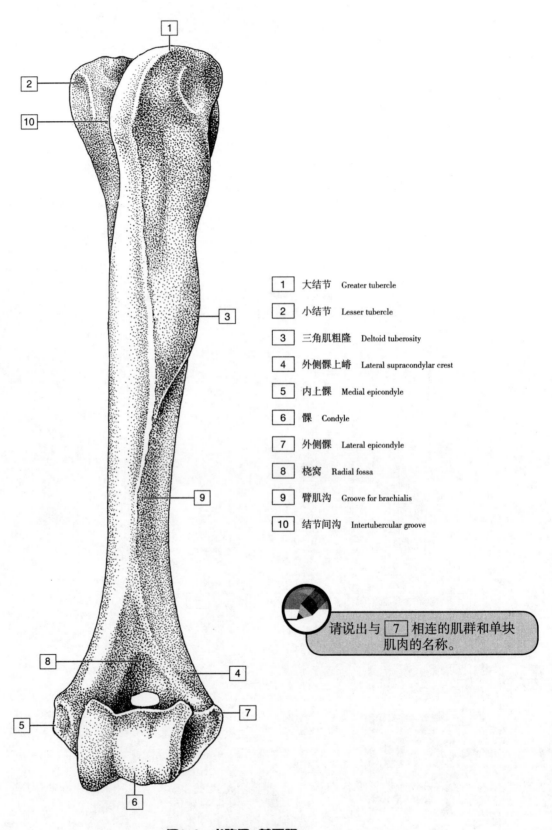

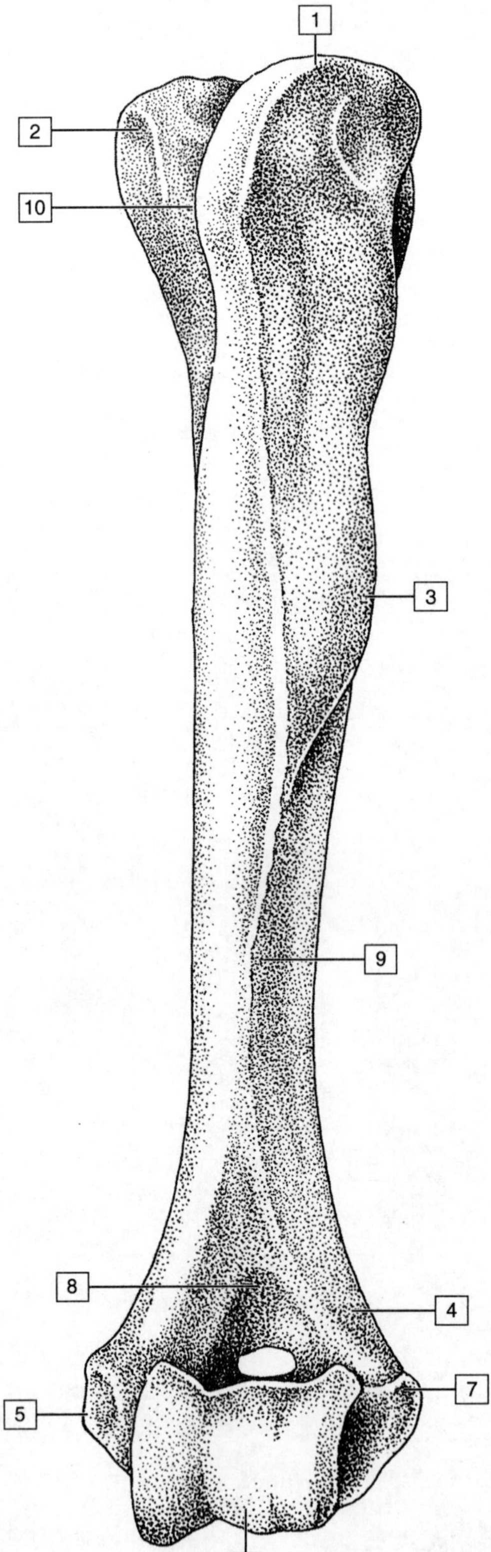

1 大结节 Greater tubercle

2 小结节 Lesser tubercle

3 三角肌粗隆 Deltoid tuberosity

4 外侧髁上嵴 Lateral supracondylar crest

5 内上髁 Medial epicondyle

6 髁 Condyle

7 外侧髁 Lateral epicondyle

8 桡窝 Radial fossa

9 臂肌沟 Groove for brachialis

10 结节间沟 Intertubercular groove

请说出与 7 相连的肌群和单块肌肉的名称。

图6-1 大肱骨，前面观

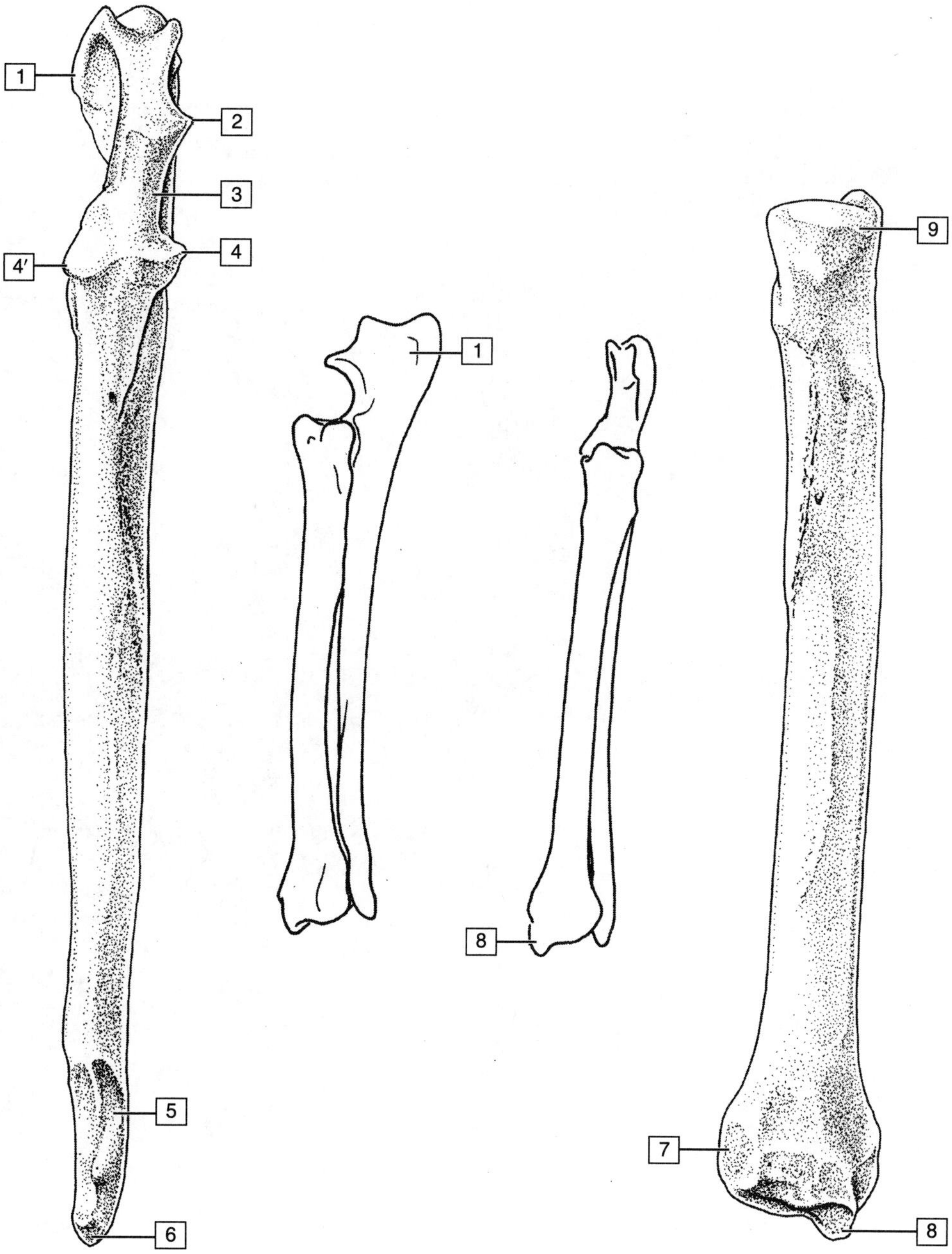

1 鹰嘴 Olecranon

2 肘突 Anconeal process

3 滑车切迹 Trochlear notch

4 外侧冠突 Lateral coronoid process

4′ 内侧冠突 Medial coronoid process

5 桡骨远端关节面 Distal articular facet for radius

6 外侧茎突(在犬尺腕骨面上) Lateral styloid process (with facet for the ulnar carpal bone in the dog)

7 尺骨关节面 Articular facet for ulna

8 内侧茎突 Medial styloid process

9 圆周面 Circumferential facet

图6-2 犬左侧尺骨和左侧桡骨

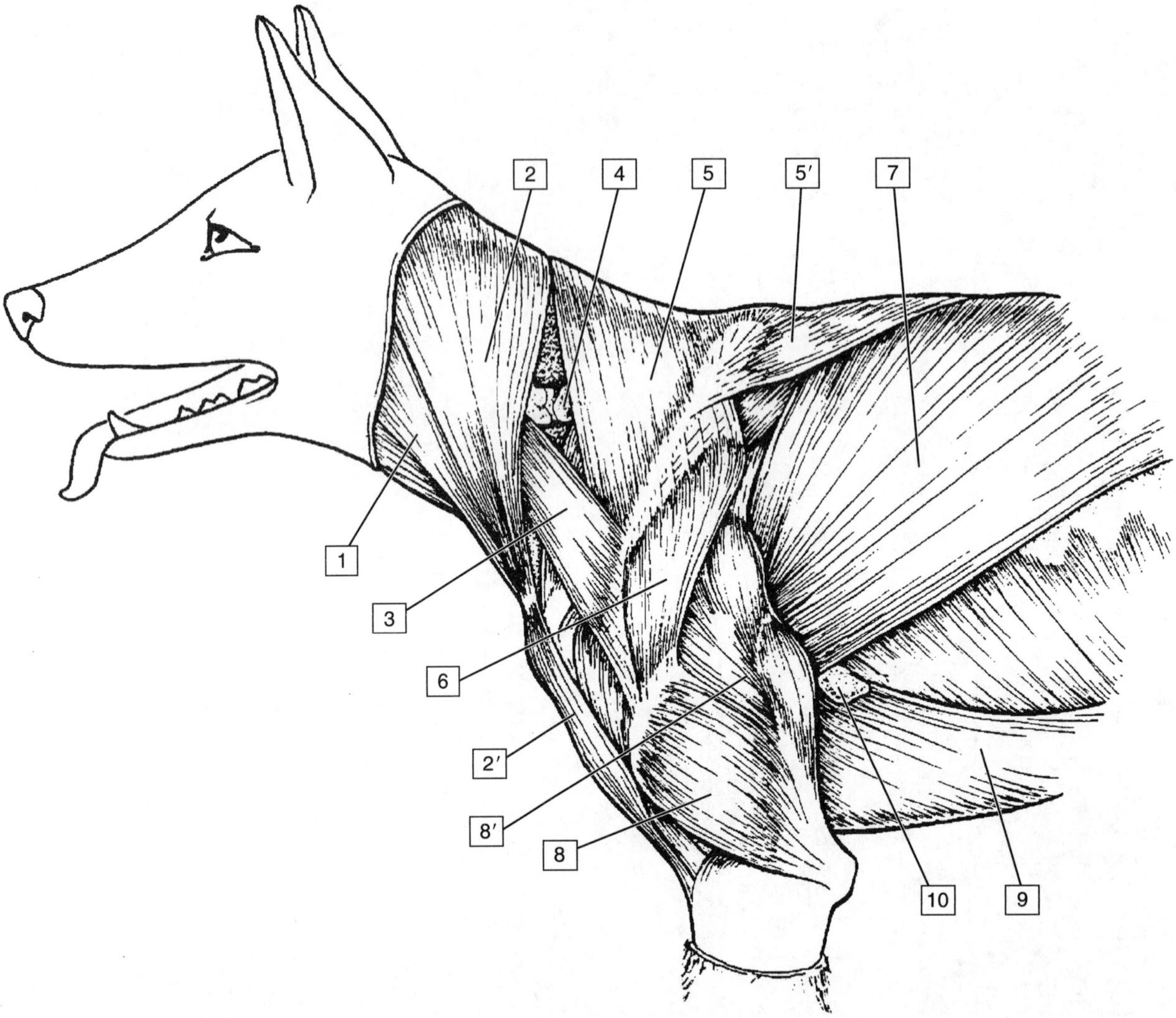

1 胸头肌 Sternocephalicus	6 三角肌 Deltoideus
2 臂头肌：锁颈肌 Brachiocephalicus: cleidocervicalis	7 背阔肌 Latissimus dorsi
2′ 臂头肌：锁臂肌 Brachiocephalicus: cleidobrachialis	8 臂三头肌外侧头 Lateral head of triceps
3 肩胛横突肌 Omotransversarius	8′ 臂三头肌长头 Long head of triceps
4 颈浅淋巴结 Superficial cervical lymph node	9 胸深（上升的）肌 Pectoralis profundus (ascendens)
5 颈斜方肌 Cervical part of trapezius	10 腋副淋巴结 Accessory axillary lymph node
5′ 胸斜方肌 Thoracic part of trapezius	

图6-3 犬肩部与臂前浅层肌肉

请说出与 7 相连的肌群或肌肉的名称。

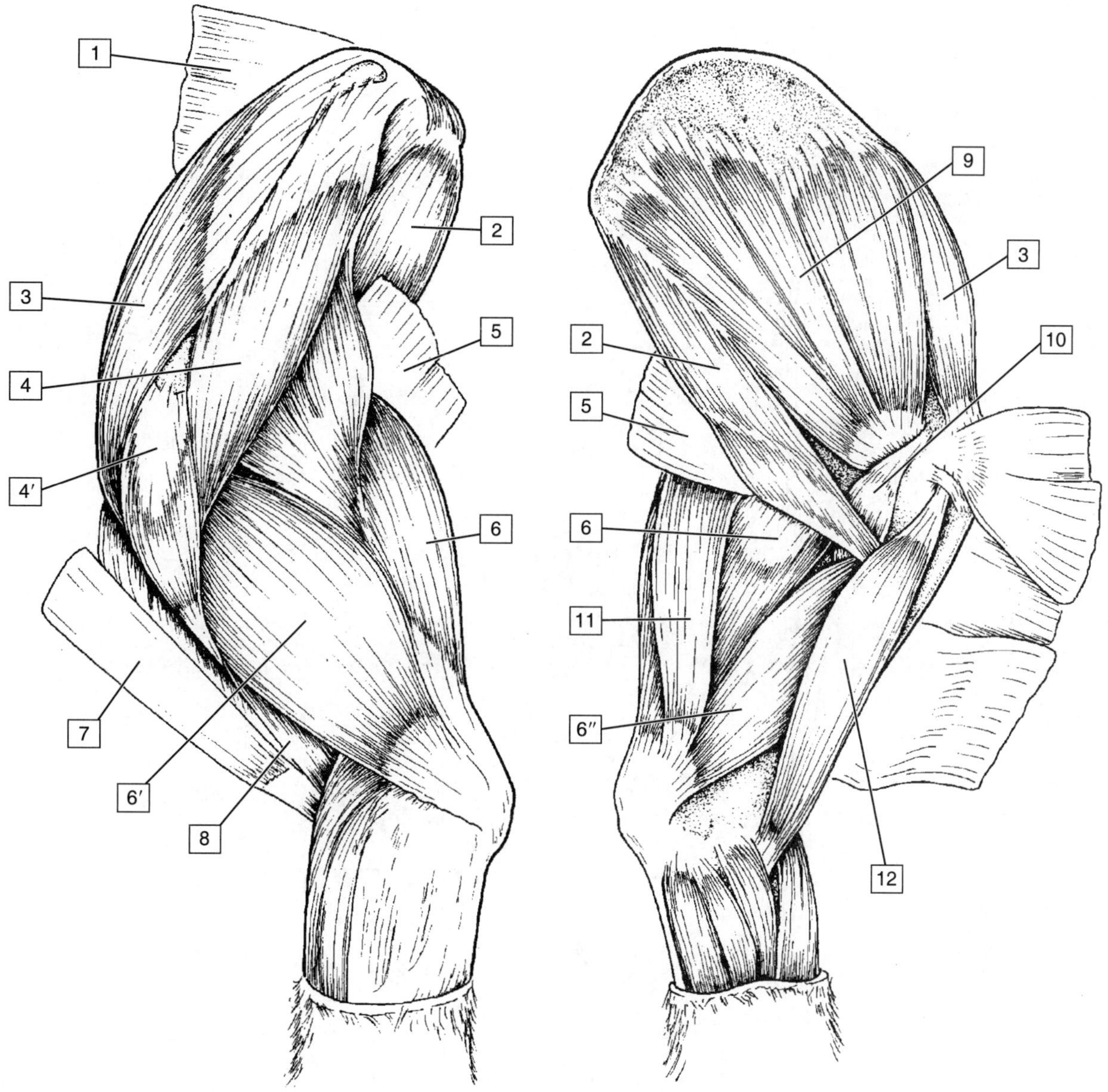

1	菱形肌 Rhomboideus
2	大圆肌 Teres major
3	冈上肌 Supraspinatus
4	三角肌肩胛部 Scapular part of deltoideus
4′	三角肌肩峰部 Acromial part of deltoideus
5	背阔肌 Latissimus dorsi
6	臂三头肌长头 Long head of triceps
6′	臂三头肌外侧头 Lateral head of triceps
6″	臂三头肌内侧头 Medial head of triceps
7	臂头肌 Brachiocephalicus
8	臂肌 Brachialis
9	肩胛下肌 Subscapularis
10	喙臂肌 Coracobrachialis
11	前臂筋膜张肌 Tensor fasciae antebrachii
12	臂二头肌 Biceps

图6-4 犬左前肢肩臂肌肉，外侧观和内侧观

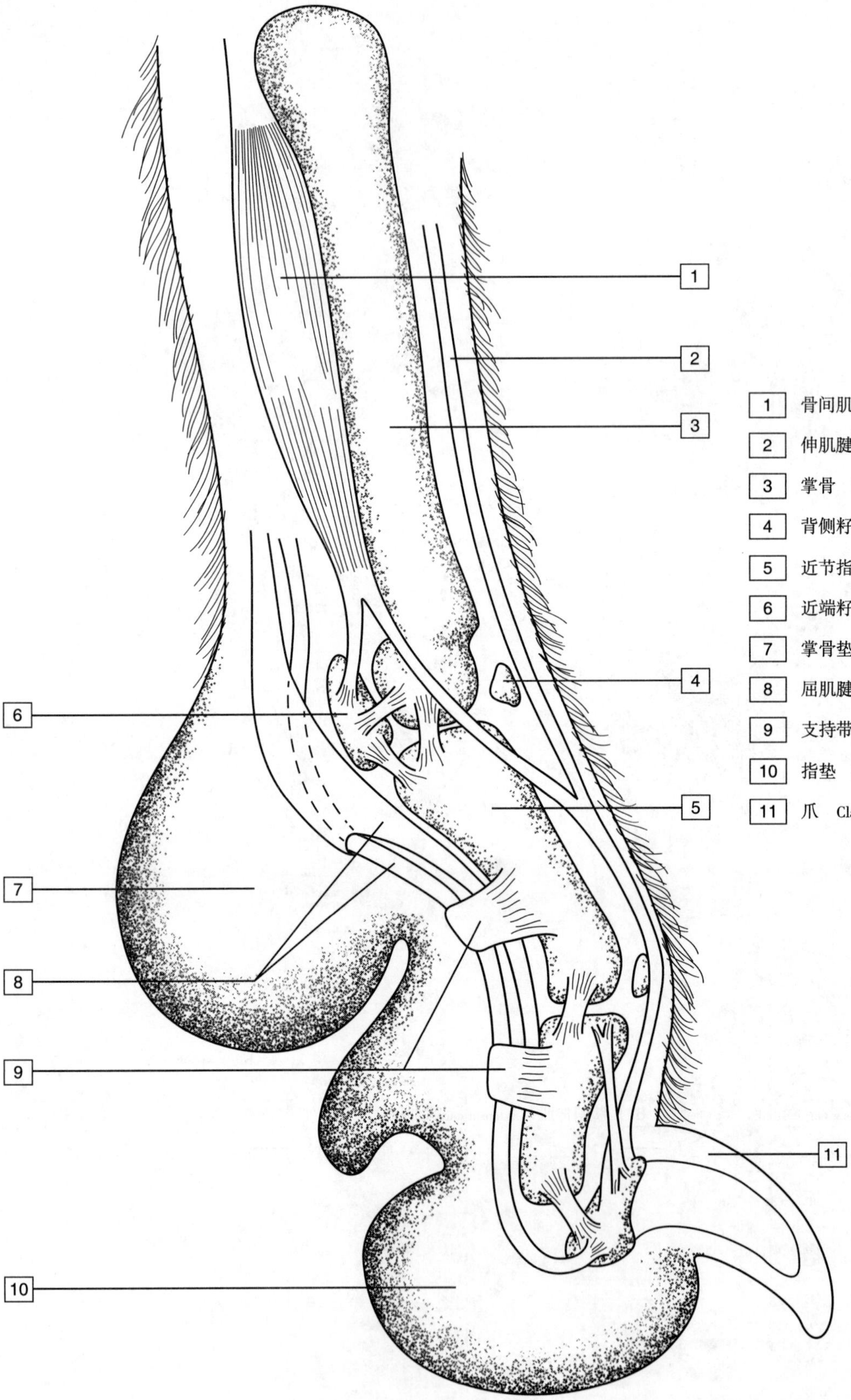

1	骨间肌	Interosseus
2	伸肌腱	Extensor tendon
3	掌骨	Metacarpal bone
4	背侧籽骨	Dorsal sesamoid bone
5	近节指骨	Proximal phalanx
6	近端籽骨	Proximal sesamoid bone
7	掌骨垫	Metacarpal pad
8	屈肌腱	Flexor tendons
9	支持带	Retinacula
10	指垫	Digital pad
11	爪	Claw

图6-5 犬爪的轴向切面

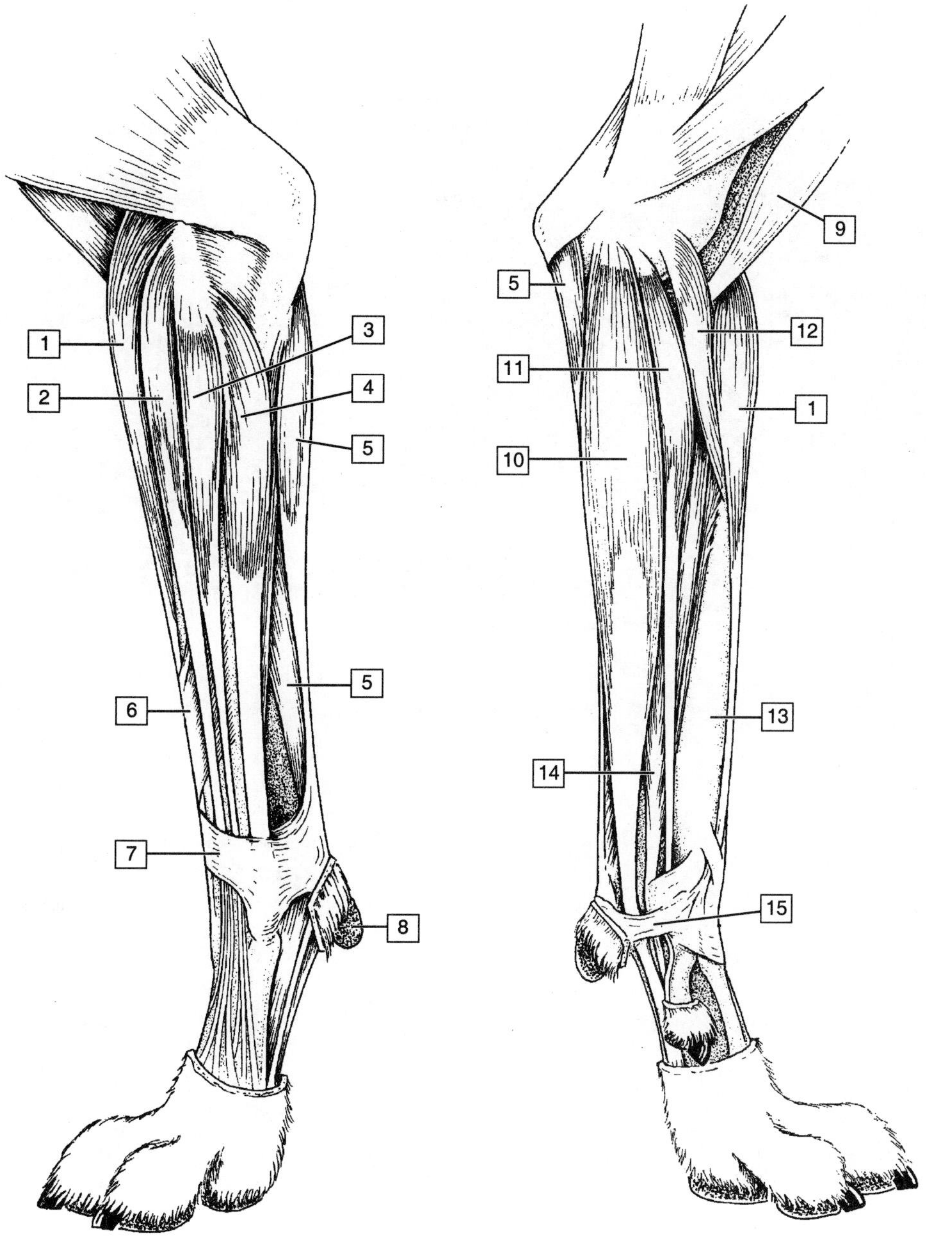

1	桡侧腕伸肌 Extensor carpi radialis		6	腕斜伸肌 Extensor carpi obliquus		11	桡侧腕屈肌 Flexor carpi radialis
2	指总伸肌 Common digital extensor		7	伸肌支持带 Extensor retinaculum		12	旋前圆肌 Pronator teres
3	指外侧伸肌 Lateral digital extensor		8	腕垫 Carpal pad		13	桡骨 Radius
4	尺骨外肌 Ulnaris lateralis		9	臂二头肌 Biceps		14	指深屈肌 Deep digital flexor
5	尺侧腕屈肌 Flexor carpi ulnaris		10	指浅屈肌 Superficial digital flexor		15	屈肌支持带 Flexor retinaculum

图6-6 犬左前肢前臂部肌肉，外侧观和内侧观

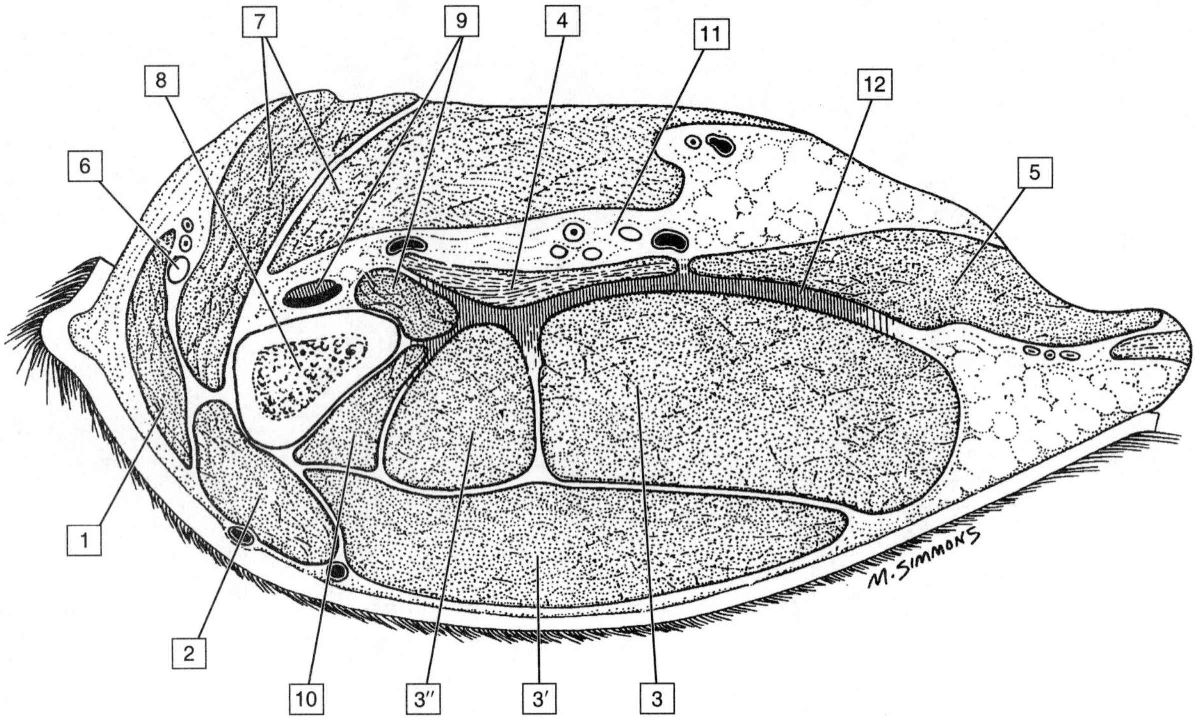

1　臂头肌　Brachiocephalicus

2　三角肌　Deltoideus

3　臂三头肌长头　Long head of triceps

3′　臂三头肌外侧头　Lateral head of triceps

3″　臂三头肌副头　Accessory head of triceps

4　大圆肌　Teres major

5　背阔肌　Latissimus dorsi

6　头静脉　Cephalic vein

7　胸肌　Pectoral muscles

8　肱骨　Humerus

9　臂二头肌肌腱和喙臂肌　Biceps tendon and coracobrachialis

10　臂肌　Brachialis

11　臂部血管和神经干　Brachial vessels and nerve trunks

12　厚的肌间筋膜　Heavy intermuscular fascia

图6-7　位于肩关节远端的犬左前肢横断面

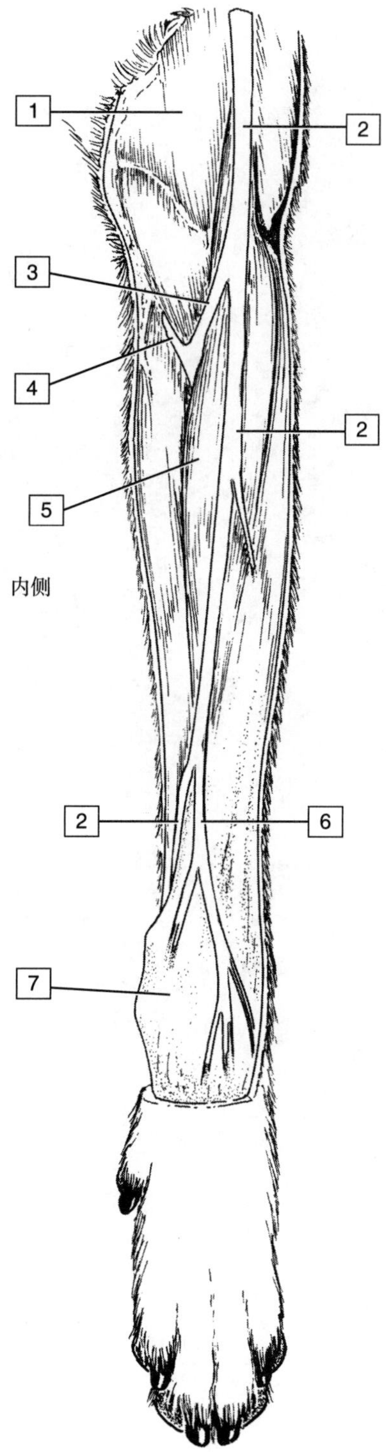

1	臂头肌 Brachiocephalicus	4	臂静脉 Brachial vein	6	副头静脉 Accessory cephalic vein
2	头静脉 Cephalic vein	5	桡侧腕伸肌 Extensor carpi radialis	7	腕部 Carpus
3	肘正中静脉 Median cubital vein				

图6-8 犬左前臂浅层静脉

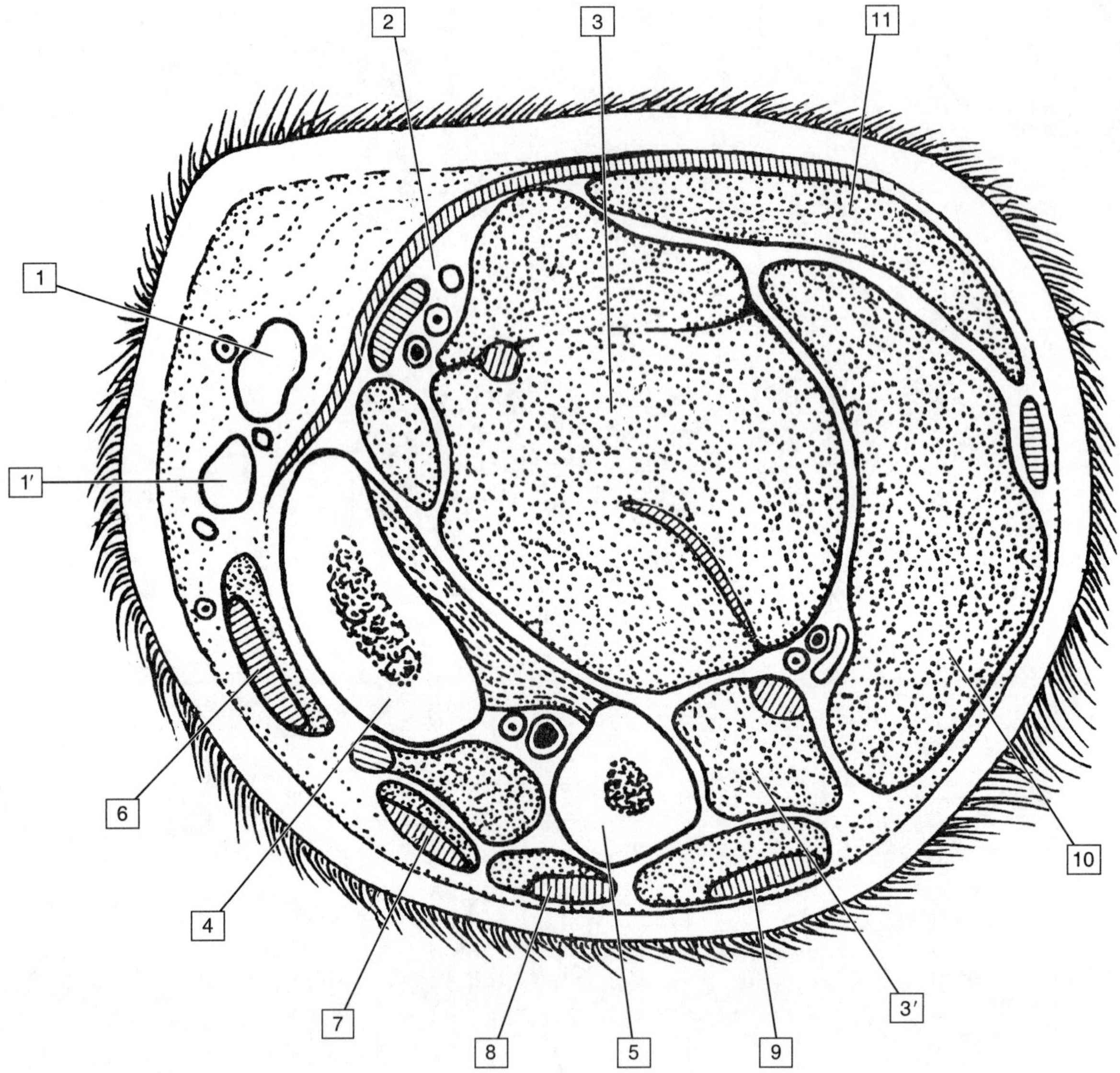

1	头静脉和桡神经浅支 Cephalic vein and branches of superficial radial nerve
1′	副头静脉 Accessory cephalic vein
2	正中血管、神经和桡侧腕屈肌 Median vessels and nerve and flexor carpi radialis
3	指深屈肌肱骨头 Humeral head of deep digital flexor
3′	指深屈肌尺骨头 Ulnar head of deep digital flexor
4	桡骨 Radius
5	尺骨 Ulna
6	桡侧腕伸肌 Extensor carpi radialis
7	指总伸肌 Common digital extensor
8	指外侧伸肌 Lateral digital extensor
9	尺外侧肌 Ulnaris lateralis
10	尺侧腕屈肌：尺侧小头位于后部，尺侧血管和神经位于前方 Flexor carpi ulnaris: its small ulnar head lies on its caudal aspects, and the ulnar vessels and nerve on its cranial aspect
11	指浅屈肌 Superficial digital flexor

图6-9 犬左前肢近腕部横断面

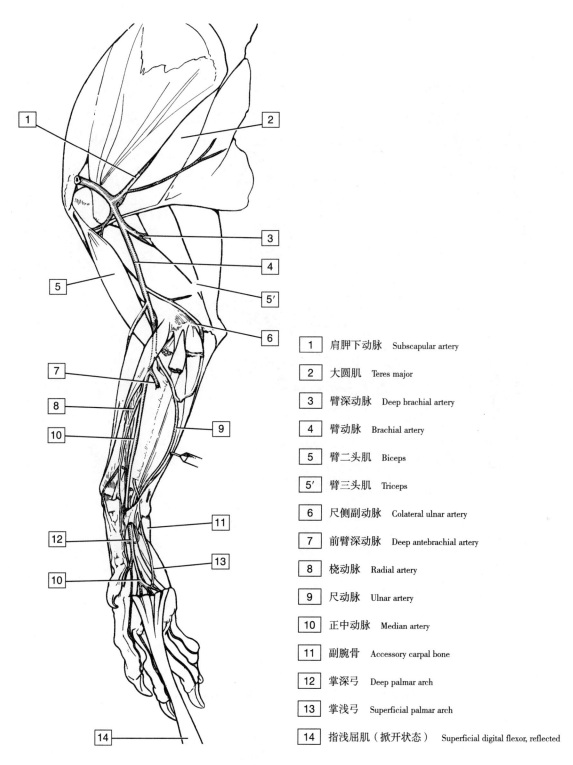

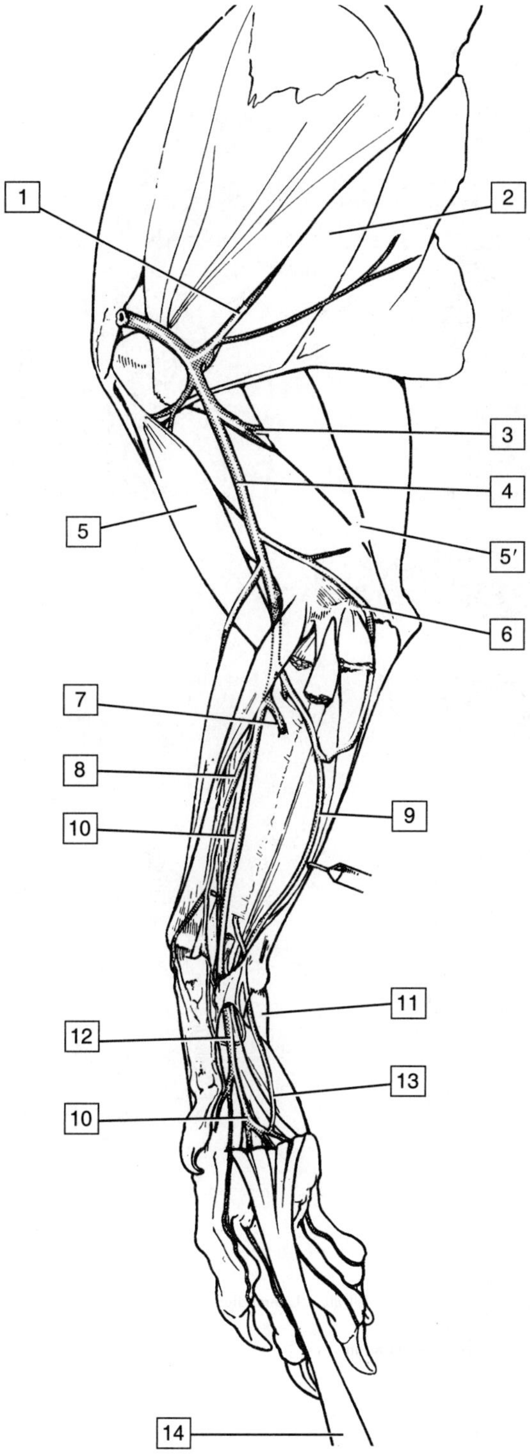

1	肩胛下动脉	Subscapular artery
2	大圆肌	Teres major
3	臂深动脉	Deep brachial artery
4	臂动脉	Brachial artery
5	臂二头肌	Biceps
5′	臂三头肌	Triceps
6	尺侧副动脉	Colateral ulnar artery
7	前臂深动脉	Deep antebrachial artery
8	桡动脉	Radial artery
9	尺动脉	Ulnar artery
10	正中动脉	Median artery
11	副腕骨	Accessory carpal bone
12	掌深弓	Deep palmar arch
13	掌浅弓	Superficial palmar arch
14	指浅屈肌（掀开状态）	Superficial digital flexor, reflected

图6-10 犬右前肢大动脉局部解剖图，内侧观

前肢后内侧肌肉已被剥离

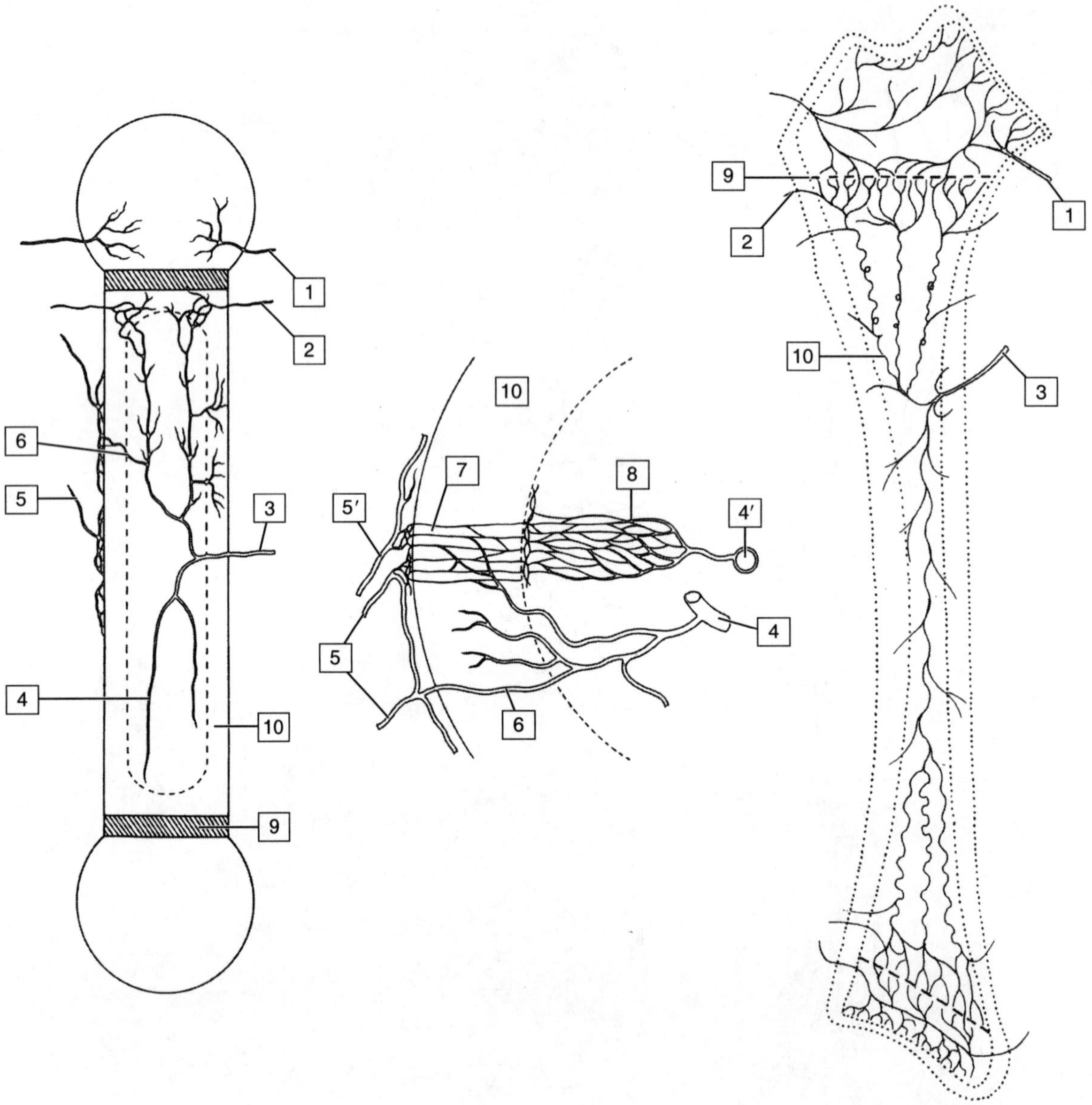

1	骨骺动脉	Epiphysial aa.
2	干骺端动脉	Metaphysial aa.
3	营养性动脉	Nutrient artery
4	骨髓动脉	Artery of the bone marrow
4′	骨髓静脉	Vein of the bone marrow
5	骨膜动脉	Periosteal aa.
5′	骨膜静脉	Periosteal vein
6	骨膜和骨髓动脉的吻合支	Anastomosis between periosteal and bone marrow aa.
7	骨密质毛细血管	Capillaries of the cortex
8	骨髓内血窦	Sinusoids in the bone marrow
9	生长软骨	Growth cartilage
10	骨密质	Cortex

图6-11　犬长骨血液供应

图中间为骨密质血供的放大

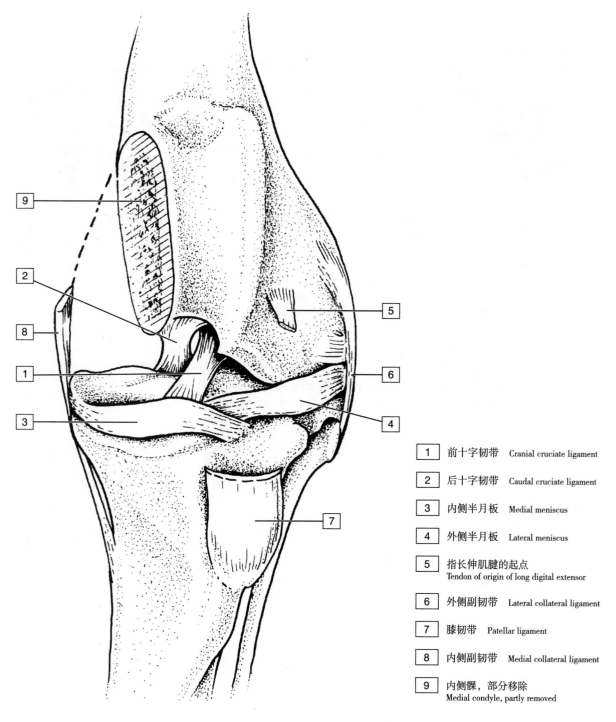

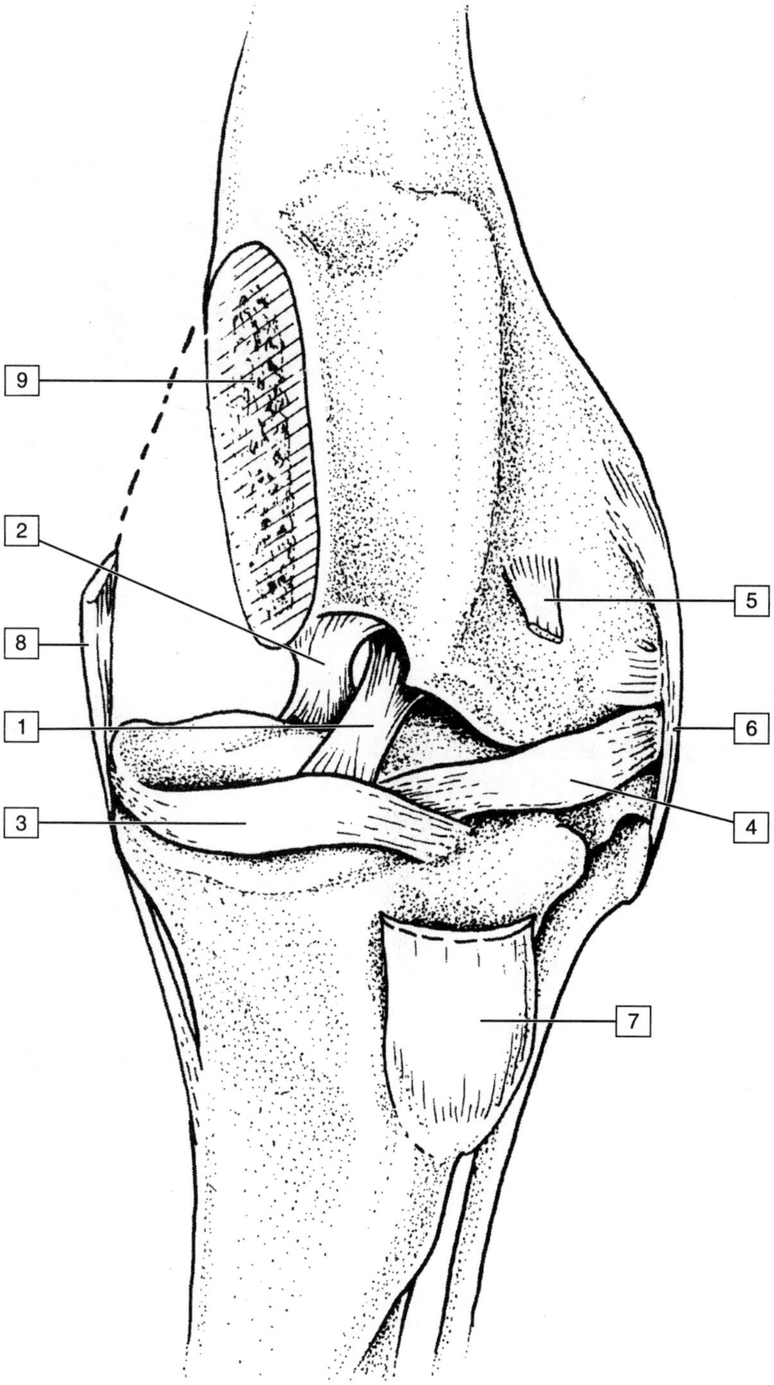

1	前十字韧带	Cranial cruciate ligament
2	后十字韧带	Caudal cruciate ligament
3	内侧半月板	Medial meniscus
4	外侧半月板	Lateral meniscus
5	指长伸肌腱的起点	Tendon of origin of long digital extensor
6	外侧副韧带	Lateral collateral ligament
7	膝韧带	Patellar ligament
8	内侧副韧带	Medial collateral ligament
9	内侧髁，部分移除	Medial condyle, partly removed

图6-12　犬左侧膝关节，前面观

切除部分关节以显示囊内和囊外韧带

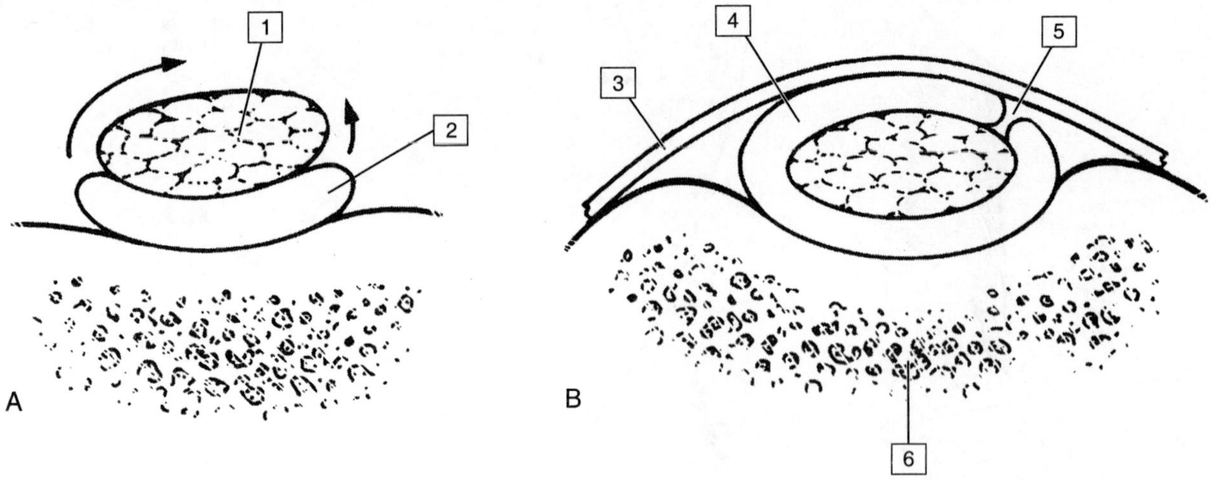

1 肌腱 Tendon

2 滑囊 Bursa

3 支持带 Retinaculum

4 腱鞘 Tendon sheath

5 腱系膜，通过它血管可到达肌腱 Mesotendon, through which blood vessels reach the tendon

6 骨骼 Bone

图6-13 犬黏液囊（A）和腱鞘（B）的横断面

滑囊允许肌腱在骨上无摩擦运动，以及在骨上和支持带下方进行腱鞘运动。箭头表示腱鞘可被看作是包裹肌腱的一个大的滑囊

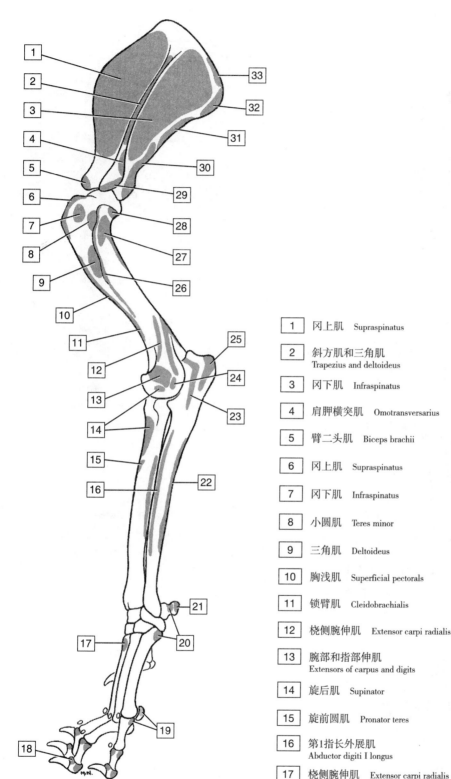

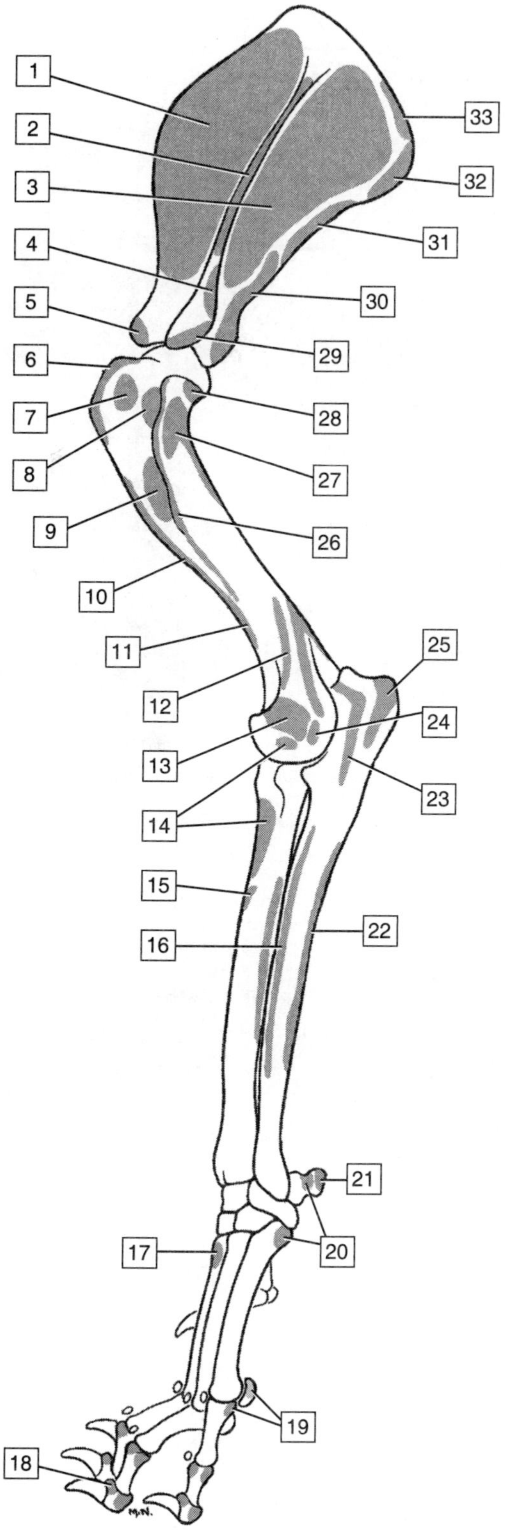

1 冈上肌 Supraspinatus
2 斜方肌和三角肌 Trapezius and deltoideus
3 冈下肌 Infraspinatus
4 肩胛横突肌 Omotransversarius
5 臂二头肌 Biceps brachii
6 冈上肌 Supraspinatus
7 冈下肌 Infraspinatus
8 小圆肌 Teres minor
9 三角肌 Deltoideus
10 胸浅肌 Superficial pectorals
11 锁臂肌 Cleidobrachialis
12 桡侧腕伸肌 Extensor carpi radialis
13 腕部和指部伸肌 Extensors of carpus and digits
14 旋后肌 Supinator
15 旋前圆肌 Pronator teres
16 第Ⅰ指长外展肌 Abductor digiti I longus
17 桡侧腕伸肌 Extensor carpi radialis
18 指外侧伸肌、指总伸肌 Lateral and common digital extensors
19 骨间肌 Interossei
20 尺外侧肌 Ulnaris lateralis
21 尺侧腕屈肌 Flexor carpi ulnaris
22 指深屈肌 Deep digital flexor
23 肘肌 Anconeus
24 尺外侧肌 Ulnaris lateralis
25 臂三头肌 Triceps
26 肘肌 Anconeus
27 臂肌 Brachialis
28 臂三头肌副头 Triceps, accessory head
29 三角肌 Deltoideus
30 臂三头肌长头和小圆肌 Triceps, long head and teres minor
31 肩胛下肌 Subscapularis
32 大圆肌 Teres major
33 菱形肌 Rhomboideus

图6-14 犬左前肢骨骼，肌肉附着，外侧观

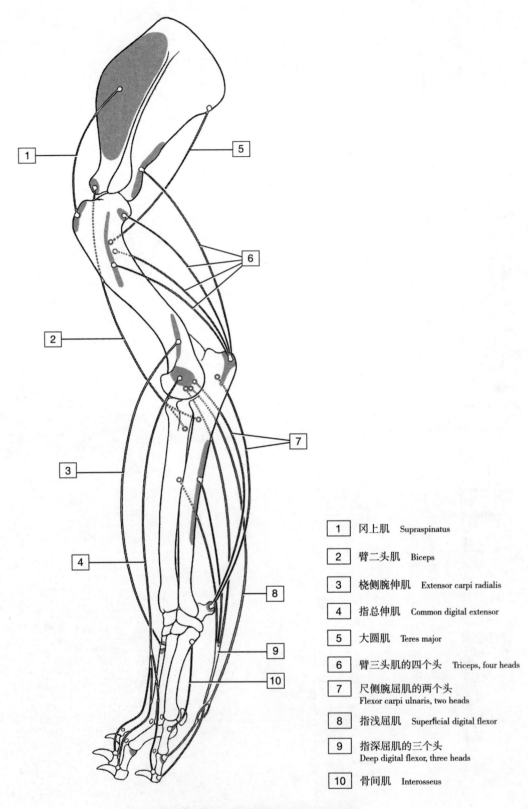

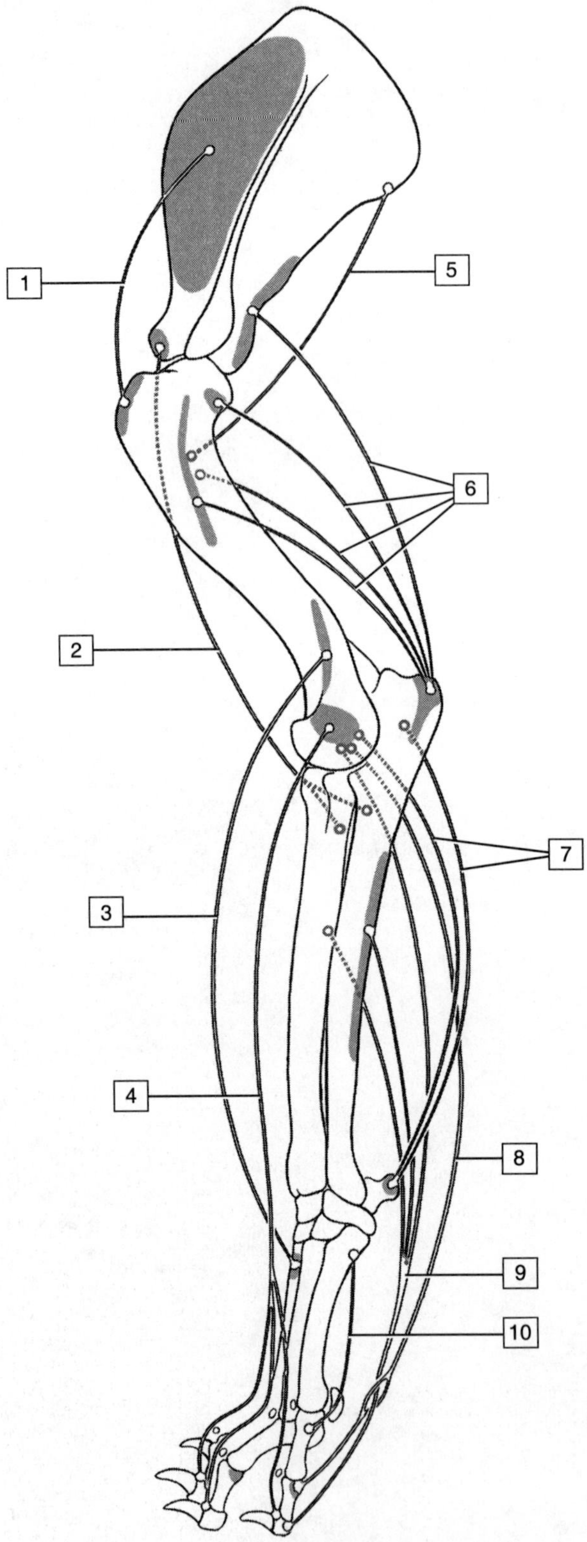

1 冈上肌 Supraspinatus

2 臂二头肌 Biceps

3 桡侧腕伸肌 Extensor carpi radialis

4 指总伸肌 Common digital extensor

5 大圆肌 Teres major

6 臂三头肌的四个头 Triceps, four heads

7 尺侧腕屈肌的两个头 Flexor carpi ulnaris, two heads

8 指浅屈肌 Superficial digital flexor

9 指深屈肌的三个头 Deep digital flexor, three heads

10 骨间肌 Interosseus

图6-15 犬前肢主要伸肌和屈肌

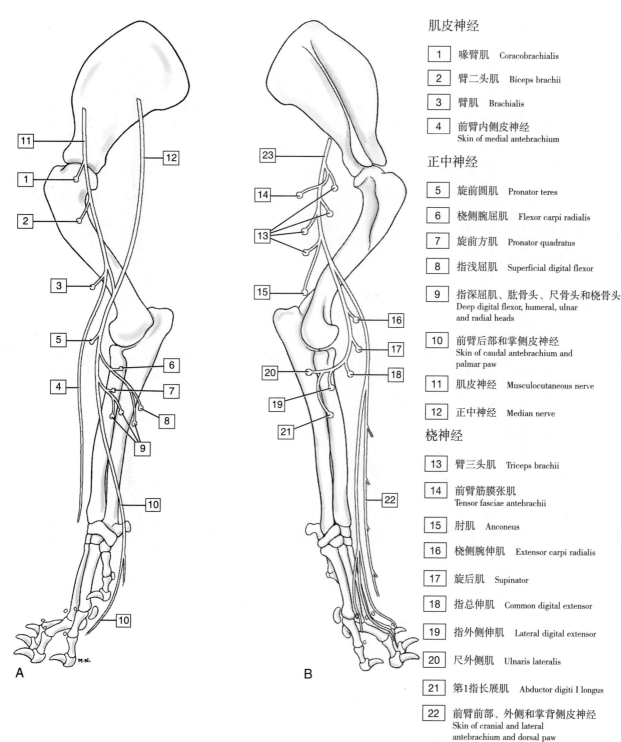

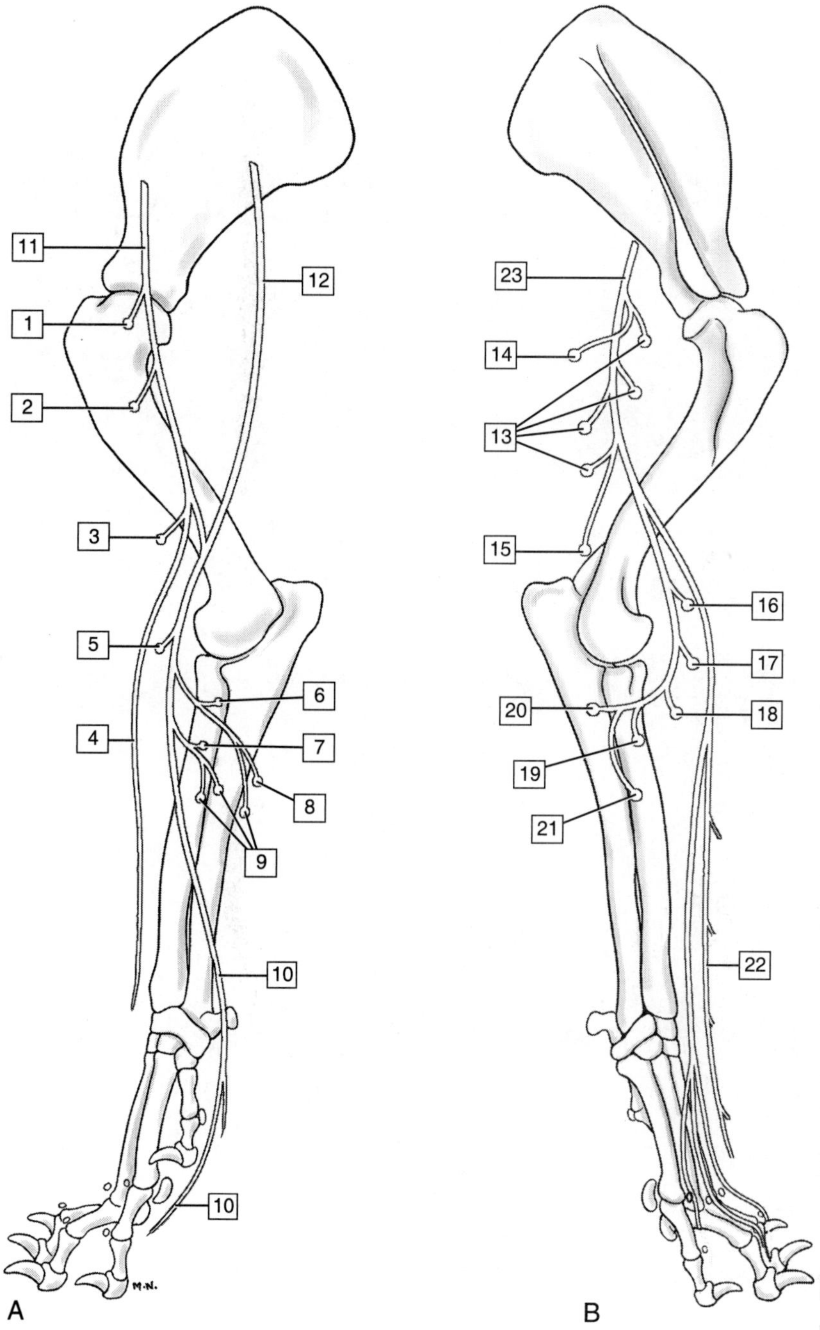

肌皮神经

1 喙臂肌 Coracobrachialis

2 臂二头肌 Biceps brachii

3 臂肌 Brachialis

4 前臂内侧皮神经 Skin of medial antebrachium

正中神经

5 旋前圆肌 Pronator teres

6 桡侧腕屈肌 Flexor carpi radialis

7 旋前方肌 Pronator quadratus

8 指浅屈肌 Superficial digital flexor

9 指深屈肌、肱骨头、尺骨头和桡骨头 Deep digital flexor, humeral, ulnar and radial heads

10 前臂后部和掌侧皮神经 Skin of caudal antebrachium and palmar paw

11 肌皮神经 Musculocutaneous nerve

12 正中神经 Median nerve

桡神经

13 臂三头肌 Triceps brachii

14 前臂筋膜张肌 Tensor fasciae antebrachii

15 肘肌 Anconeus

16 桡侧腕伸肌 Extensor carpi radialis

17 旋后肌 Supinator

18 指总伸肌 Common digital extensor

19 指外侧伸肌 Lateral digital extensor

20 尺外侧肌 Ulnaris lateralis

21 第1指长展肌 Abductor digiti I longus

22 前臂前部、外侧和掌背侧皮神经 Skin of cranial and lateral antebrachium and dorsal paw

23 桡神经 Radial nerve

图6-16 犬右前肢神经分布

A. 犬右前肢肌皮神经、正中神经的分布（内侧观）；B. 犬右前肢桡神经分布（外侧观）

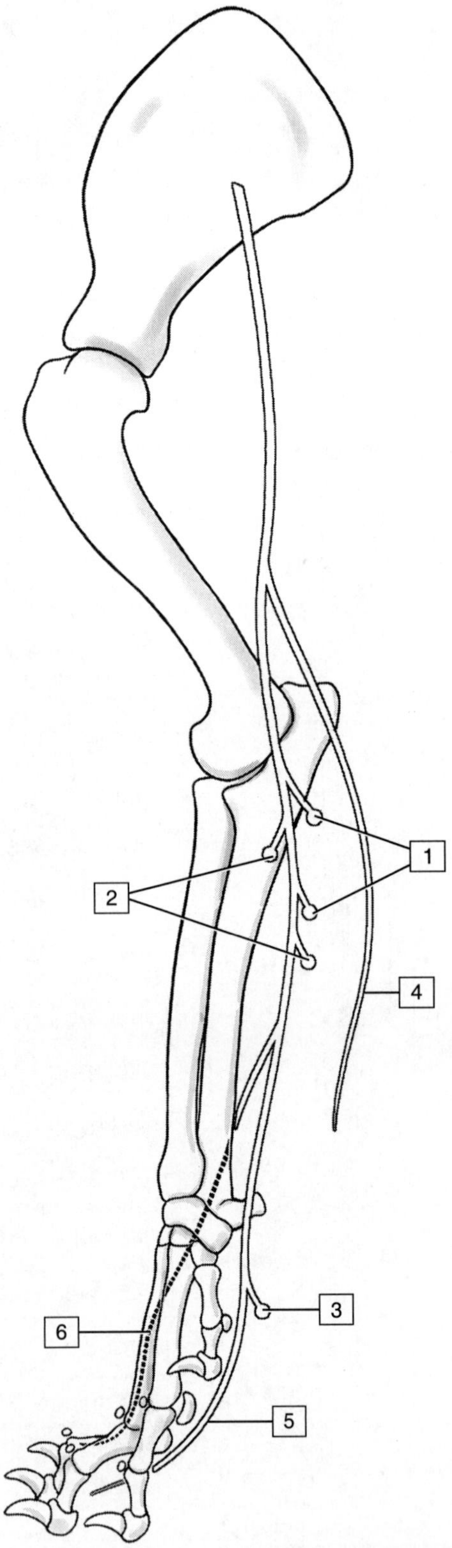

1 尺侧腕屈肌尺骨头和肱骨头 Flexor carpi ulnaris, ulnar and humeral heads

2 指深屈肌尺骨头和肱骨头 Deep digital flexor, ulnar and humeral heads

3 骨间肌 Interossei

4 前臂后皮神经 Skin of caudal antebrachium

5 掌侧皮神经 Skin of palmar paw

6 第5掌侧和指外侧皮神经 Skin of fifth metacarpal, lateral surface of digit

图6-17 犬右前肢尺神经分布，内侧观

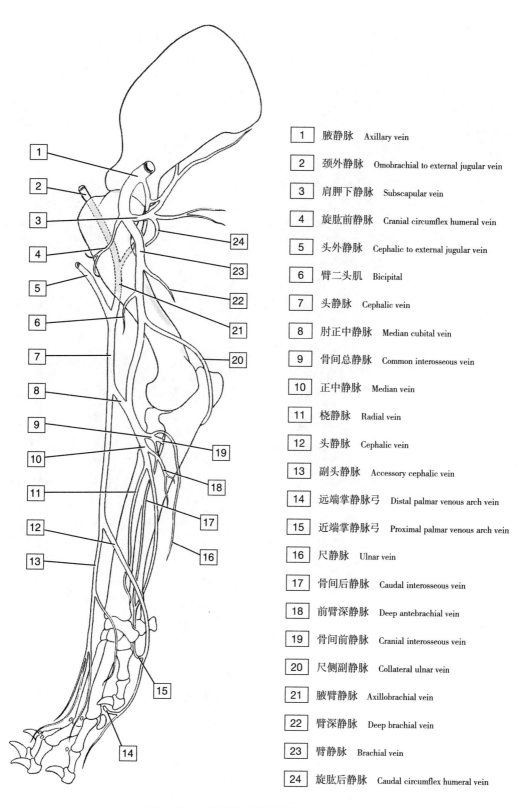

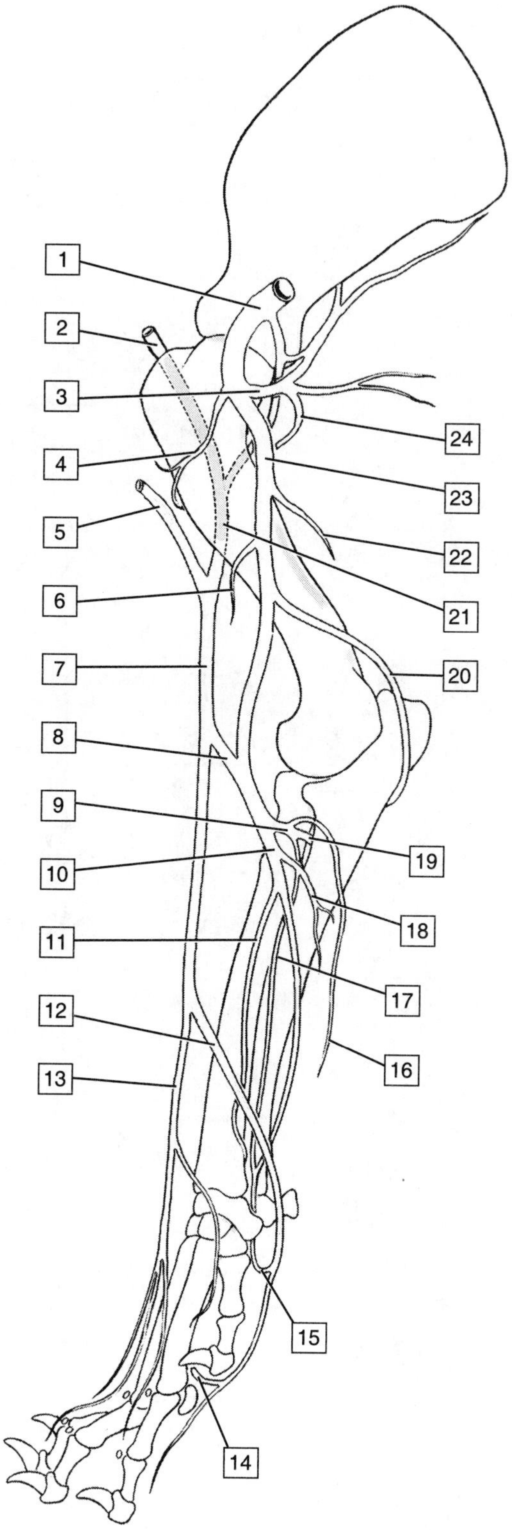

1	腋静脉	Axillary vein
2	颈外静脉	Omobrachial to external jugular vein
3	肩胛下静脉	Subscapular vein
4	旋肱前静脉	Cranial circumflex humeral vein
5	头外静脉	Cephalic to external jugular vein
6	臂二头肌	Bicipital
7	头静脉	Cephalic vein
8	肘正中静脉	Median cubital vein
9	骨间总静脉	Common interosseous vein
10	正中静脉	Median vein
11	桡静脉	Radial vein
12	头静脉	Cephalic vein
13	副头静脉	Accessory cephalic vein
14	远端掌静脉弓	Distal palmar venous arch vein
15	近端掌静脉弓	Proximal palmar venous arch vein
16	尺静脉	Ulnar vein
17	骨间后静脉	Caudal interosseous vein
18	前臂深静脉	Deep antebrachial vein
19	骨间前静脉	Cranial interosseous vein
20	尺侧副静脉	Collateral ulnar vein
21	腋臂静脉	Axillobrachial vein
22	臂深静脉	Deep brachial vein
23	臂静脉	Brachial vein
24	旋肱后静脉	Caudal circumflex humeral vein

图6-18 犬右前肢静脉示意图，内侧观

马

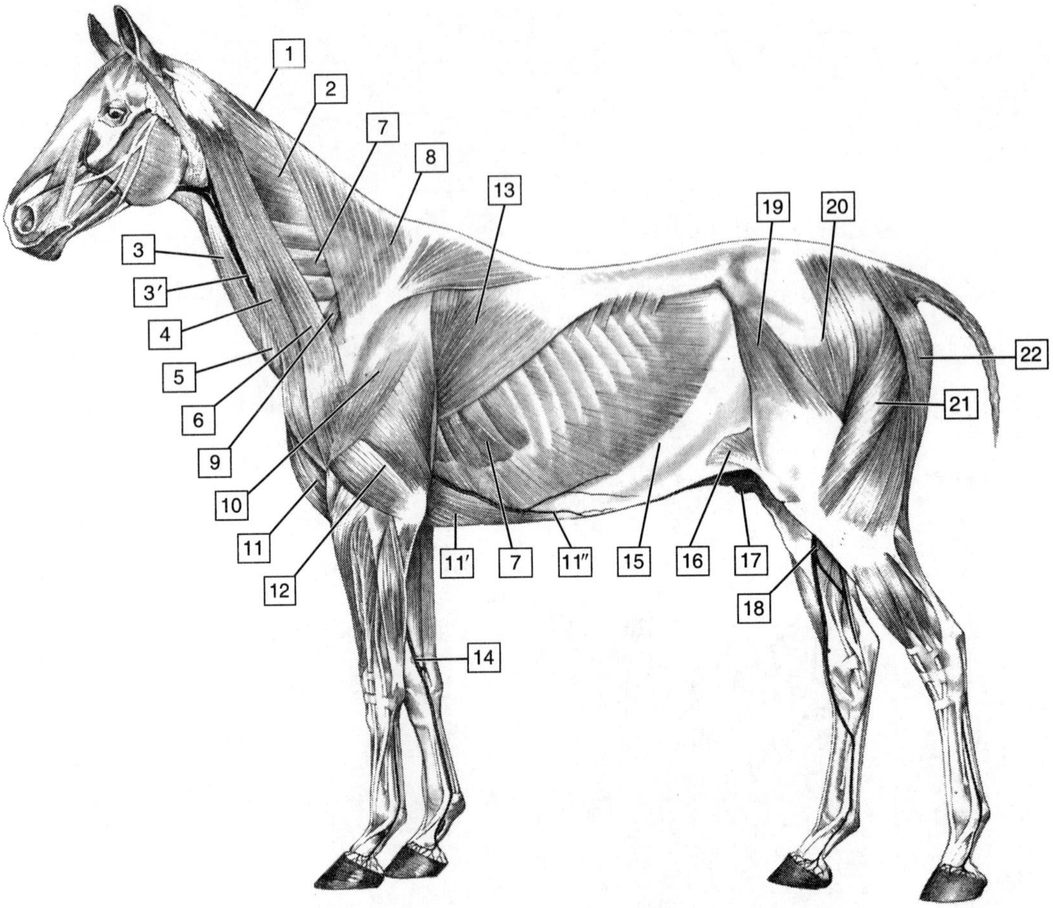

1	菱形肌	Rhomboideus
2	夹肌	Splenius
3	胸头肌	Sternocephalicus
3′	颈静脉	Jugular vein
4	臂头肌	Brachiocephalicus
5	颈皮肌	Cutaneous colli
6	锁枕肌	Omotransversarius
7	腹侧锯肌	Serratus ventralis
8	斜方肌	Trapezius
9	锁骨下肌	Subclavius
10	三角肌	Deltoideus
11	胸降肌	Pectoralis descendens
11′	胸深肌	Pectoralis ascendens
11″	胸浅静脉	Superficial thoracic vein
12	臂三头肌	Triceps
13	背阔肌	Latissimus dorsi
14	头静脉	Cephalic vein
15	腹外斜肌	External abdominal oblique
16	皮干残余形成侧面皱褶	Stump of cutaneus trunci forming flank fold
17	鞘	Sheath
18	内侧隐静脉	Medial saphenous vein
19	阔筋膜张肌	Tensor fasciae latae
20	臀浅肌	Gluteus superficialis
21	股二头肌	Biceps femoris
22	半腱肌	Semitendinosus

图6-19 马浅层肌肉和静脉

除了颈皮肌以外的其他皮肌已被剥离

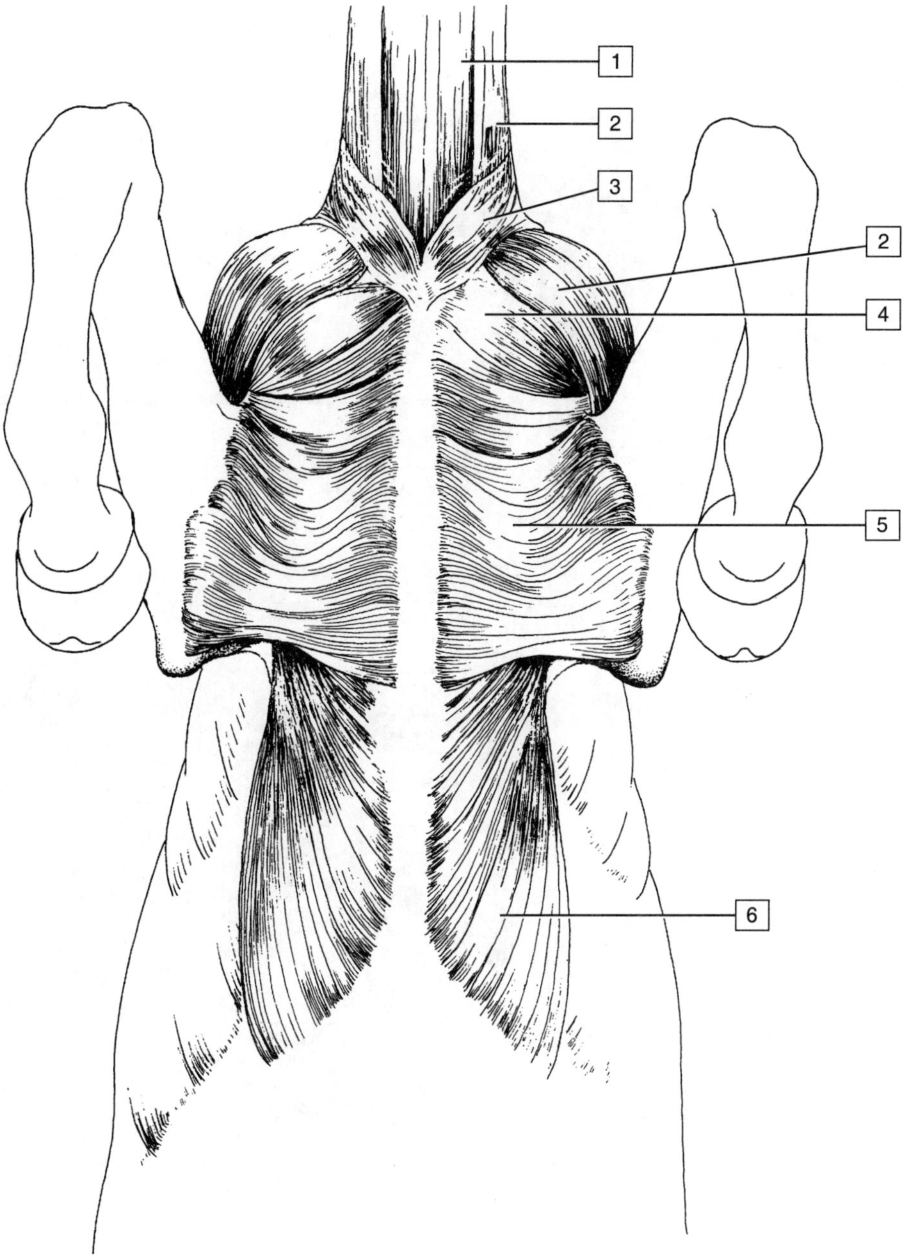

1	胸头肌 Sternocephalicus	4	胸降肌 Pectoralis descendens
2	臂头肌 Brachiocephalicus	5	胸横肌 Pectoralis transversus
3	颈皮肌 Cutaneous colli	6	胸深肌 Pectoralis profundus

图6-20 马胸部腹侧浅层肌肉

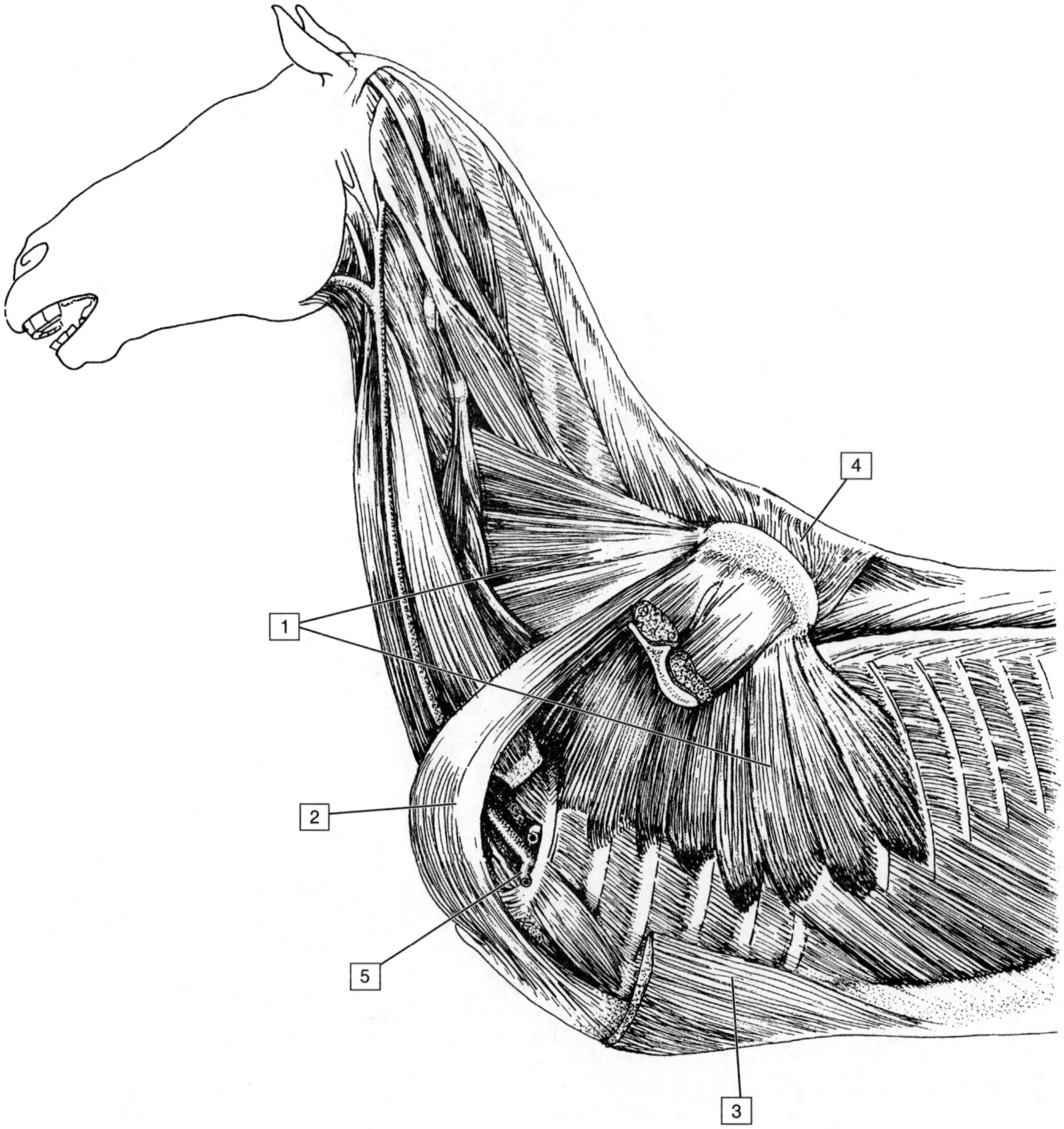

1 腹侧锯肌 Serratus ventralis

2 锁骨下肌 Subclavius

3 胸深肌 Pectoralis profundus

4 菱形肌 Rhomboideus

5 腋血管绕第1肋返回至前肢 Axillary vessels turning around first rib into limb

图6-21 马躯干与前肢相连的深层肌肉

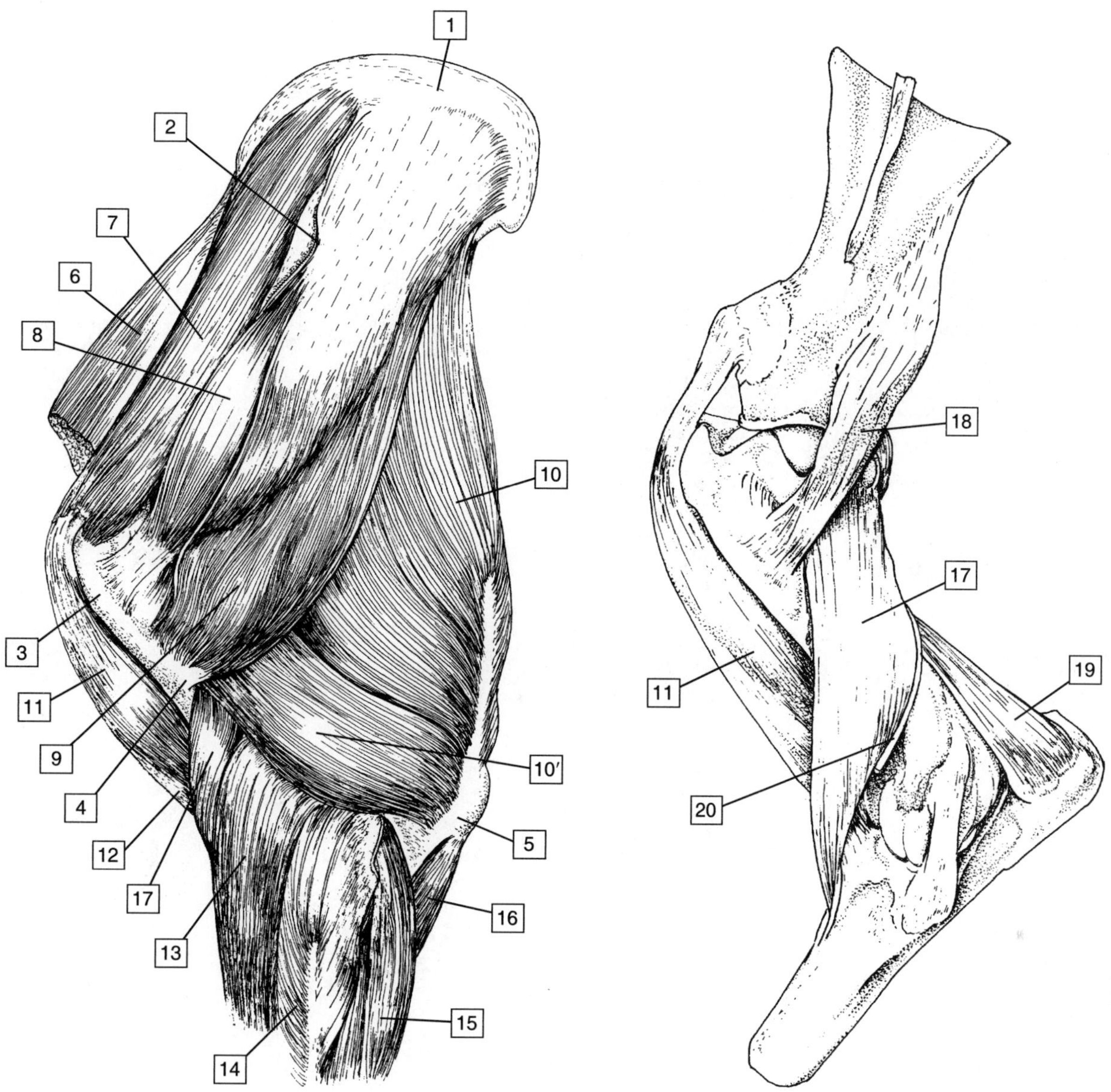

1	肩胛软骨 Scapular cartilage	6	锁骨下肌 Subclavius	11	臂二头肌 Biceps	16	指深屈肌的尺骨头 Ulnar head of deep digital flexor
2	肩胛冈 Scapular spine	7	冈上肌 Supraspinatus	12	臂二头肌腱膜 Lacertus fibrosus	17	臂肌 Brachialis
3	肱骨大结节 Greater tubercle of humerus	8	冈下肌 Infraspinatus	13	桡侧腕伸肌 Extensor carpi radialis	18	小圆肌 Teres minor
4	肱骨三角肌粗隆 Deltoid tuberosity of humerus	9	三角肌 Deltoideus	14	指总伸肌 Common digital extensor	19	肘肌 Anconeus
5	鹰嘴 Olecranon	10	臂三头肌长头 Long head of triceps	15	尺外侧肌 Ulnaris lateralis	20	桡神经 Radial nerve
		10′	臂三头肌外侧头 Lateral head of triceps				

图6-22 马肩部和肘关节相关肌肉，外侧观

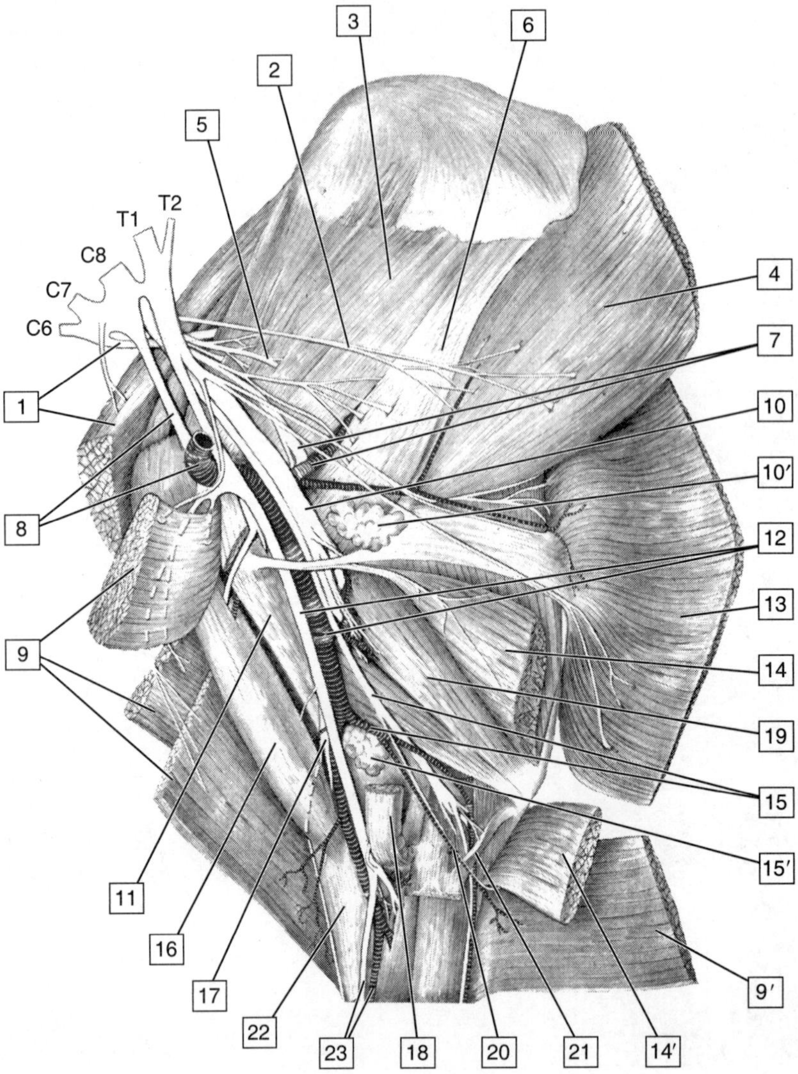

1 肩胛上神经和锁骨下肌 Suprascapular nerve and subclavius

2 胸背神经 Thoracodorsal nerve

3 肩胛下肌 Subscapularis

4 背阔肌 Latissimus dorsi

5 肩胛下神经 Subscapular nerve

6 大圆肌 Teres major

7 腋神经和肩胛下动脉 Axillary nerve and subscapular artery

8 肌皮神经和腋动脉 Musculocutaneous nerve and axillary artery

9 胸深肌 Pectoralis profundus

9′ 胸降肌 Pectoralis descendens

10 桡神经 Radial nerve

10′ 腋淋巴结 Axillary lymph nodes

11 喙臂肌 Coracobrachialis

12 正中神经和臂动脉 Median nerve and brachial artery

13 皮干 Cutaneous trunci

14 前臂筋膜张肌断端 Stump of tensor fasciae antebrachii

14′ 前臂筋膜张肌断端 Stump of tensor fasciae antebrachii

15 尺神经和尺侧副动脉 Ulnar nerve and collateral ulnar artery

15′ 肘淋巴结 Cubital lymph nodes

16 臂二头肌 Biceps

17 肌皮神经和前臂内侧皮神经 Musculocutaneous and medial cutaneous antebrachial nerve

18 桡侧腕屈肌 Flexor carpi radialis

19 臂三头肌 Triceps

20 前臂后支神经 Caudal cutaneous antebrachial nerve

21 尺侧腕屈肌 Flexor carpi ulnaris

22 臂二头肌腱膜 Lacertus fibrosus

23 正中神经和正中动脉 Median nerve and artery

图6-23 马右肩和右臂内侧面的神经、血管和肌肉

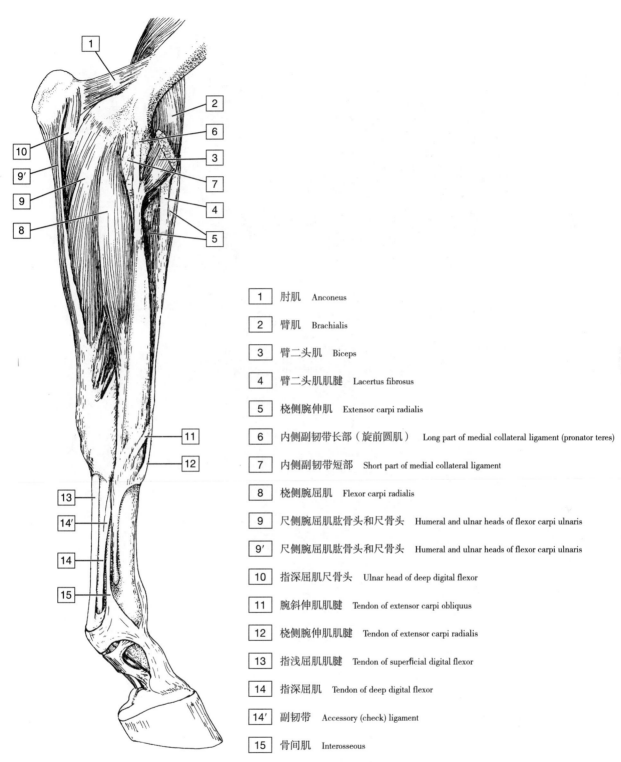

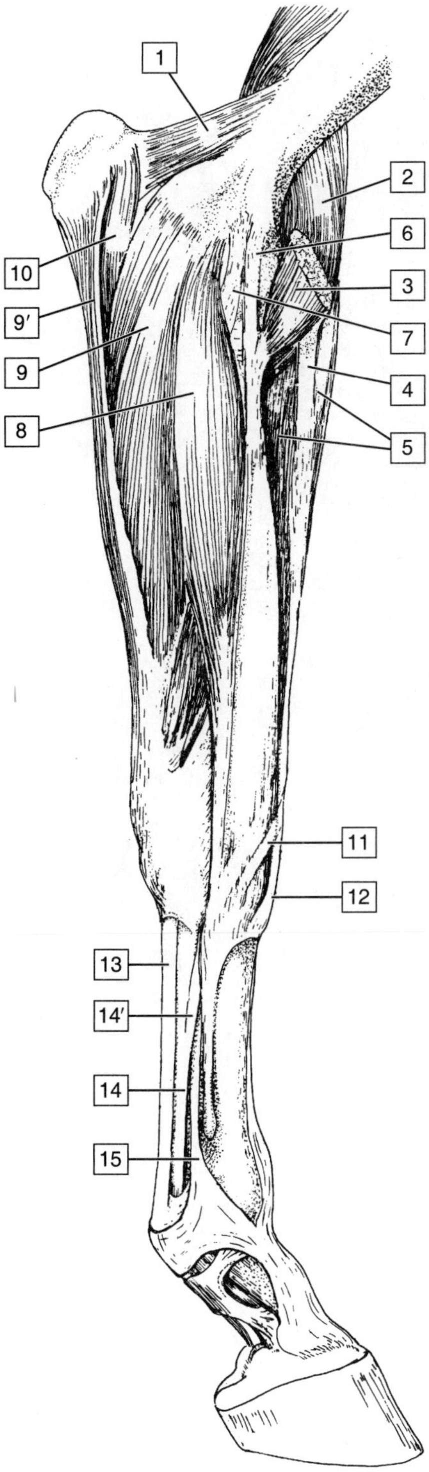

1	肘肌	Anconeus
2	臂肌	Brachialis
3	臂二头肌	Biceps
4	臂二头肌肌腱	Lacertus fibrosus
5	桡侧腕伸肌	Extensor carpi radialis
6	内侧副韧带长部（旋前圆肌）	Long part of medial collateral ligament (pronator teres)
7	内侧副韧带短部	Short part of medial collateral ligament
8	桡侧腕屈肌	Flexor carpi radialis
9	尺侧腕屈肌肱骨头和尺骨头	Humeral and ulnar heads of flexor carpi ulnaris
9′	尺侧腕屈肌肱骨头和尺骨头	Humeral and ulnar heads of flexor carpi ulnaris
10	指深屈肌尺骨头	Ulnar head of deep digital flexor
11	腕斜伸肌肌腱	Tendon of extensor carpi obliquus
12	桡侧腕伸肌肌腱	Tendon of extensor carpi radialis
13	指浅屈肌肌腱	Tendon of superficial digital flexor
14	指深屈肌	Tendon of deep digital flexor
14′	副韧带	Accessory (check) ligament
15	骨间肌	Interosseous

图6-24　马左前肢深层肌肉，内侧观

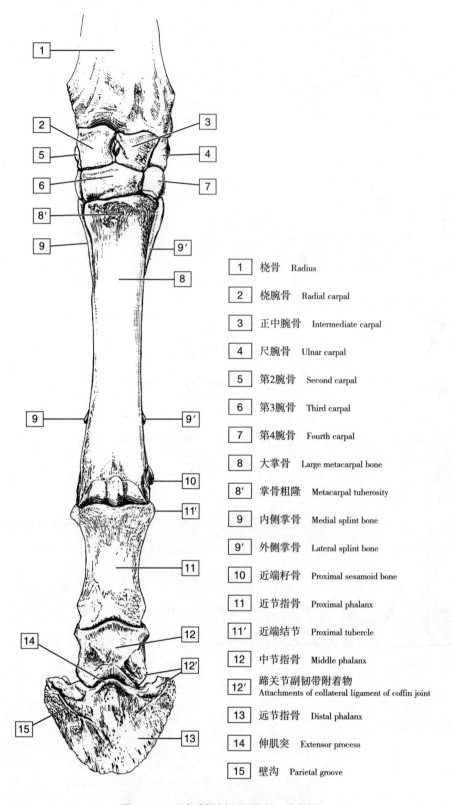

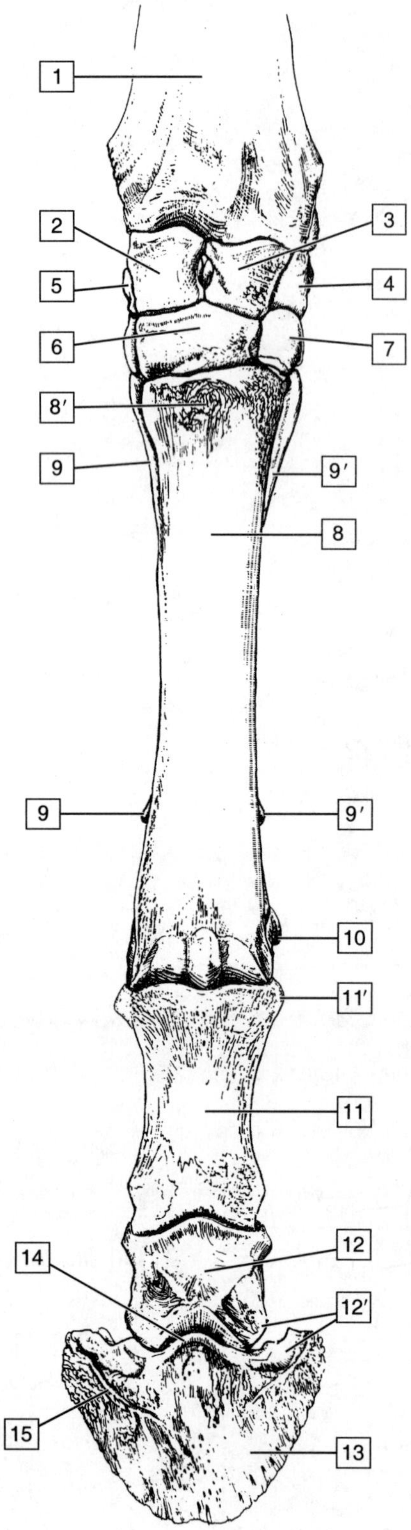

1 桡骨 Radius

2 桡腕骨 Radial carpal

3 正中腕骨 Intermediate carpal

4 尺腕骨 Ulnar carpal

5 第2腕骨 Second carpal

6 第3腕骨 Third carpal

7 第4腕骨 Fourth carpal

8 大掌骨 Large metacarpal bone

8′ 掌骨粗隆 Metacarpal tuberosity

9 内侧掌骨 Medial splint bone

9′ 外侧掌骨 Lateral splint bone

10 近端籽骨 Proximal sesamoid bone

11 近节指骨 Proximal phalanx

11′ 近端结节 Proximal tubercle

12 中节指骨 Middle phalanx

12′ 蹄关节副韧带附着物 Attachments of collateral ligament of coffin joint

13 远节指骨 Distal phalanx

14 伸肌突 Extensor process

15 壁沟 Parietal groove

图6-25 马左前肢远端骨骼，背侧观

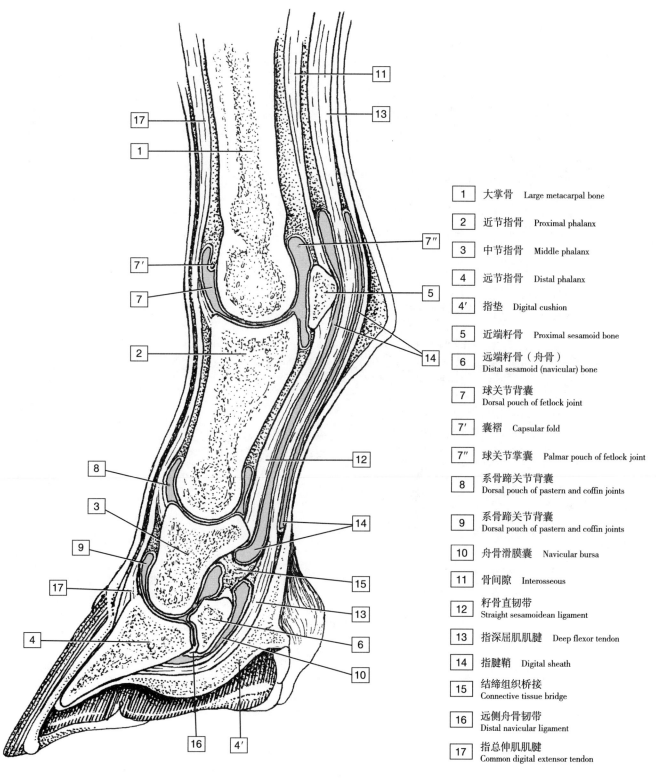

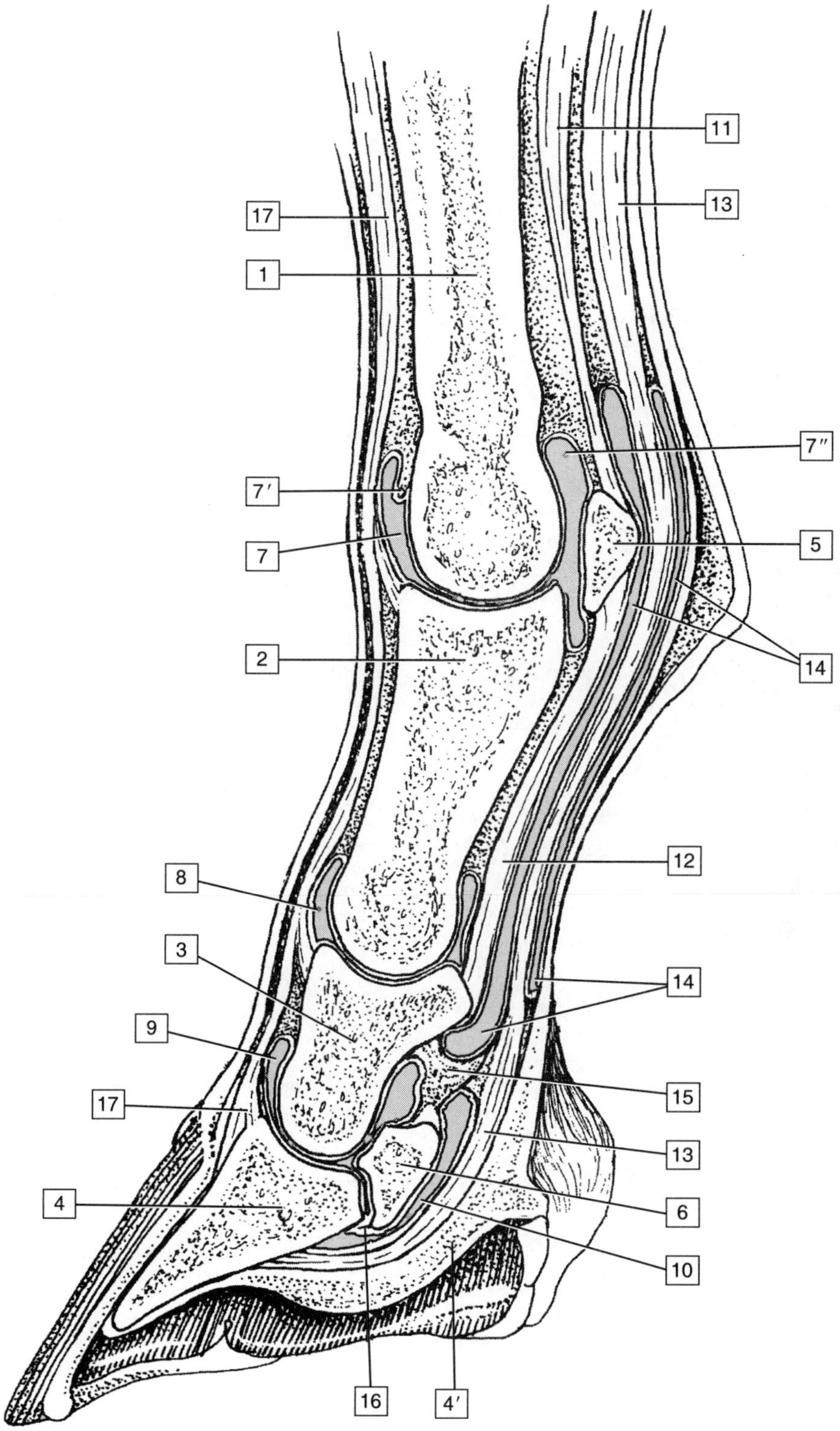

1 大掌骨 Large metacarpal bone

2 近节指骨 Proximal phalanx

3 中节指骨 Middle phalanx

4 远节指骨 Distal phalanx

4′ 指垫 Digital cushion

5 近端籽骨 Proximal sesamoid bone

6 远端籽骨（舟骨） Distal sesamoid (navicular) bone

7 球关节背囊 Dorsal pouch of fetlock joint

7′ 囊褶 Capsular fold

7″ 球关节掌囊 Palmar pouch of fetlock joint

8 系骨蹄关节背囊 Dorsal pouch of pastern and coffin joints

9 系骨蹄关节背囊 Dorsal pouch of pastern and coffin joints

10 舟骨滑膜囊 Navicular bursa

11 骨间隙 Interosseous

12 籽骨直韧带 Straight sesamoidean ligament

13 指深屈肌肌腱 Deep flexor tendon

14 指腱鞘 Digital sheath

15 结缔组织桥接 Connective tissue bridge

16 远侧舟骨韧带 Distal navicular ligament

17 指总伸肌肌腱 Common digital extensor tendon

图6-26 马蹄部轴面

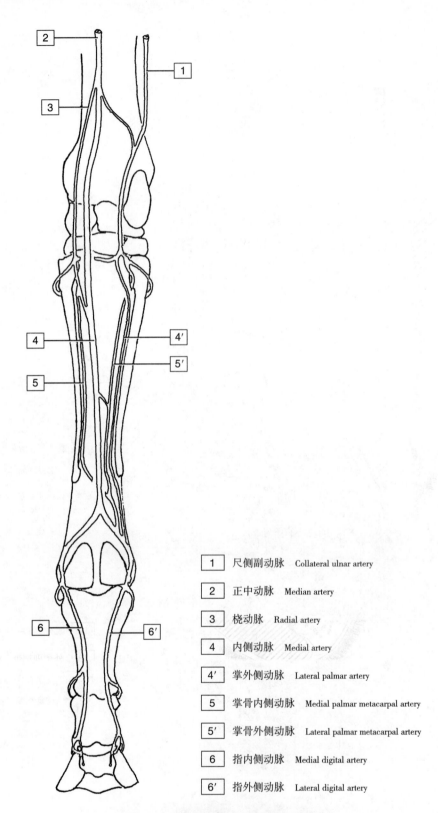

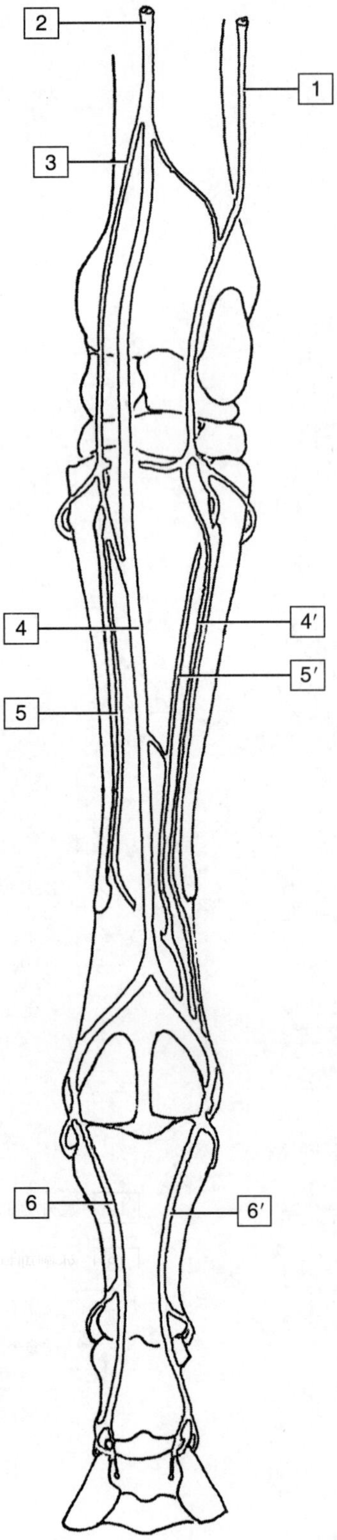

1 尺侧副动脉 Collateral ulnar artery

2 正中动脉 Median artery

3 桡动脉 Radial artery

4 内侧动脉 Medial artery

4′ 掌外侧动脉 Lateral palmar artery

5 掌骨内侧动脉 Medial palmar metacarpal artery

5′ 掌骨外侧动脉 Lateral palmar metacarpal artery

6 指内侧动脉 Medial digital artery

6′ 指外侧动脉 Lateral digital artery

图6-27 马右前肢主要动脉，掌侧观

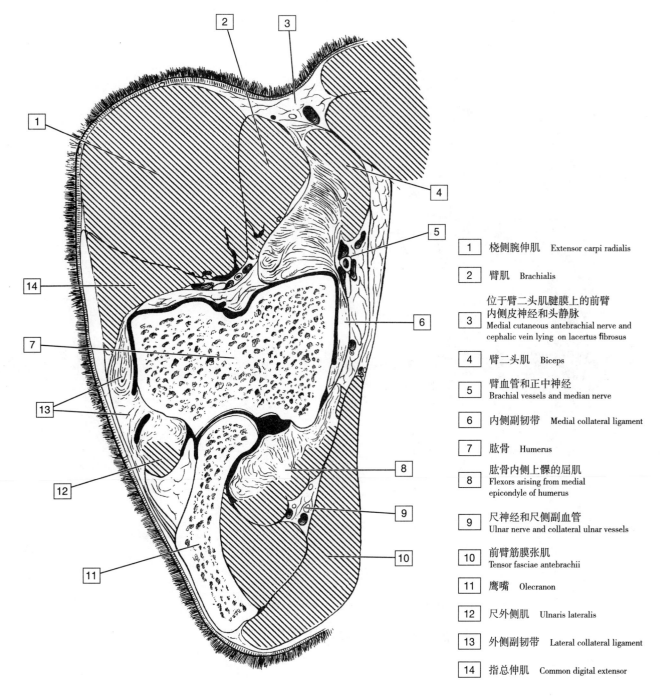

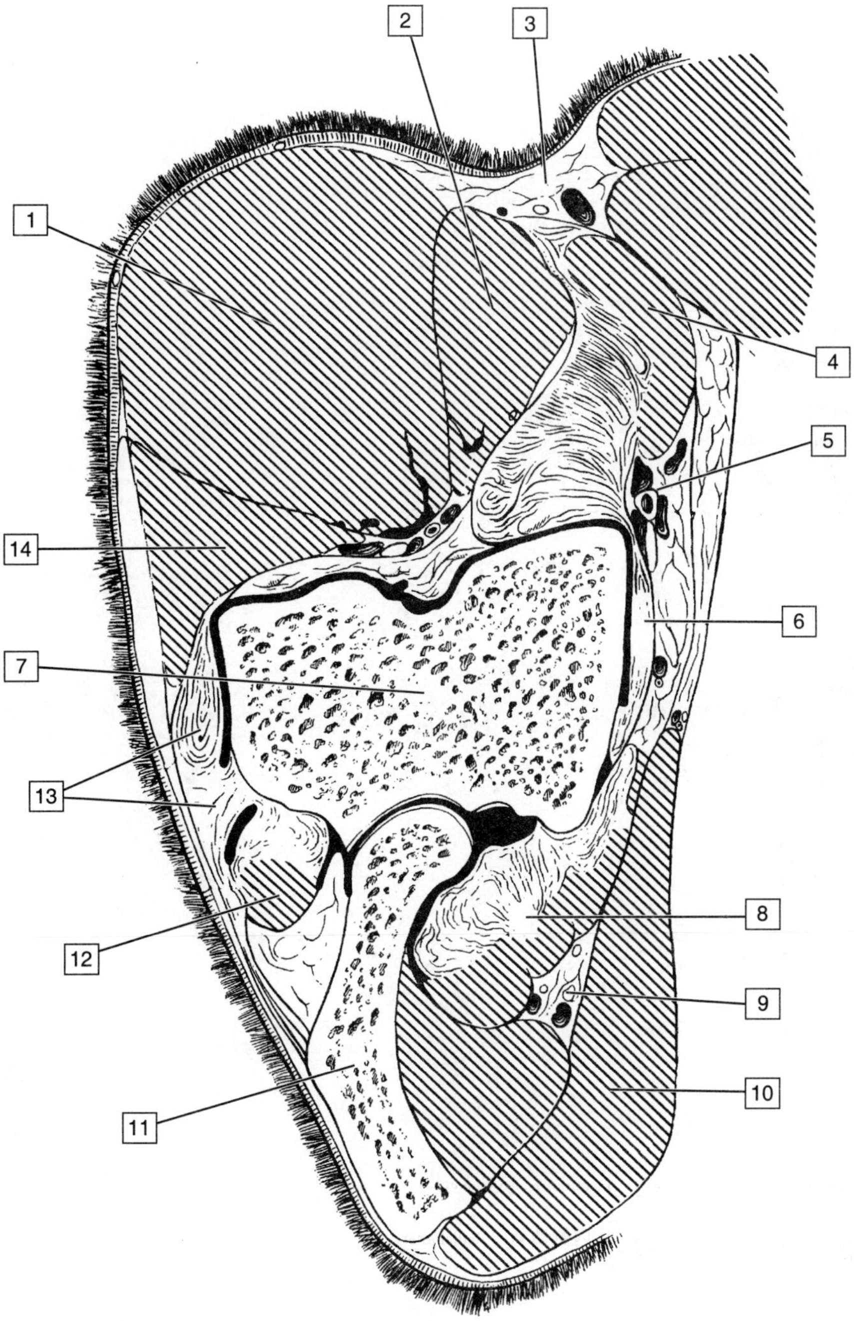

1	桡侧腕伸肌	Extensor carpi radialis
2	臂肌	Brachialis
3	位于臂二头肌腱膜上的前臂内侧皮神经和头静脉	Medial cutaneous antebrachial nerve and cephalic vein lying on lacertus fibrosus
4	臂二头肌	Biceps
5	臂血管和正中神经	Brachial vessels and median nerve
6	内侧副韧带	Medial collateral ligament
7	肱骨	Humerus
8	肱骨内侧上髁的屈肌	Flexors arising from medial epicondyle of humerus
9	尺神经和尺侧副血管	Ulnar nerve and collateral ulnar vessels
10	前臂筋膜张肌	Tensor fasciae antebrachii
11	鹰嘴	Olecranon
12	尺外侧肌	Ulnaris lateralis
13	外侧副韧带	Lateral collateral ligament
14	指总伸肌	Common digital extensor

图6-28 马左肘横断面

牛

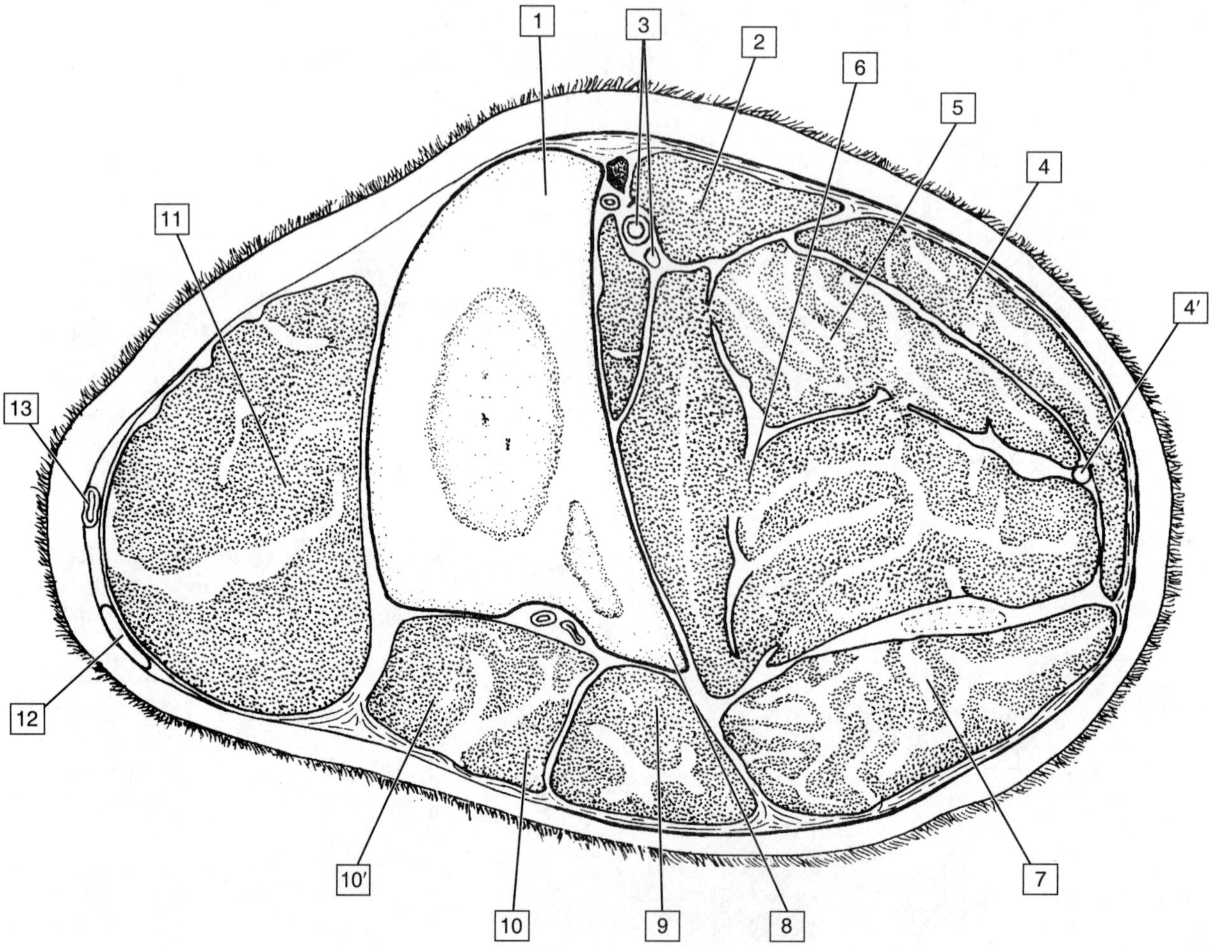

1	桡骨 Radius	8	尺骨 Ulna
2	桡侧腕屈肌 Flexor carpi radialis	9	指外侧伸肌 Lateral digital extensor
3	正中血管和神经 Median vessels and nerve	10	指总伸肌 Common digital extensor
4	尺侧腕屈肌 Flexor carpi ulnaris	10′	指总伸肌 Common digital extensor
4′	尺神经 Ulnar nerve	11	桡侧腕伸肌 Extensor carpi radialis
5	指浅屈肌 Superficial digital flexor	12	桡神经浅支 Superficial branch of radial nerve
6	指深屈肌 Deep digital flexor	13	头静脉 Cephalic vein
7	尺骨外侧肌 Ulnaris lateralis		

图6-29 牛左前臂正中横断面

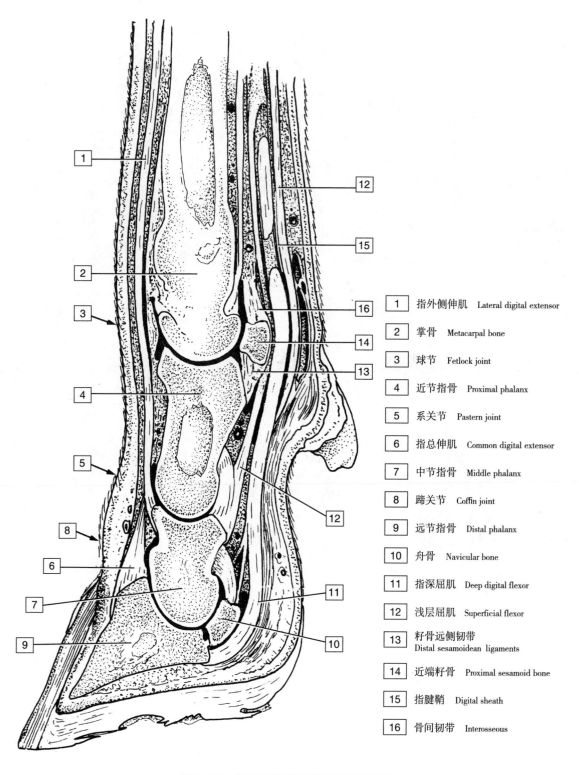

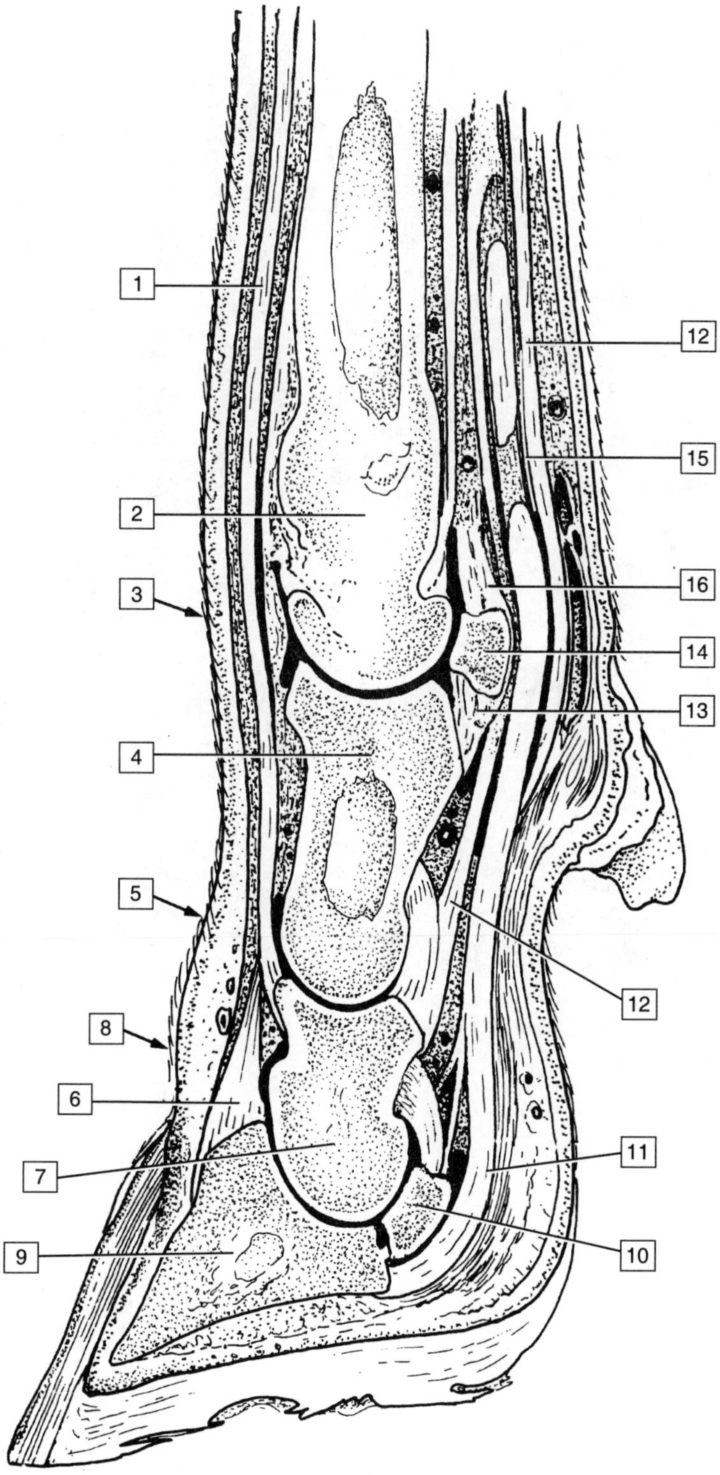

1	指外侧伸肌	Lateral digital extensor
2	掌骨	Metacarpal bone
3	球节	Fetlock joint
4	近节指骨	Proximal phalanx
5	系关节	Pastern joint
6	指总伸肌	Common digital extensor
7	中节指骨	Middle phalanx
8	蹄关节	Coffin joint
9	远节指骨	Distal phalanx
10	舟骨	Navicular bone
11	指深屈肌	Deep digital flexor
12	浅层屈肌	Superficial flexor
13	籽骨远侧韧带	Distal sesamoidean ligaments
14	近端籽骨	Proximal sesamoid bone
15	指腱鞘	Digital sheath
16	骨间韧带	Interosseous

图6-30 经外侧趾劈开的牛蹄矢状面

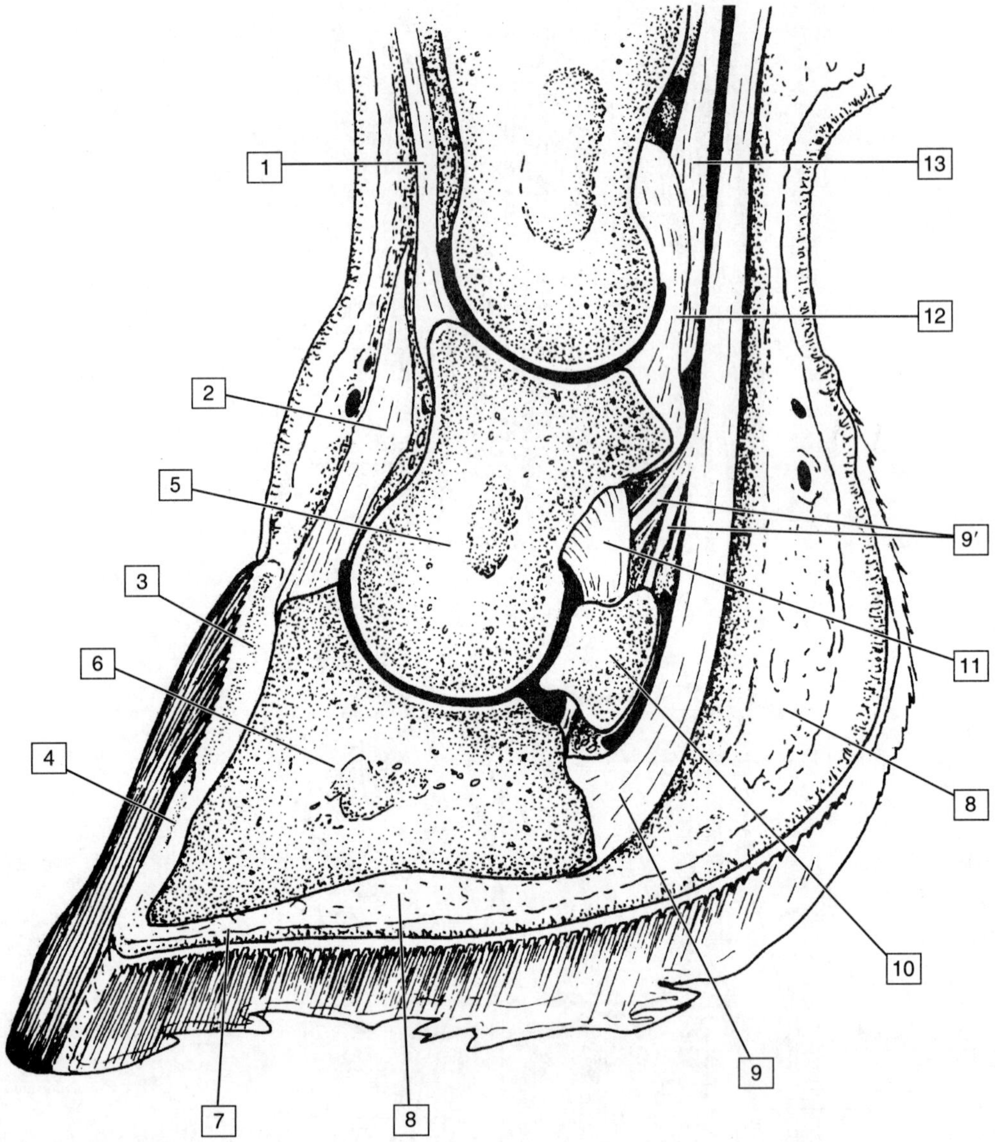

1	固有（内侧）指伸肌 Proper (medial) digital extensor
2	指总伸肌 Common digital extensor
3	蹄冠真皮 Coronary dermis
4	层状真皮 Laminar dermis
5	中节指骨 Middle phalanx
6	远节指骨 Distal phalanx
7	带有指垫蹄底真皮 Sole dermis covered by sole
8	蹄垫 Digital cushion
9	指深屈肌 Deep digital flexor
9′	至中节指骨和舟骨的指深屈肌纤维 Fibers of deep digital flexor to the middle phalanx and navicular bone
10	舟骨 Navicular bone
11	舟骨副韧带 Collateral navicular ligament
12	骰关节掌侧韧带 Palmar ligaments of pastern joint
13	指浅屈肌 Superficial digital flexor

图6-31 牛前蹄内侧趾矢状面

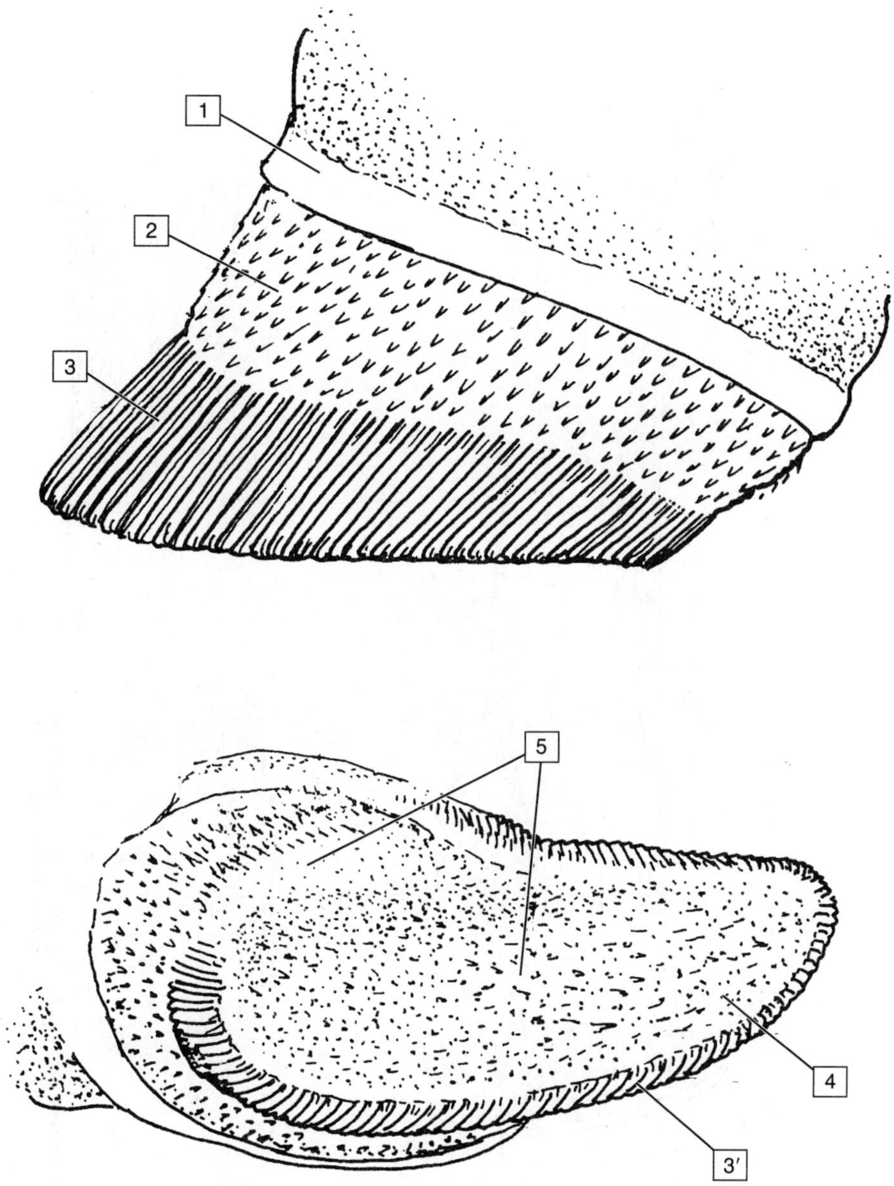

1	蹄缘真皮	Perioplic dermis
2	蹄冠真皮	Coronary dermis
3	层状真皮	Laminar dermis
3′	远端的终末乳头	Terminal papillae at the distal ends of the laminae
4	蹄底真皮	Sole dermis
5	蹄球真皮	Dermis of the bulb

图6-32 产生角质的牛蹄真皮，远轴面及底面观

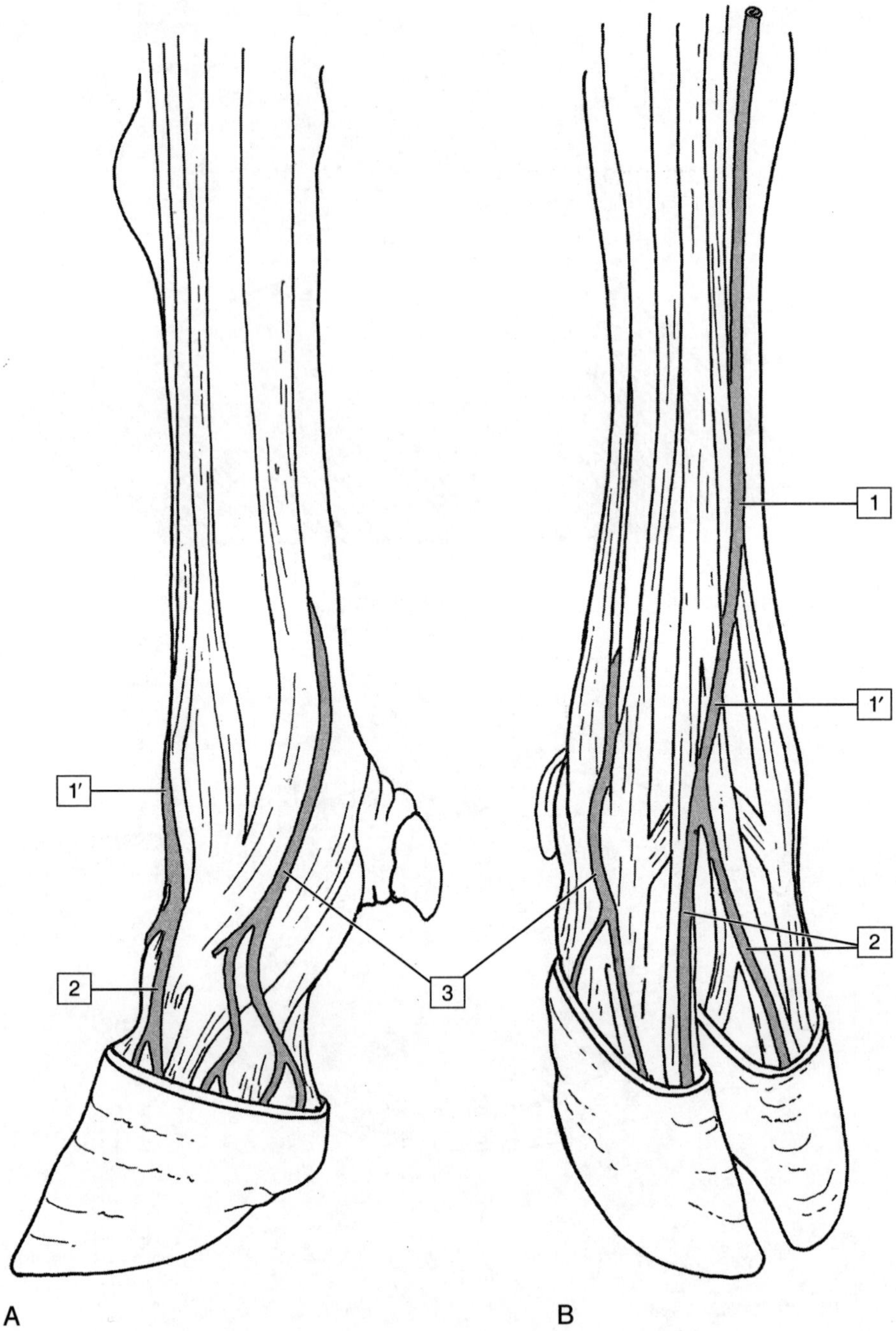

1	副头静脉 Accessory cephalic vein	2	指背侧静脉 Dorsal digital vein
1′	第3指背侧总静脉 Dorsal common digital vein Ⅲ	3	掌轴侧指静脉 Abaxial palmar digital vein

图6-33 牛前肢的主要静脉

A. 左前肢，外侧观；B. 右前肢，背侧观

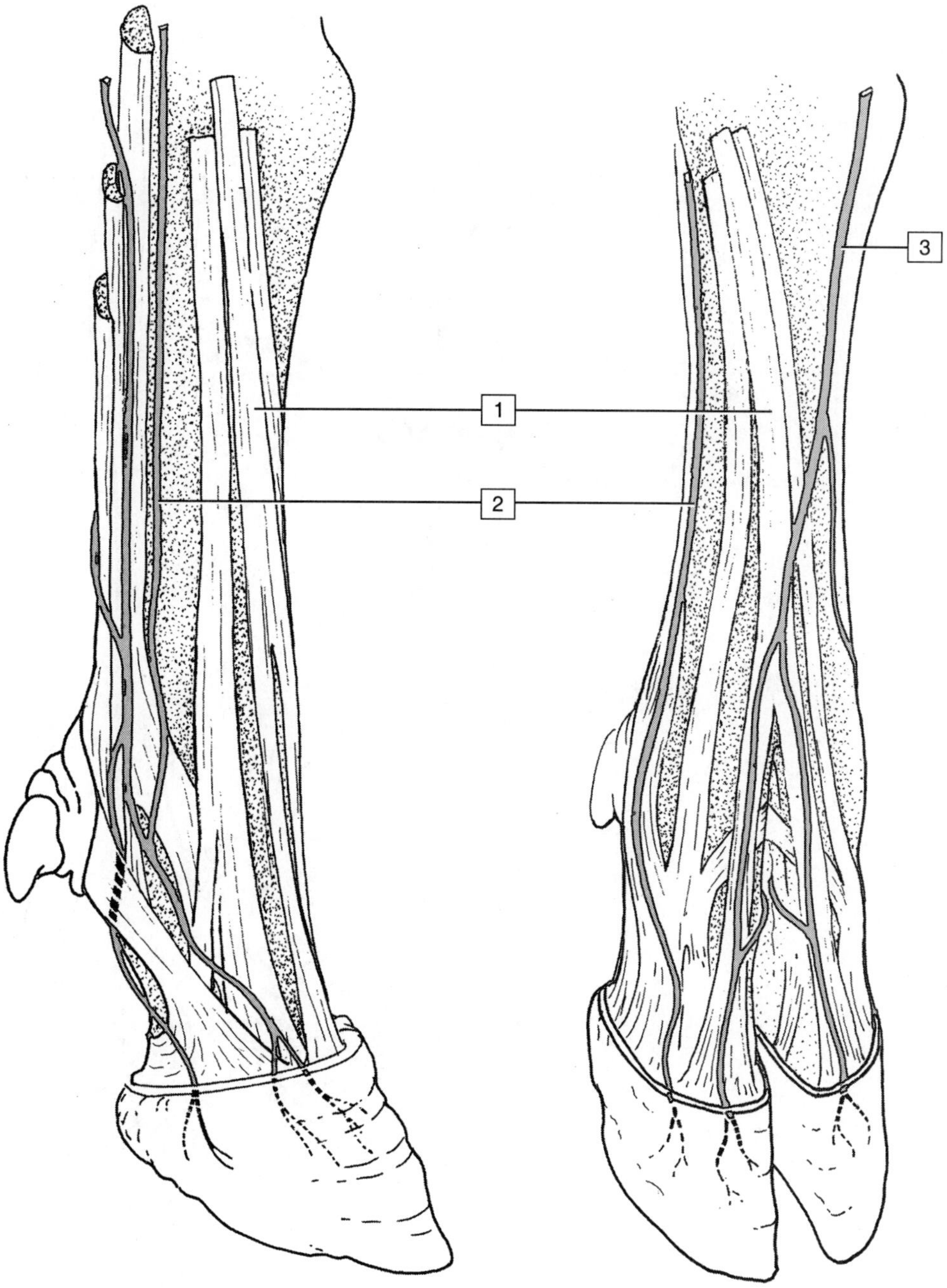

1 趾伸肌肌腱 Digital extensor tendons

2 尺神经背侧支 Dorsal branch of ulnar nerve

3 桡神经浅支 Superficial branch of radial nerve

图6-34 牛右前蹄主要神经，外侧观和背侧观

猪

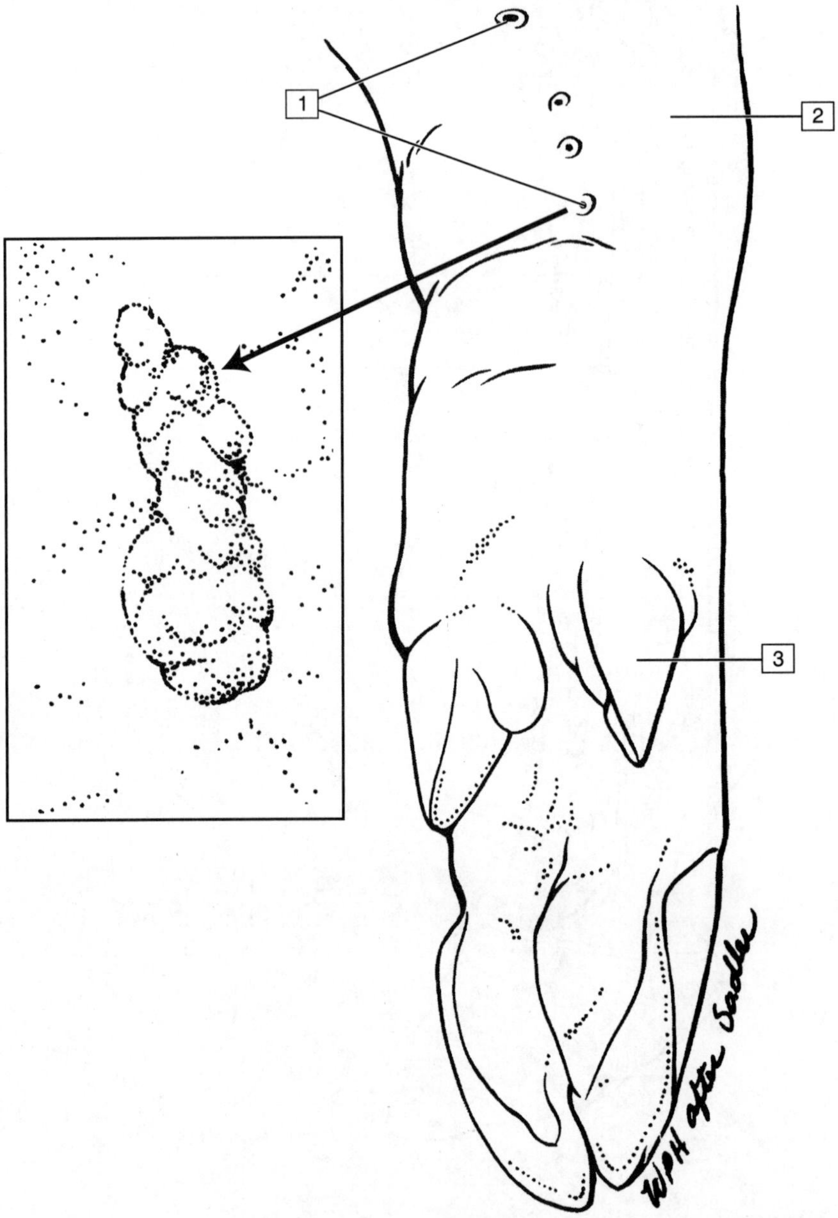

1	腕腺 Carpal glands	3	内侧悬蹄 Hoof of medial dewclaw
2	腕内侧表面 Medial surface of carpus		

图6-35 猪左前脚，后内侧观

鸟类

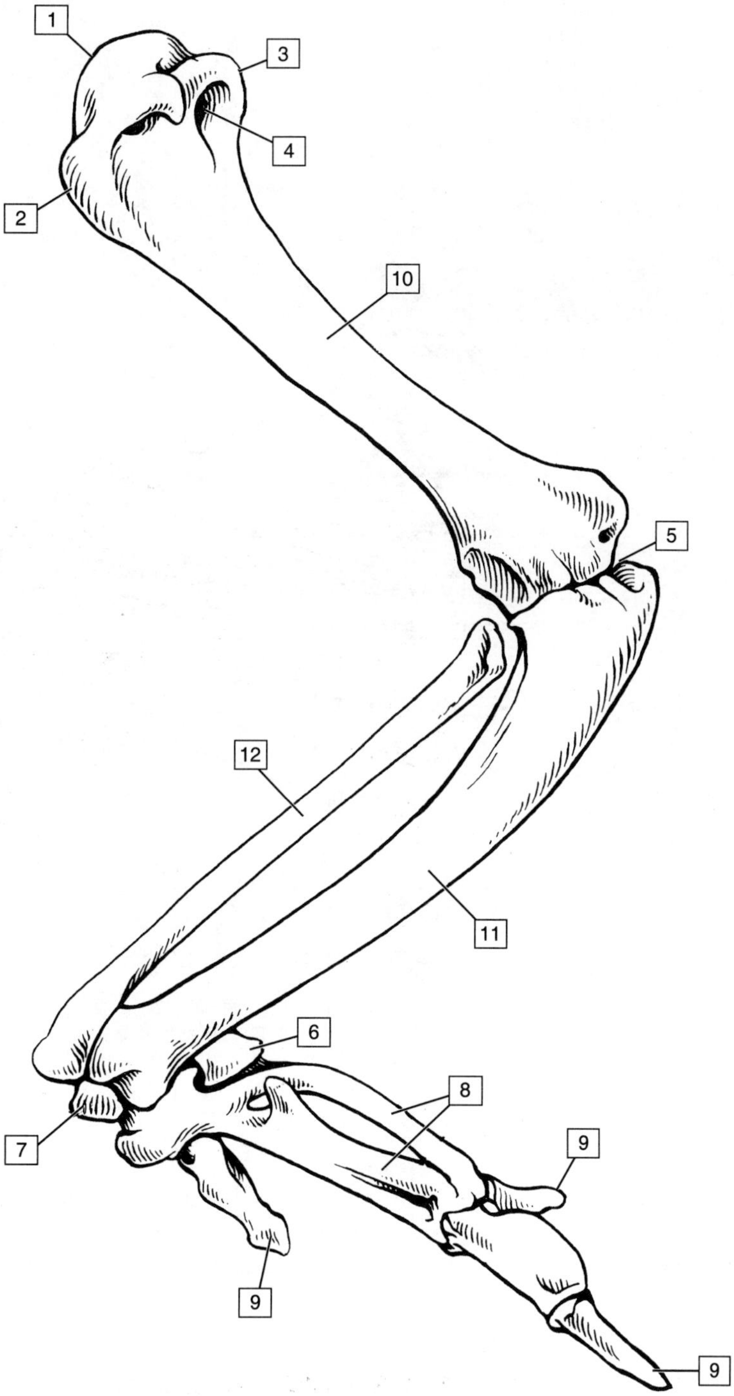

1	肱骨头 Head of humerus	4	气孔 Pneumatic foramen	7	桡腕骨 Radial carpal	10	肱骨 Humerus
2	背侧结节 Dorsal tubercle	5	肘关节 Elbow joint	8	腕掌骨 Carpometacarpals	11	尺骨 Ulna
3	腹侧结节 Ventral tubercle	6	尺腕骨 Ulnar carpal	9	趾 Digits	12	桡骨 Radius

图6-36 家禽左翼部骨骼，部分向外侧伸展

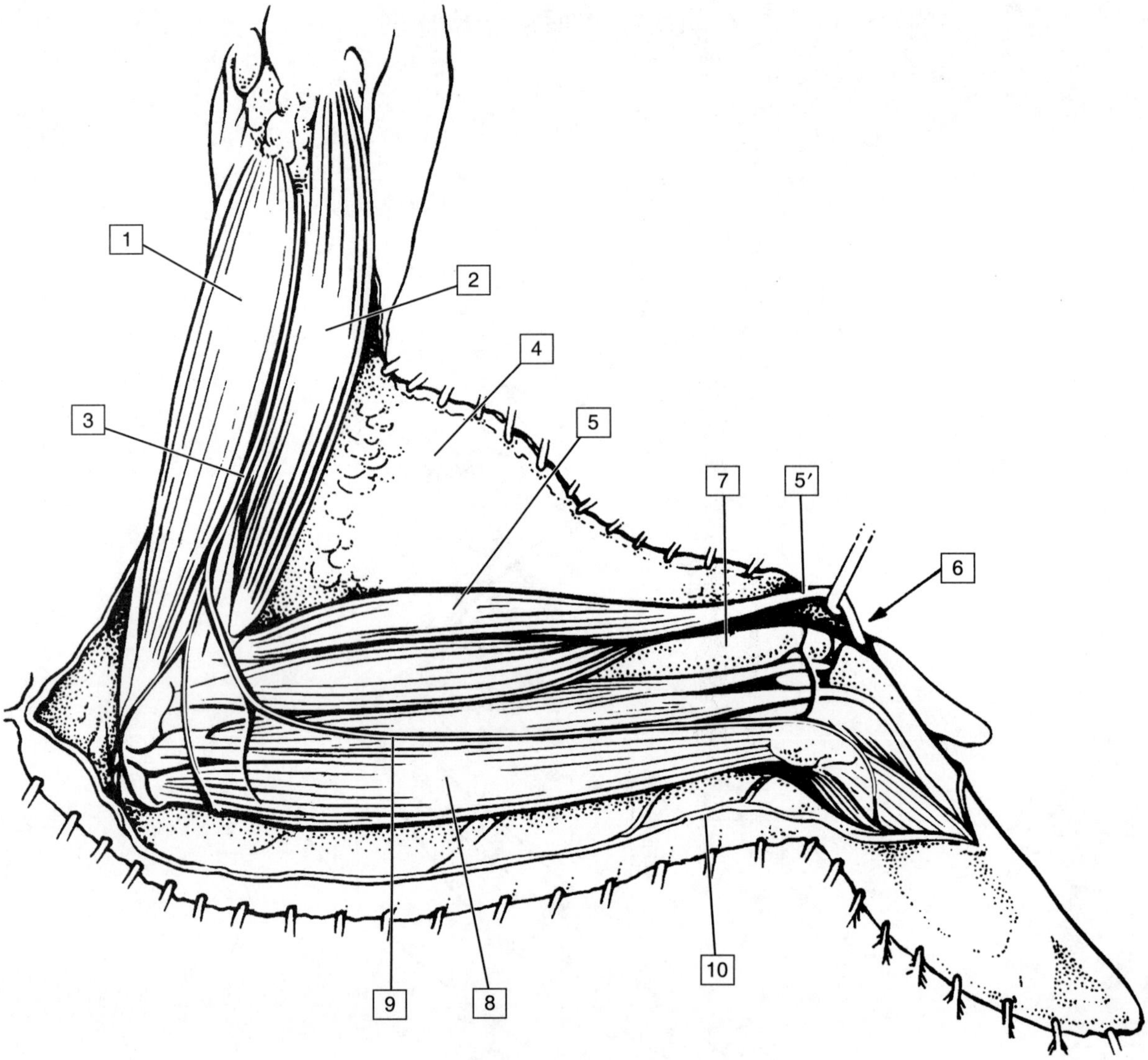

1 臂三头肌 Triceps

2 臂二头肌 Biceps

3 臂静脉 Brachial vein

4 皮褶（前膜） Skin fold (propatagium)

5 桡侧腕伸肌 Extensor carpi radialis

5′ 桡侧腕伸肌肌腱 Tendon of extensor carpi radialis

6 腕关节 Carpal joint

7 桡骨皮下部 Subcutaneous part of radius

8 尺侧腕屈肌 Flexor carpi ulnaris

9 尺侧翼（皮）静脉 Cutaneous ulnar (wing) vein

10 掀开的皮肤 Reflected skin

图6-37 侧向伸展的家禽左翼部腹侧浅层结构

不同动物的解剖结构比较

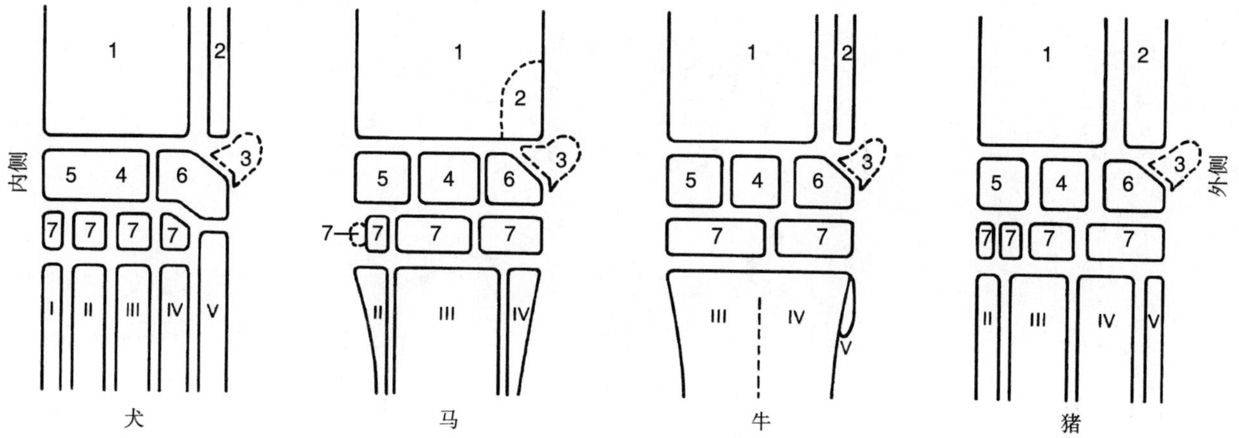

1	桡骨 Radius	5	桡腕骨 Radial carpal bone
2	尺骨 Ulna	6	尺腕骨 Ulnar carpal bone
3	副腕骨 Accessory carpal bone	7	腕骨远端 Distal carpal bones
4	中间腕骨 Intermediate carpal bone		

图6–38 腕骨

犬、马、牛和猪的腕骨模式图，

罗马数字表示掌骨

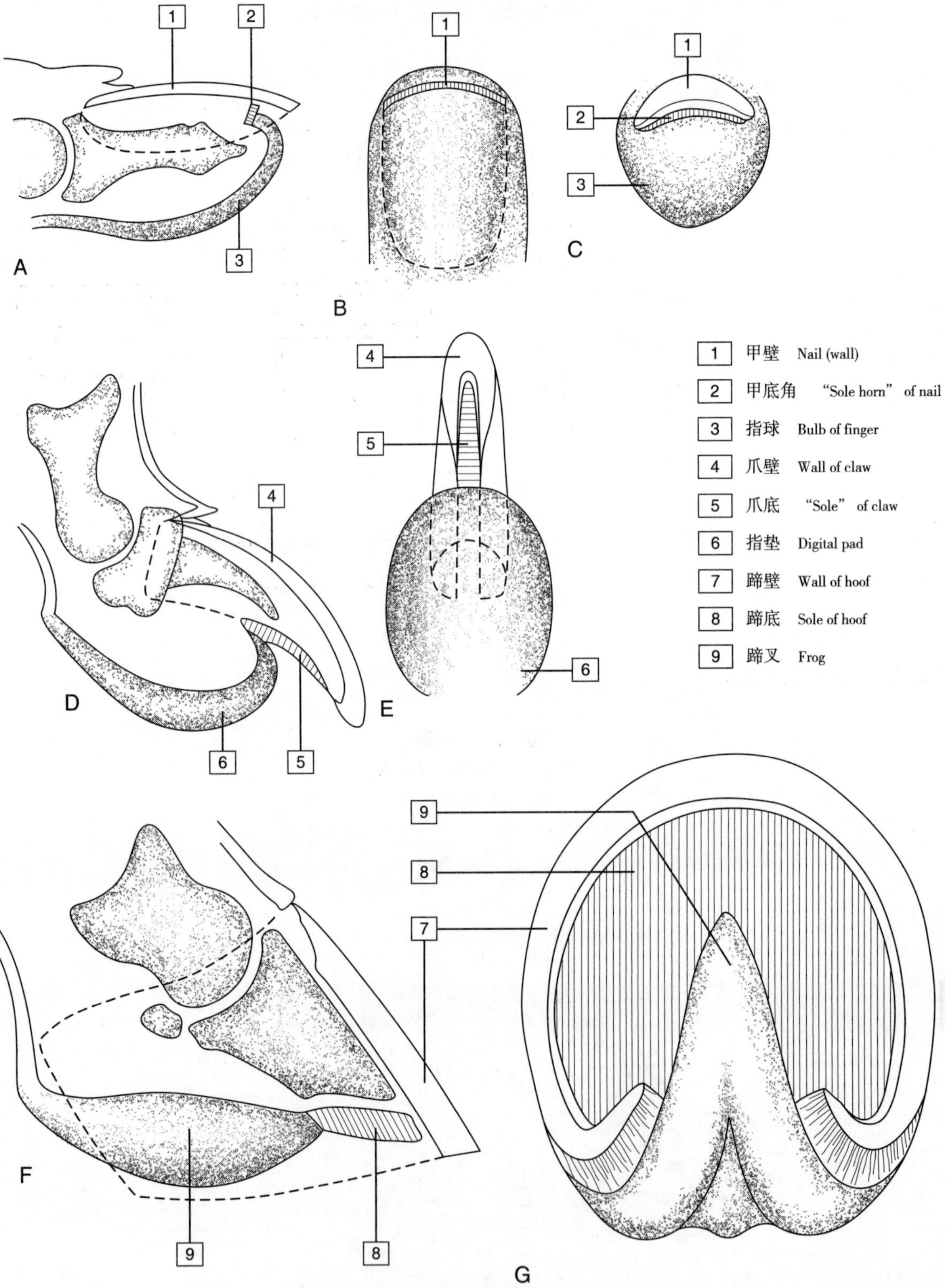

图6-39 甲、爪和蹄示意图

A~C. 人指尖的纵切面、掌侧面和前面观；D、E. 犬爪的纵切面和掌侧面；F、G. 马蹄的纵切面和底面

7

后 肢

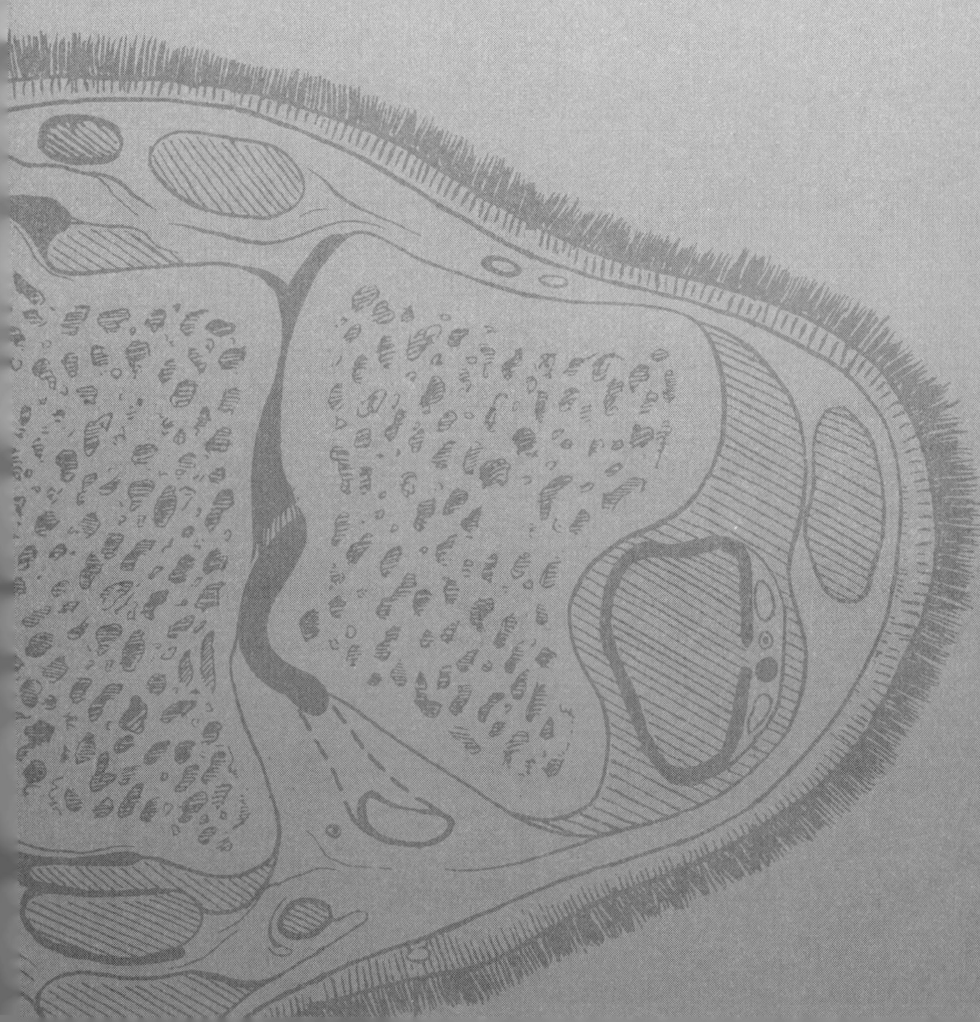

犬

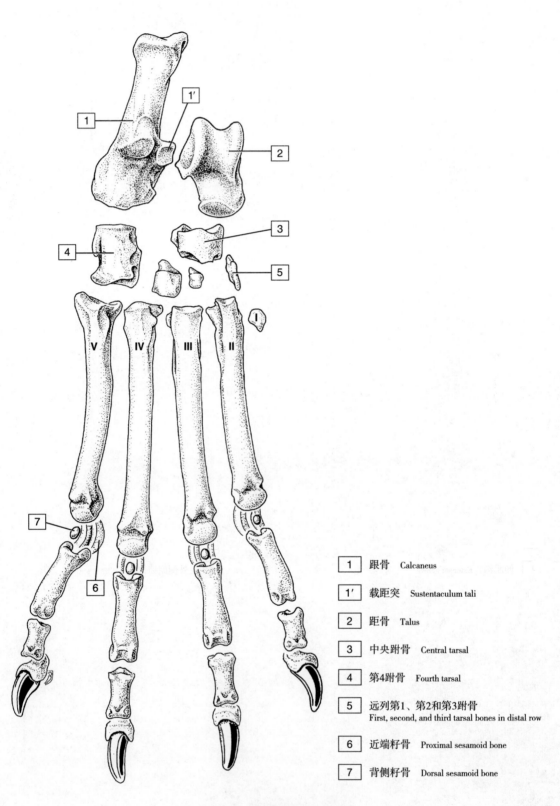

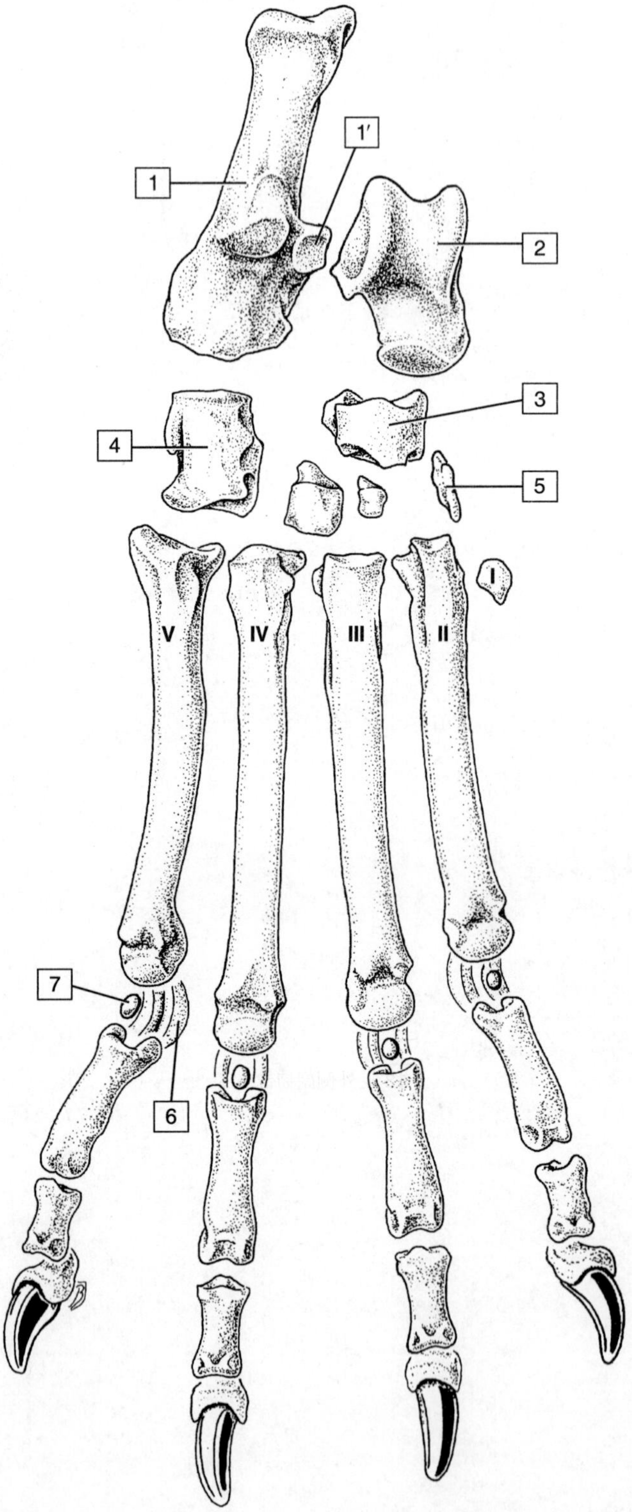

1	跟骨	Calcaneus
1′	载距突	Sustentaculum tali
2	距骨	Talus
3	中央跗骨	Central tarsal
4	第4跗骨	Fourth tarsal
5	远列第1、第2和第3跗骨	First, second, and third tarsal bones in distal row
6	近端籽骨	Proximal sesamoid bone
7	背侧籽骨	Dorsal sesamoid bone

图7-1 犬右脚骨骼，背侧观

罗马数字表示跖骨

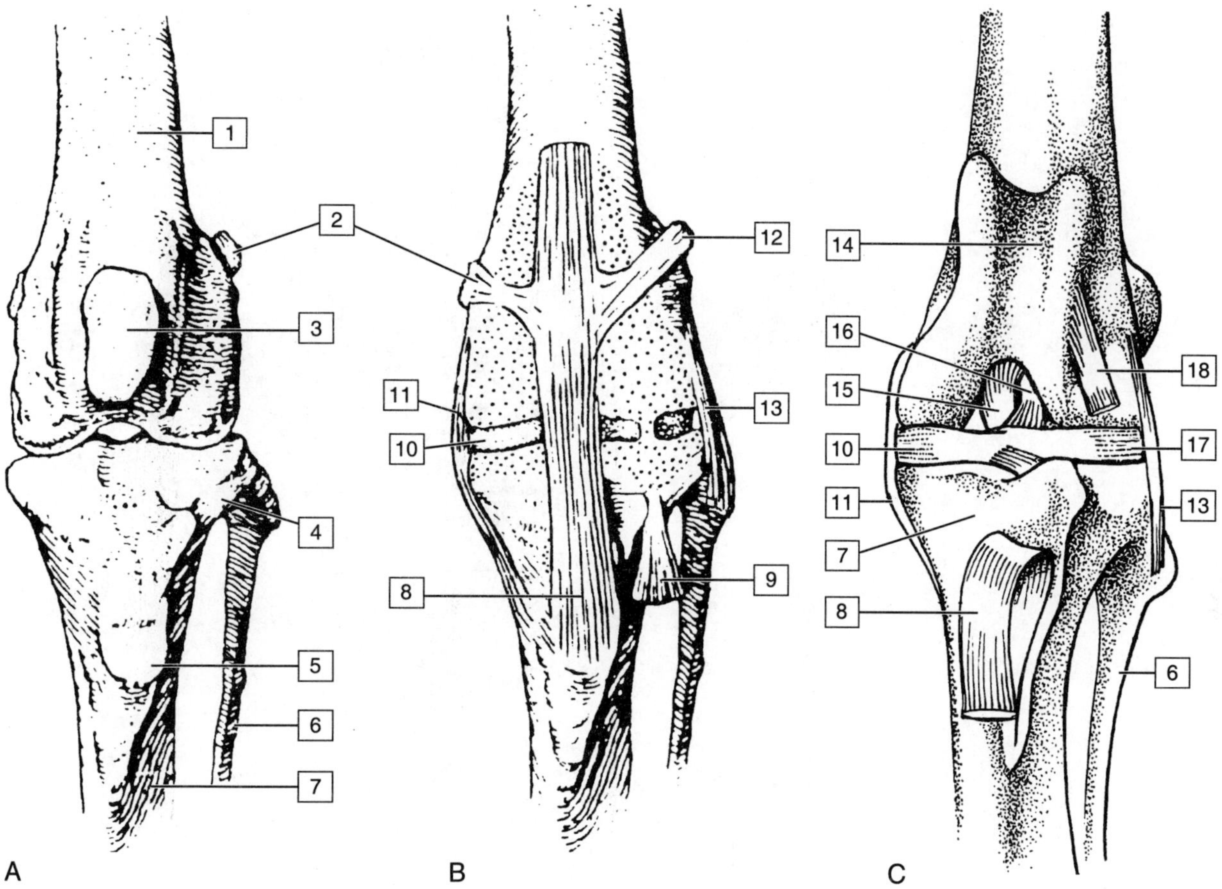

1 股骨 Femur

2 腓肠肌内的籽骨 Sesamoids in gastrocnemius

3 膝盖骨 Patella

4 伸肌沟 Extensor groove

5 胫骨粗隆 Tibial tuberosity

6 腓骨 Fibula

7 胫骨 Tibia

8 膝韧带 Patellar ligament

9 指长伸肌腱穿过伸肌沟 Tendon of long digital extensor passing through extensor groove

10 内侧半月板 Medial meniscus

11 内侧副韧带 Medial collateral ligament

12 外侧股膝韧带 Lateral femoropatellar ligament

13 外侧副韧带 Lateral collateral ligament

14 滑车 Trochlea

15 后交叉韧带 Caudal cruciate ligament

16 前交叉韧带 Cranial cruciate ligament

17 外侧半月板 Lateral meniscus

18 9的断端 Stump of 9

图7-2 犬左膝关节，前面观（A~C）

关节囊的范围如B所示，图C中的膝盖骨已被剥离

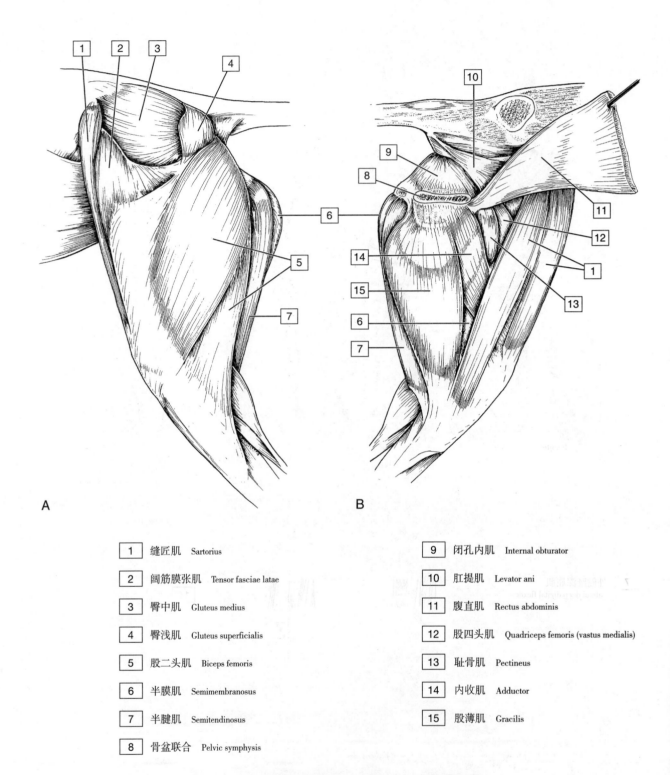

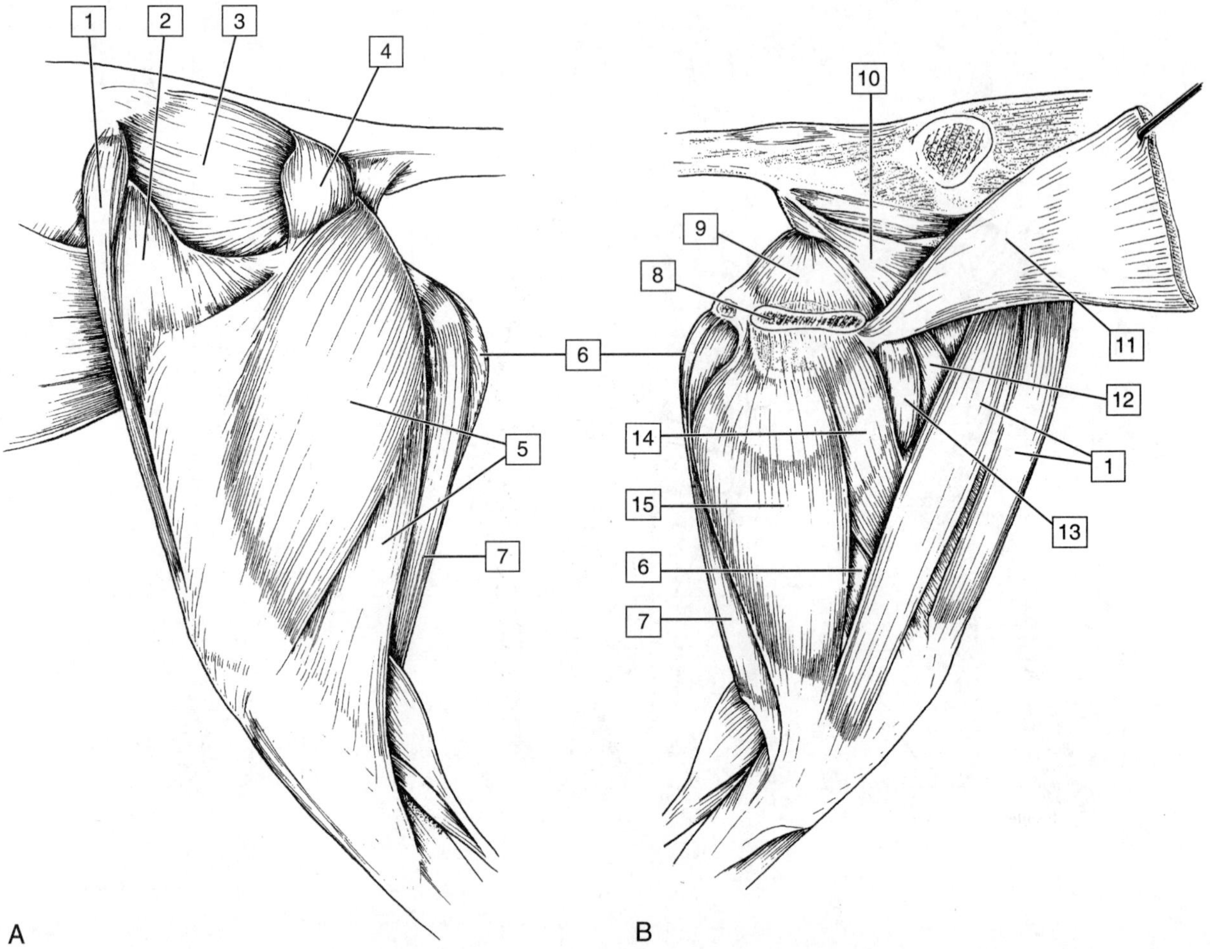

1	缝匠肌 Sartorius	9	闭孔内肌 Internal obturator
2	阔筋膜张肌 Tensor fasciae latae	10	肛提肌 Levator ani
3	臀中肌 Gluteus medius	11	腹直肌 Rectus abdominis
4	臀浅肌 Gluteus superficialis	12	股四头肌 Quadriceps femoris (vastus medialis)
5	股二头肌 Biceps femoris	13	耻骨肌 Pectineus
6	半膜肌 Semimembranosus	14	内收肌 Adductor
7	半腱肌 Semitendinosus	15	股薄肌 Gracilis
8	骨盆联合 Pelvic symphysis		

图7-3 犬后肢和大腿肌肉，外侧观（A）和内侧观（B）

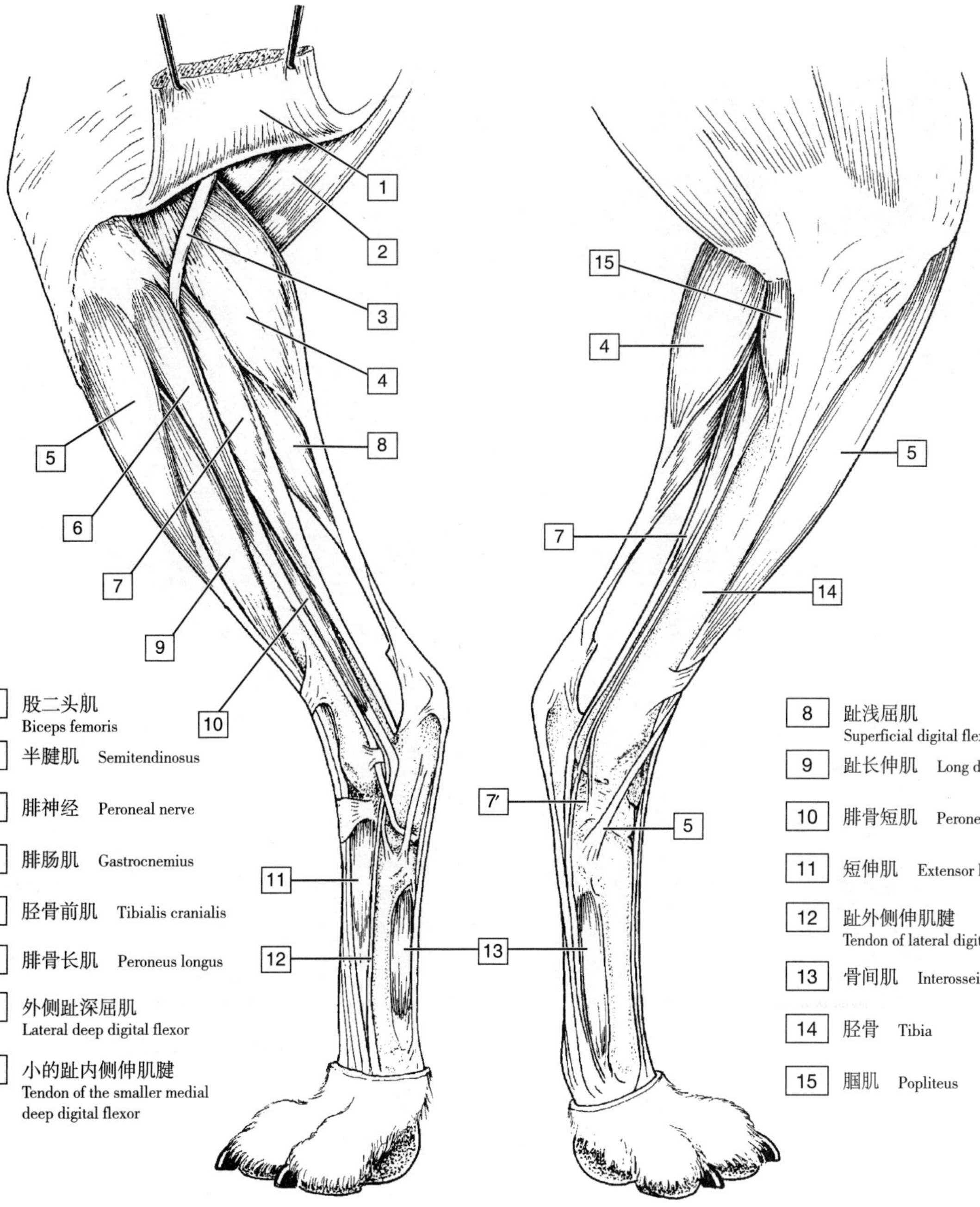

1 股二头肌 Biceps femoris
2 半腱肌 Semitendinosus
3 腓神经 Peroneal nerve
4 腓肠肌 Gastrocnemius
5 胫骨前肌 Tibialis cranialis
6 腓骨长肌 Peroneus longus
7 外侧趾深屈肌 Lateral deep digital flexor
7′ 小的趾内侧伸肌腱 Tendon of the smaller medial deep digital flexor
8 趾浅屈肌 Superficial digital flexor
9 趾长伸肌 Long digital extensor
10 腓骨短肌 Peroneus brevis
11 短伸肌 Extensor brevis
12 趾外侧伸肌腱 Tendon of lateral digital extensor
13 骨间肌 Interossei
14 胫骨 Tibia
15 腘肌 Popliteus

图7-4 犬左小腿肌肉，外侧观和内侧观

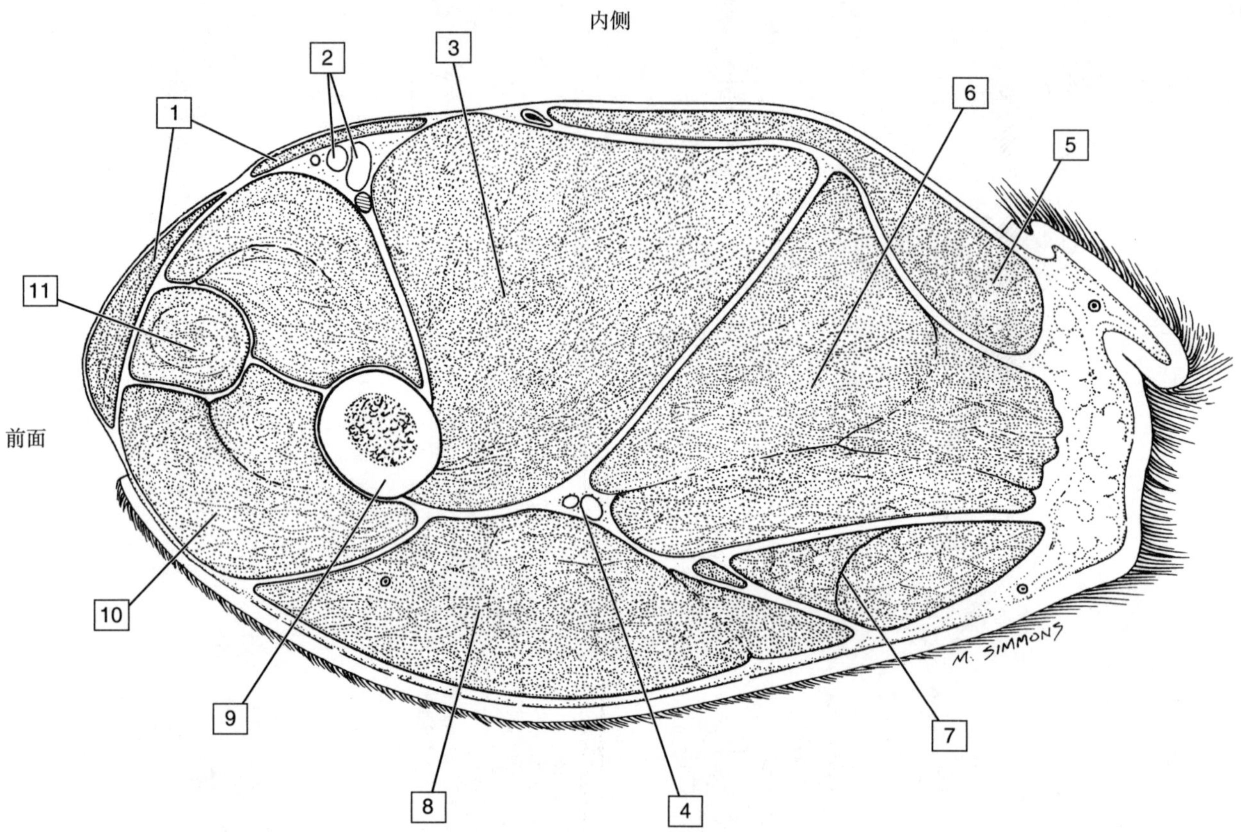

1 缝匠肌 Sartorius	7 半腱肌 Semitendinosus
2 股血管 Femoral vessels	8 股二头肌 Biceps femoris
3 内收肌 Adductor	9 股骨 Femur
4 坐骨神经 Sciatic nerve	10 股外侧股（股四头肌）Vastus lateralis (of quadriceps)
5 股薄肌 Gracilis	11 股直肌 Rectus femoris
6 半膜肌 Semimembranosus	

图7-5 犬左大腿横断面

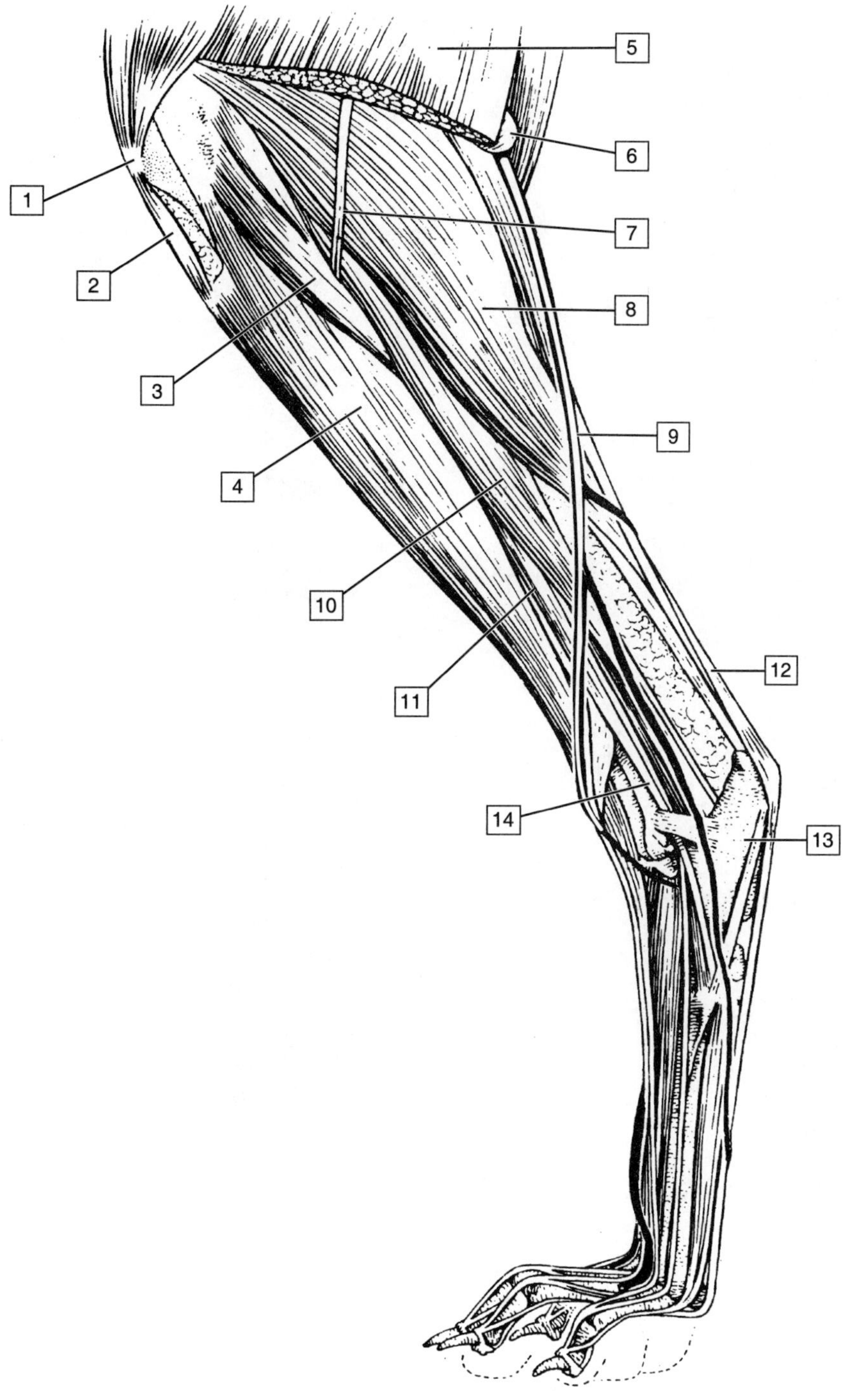

1 膝盖骨 Patella

2 膝韧带 Patellar ligament

3 腓骨长肌 Peroneus longus

4 胫骨前肌 Tibialis cranialis

5 股二头肌 Biceps femoris

6 腘淋巴结 Popliteal lymph node

7 腓总神经 Common peroneal nerve

8 腓肠肌外侧头 Lateral head of gastrocnemius

9 外侧隐静脉 Lateral saphenous vein

10 趾深屈肌 Deep digital flexor

11 腓浅神经 Superficial peroneal nerve

12 跟腱 Calcaneal tendon

13 跟骨 Calcaneus

14 腓骨长肌腱 Peroneus longus tendon

图7-6 犬左后肢

图示外侧隐静脉，外侧观

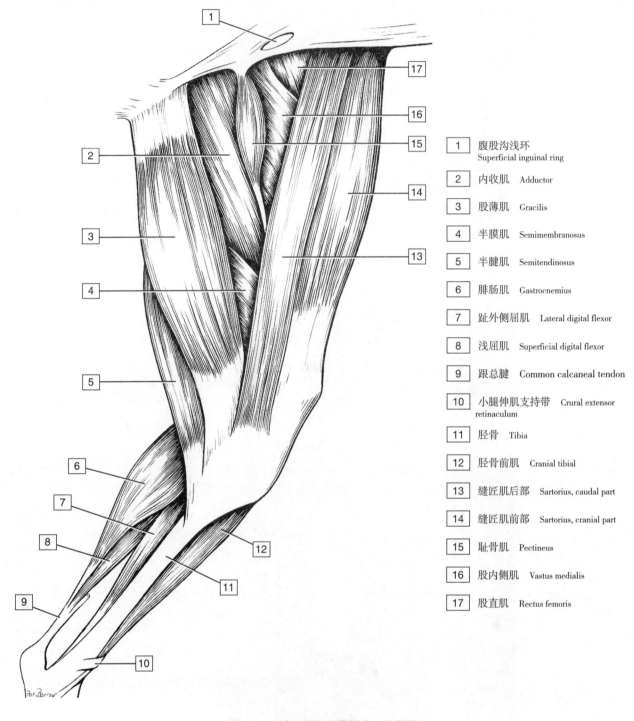

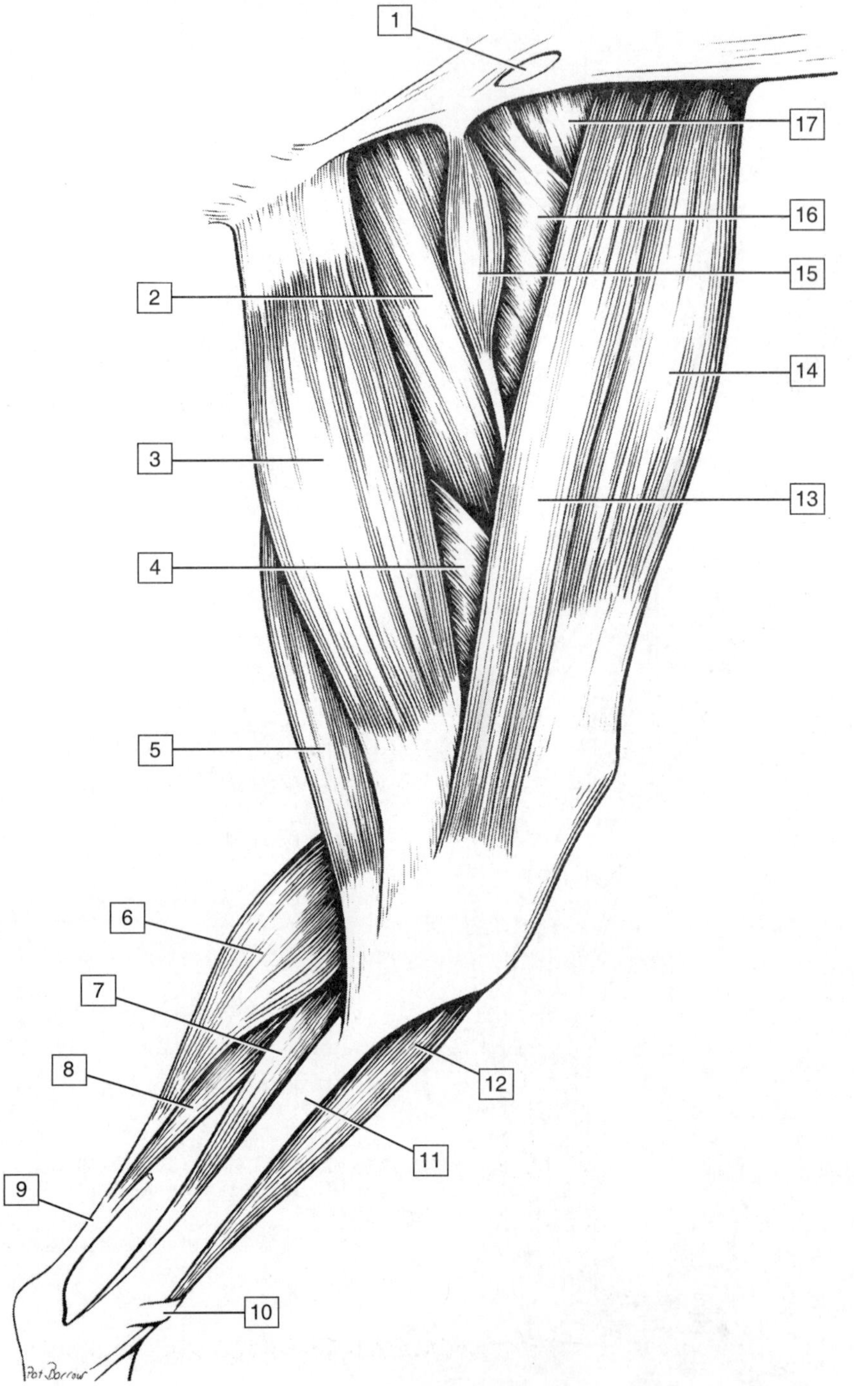

1 腹股沟浅环 Superficial inguinal ring

2 内收肌 Adductor

3 股薄肌 Gracilis

4 半膜肌 Semimembranosus

5 半腱肌 Semitendinosus

6 腓肠肌 Gastrocnemius

7 趾外侧屈肌 Lateral digital flexor

8 浅屈肌 Superficial digital flexor

9 跟总腱 Common calcaneal tendon

10 小腿伸肌支持带 Crural extensor retinaculum

11 胫骨 Tibia

12 胫骨前肌 Cranial tibial

13 缝匠肌后部 Sartorius, caudal part

14 缝匠肌前部 Sartorius, cranial part

15 耻骨肌 Pectineus

16 股内侧肌 Vastus medialis

17 股直肌 Rectus femoris

图7-7 犬左后肢浅层肌肉，内侧观

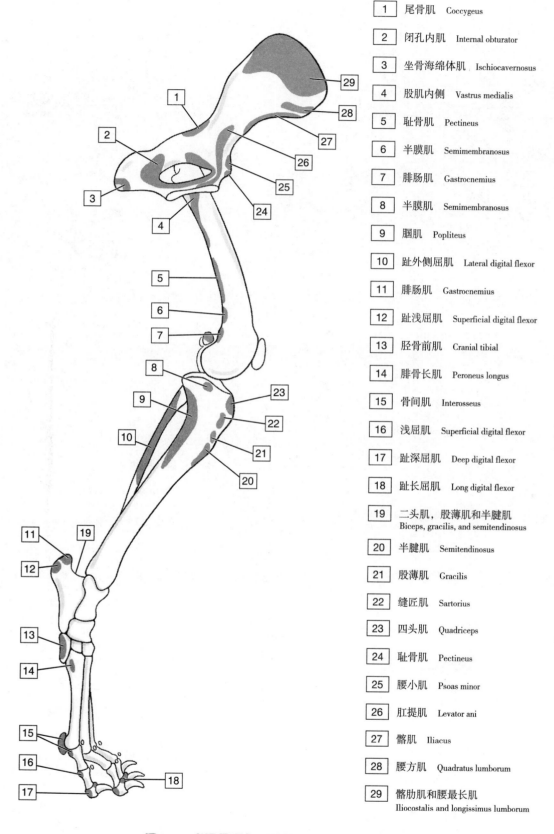

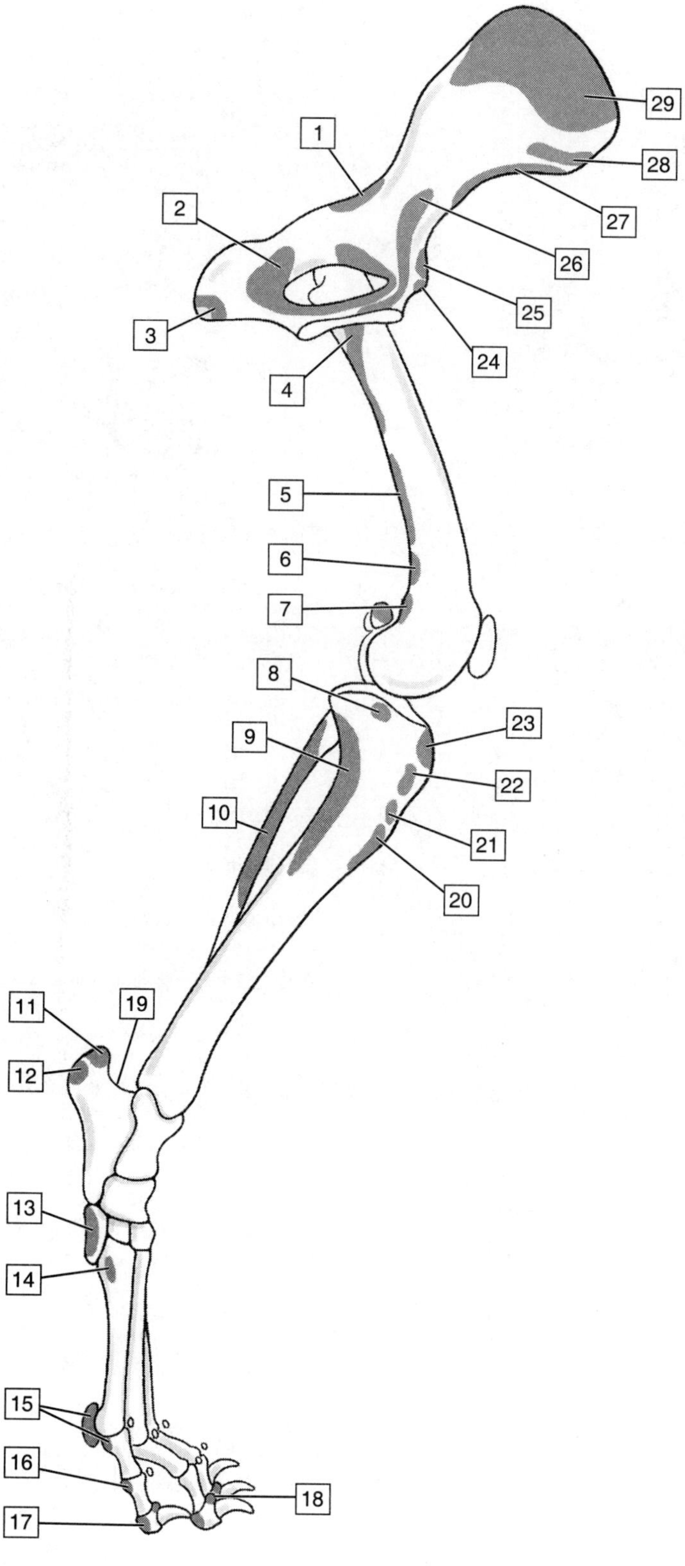

1 尾骨肌 Coccygeus
2 闭孔内肌 Internal obturator
3 坐骨海绵体肌 Ischiocavernosus
4 股肌内侧 Vastrus medialis
5 耻骨肌 Pectineus
6 半膜肌 Semimembranosus
7 腓肠肌 Gastrocnemius
8 半膜肌 Semimembranosus
9 腘肌 Popliteus
10 趾外侧屈肌 Lateral digital flexor
11 腓肠肌 Gastrocnemius
12 趾浅屈肌 Superficial digital flexor
13 胫骨前肌 Cranial tibial
14 腓骨长肌 Peroneus longus
15 骨间肌 Interosseus
16 浅屈肌 Superficial digital flexor
17 趾深屈肌 Deep digital flexor
18 趾长屈肌 Long digital flexor
19 二头肌，股薄肌和半腱肌 Biceps, gracilis, and semitendinosus
20 半腱肌 Semitendinosus
21 股薄肌 Gracilis
22 缝匠肌 Sartorius
23 四头肌 Quadriceps
24 耻骨肌 Pectineus
25 腰小肌 Psoas minor
26 肛提肌 Levator ani
27 髂肌 Iliacus
28 腰方肌 Quadratus lumborum
29 髂肋肌和腰最长肌 Iliocostalis and longissimus lumborum

图7-8 犬骨盆和左后肢的肌肉附着，内侧观

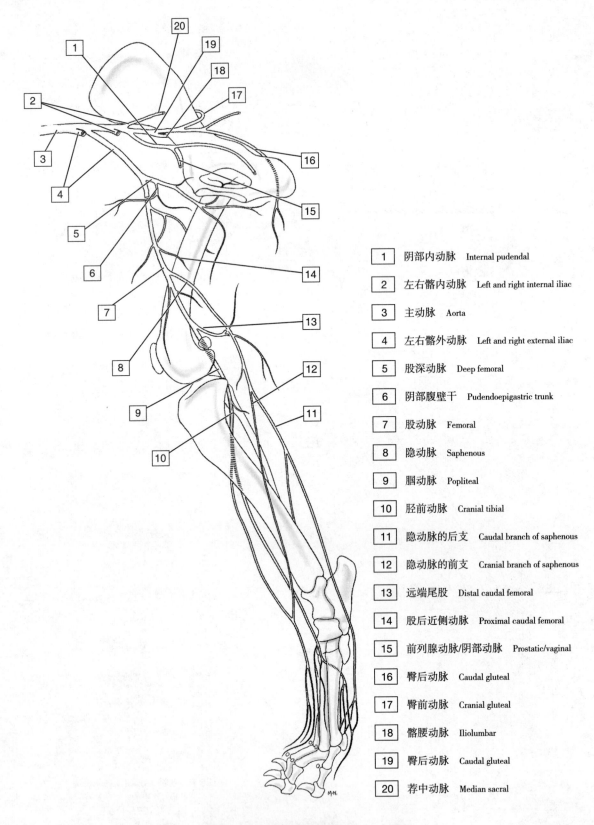

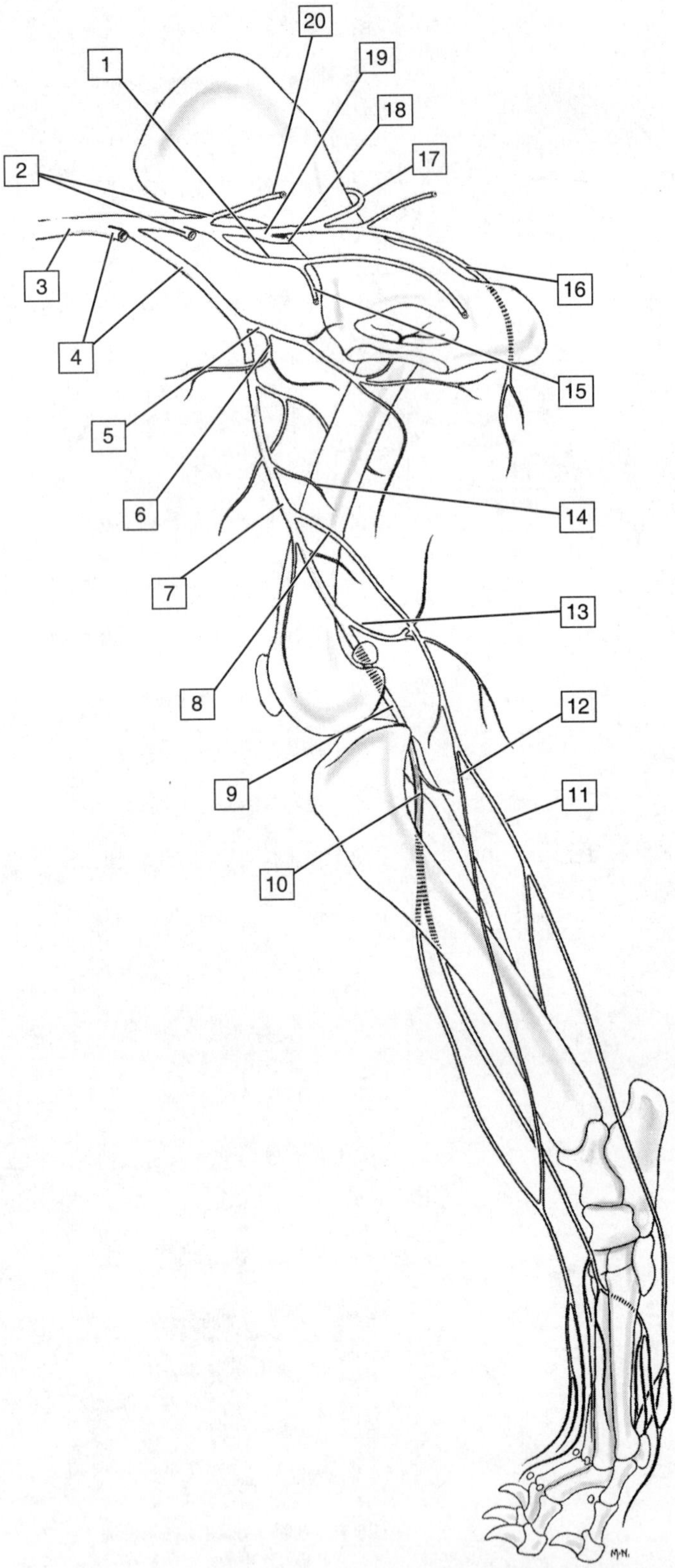

1	阴部内动脉	Internal pudendal
2	左右髂内动脉	Left and right internal iliac
3	主动脉	Aorta
4	左右髂外动脉	Left and right external iliac
5	股深动脉	Deep femoral
6	阴部腹壁干	Pudendoepigastric trunk
7	股动脉	Femoral
8	隐动脉	Saphenous
9	腘动脉	Popliteal
10	胫前动脉	Cranial tibial
11	隐动脉的后支	Caudal branch of saphenous
12	隐动脉的前支	Cranial branch of saphenous
13	远端尾股	Distal caudal femoral
14	股后近侧动脉	Proximal caudal femoral
15	前列腺动脉/阴部动脉	Prostatic/vaginal
16	臀后动脉	Caudal gluteal
17	臀前动脉	Cranial gluteal
18	髂腰动脉	Iliolumbar
19	臀后动脉	Caudal gluteal
20	荐中动脉	Median sacral

图7-9 犬右后肢动脉模式图，内侧观

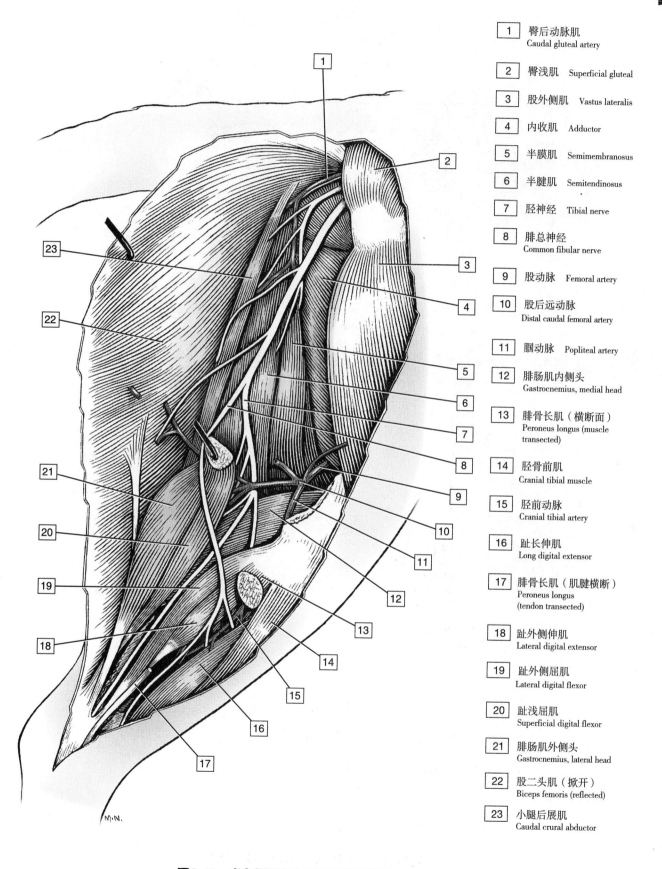

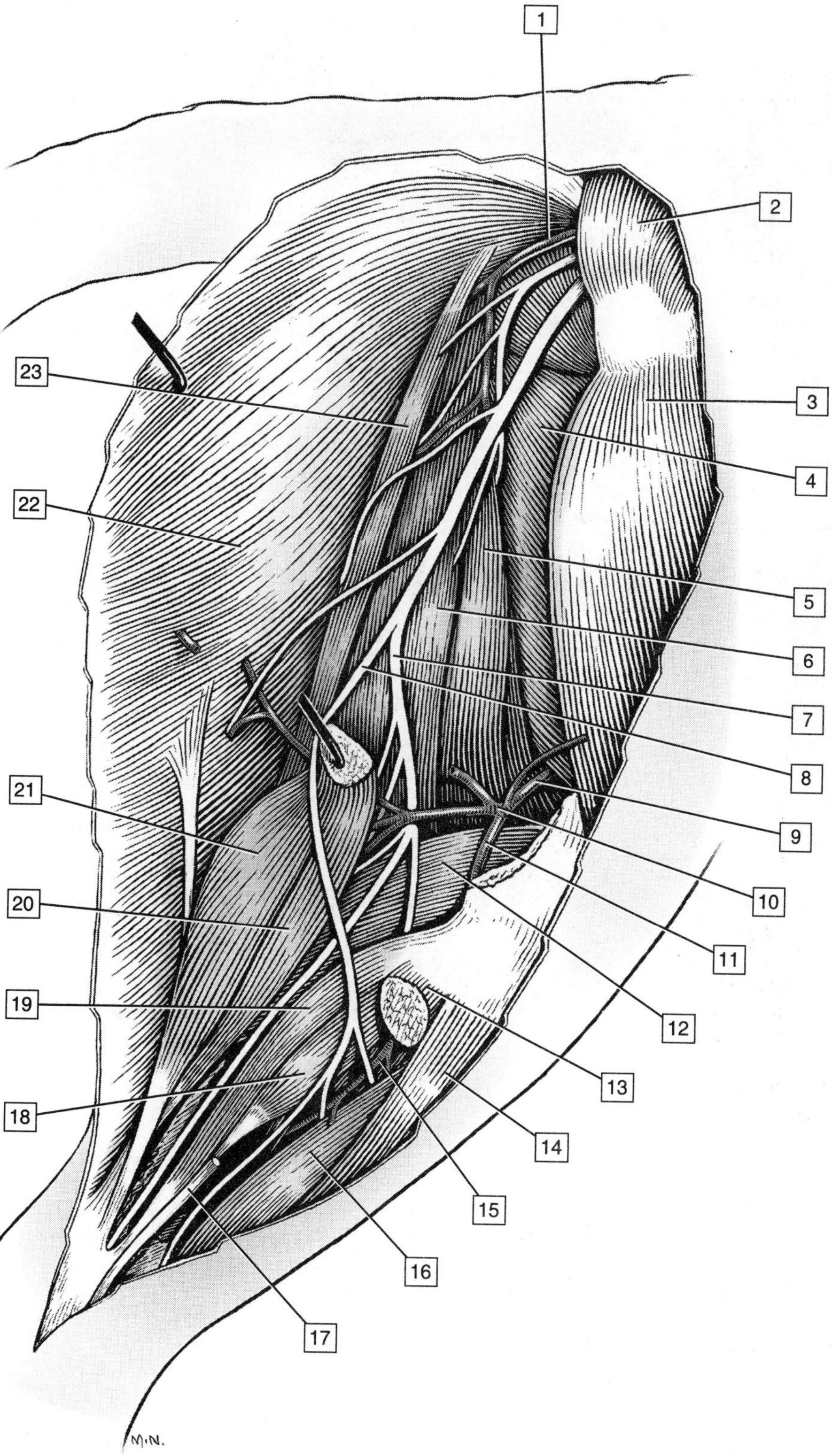

1 臀后动脉肌 Caudal gluteal artery

2 臀浅肌 Superficial gluteal

3 股外侧肌 Vastus lateralis

4 内收肌 Adductor

5 半膜肌 Semimembranosus

6 半腱肌 Semitendinosus

7 胫神经 Tibial nerve

8 腓总神经 Common fibular nerve

9 股动脉 Femoral artery

10 股后远动脉 Distal caudal femoral artery

11 腘动脉 Popliteal artery

12 腓肠肌内侧头 Gastrocnemius, medial head

13 腓骨长肌（横断面） Peroneus longus (muscle transected)

14 胫骨前肌 Cranial tibial muscle

15 胫前动脉 Cranial tibial artery

16 趾长伸肌 Long digital extensor

17 腓骨长肌（肌腱横断） Peroneus longus (tendon transected)

18 趾外侧伸肌 Lateral digital extensor

19 趾外侧屈肌 Lateral digital flexor

20 趾浅屈肌 Superficial digital flexor

21 腓肠肌外侧头 Gastrocnemius, lateral head

22 股二头肌（掀开） Biceps femoris (reflected)

23 小腿后展肌 Caudal crural abductor

图7-10 犬右侧股部与小腿动脉和神经，外侧观

马

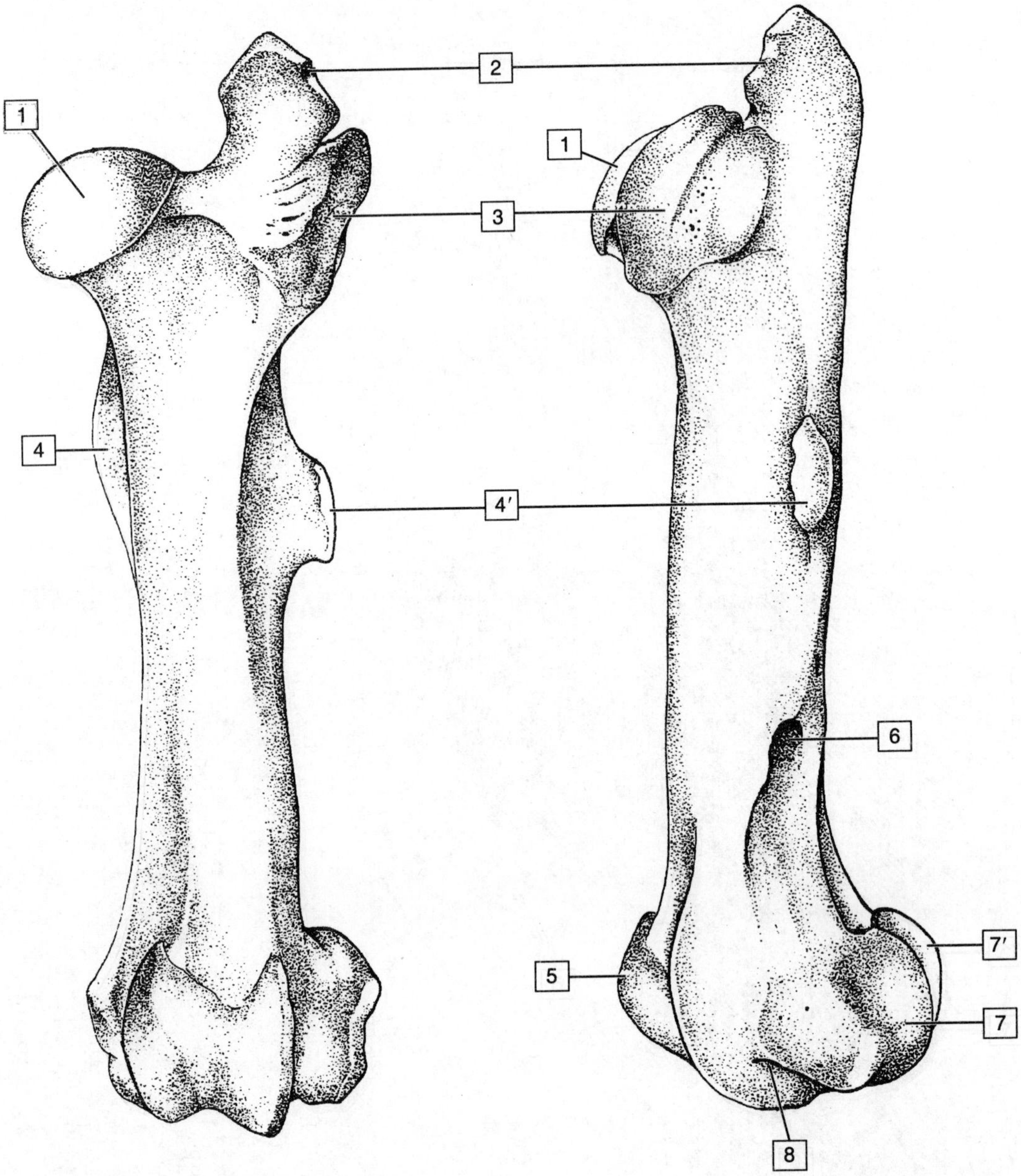

1	股骨头	Head of femur
2	大转子前部	Cranial part of greater trochanter
3	大转子后部	Caudal part of greater trochanter
4	小转子	Lesser trochanter
4′	第3转子	Third trochanter
5	内侧滑车嵴的近端扩大	Enlarged proximal end of medial trochlear ridge
6	髁上窝	Supracondylar fossa
7	外侧髁	Lateral condyle
7′	内侧髁	Medial condyle
8	伸肌窝	Extensor fossa

图7-11　马左侧股骨，前面观和外侧观

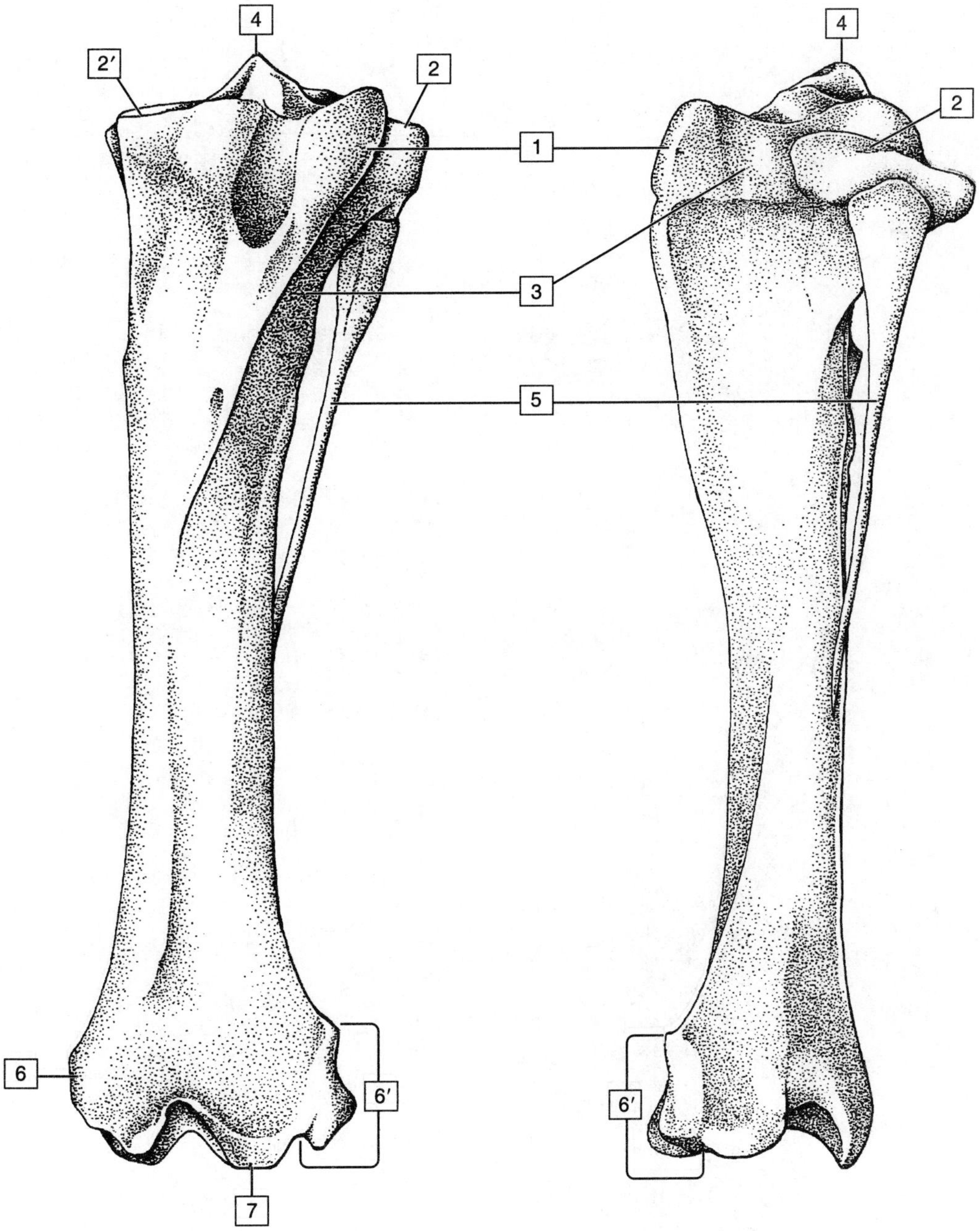

1	胫骨粗隆	Tibial tuberosity
2	外侧髁	Lateral condyle
2′	内侧髁	Medial condyle
3	伸肌沟	Extensor groove
4	髁间隆起	Intercondylar eminence
5	腓骨	Fibula
6	内踝	Medial malleolus
6′	马的外踝（代表腓骨远端）	Lateral malleolus in the horse (representing distal end of fibula)
7	内侧踝	Cochlea

图7-12　马左侧胫骨和腓骨，前面观和外侧观

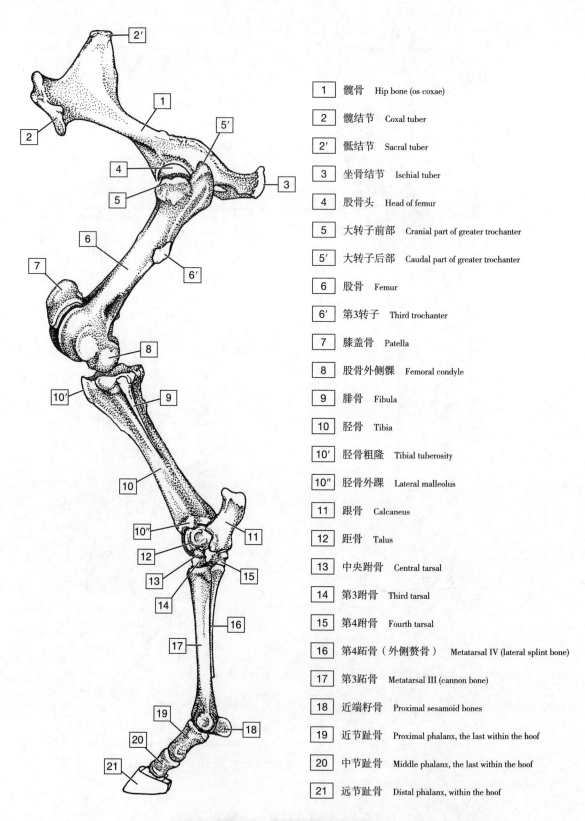

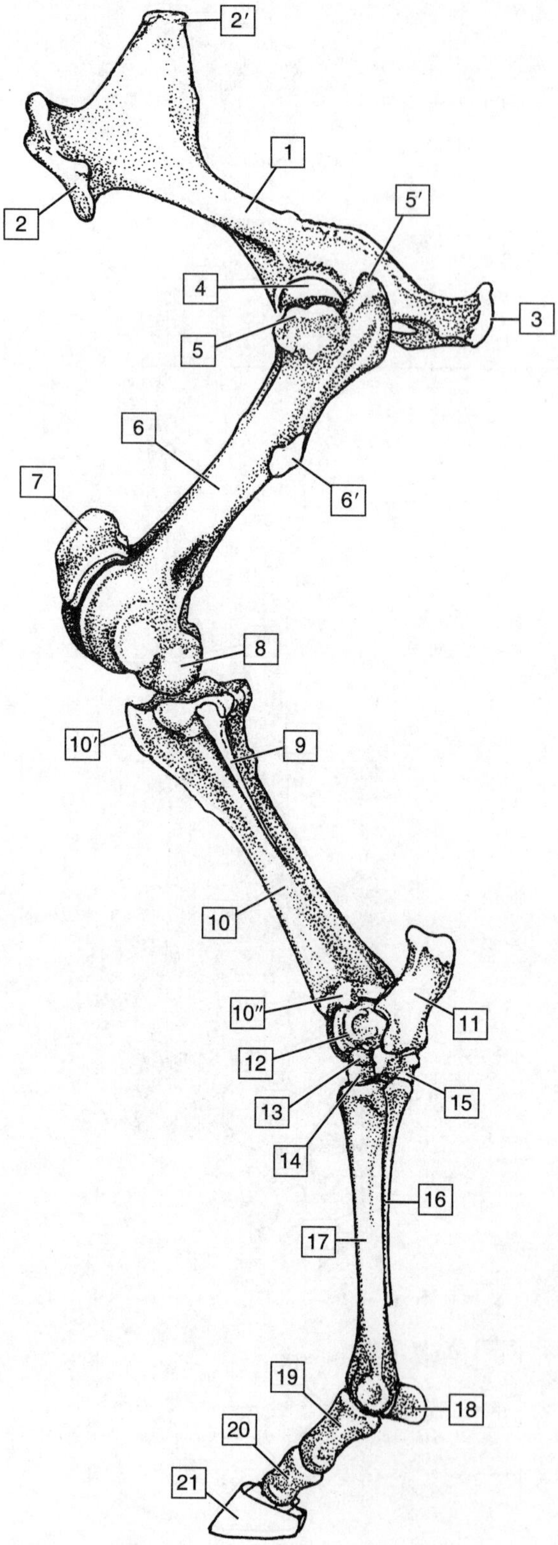

1	髋骨	Hip bone (os coxae)
2	髋结节	Coxal tuber
2′	骶结节	Sacral tuber
3	坐骨结节	Ischial tuber
4	股骨头	Head of femur
5	大转子前部	Cranial part of greater trochanter
5′	大转子后部	Caudal part of greater trochanter
6	股骨	Femur
6′	第3转子	Third trochanter
7	膝盖骨	Patella
8	股骨外侧髁	Femoral condyle
9	腓骨	Fibula
10	胫骨	Tibia
10′	胫骨粗隆	Tibial tuberosity
10″	胫骨外踝	Lateral malleolus
11	跟骨	Calcaneus
12	距骨	Talus
13	中央跗骨	Central tarsal
14	第3跗骨	Third tarsal
15	第4跗骨	Fourth tarsal
16	第4跖骨（外侧赘骨）	Metatarsal IV (lateral splint bone)
17	第3跖骨	Metatarsal III (cannon bone)
18	近端籽骨	Proximal sesamoid bones
19	近节趾骨	Proximal phalanx, the last within the hoof
20	中节趾骨	Middle phalanx, the last within the hoof
21	远节趾骨	Distal phalanx, within the hoof

图7-13 马左后肢骨，外侧观

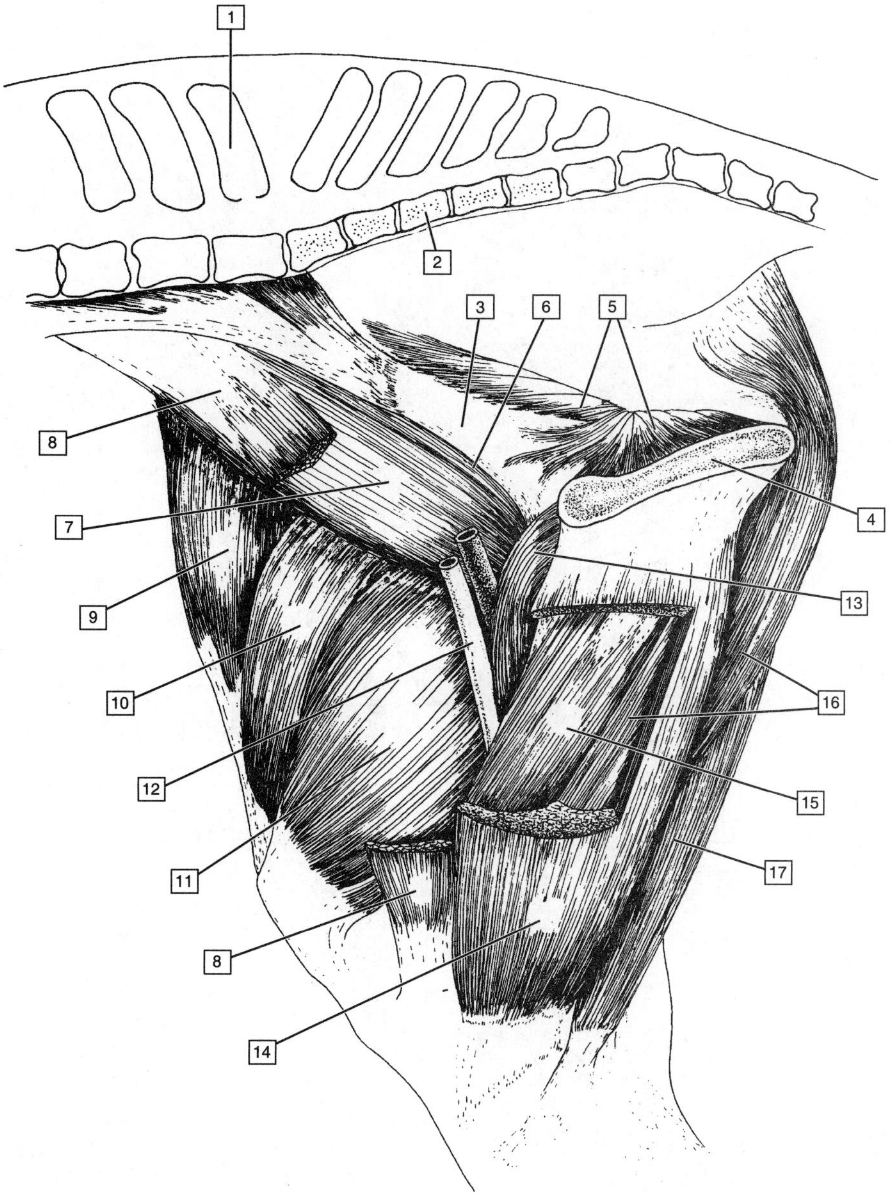

1	最后腰椎 Last lumbar vertebra
2	荐骨 Sacrum
3	髂骨干 Shaft of ilium
4	骨盆联合 Pelvic symphysis
5	闭孔内肌 Internal obturator
6	腰小肌 Psoas minor
7	髂腰肌 Iliopsoas
8	缝匠肌，切除 Sartorius, resected
9	阔筋膜张肌 Tensor fasciae latae
10	股直肌 Rectus femoris
11	股内侧肌 Vastus medialis
12	股三角中的股血管 Femoral vessels infemoral triangle
13	耻骨肌 Pectineus
14	股薄肌，有孔 Gracilis, fenestrated
15	内收肌 Adductor
16	半膜肌 Semimembranosus
17	半腱肌 Semitendinosus

图7-14 马股，内侧观

请写出支配 13 ~ 15 的神经。

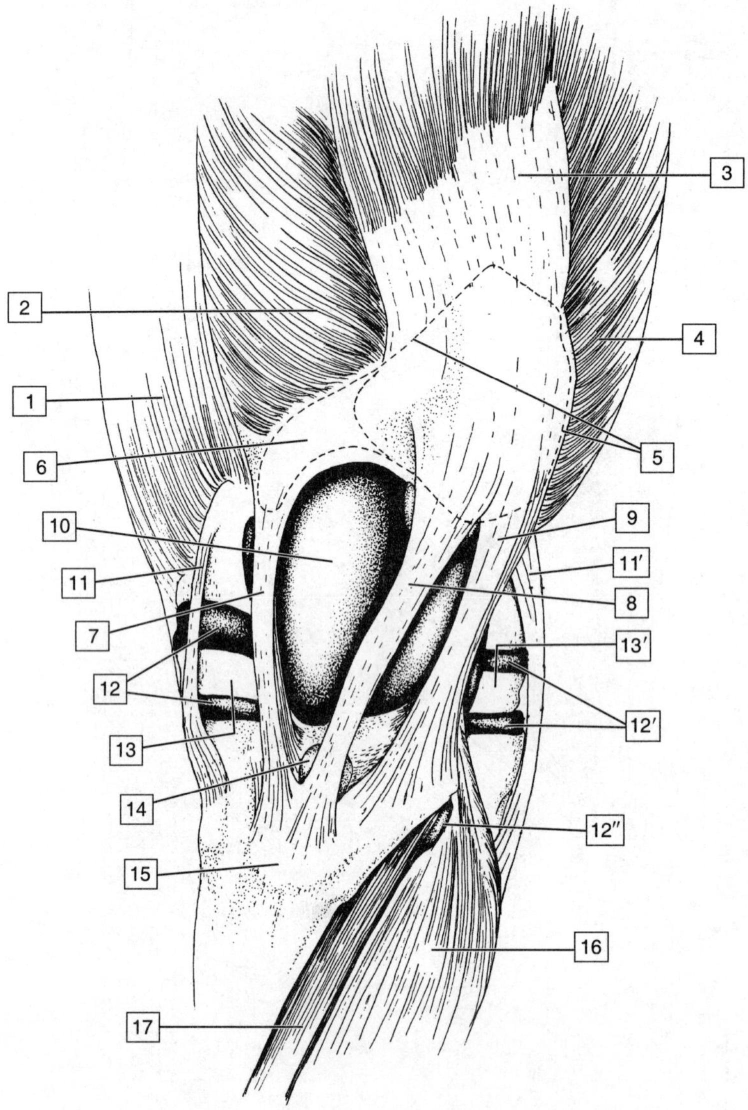

1	内收肌	Adductor
2	股内侧肌	Vastus medialis
3	股直肌	Rectus femoris
4	股外侧肌	Vastus lateralis
5	膝盖骨轮廓	Outline of patella
6	膝盖骨纤维软骨轮廓	Outline of patellar fibrocartilage
7	内侧膝韧带	Medial patellar ligament
8	中间膝韧带	Intermediate patellar ligament
9	外侧膝韧带	Lateral patellar ligament
10	股骨滑车内侧嵴上的关节囊	Joint capsule over medial ridge of femoral trochlea
11	内侧副韧带	Medial collateral ligament
11′	外侧副韧带	Lateral collateral ligaments
12	内侧股胫关节囊	Medial femorotibial joint capsule
12′	外侧股胫关节囊	Lateral femorotibial joint capsule
12″	第3腓骨肌和趾长伸肌的组合腱在12的凹陷处	Recess of 12 under combined tendon of peroneus tertius and long digital extensor
13	内侧半月板	Medial meniscus
13′	外侧半月板	Lateral menisci
14	膝下深囊	Distal infrapatellar bursa
15	胫骨粗隆	Tibial tuberosity
16	趾长伸肌	Long digital extensor
17	胫骨前肌	Tibialis cranialis

图7-15　马左膝关节，前面观

请写出支配 2 ~ 4 的神经。

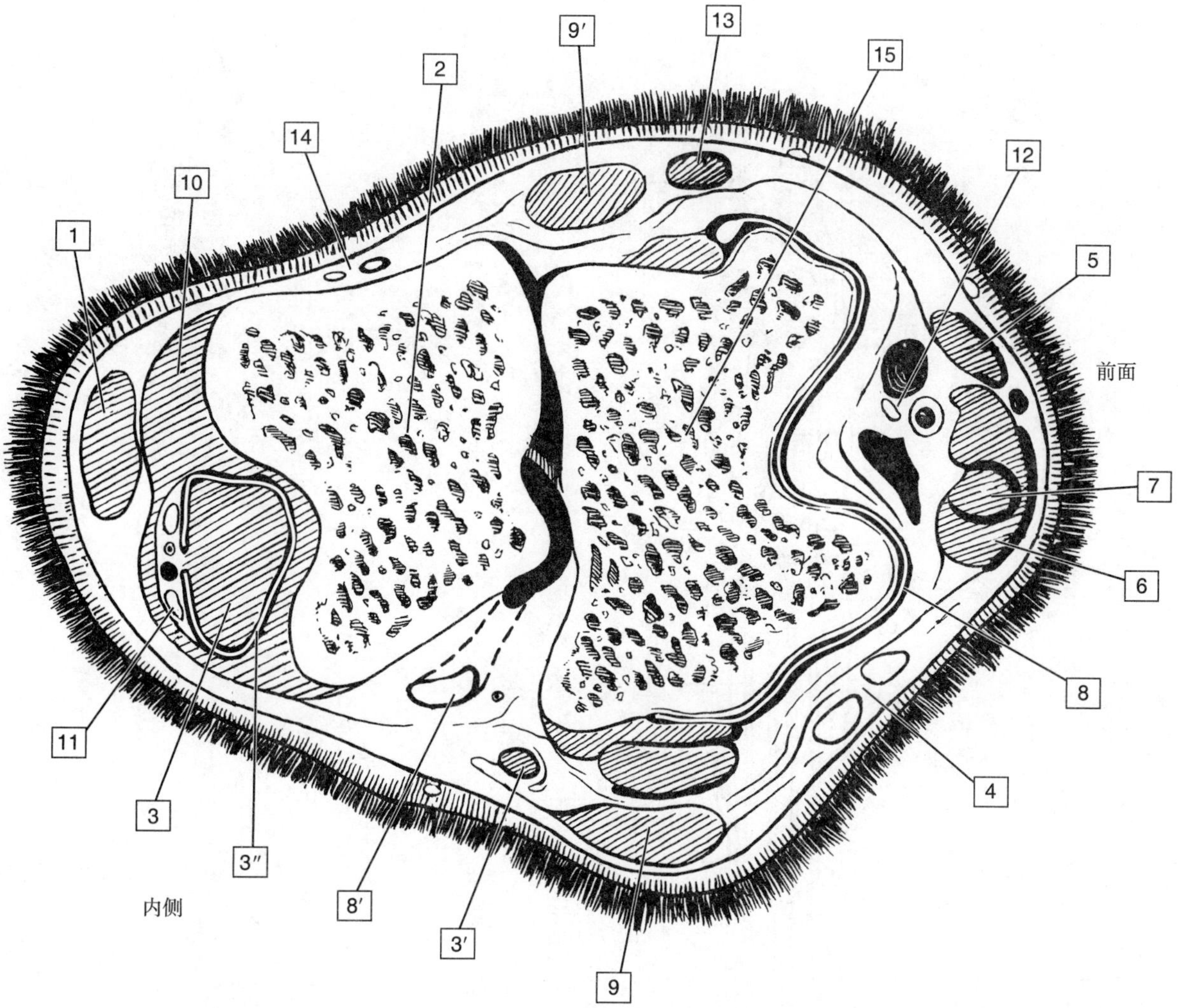

1	趾浅屈肌　Superficial	8′	跗关节的跖内侧囊　Medioplantar pouch of tarsocrural joint
2	跟骨　Calcaneus	9	内侧副韧带（浅部）　Medial collateral ligament (superficial part)
3	外侧趾深屈肌和胫骨后肌 Lateral deep digital flexor and tibialis caudalis	9′	外侧副韧带（浅部）　Lateral collateral ligament (superficial part)
3′	内侧趾深屈肌腱　Tendon of medial deep digital flexor	10	跖侧长韧带　Long plantar ligament
3″	跗骨鞘　Tarsal sheath	11	足底神经和隐静脉　Plantar nerves and saphenous vessels
4	内侧隐静脉前支 Cranial branch of medial saphenous vein	12	胫前血管和腓深神经　Cranial tibial vessels and deep peroneal nerve
5	趾长伸肌　Long digital extensor	13	趾外侧伸肌　Lateral digital extensor
6	第3腓骨肌　Peroneus tertius	14	小腿后侧皮神经和外侧隐静脉 Caudal cutaneous sural nerve and lateral saphenous vein
7	胫骨前肌　Tibialis cranialis	15	距骨　Talus
8	跗关节的背侧和跖内侧囊 Dorsal and medioplantar pouches of tarsocrural joint		

图7-16　马的跗关节黏液囊、腱鞘和关节囊

本图为横断面的近表面

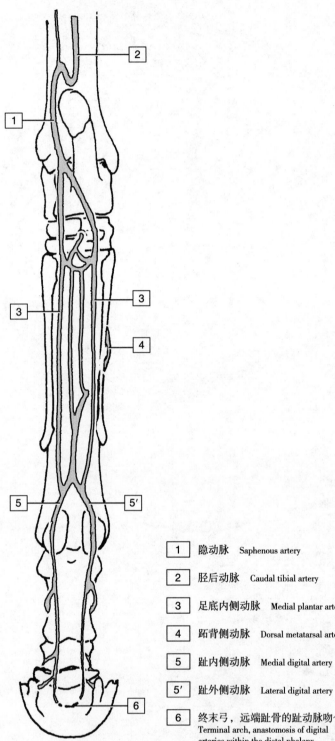

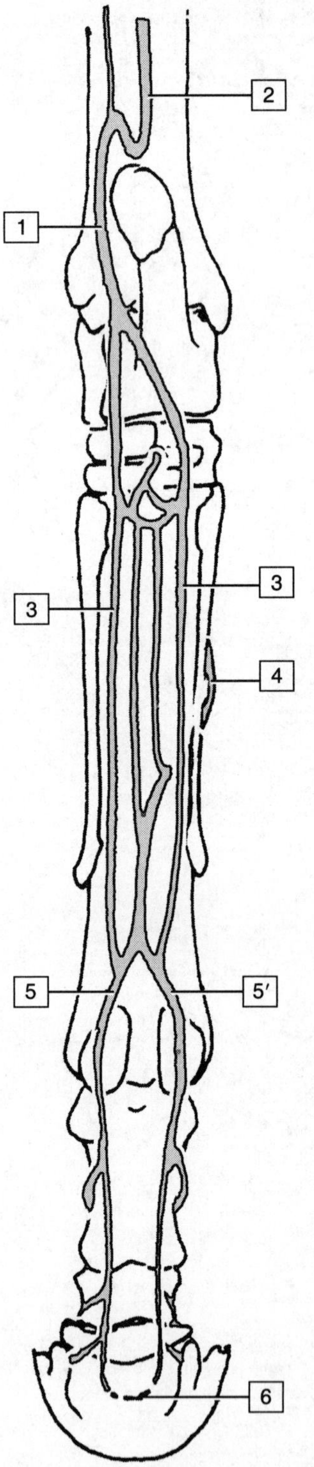

1	隐动脉	Saphenous artery
2	胫后动脉	Caudal tibial artery
3	足底内侧动脉	Medial plantar artery
4	跖背侧动脉	Dorsal metatarsal artery
5	趾内侧动脉	Medial digital artery
5′	趾外侧动脉	Lateral digital artery
6	终末弓，远端趾骨的趾动脉吻合	Terminal arch, anastomosis of digital arteries within the distal phalanx

图7-17　马右后肢主要动脉，跖侧观

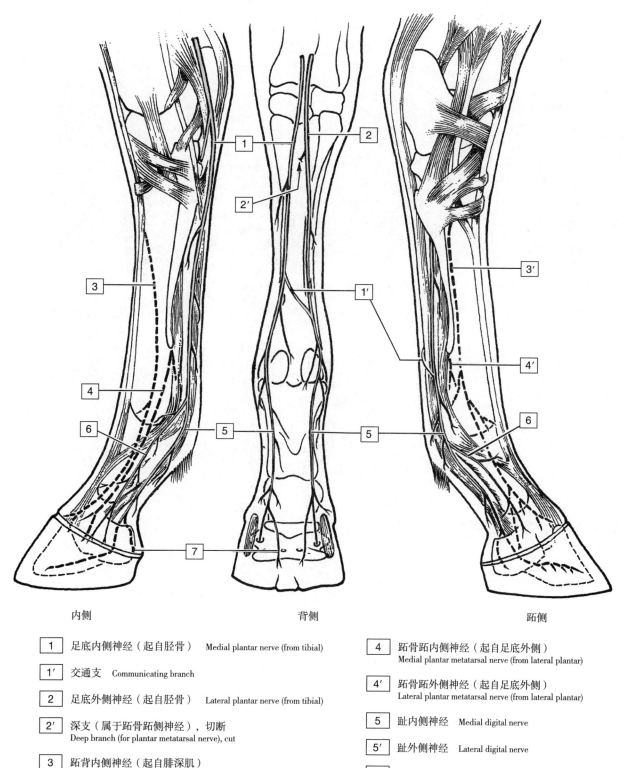

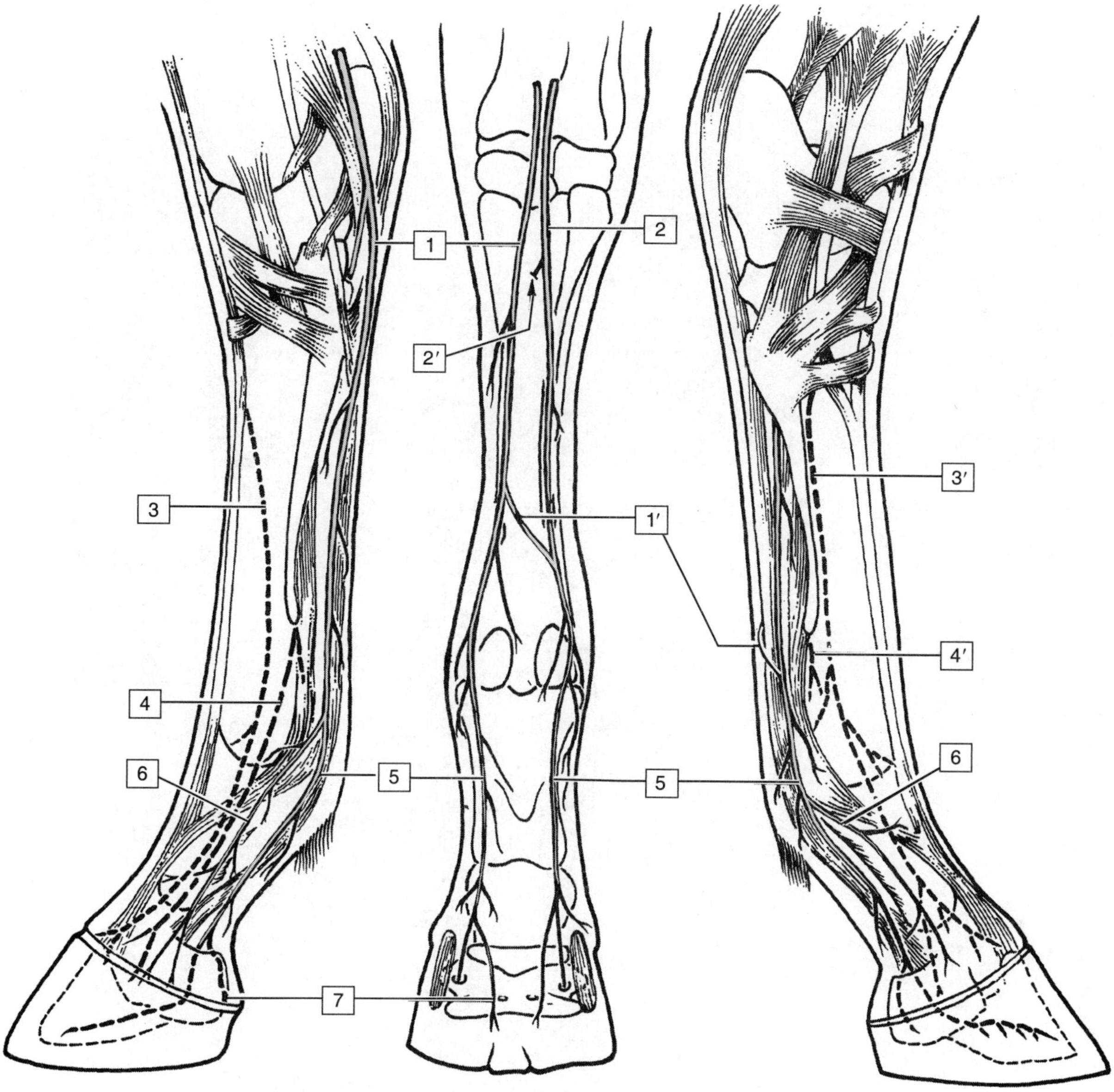

1	足底内侧神经（起自胫骨）	Medial plantar nerve (from tibial)
1′	交通支	Communicating branch
2	足底外侧神经（起自胫骨）	Lateral plantar nerve (from tibial)
2′	深支（属于跖骨跖侧神经），切断	Deep branch (for plantar metatarsal nerve), cut
3	跖背内侧神经（起自腓深肌）	Medial dorsal metatarsal nerve (from deep peroneal)
3′	跖背外侧神经（起自腓深肌）	Lateral dorsal metatarsal nerve (from deep peroneal)
4	跖骨跖内侧神经（起自足底外侧）	Medial plantar metatarsal nerve (from lateral plantar)
4′	跖骨跖外侧神经（起自足底外侧）	Lateral plantar metatarsal nerve (from lateral plantar)
5	趾内侧神经	Medial digital nerve
5′	趾外侧神经	Lateral digital nerve
6	趾神经背侧支	Dorsal branch of digital nerve
7	分支至趾垫	Branch to digital cushion

图7-18 马右后蹄部神经

牛

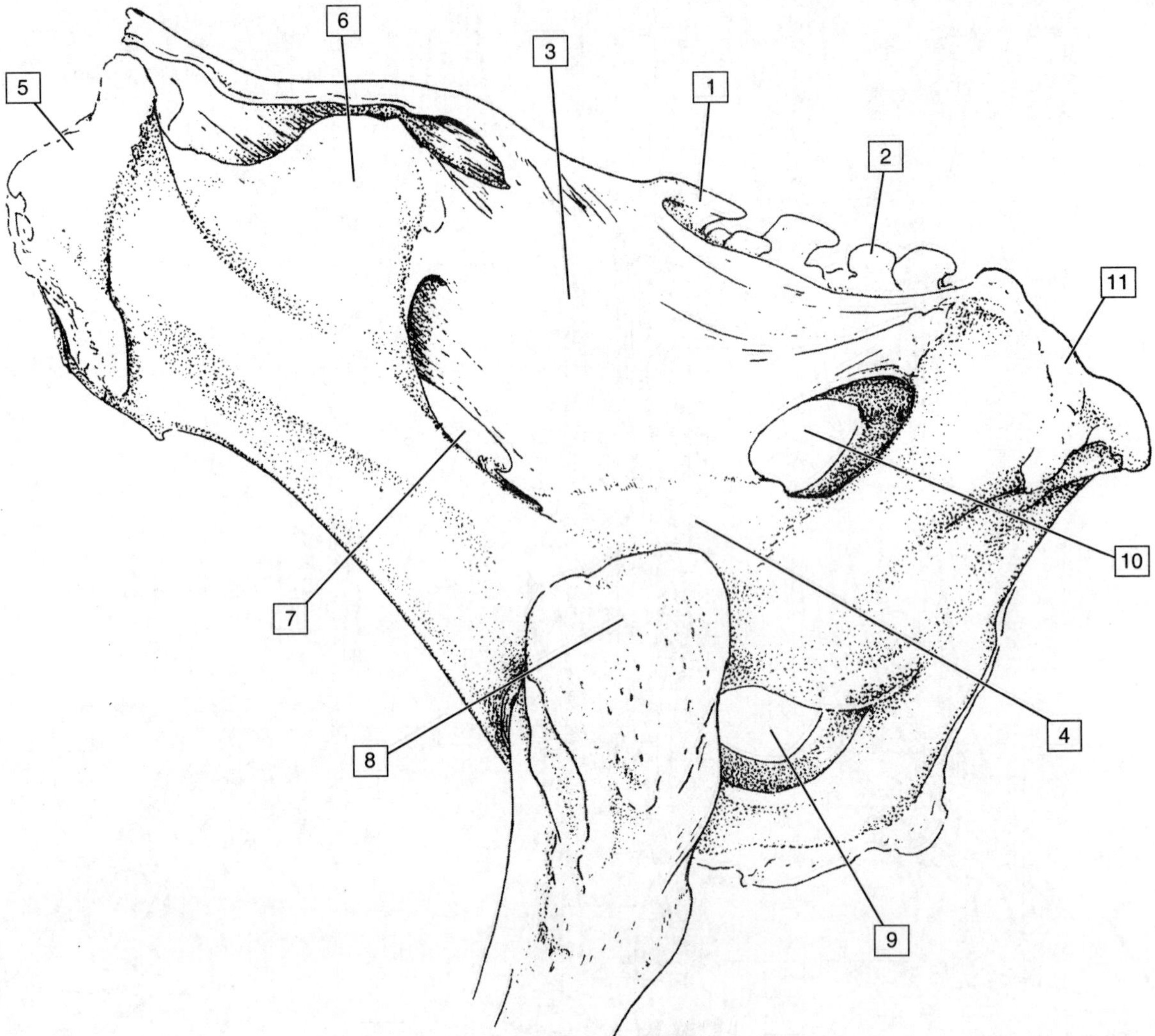

1 骶骨 Sacrum
2 尾椎 Caudal vertebra(e)
3 荐结节韧带 Sacrotuberous ligament
4 坐骨棘 Ischial spine
5 髋结节 Coxal tuber
6 荐结节 Sacral tuber
7 坐骨大孔 Greater sciatic foramen
8 大转子 Greater trochanter
9 闭孔 Obturator foramen
10 坐骨小孔 Lesser sciatic foramen
11 坐骨管 Ischial tube

图7-19 牛荐坐韧带，左外侧观

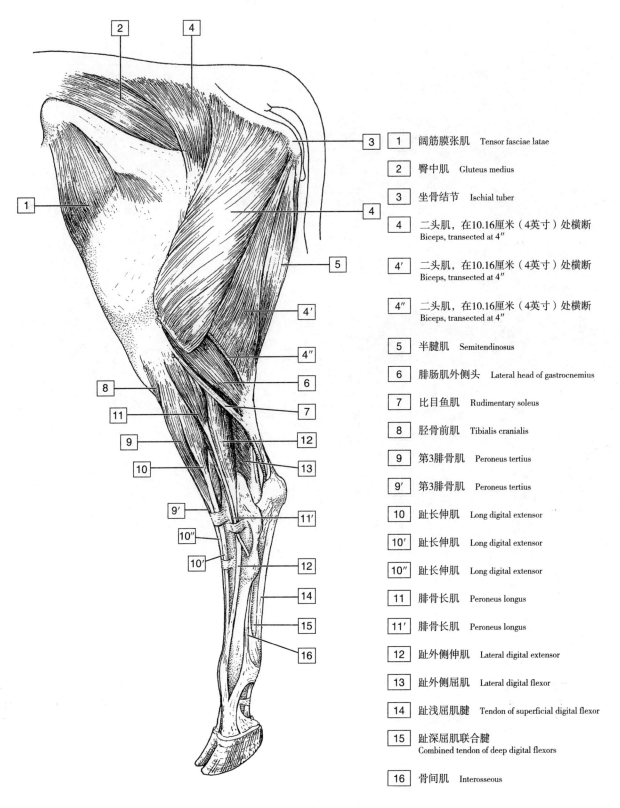

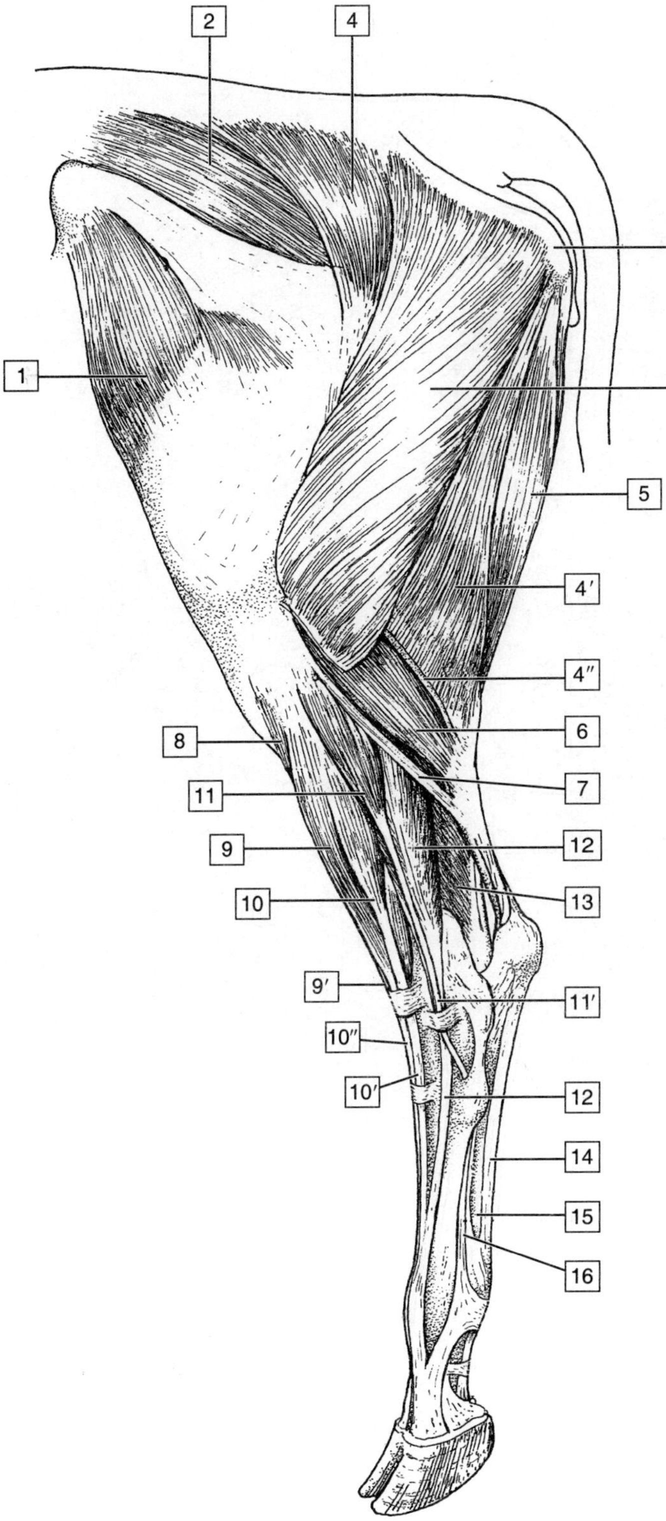

1 阔筋膜张肌 Tensor fasciae latae

2 臀中肌 Gluteus medius

3 坐骨结节 Ischial tuber

4 二头肌，在10.16厘米（4英寸）处横断 Biceps, transected at 4″

4′ 二头肌，在10.16厘米（4英寸）处横断 Biceps, transected at 4″

4″ 二头肌，在10.16厘米（4英寸）处横断 Biceps, transected at 4″

5 半腱肌 Semitendinosus

6 腓肠肌外侧头 Lateral head of gastrocnemius

7 比目鱼肌 Rudimentary soleus

8 胫骨前肌 Tibialis cranialis

9 第3腓骨肌 Peroneus tertius

9′ 第3腓骨肌 Peroneus tertius

10 趾长伸肌 Long digital extensor

10′ 趾长伸肌 Long digital extensor

10″ 趾长伸肌 Long digital extensor

11 腓骨长肌 Peroneus longus

11′ 腓骨长肌 Peroneus longus

12 趾外侧伸肌 Lateral digital extensor

13 趾外侧屈肌 Lateral digital flexor

14 趾浅屈肌腱 Tendon of superficial digital flexor

15 趾深屈肌联合腱 Combined tendon of deep digital flexors

16 骨间肌 Interosseous

图7-20 牛左后肢肌肉，外侧观

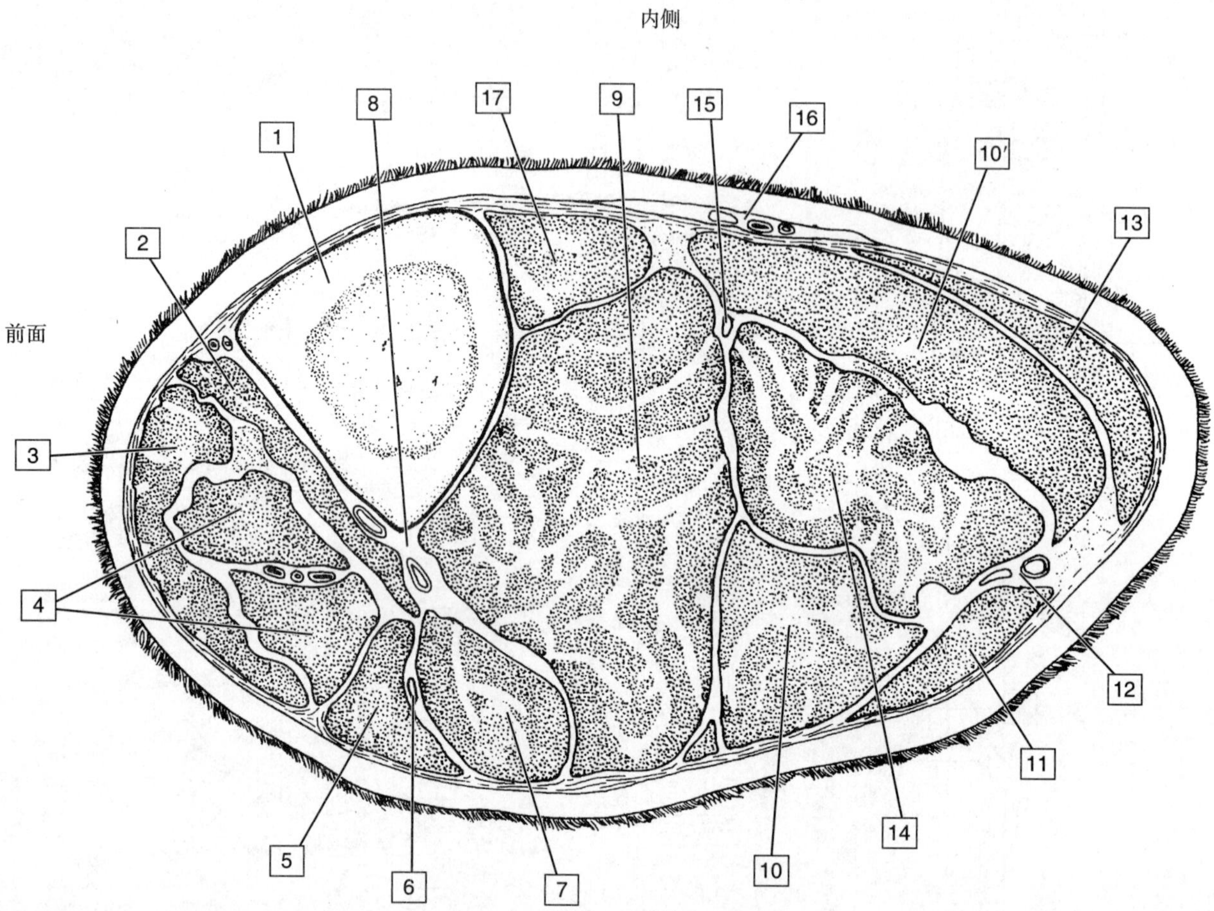

1	胫骨	Tibia
2	胫骨前肌	Tibialis cranialis
3	第3腓肠肌	Peroneus tertius
4	趾长伸肌	Long digital extensor
5	腓骨长肌	Peroneus longus
6	腓神经	Peroneal nerve
7	趾外侧伸肌	Lateral digital extensor
8	胫前血管	Cranial tibial vessels
9	趾深屈肌	Deep digital flexors
10	腓肠肌外侧头	Lateral head gastrocnemius
10′	腓肠肌内侧头	Medial head of gastrocnemius
11	股二头肌	Biceps femoris
12	腓肠肌后皮神经和外侧隐静脉	Caudal cutaneous sural nerve and lateral saphenous vein
13	半腱肌	Semitendinosus
14	趾浅屈肌	Superficial digital flexor
15	胫神经	Tibial nerve
16	隐血管和神经	Saphenous vessels and nerve
17	腘肌	Popliteus

图7-21 牛左腿的横断面

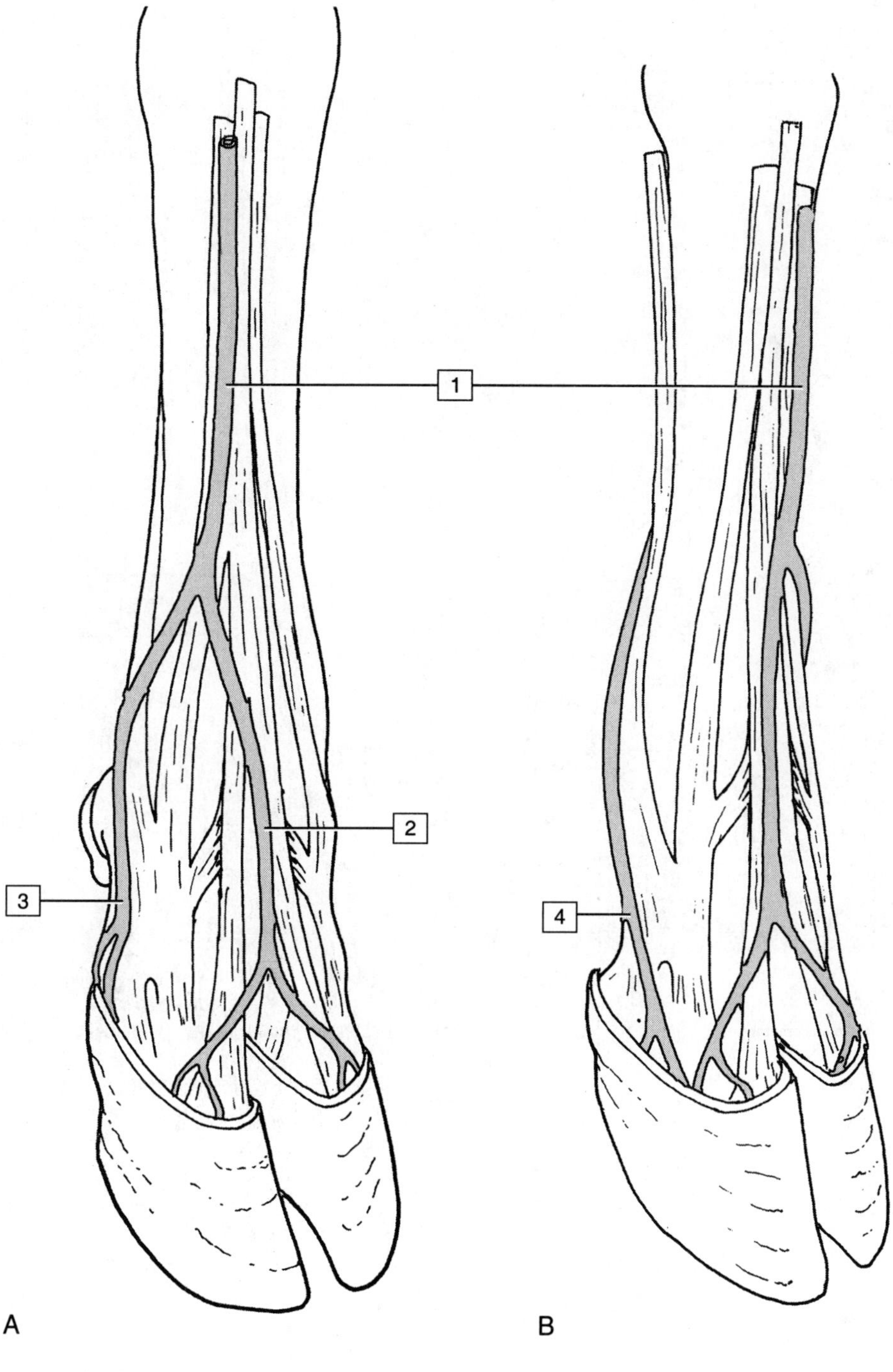

1	侧隐静脉前支　Cranial tributary of lateral saphenous vein	3	外侧趾足底静脉　Plantar vein of lateral digit
2	第3趾总背侧静脉　Dorsal common digital vein III	4	内侧趾足底静脉　Plantar vein of medial digit

图7-22　牛右后肢大静脉

A. 右后脚，背外侧观；B. 左后脚，背内侧观

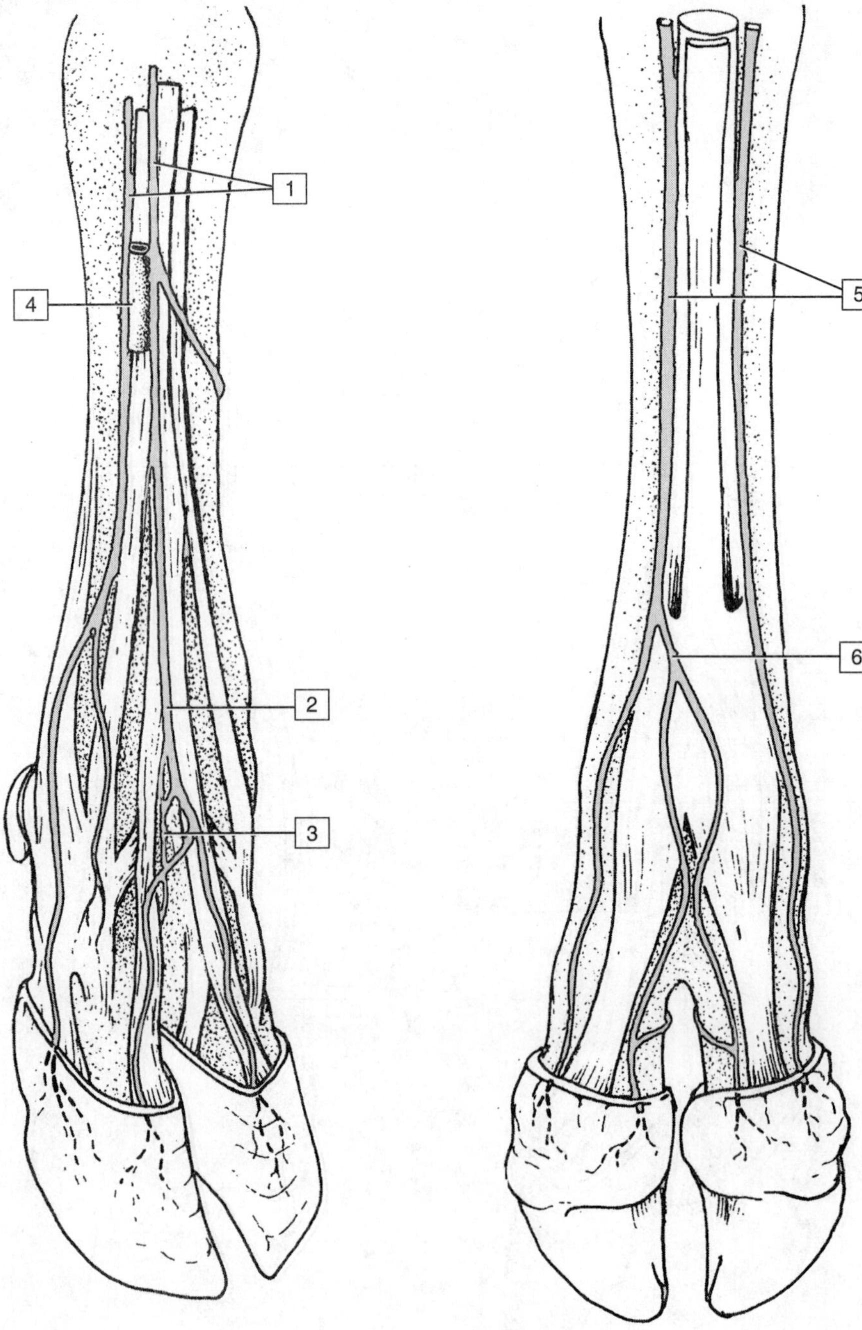

1 腓总神经的内侧和中间支 Lateral and middle branches of superficial peroneal nerve	4 外侧隐静脉前支 Cranial tributary of lateral saphenous vein
2 第3趾总背侧神经 Dorsal common digital nerve III	5 内侧和外侧足底神经 Medial and lateral plantar nn.
3 腓深神经 Deep peroneal nerve	6 第3趾跖总神经 Plantar common digital nerve III

图7-23 牛右后肢神经

A. 右后脚，背外侧观；B. 右后脚，足底面

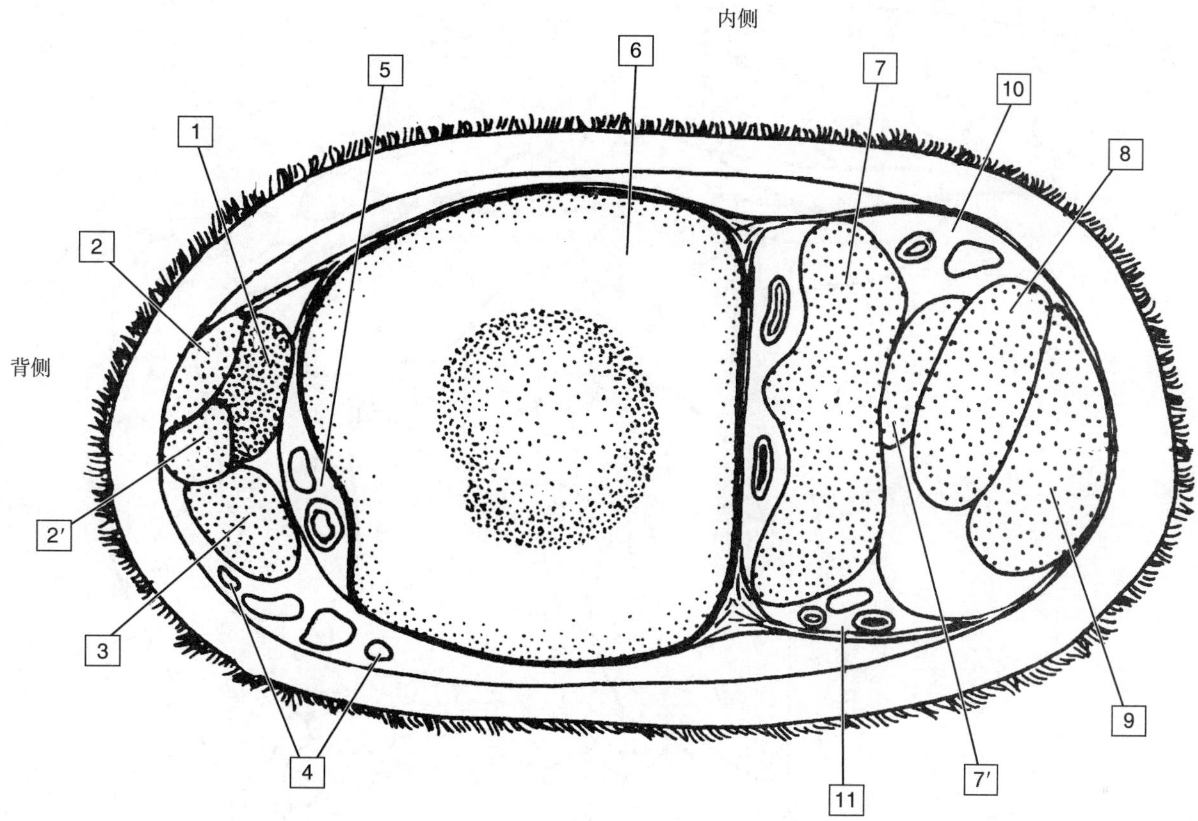

1	短伸肌 Extensor brevis	6	跖骨 Metatarsal bone
2	趾长伸肌 Long digital extensor	7	骨间肌 Interosseous
2′	趾长伸肌 Long digital extensor	7′	从骨间肌到趾浅屈肌 Band from interosseous to superficial digital flexor
3	趾外侧伸肌 Lateral digital extensor	8	趾深屈肌 Deep digitalflexor
4	腓浅神经的分支和侧隐静脉的前支 Branches of superficial peroneal nerve and cranial tributary of lateral saphenous vein	9	趾浅屈肌 Superficial digital flexor
5	腓深神经与跖背动脉（胫骨前肌前延续） Deep peroneal nerve and dorsal metatarsal artery (continuation of cranial tibial)	10	足底内侧神经和血管 Medial plantar nerve and vessel
		11	足底外侧神经和血管 Lateral plantar nerve and vessel

图7-24 牛左跖骨横断面

猪

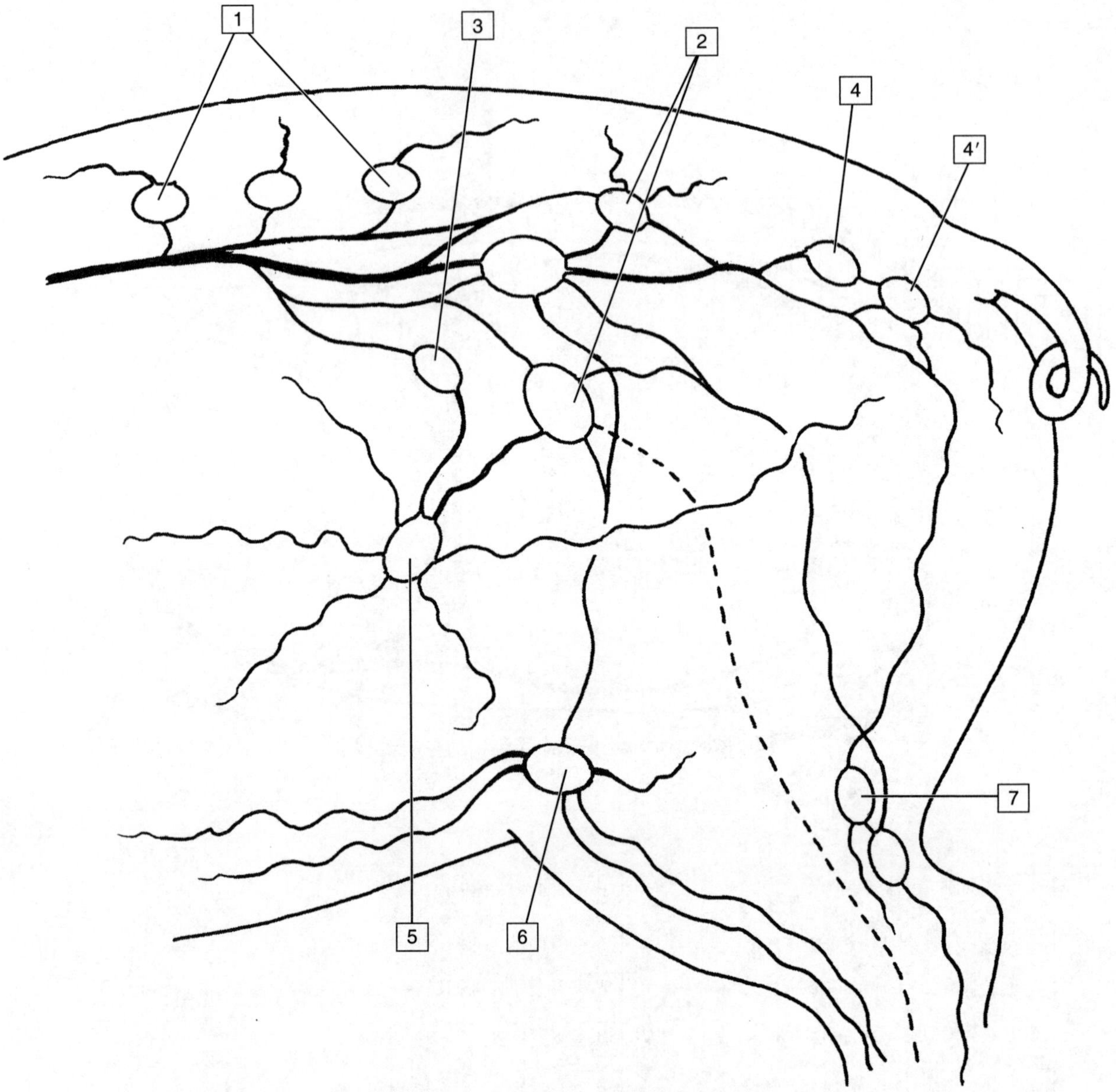

1	腰主动脉淋巴结 Lumbar aortic nodes	4′	臀淋巴结 Gluteal nodes
2	髂内淋巴结 Medial iliac nodes	5	髂下淋巴结 Subiliac nodes
3	髂外淋巴结 Lateral iliac node	6	腹股沟浅淋巴结 Superficial inguinal nodes
4	坐骨结节 Ischial node	7	腘淋巴结 Popliteal nodes

图7-25 猪后肢淋巴结，外侧观

不同动物的解剖结构比较

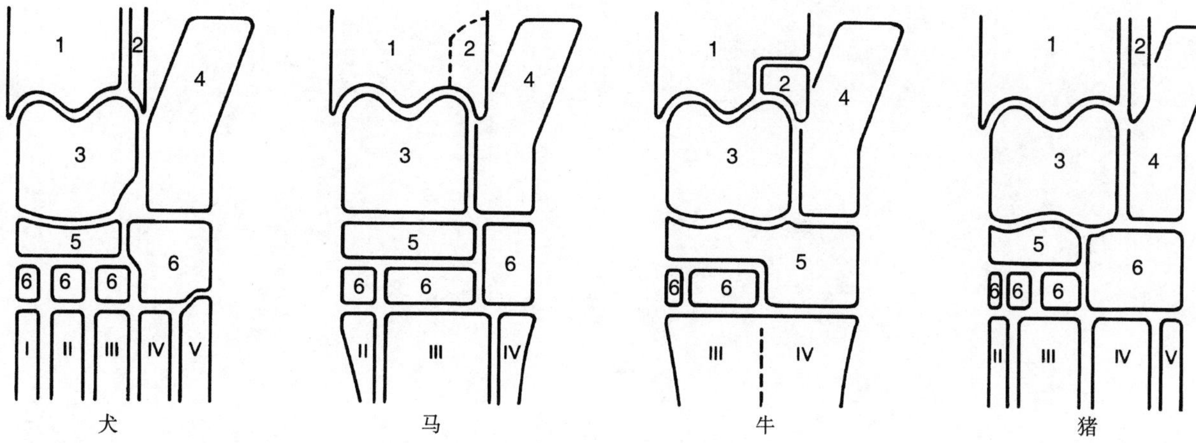

1	桡骨 Radius		4	跟骨 Calcaneus	
2	腓骨 Fibula		5	跗骨中央 Central tarsal bone	
3	距骨 Talus		6	跗骨远端 Distal tarsal bones	

图7-26 跗骨示意图

罗马数字表示跖骨，阿拉伯数字表示跗骨前端

8

异宠

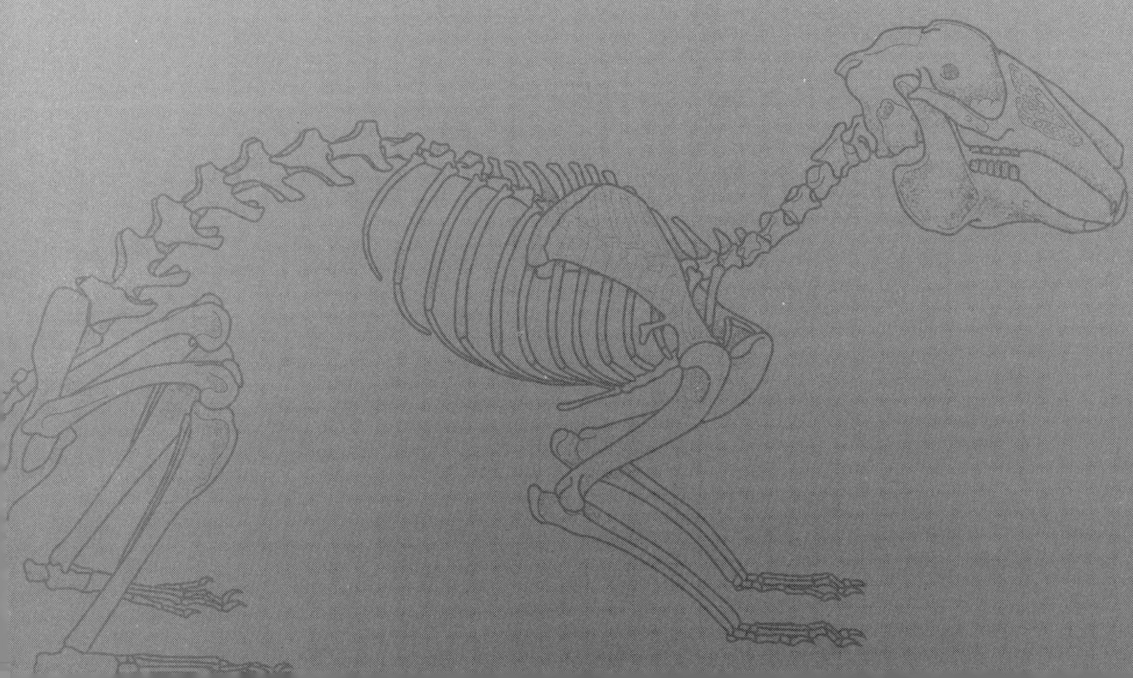

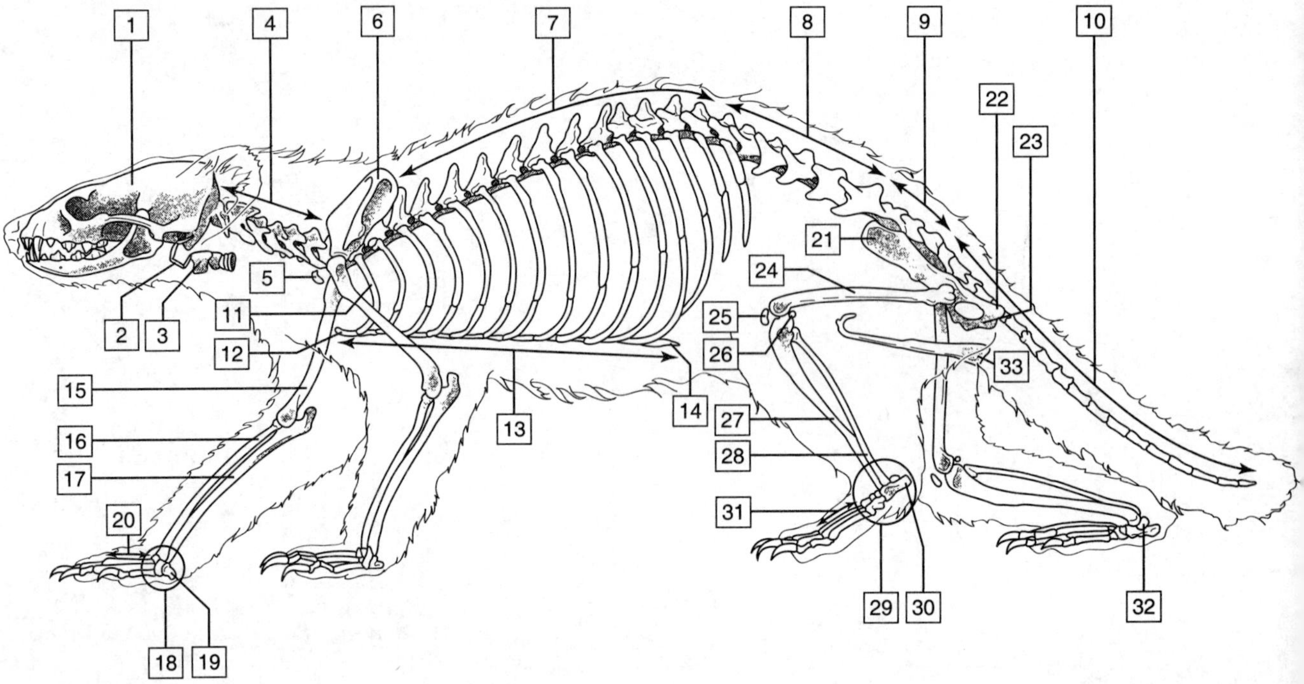

1 颅骨 Calvaria
2 舌骨 Hyoid apparatus
3 喉 Larynx
4 7枚颈椎 Seven cervical vertebrae
5 锁骨 Clavicle
6 肩胛骨 Scapula
7 15枚胸椎 Fifteen thoracic vertebrae
8 5枚腰椎 Five lumbar vertebrae
9 3枚荐椎 Three sacral vertebrae
10 18枚尾椎 Eighteen caudal vertebrae
11 第1肋 First rib
12 胸骨柄 Manubrium
13 胸骨 Sternum
14 剑突 Xiphoid process
15 肱骨 Humerus
16 桡骨 Radius
17 尺骨 Ulna
18 腕骨 Carpal bones
19 副腕骨 Accessory carpal bone
20 掌骨 Metacarpal bones
21 髂骨 Ilium
22 坐骨 Ischium
23 耻骨 Pubis
24 股骨 Femur
25 膝盖骨 Patella
26 腓肠豆 Fabella
27 胫骨 Tibia
28 腓骨 Fibula
29 跗骨 Tarsal bones
30 跟骨 Calcaneus
31 跖骨 Metatarsal bon
32 距骨 Talus
33 阴茎骨 Os penis

图8-1 鼬科动物白鼬的骨骼

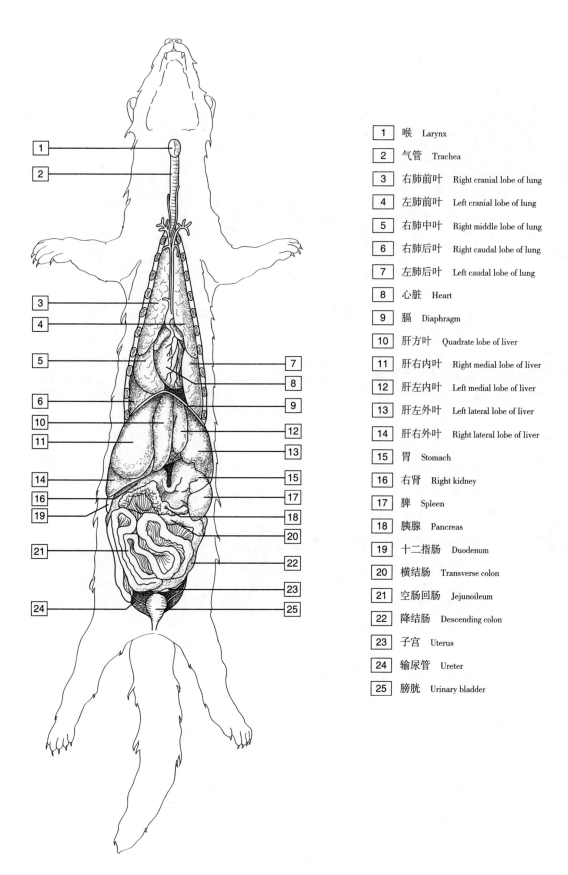

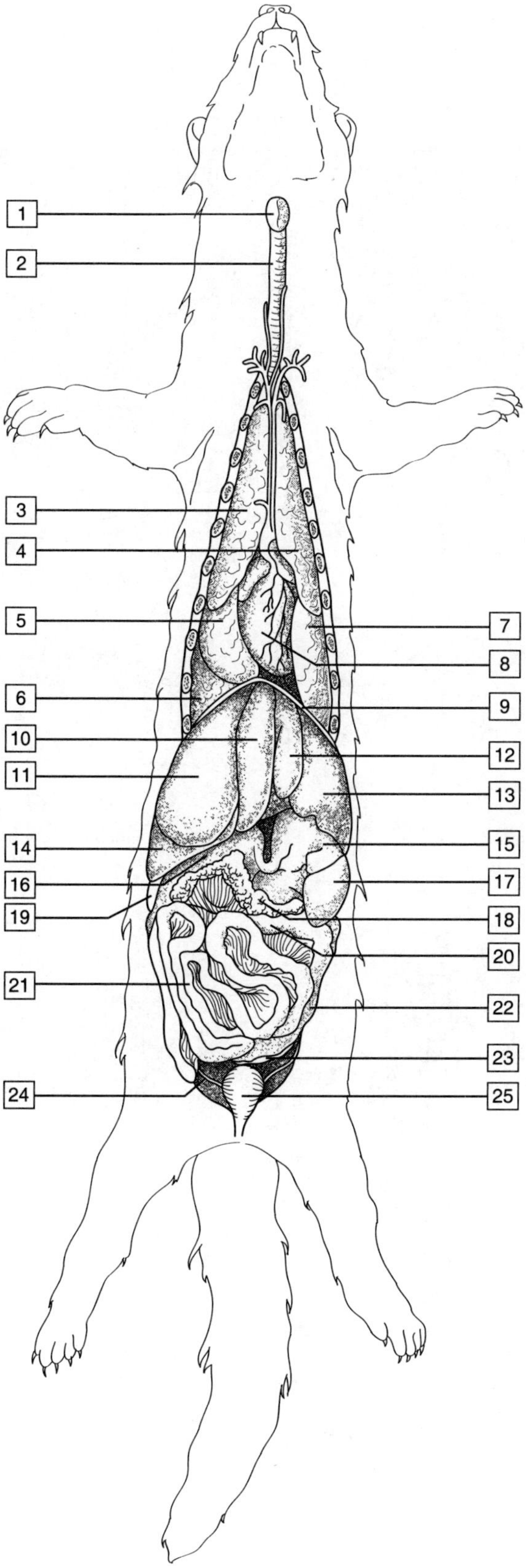

1 喉 Larynx
2 气管 Trachea
3 右肺前叶 Right cranial lobe of lung
4 左肺前叶 Left cranial lobe of lung
5 右肺中叶 Right middle lobe of lung
6 右肺后叶 Right caudal lobe of lung
7 左肺后叶 Left caudal lobe of lung
8 心脏 Heart
9 膈 Diaphragm
10 肝方叶 Quadrate lobe of liver
11 肝右内叶 Right medial lobe of liver
12 肝左内叶 Left medial lobe of liver
13 肝左外叶 Left lateral lobe of liver
14 肝右外叶 Right lateral lobe of liver
15 胃 Stomach
16 右肾 Right kidney
17 脾 Spleen
18 胰腺 Pancreas
19 十二指肠 Duodenum
20 横结肠 Transverse colon
21 空肠回肠 Jejunoileum
22 降结肠 Descending colon
23 子宫 Uterus
24 输尿管 Ureter
25 膀胱 Urinary bladder

图8-2 鼬科动物白鼬的内部解剖图，腹侧观

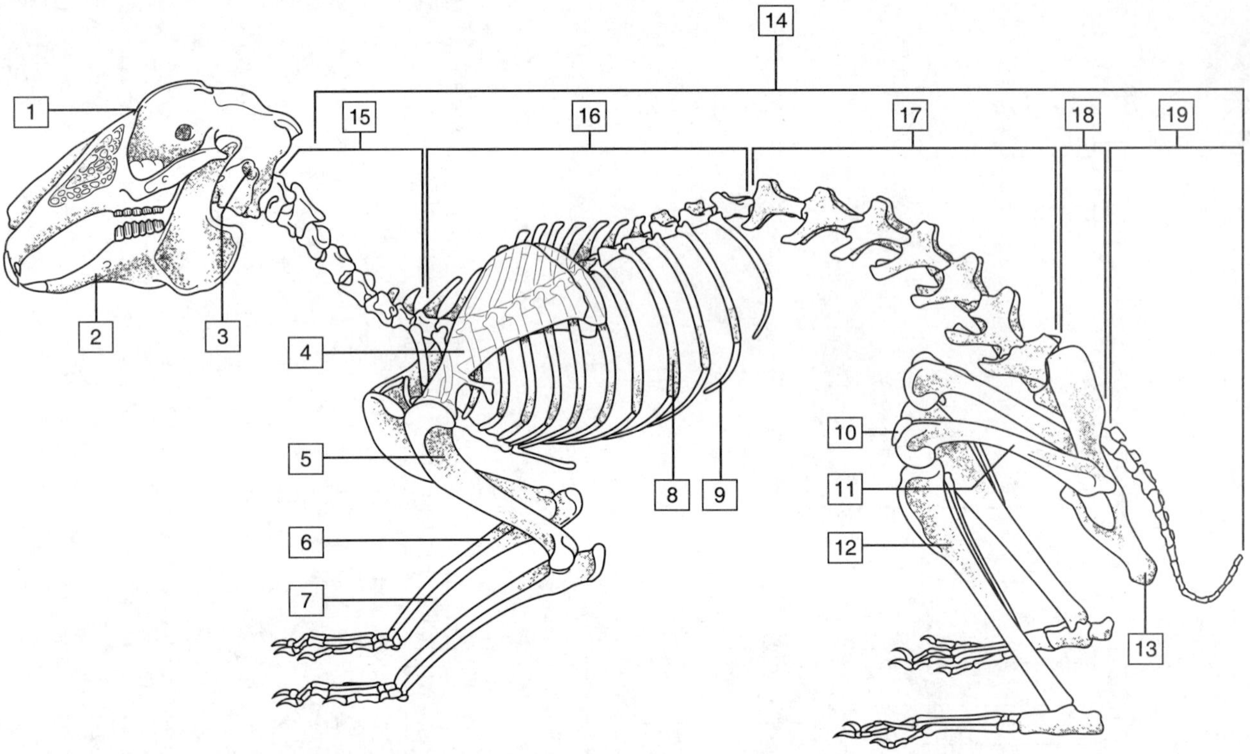

1	颅骨	Skull	6	桡骨	Radius	11	股骨	Femur	16	胸椎	Thoracic
2	下颌骨	Mandible	7	尺骨	Ulna	12	胫骨	Tibia	17	腰椎	Lumbar
3	鼓泡	Tympanic bulla	8	肋骨	Rib	13	骨盆	Pelvis	18	荐椎	Sacral
4	肩胛骨	Scapula	9	肋软骨	Cartilage	14	脊椎	Vertebrae	19	尾椎	Coccygeal
5	肱骨	Humerus	10	膝盖骨	Patella	15	颈椎	Cervical			

图8-3　兔的骨骼

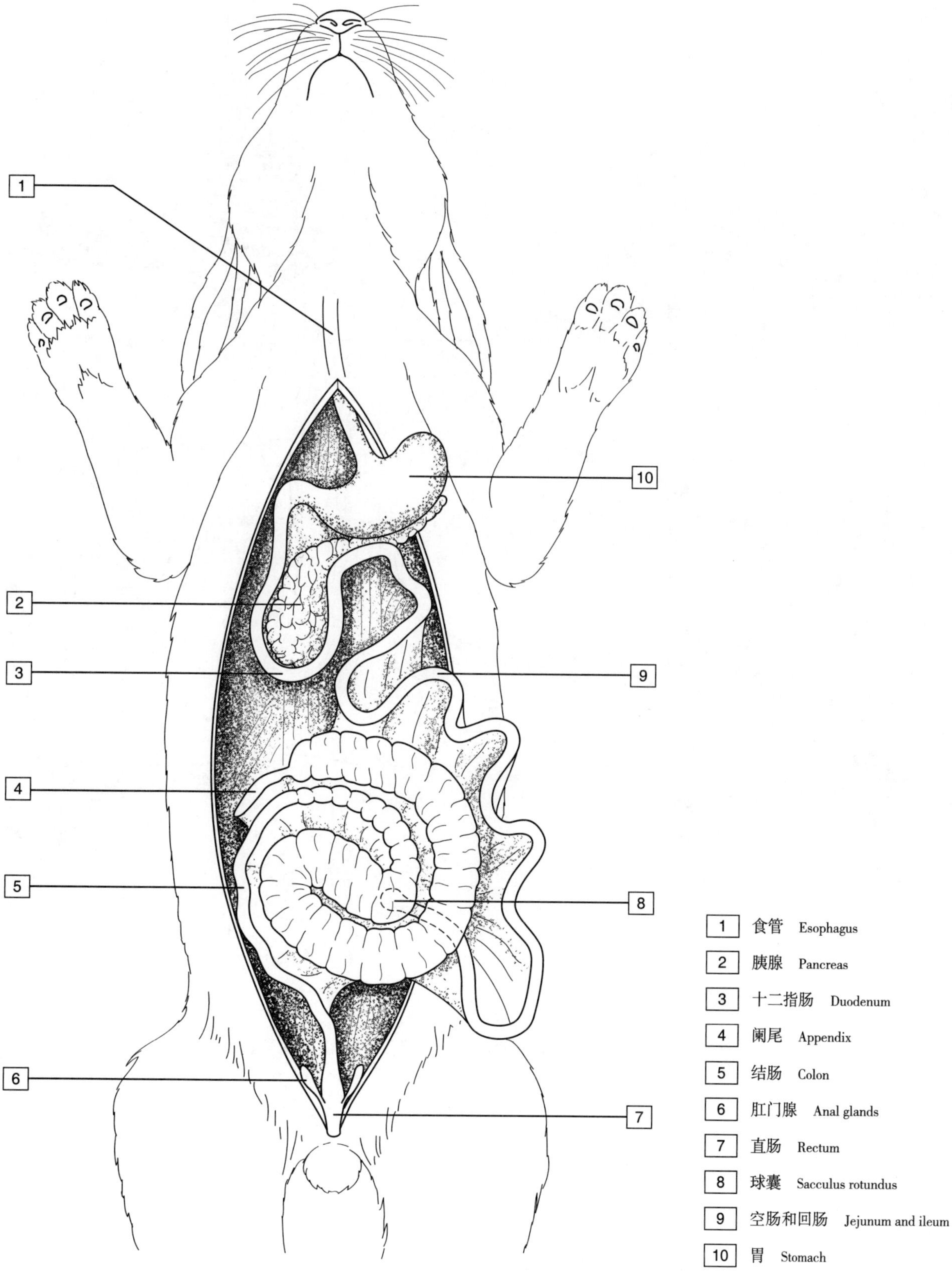
1
2
3
4
5
6
7
8
9
10
1 食管 Esophagus
2 胰腺 Pancreas
3 十二指肠 Duodenum
4 阑尾 Appendix
5 结肠 Colon
6 肛门腺 Anal glands
7 直肠 Rectum
8 球囊 Sacculus rotundus
9 空肠和回肠 Jejunum and ileum
10 胃 Stomach

图8-4 雌性野兔内脏解剖图，腹侧观

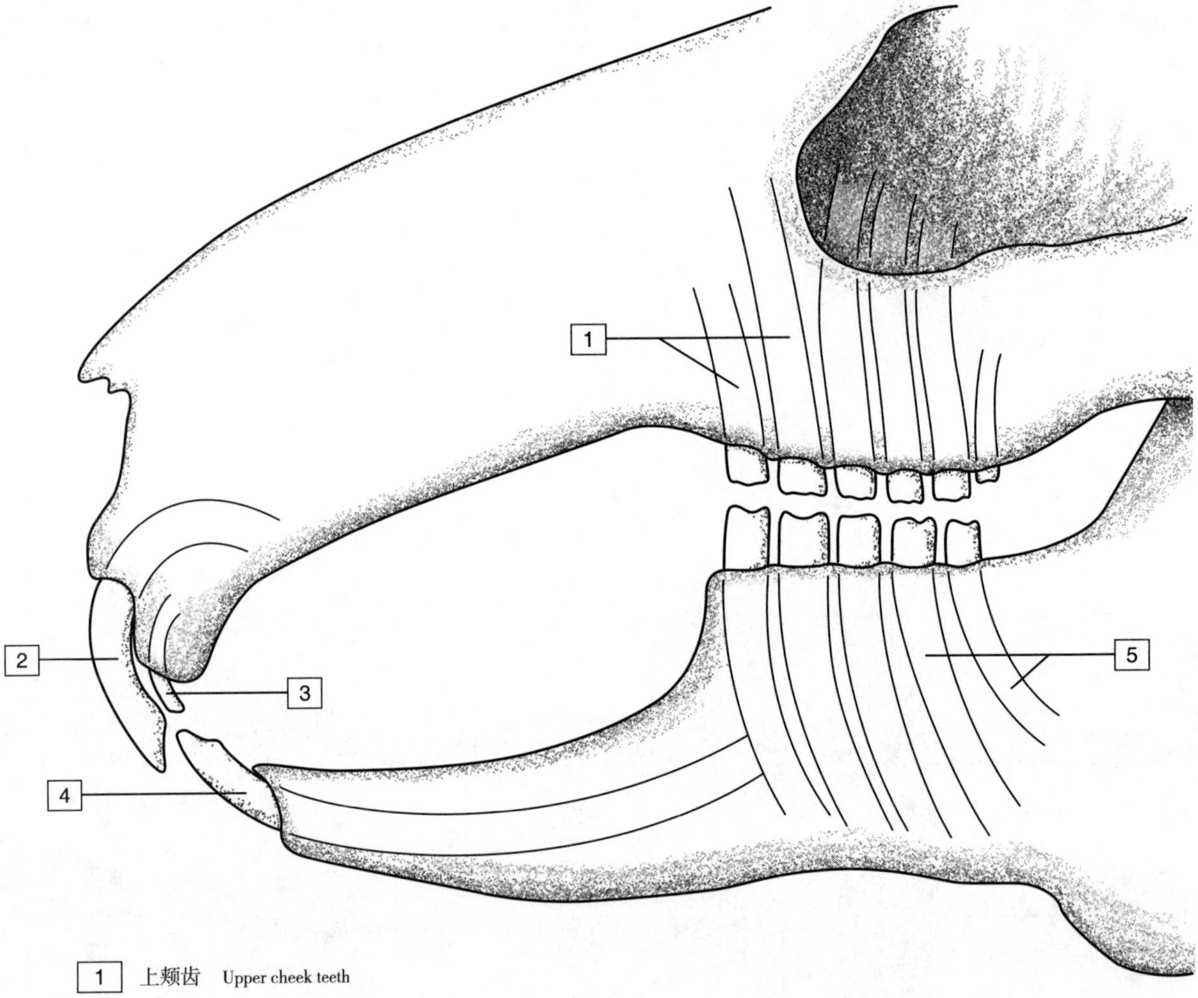

1	上颊齿	Upper cheek teeth
2	上切齿	Upper incisor
3	三角齿	Peg tooth
4	下切齿	Lower incisor
5	下颊齿	Lower cheek teeth

图8-5　野兔头骨横断面显示的下颌、上颌及单个牙齿

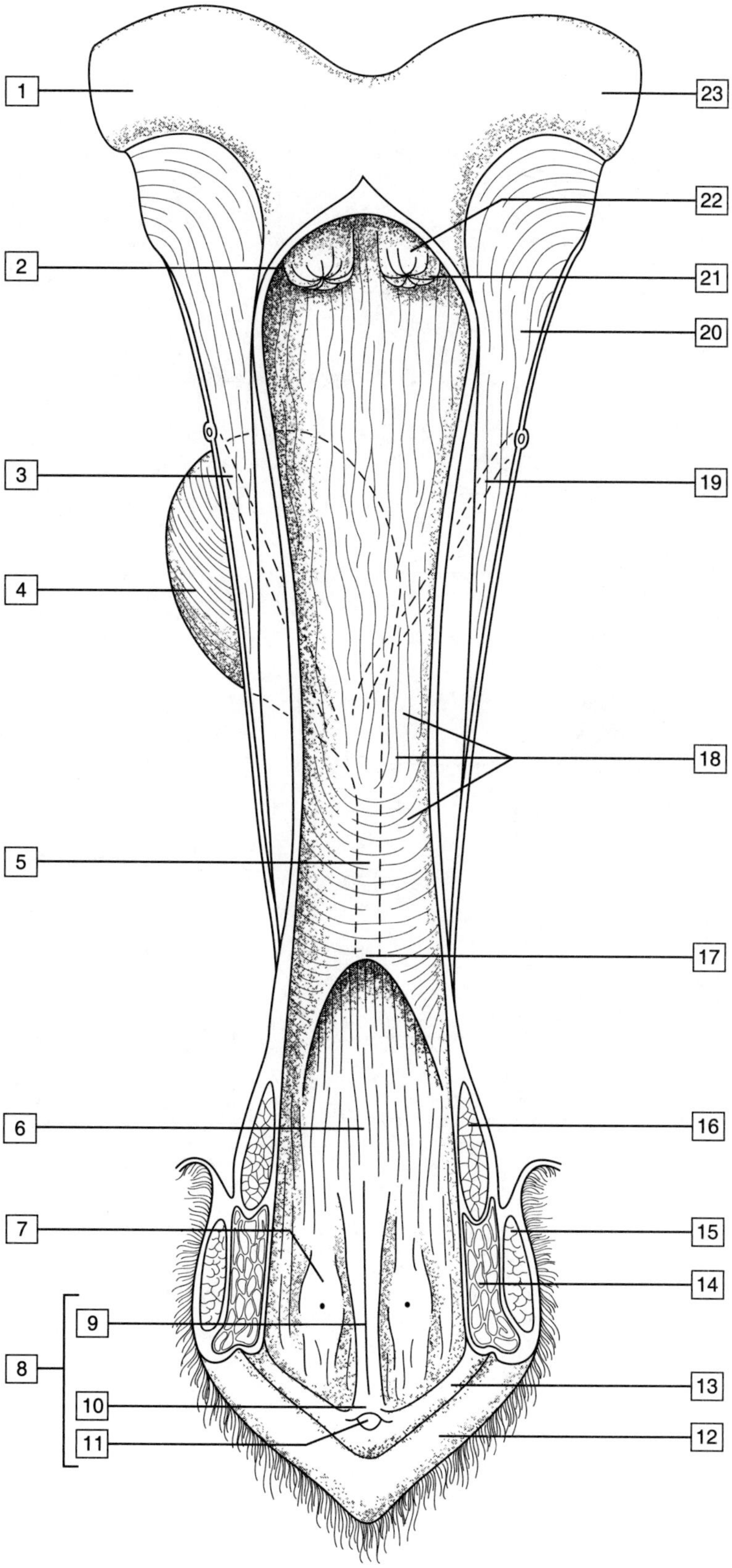

1 左侧子宫角 Left uterine horn
2 阴道穹隆 Vaginal fornix
3 左侧输尿管 Left ureter
4 膀胱 Urinary bladder
5 尿道 Urethra
6 阴道前庭 Vaginal vestibule
7 前庭大腺 Greater vestibular gland
8 阴蒂 Clitoris
9 阴蒂前部 Corpus of the clitoris
10 包皮 Preputium of the clitoris
11 阴蒂头 Glans of the clitoris
12 大阴唇 Greater lip of pudenda
13 小阴唇 Lesser lip of pudenda
14 前庭收缩肌 Constrictor muscle of vestibule
15 包皮腺 Preputial gland
16 前庭球 Vestibular bulb
17 尿道外口 External urethral opening
18 阴道（腹底壁） Vagina (ventral floor)
19 右侧输尿管 Right ureter
20 子宫阔韧带 Uterine broad ligament
21 子宫外口 External uterine ostium
22 右侧子宫颈阴道部 Right vaginal cervix
23 右侧子宫角 Right uterine horn

图8-6 野兔的阴道

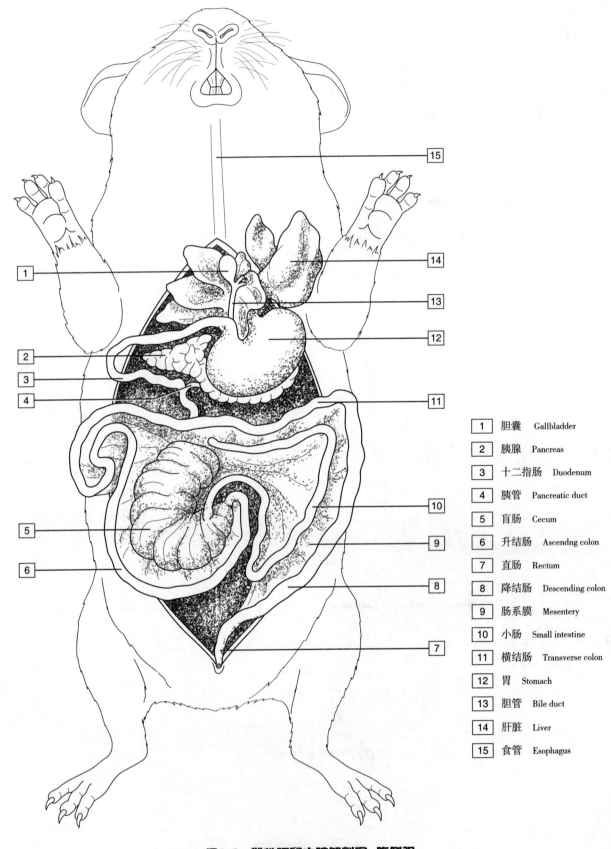

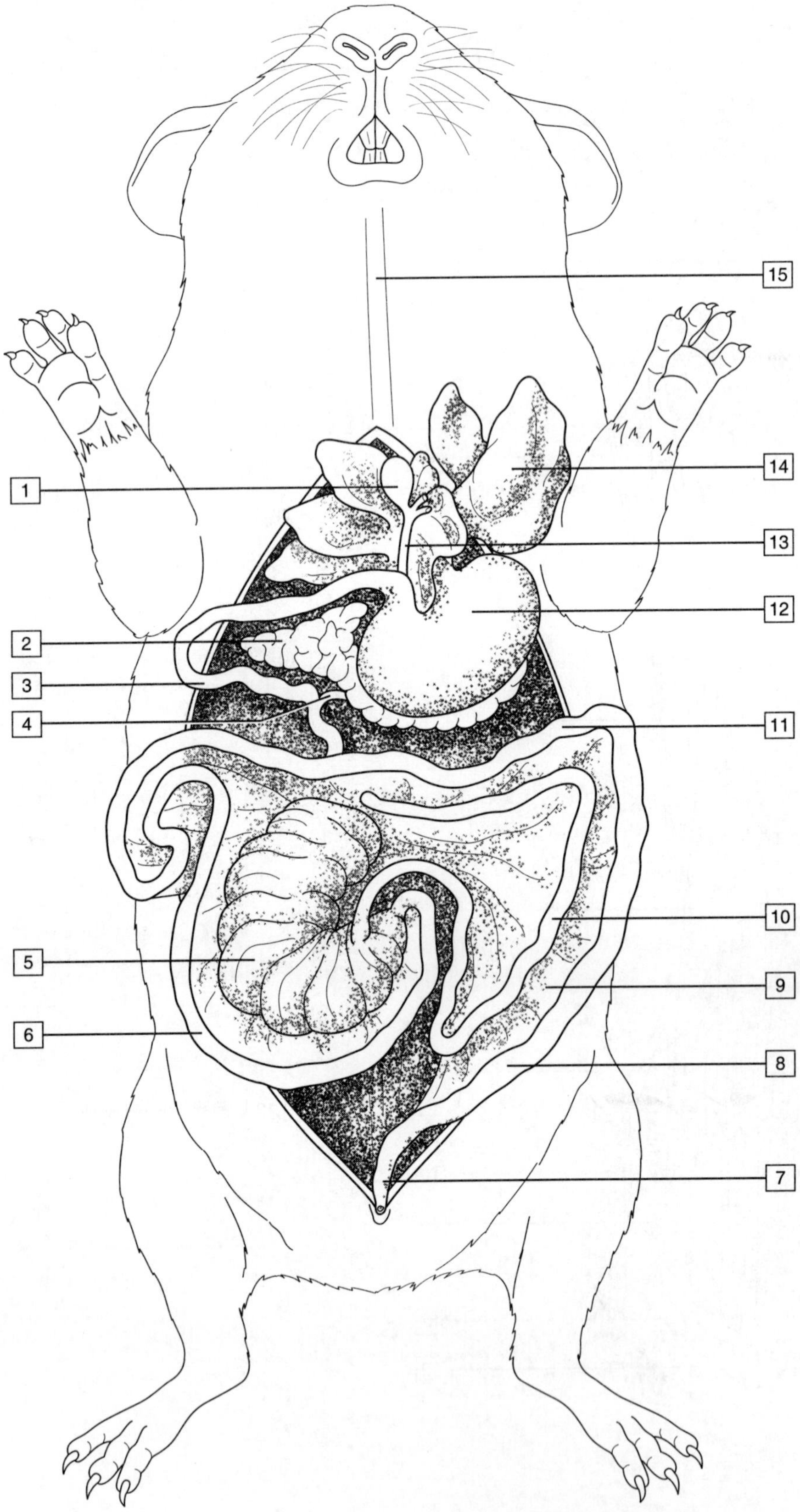

1	胆囊	Gallbladder
2	胰腺	Pancreas
3	十二指肠	Duodenum
4	胰管	Pancreatic duct
5	盲肠	Cecum
6	升结肠	Ascendng colon
7	直肠	Rectum
8	降结肠	Descending colon
9	肠系膜	Mesentery
10	小肠	Small intestine
11	横结肠	Transverse colon
12	胃	Stomach
13	胆管	Bile duct
14	肝脏	Liver
15	食管	Esophagus

图8-7　雌性豚鼠内脏解剖图，腹侧观

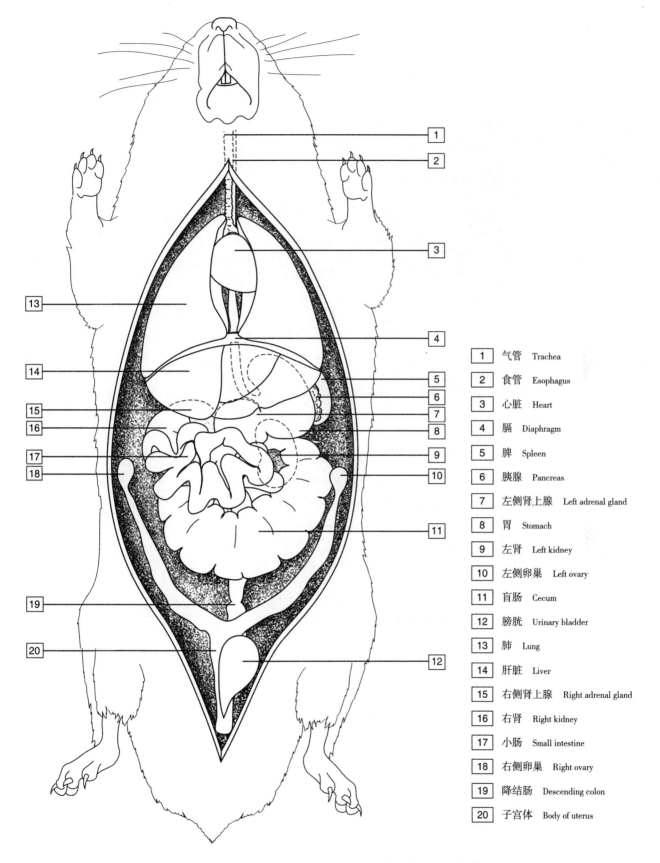

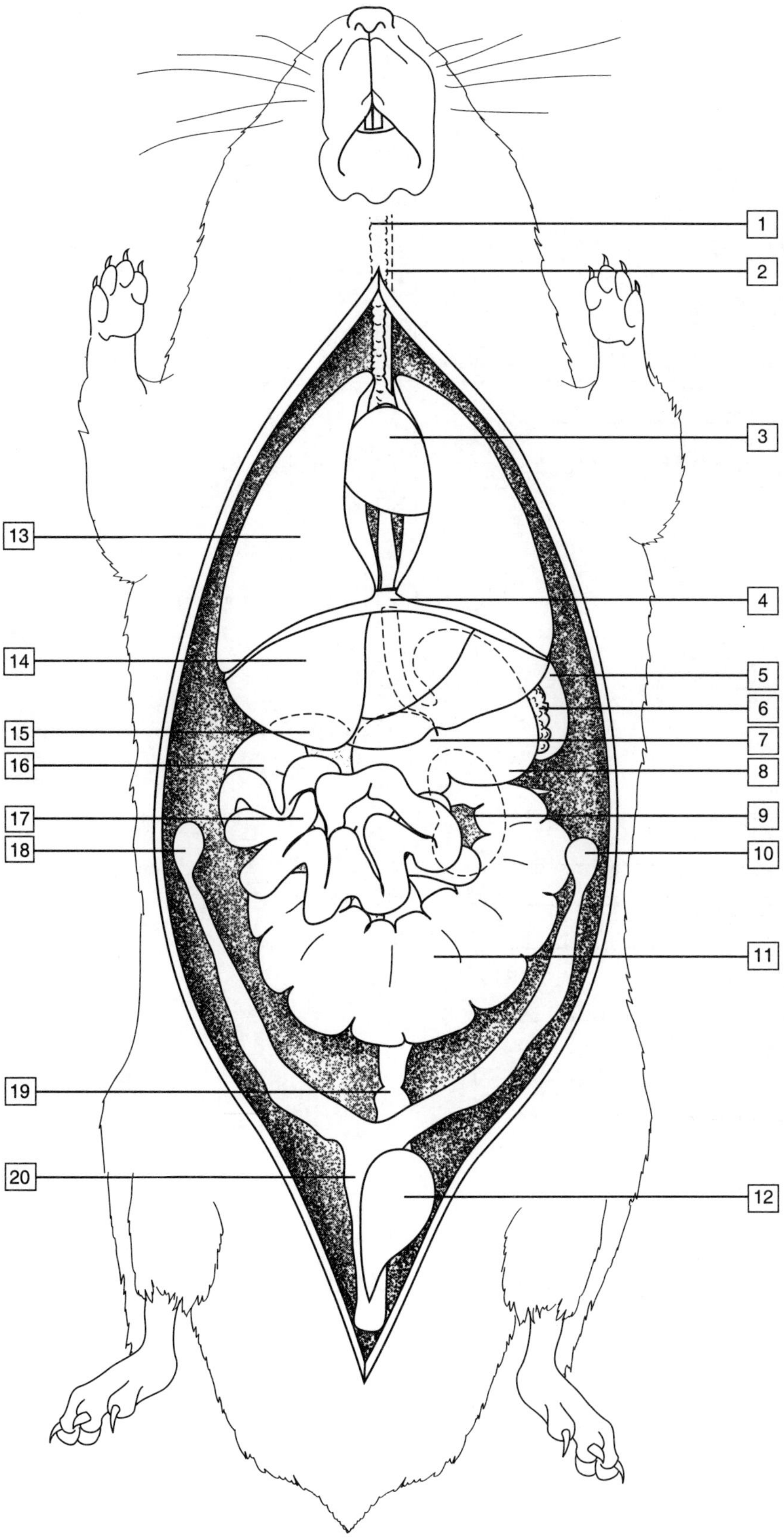

1 气管 Trachea
2 食管 Esophagus
3 心脏 Heart
4 膈 Diaphragm
5 脾 Spleen
6 胰腺 Pancreas
7 左侧肾上腺 Left adrenal gland
8 胃 Stomach
9 左肾 Left kidney
10 左侧卵巢 Left ovary
11 盲肠 Cecum
12 膀胱 Urinary bladder
13 肺 Lung
14 肝脏 Liver
15 右侧肾上腺 Right adrenal gland
16 右肾 Right kidney
17 小肠 Small intestine
18 右侧卵巢 Right ovary
19 降结肠 Descending colon
20 子宫体 Body of uterus

图8-8 仓鼠科动物仓鼠（雌性）内脏解剖图，腹侧观

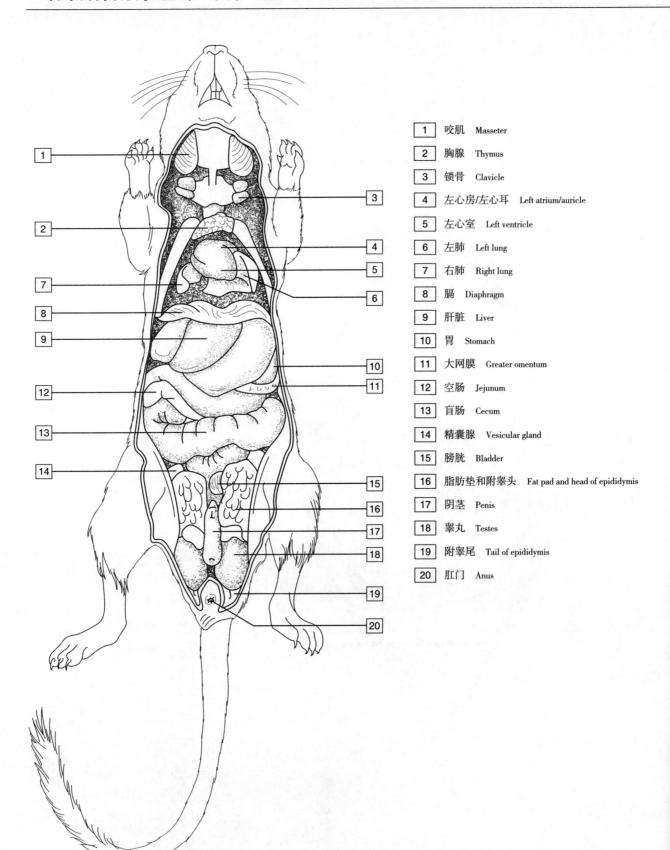

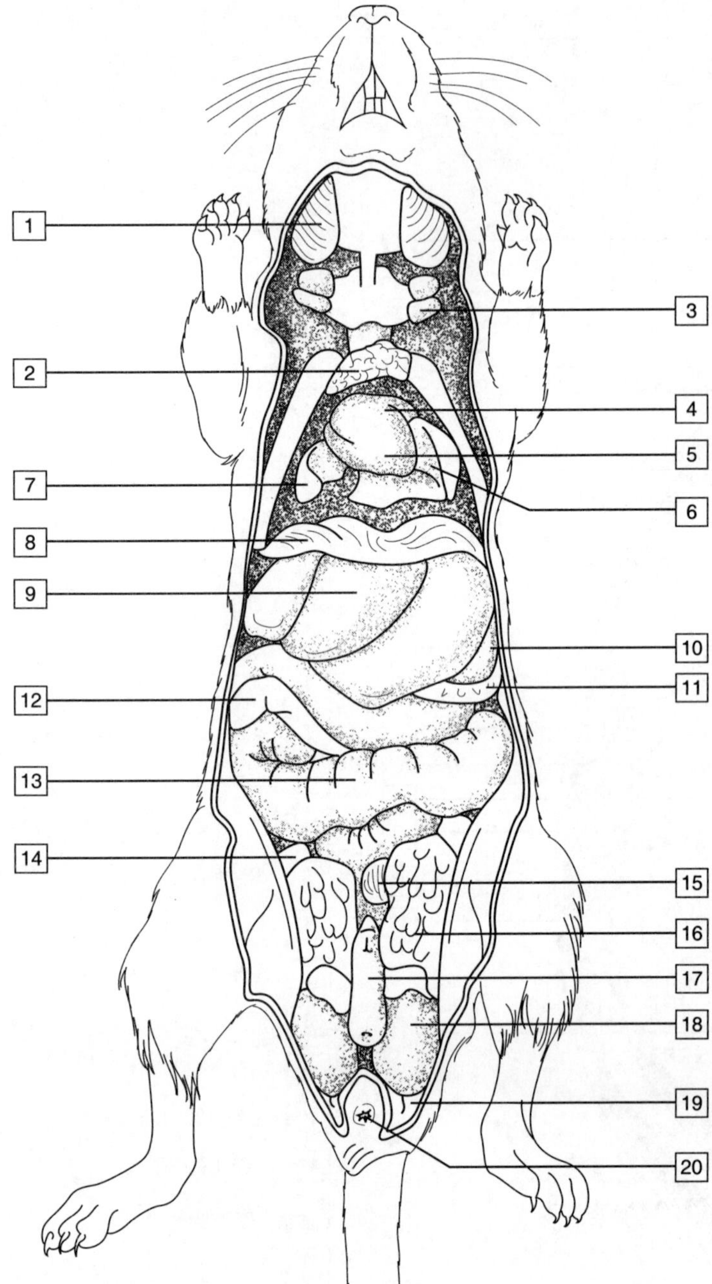

1 咬肌 Masseter
2 胸腺 Thymus
3 锁骨 Clavicle
4 左心房/左心耳 Left atrium/auricle
5 左心室 Left ventricle
6 左肺 Left lung
7 右肺 Right lung
8 膈 Diaphragm
9 肝脏 Liver
10 胃 Stomach
11 大网膜 Greater omentum
12 空肠 Jejunum
13 盲肠 Cecum
14 精囊腺 Vesicular gland
15 膀胱 Bladder
16 脂肪垫和附睾头 Fat pad and head of epididymis
17 阴茎 Penis
18 睾丸 Testes
19 附睾尾 Tail of epididymis
20 肛门 Anus

图8-9 仓鼠科动物蒙古沙鼠（雄性）内脏解剖图，腹侧观

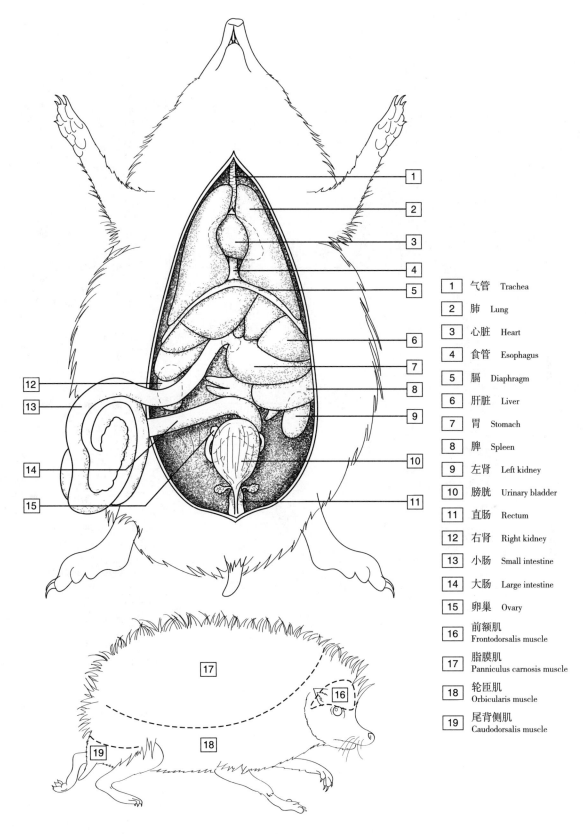

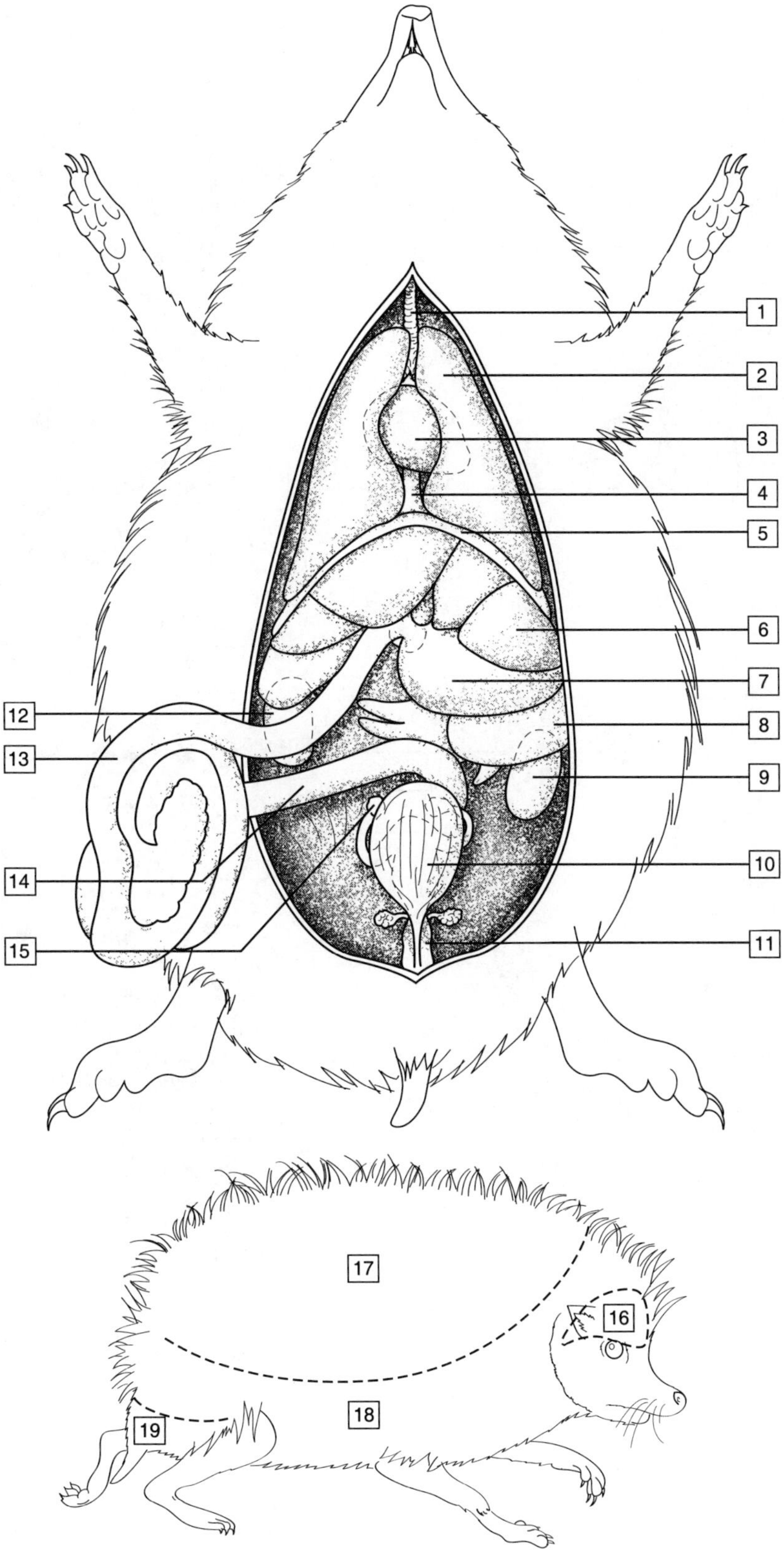

1 气管 Trachea
2 肺 Lung
3 心脏 Heart
4 食管 Esophagus
5 膈 Diaphragm
6 肝脏 Liver
7 胃 Stomach
8 脾 Spleen
9 左肾 Left kidney
10 膀胱 Urinary bladder
11 直肠 Rectum
12 右肾 Right kidney
13 小肠 Small intestine
14 大肠 Large intestine
15 卵巢 Ovary
16 前额肌 Frontodorsalis muscle
17 脂膜肌 Panniculus carnosis muscle
18 轮匝肌 Orbicularis muscle
19 尾背侧肌 Caudodorsalis muscle

图8-10 猬科动物非洲刺猬（雌性）内脏解剖图（腹侧观，上图），以及用于卷曲身体和控制脊柱的肌肉（下图）

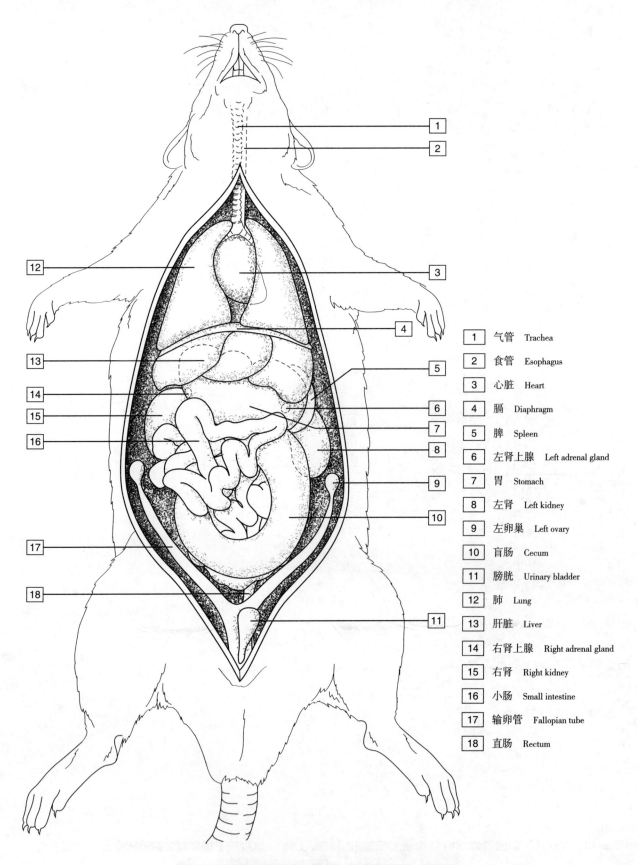

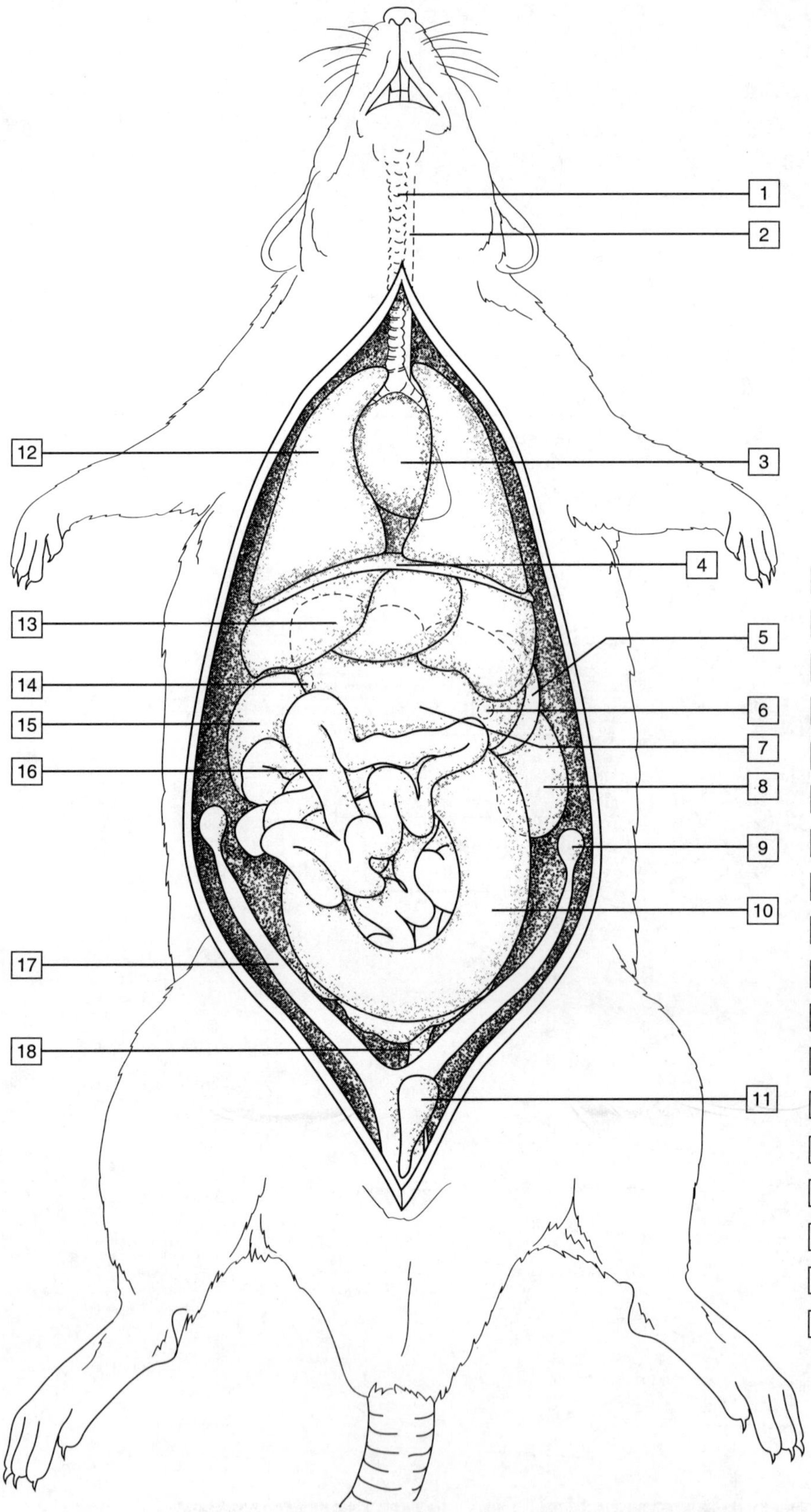

1	气管	Trachea
2	食管	Esophagus
3	心脏	Heart
4	膈	Diaphragm
5	脾	Spleen
6	左肾上腺	Left adrenal gland
7	胃	Stomach
8	左肾	Left kidney
9	左卵巢	Left ovary
10	盲肠	Cecum
11	膀胱	Urinary bladder
12	肺	Lung
13	肝脏	Liver
14	右肾上腺	Right adrenal gland
15	右肾	Right kidney
16	小肠	Small intestine
17	输卵管	Fallopian tube
18	直肠	Rectum

图8-11　鼠科动物大鼠（雌性）内脏解剖图，腹侧观

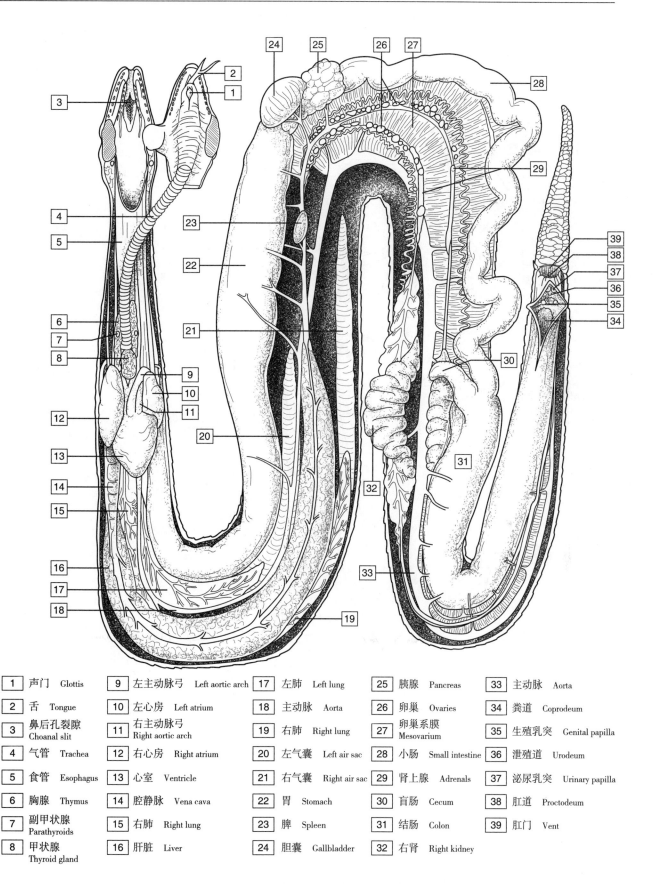

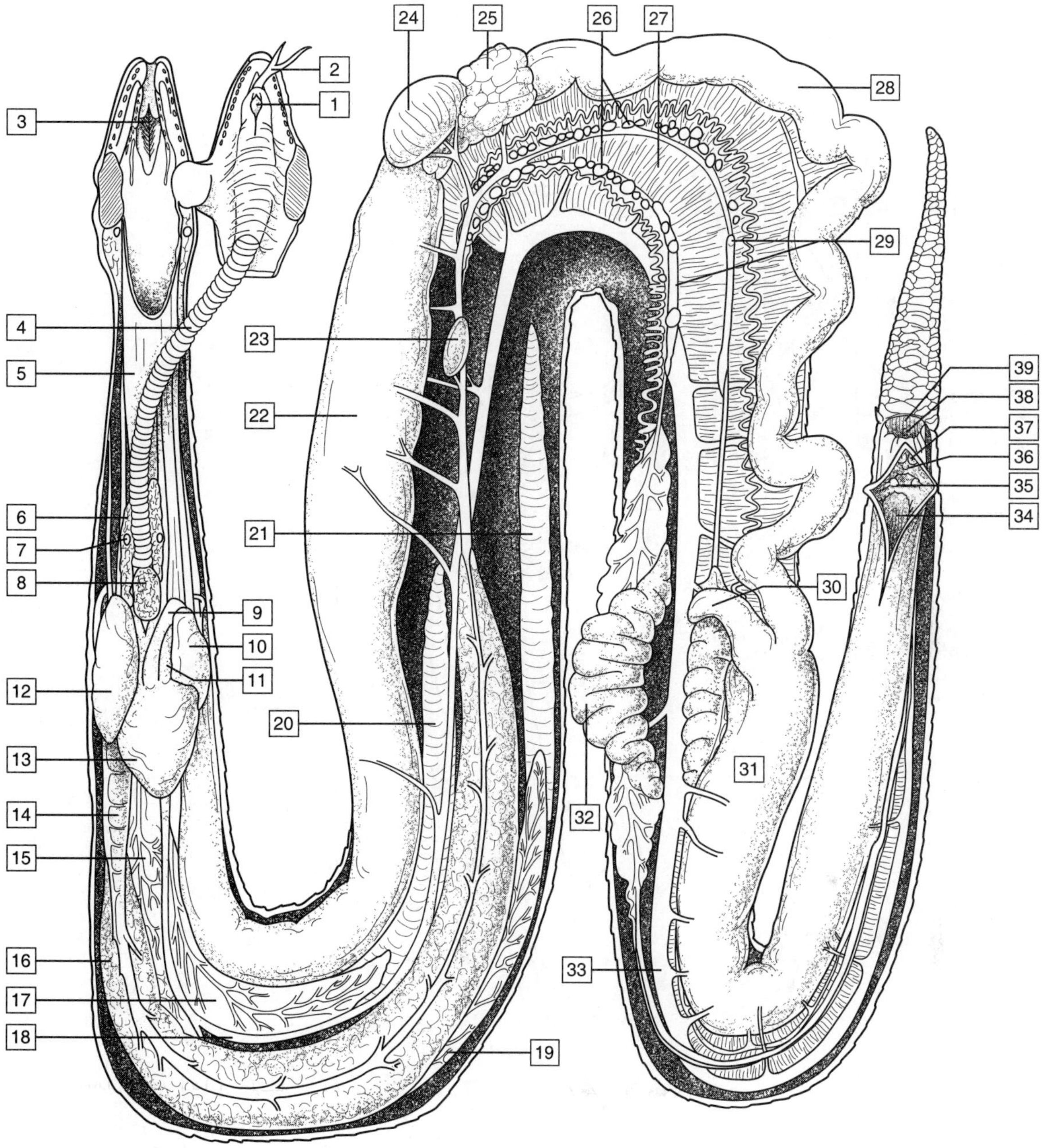

1	声门 Glottis	9	左主动脉弓 Left aortic arch	17	左肺 Left lung	25	胰腺 Pancreas	33	主动脉 Aorta
2	舌 Tongue	10	左心房 Left atrium	18	主动脉 Aorta	26	卵巢 Ovaries	34	粪道 Coprodeum
3	鼻后孔裂隙 Choanal slit	11	右主动脉弓 Right aortic arch	19	右肺 Right lung	27	卵巢系膜 Mesovarium	35	生殖乳突 Genital papilla
4	气管 Trachea	12	右心房 Right atrium	20	左气囊 Left air sac	28	小肠 Small intestine	36	泄殖道 Urodeum
5	食管 Esophagus	13	心室 Ventricle	21	右气囊 Right air sac	29	肾上腺 Adrenals	37	泌尿乳突 Urinary papilla
6	胸腺 Thymus	14	腔静脉 Vena cava	22	胃 Stomach	30	盲肠 Cecum	38	肛道 Proctodeum
7	副甲状腺 Parathyroids	15	右肺 Right lung	23	脾 Spleen	31	结肠 Colon	39	肛门 Vent
8	甲状腺 Thyroid gland	16	肝脏 Liver	24	胆囊 Gallbladder	32	右肾 Right kidney		

图8-12　蛇科动物蛇内脏解剖图，腹侧观

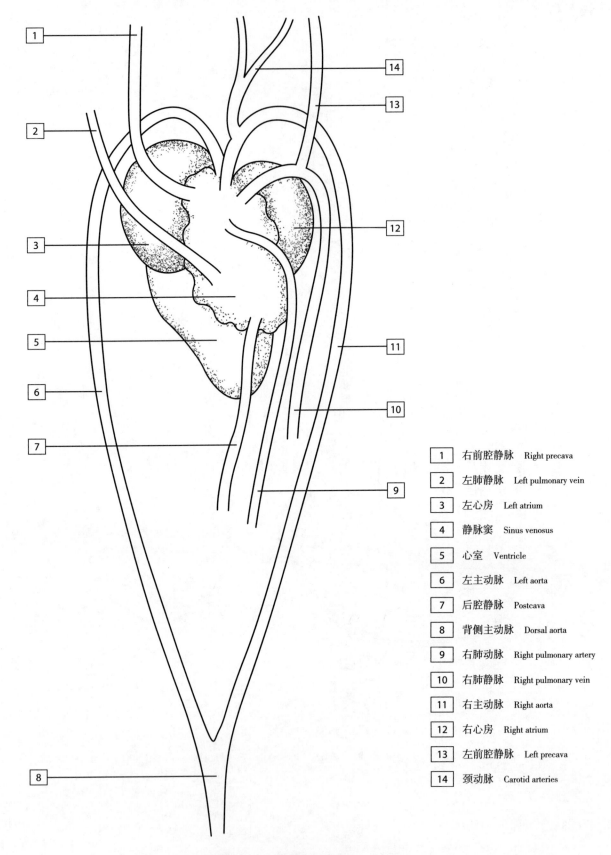

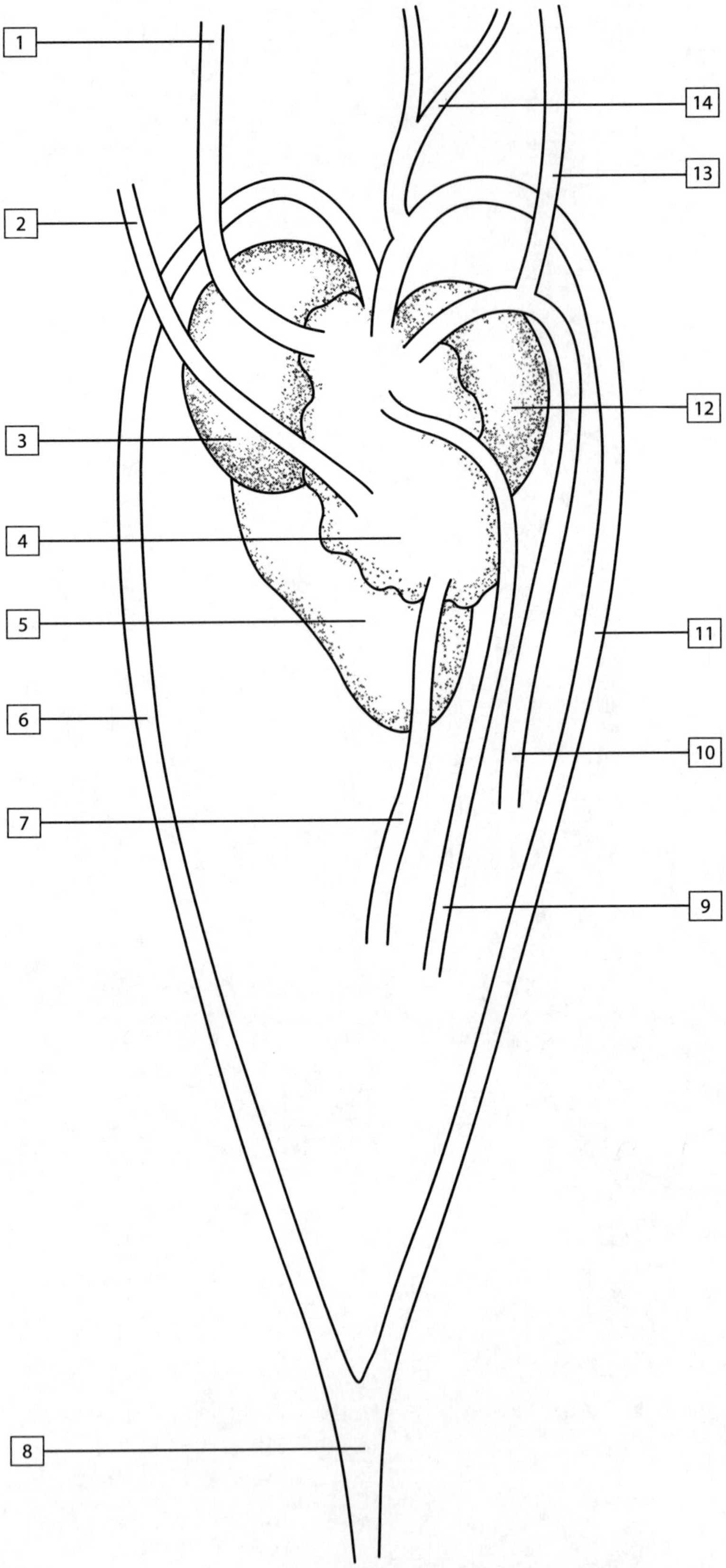

1 右前腔静脉 Right precava

2 左肺静脉 Left pulmonary vein

3 左心房 Left atrium

4 静脉窦 Sinus venosus

5 心室 Ventricle

6 左主动脉 Left aorta

7 后腔静脉 Postcava

8 背侧主动脉 Dorsal aorta

9 右肺动脉 Right pulmonary artery

10 右肺静脉 Right pulmonary vein

11 右主动脉 Right aorta

12 右心房 Right atrium

13 左前腔静脉 Left precava

14 颈动脉 Carotid arteries

图8-13 蛇科动物蛇心脏，背侧观

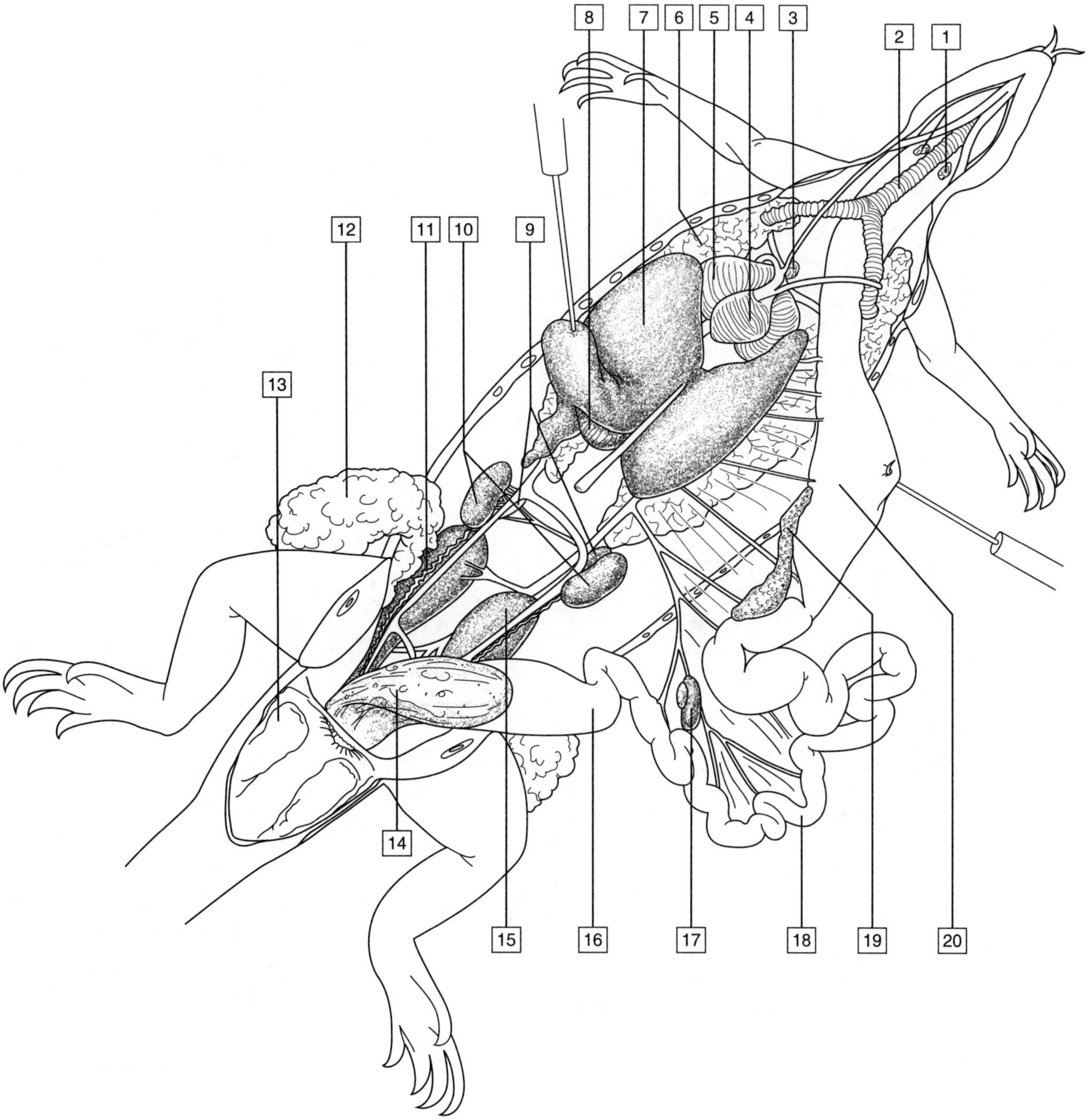

1 甲状旁腺 Parathyroids	6 肺 Lung	11 输精管 Vas deferens	16 结肠 Colon
2 气管 Trachea	7 肝脏 Liver	12 脂肪垫 Fat pad	17 脾 Spleen
3 甲状腺 Thyroid gland	8 胆囊 Gallbladder	13 半阴茎囊 Hemipenis sac	18 小肠 Small intestine
4 心室 Ventricle	9 肾上腺 Adrenal glands	14 膀胱 Bladder	19 胰腺 Pancreas
5 右心房 Right atrium	10 睾丸 Testes	15 肾脏 Kidney	20 胃 Stomach

图8-14 蜥蜴科动物草原巨蜥内脏解剖图，腹侧观

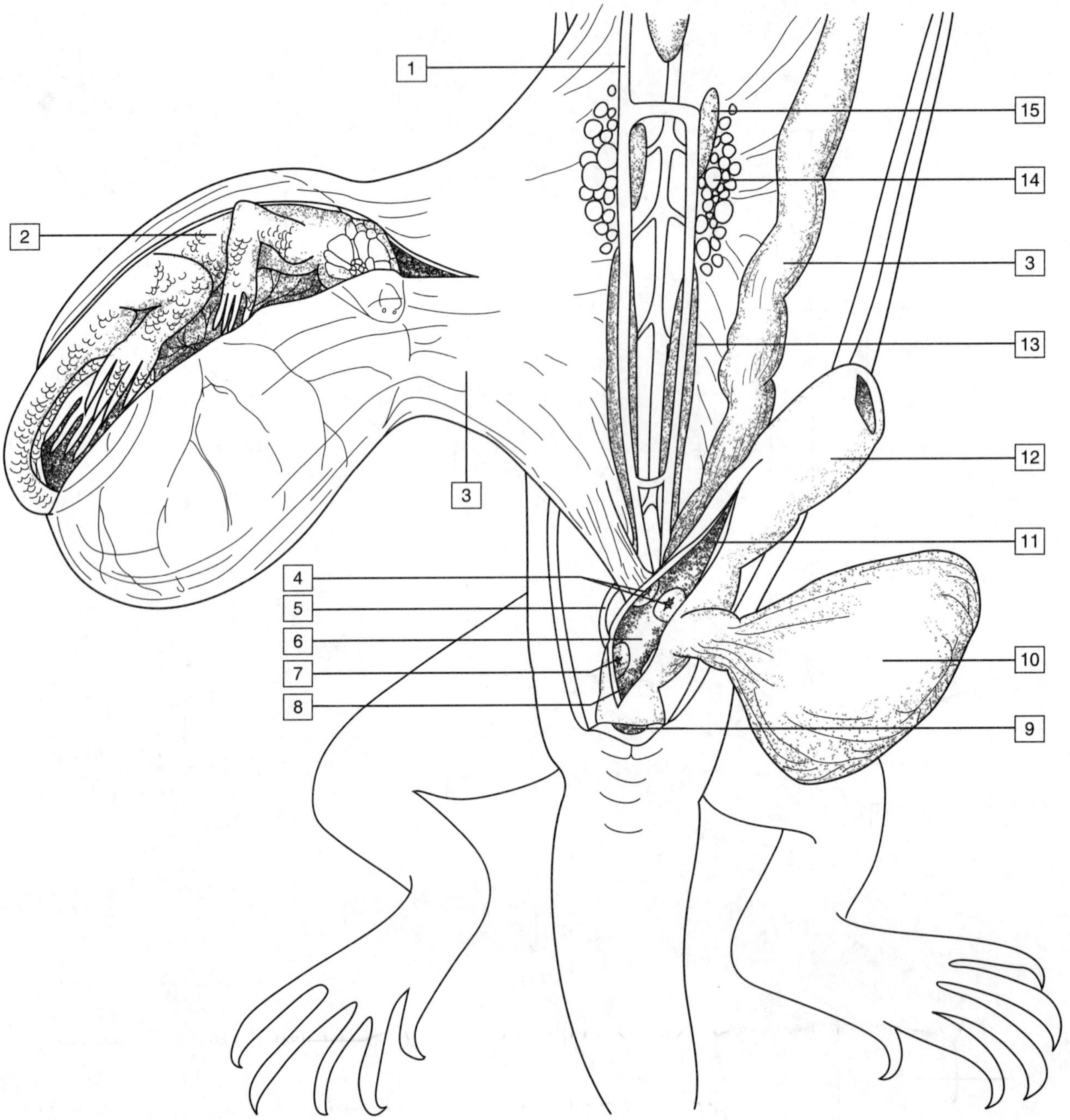

1	腔静脉 Vena cava	5	输尿管 Ureter	9	肛门 Vent	13	肾脏 Kidney
2	胎儿 fetus	6	泄殖道 Urodeum	10	膀胱 Urinary bladder	14	卵巢 Ovary
3	壳腺 Shell gland	7	尿孔 Urinary pore	11	粪道 Coprodeum	15	左肾上腺 Left adrenal gland
4	生殖孔 Genital pores	8	肛道 Proctodeum	12	远端结肠 Distal colon		

图8-15 蜥蜴科动物蜥蜴（雌性）的泄殖腔解剖图

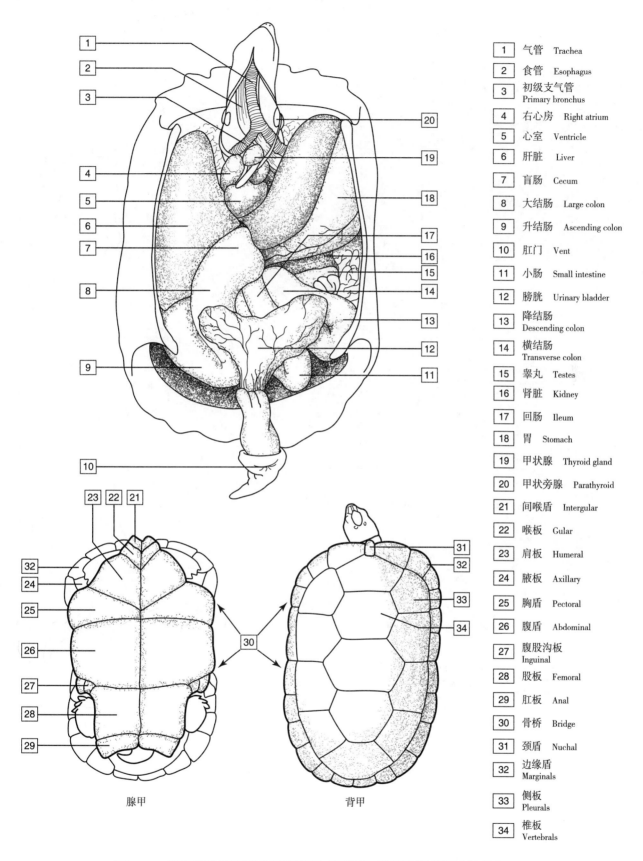

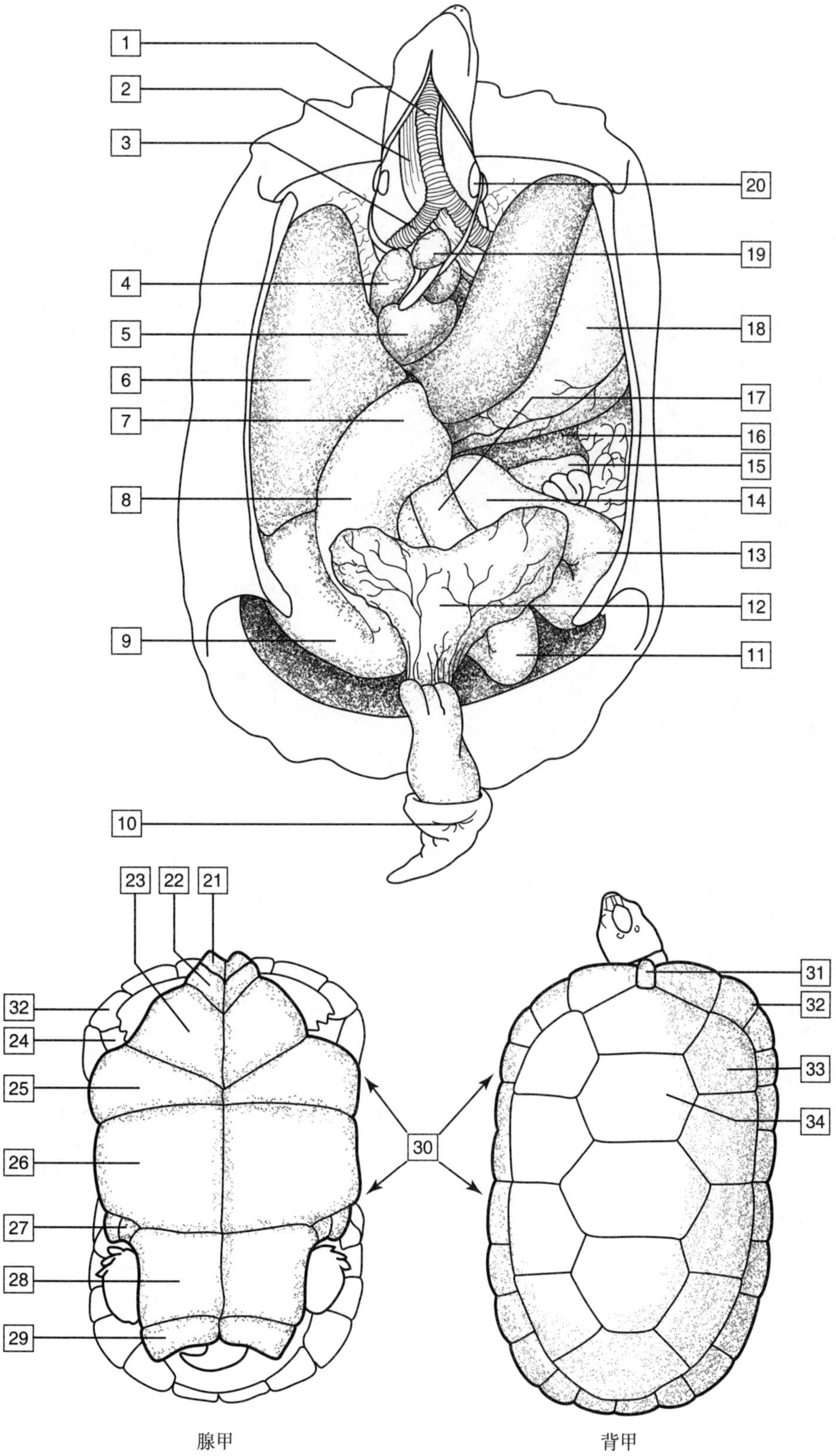

1 气管 Trachea
2 食管 Esophagus
3 初级支气管 Primary bronchus
4 右心房 Right atrium
5 心室 Ventricle
6 肝脏 Liver
7 盲肠 Cecum
8 大结肠 Large colon
9 升结肠 Ascending colon
10 肛门 Vent
11 小肠 Small intestine
12 膀胱 Urinary bladder
13 降结肠 Descending colon
14 横结肠 Transverse colon
15 睾丸 Testes
16 肾脏 Kidney
17 回肠 Ileum
18 胃 Stomach
19 甲状腺 Thyroid gland
20 甲状旁腺 Parathyroid
21 间喉盾 Intergular
22 喉板 Gular
23 肩板 Humeral
24 腋板 Axillary
25 胸盾 Pectoral
26 腹盾 Abdominal
27 腹股沟板 Inguinal
28 股板 Femoral
29 肛板 Anal
30 骨桥 Bridge
31 颈盾 Nuchal
32 边缘盾 Marginals
33 侧板 Pleurals
34 椎板 Vertebrals

图8-16 海龟科动物乌龟内脏解剖图，腹侧观，以及胸甲和甲壳的命名

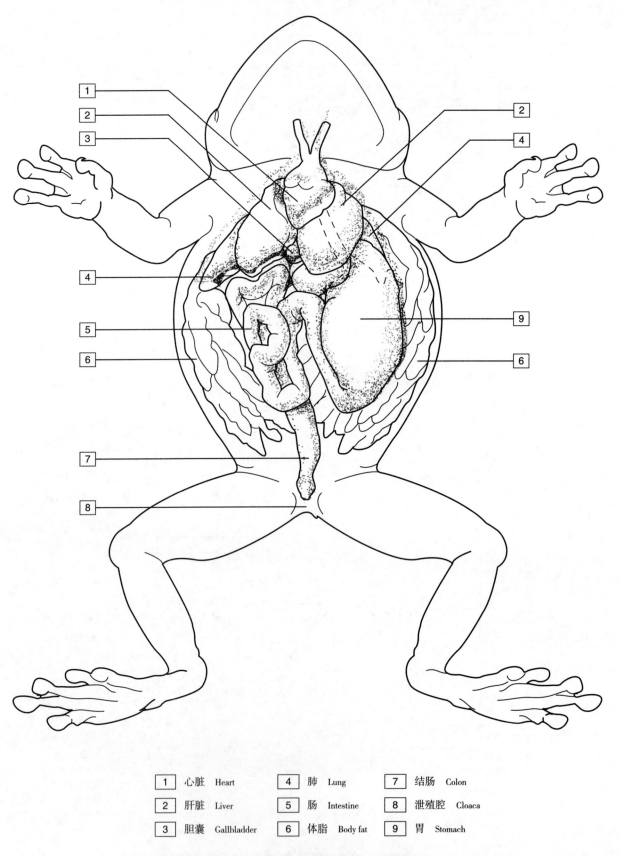

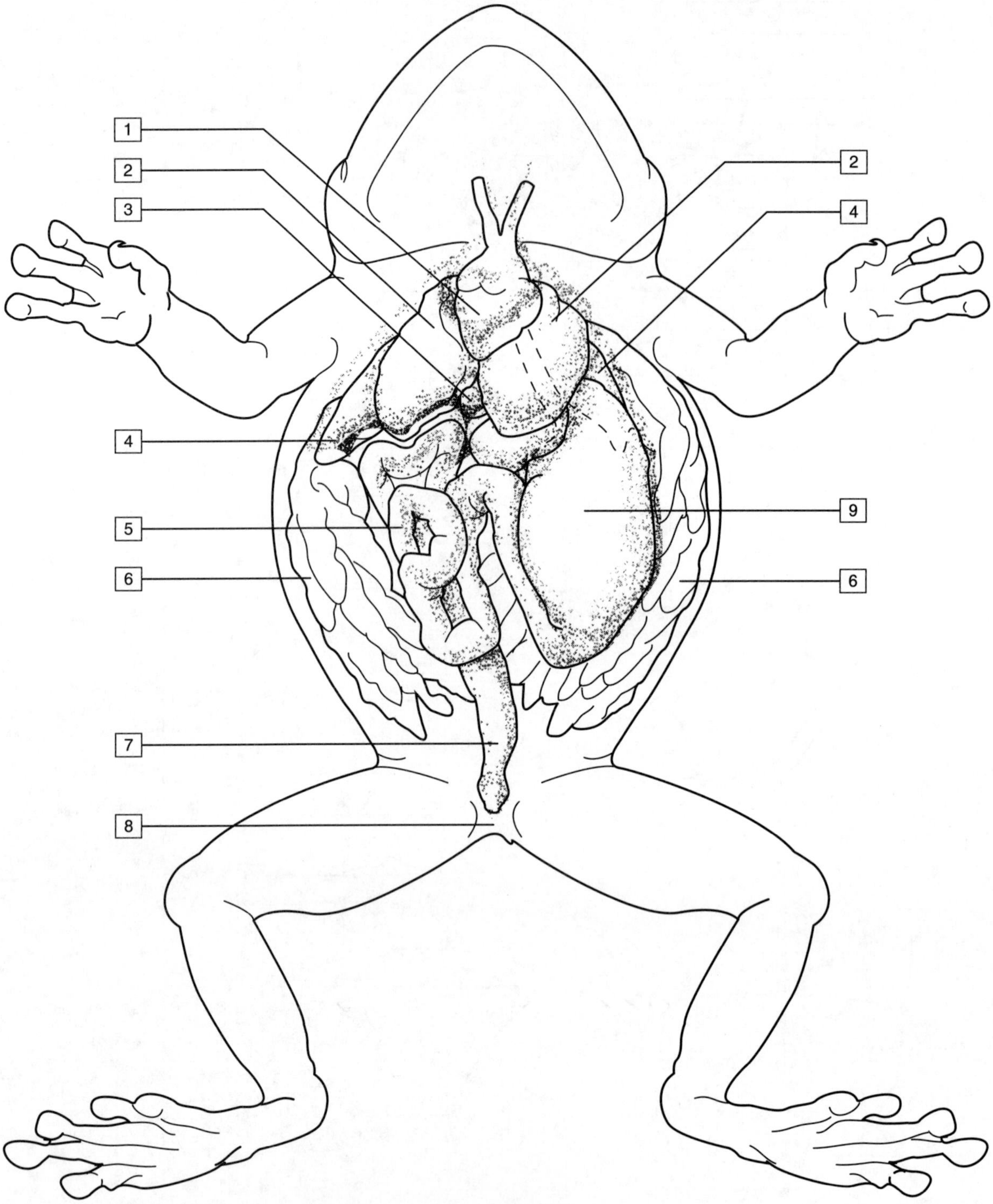

1	心脏	Heart	4	肺	Lung	7	结肠	Colon
2	肝脏	Liver	5	肠	Intestine	8	泄殖腔	Cloaca
3	胆囊	Gallbladder	6	体脂	Body fat	9	胃	Stomach

图8-17　青蛙的内脏解剖图，腹侧观

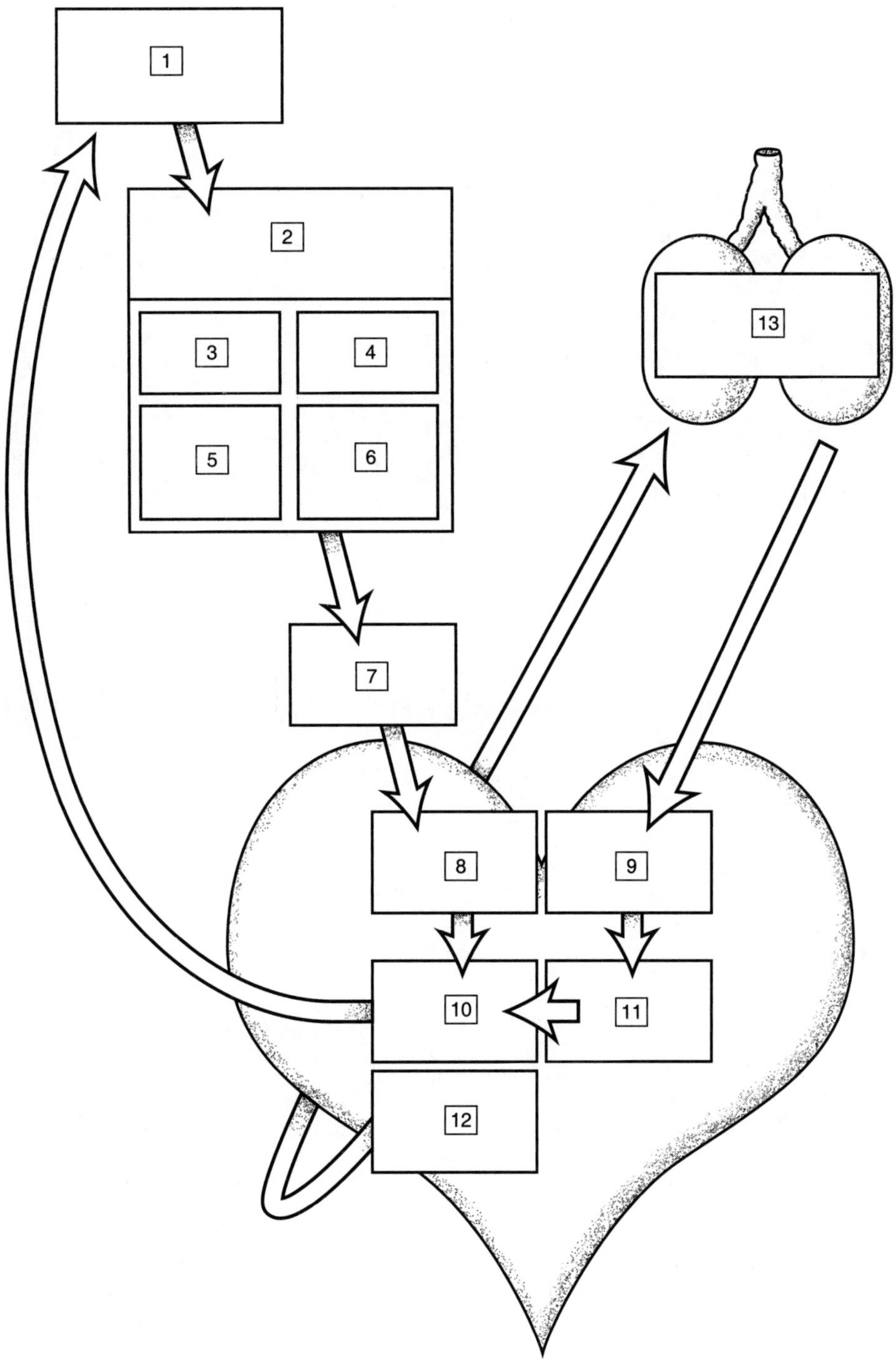

1 大循环 General circulation

2 返回心脏的主要血管 Major vessels returning to the heart

3 左前腔静脉 Left precava

4 右前腔静脉 Right precava

5 肝左静脉 Left hepatic vein

6 后腔静脉 Postcava

7 静脉窦 Sinus venosus

8 右心房 Right atrium

9 左心房 Left atrium

10 静脉腔 Cavum venosum

11 动脉腔 Cavum arteriosum

12 肺腔 Cavum pulmonade

13 肺循环 Pulmonary circulation

图8-18 非鳄目爬行动物的一般循环